# PRINCIPLES OF HORMONE/ BEHAVIOR RELATIONS

T0297124

# PRINCIPLES OF HORMONE/ BEHAVIOR RELATIONS

## SECOND EDITION

DONALD W. PFAFF
*The Rockefeller University, New York, NY, United States*

ROBERT T. RUBIN
*Department of Psychiatry and Biobehavioral Sciences*
*David Geffen School of Medicine at UCLA*
*Los Angeles, CA, United States*

JILL E. SCHNEIDER
*Lehigh University, Bethlehem, PA, United States*

GEOFFREY A. HEAD
*Neuropharmacology Laboratory, Baker IDI, Melbourne, VIC, Australia*

ACADEMIC PRESS
An imprint of Elsevier

Academic Press is an imprint of Elsevier
125 London Wall, London EC2Y 5AS, United Kingdom
525 B Street, Suite 1800, San Diego, CA 92101-4495, United States
50 Hampshire Street, 5th Floor, Cambridge, MA 02139, United States
The Boulevard, Langford Lane, Kidlington, Oxford OX5 1GB, United Kingdom

**Notices**
Knowledge and best practice in this field are constantly changing. As new research and
experience broaden our understanding, changes in research methods, professional
practices, or medical treatment may become necessary.

Practitioners and researchers must always rely on their own experience and knowledge in
evaluating and using any information, methods, compounds, or experiments described
herein. In using such information or methods they should be mindful of their own safety
and the safety of others, including parties for whom they have a professional
responsibility.

To the fullest extent of the law, neither the Publisher nor the authors, contributors, or
editors, assume any liability for any injury and/or damage to persons or property as a
matter of products liability, negligence or otherwise, or from any use or operation of any
methods, products, instructions, or ideas contained in the material herein.

**Library of Congress Cataloging-in-Publication Data**
A catalog record for this book is available from the Library of Congress

**British Library Cataloguing-in-Publication Data**
A catalogue record for this book is available from the British Library

ISBN: 978-0-12-811371-4

For information on all Academic Press publications visit our
website at https://www.elsevier.com/books-and-journals

Working together
to grow libraries in
developing countries

www.elsevier.com • www.bookaid.org

*Publisher:* Nikki Levy
*Acquisition Editor:* April Farr
*Editorial Project Manager:* Timothy Bennett
*Production Project Manager:* Priya Kumaraguruparan
*Cover Designer:* Vicky Pearson Esser

Typeset by TNQ Books and Journals

# Contents

# II

# HISTORY: HORMONE EFFECTS CAN DEPEND ON FAMILY, GENDER, AND DEVELOPMENT

## 9. Sex Differences Can Influence Behavioral Responses

## 10. Hormone Actions Early in Development Can Influence Hormone Responsiveness in the CNS During Adulthood

## 11. Epigenetic Changes Mediate Effects of Hormones on Behavior

## 12. Puberty Alters Hormone Secretion and Hormone Responsivity and Heralds Sex Differences

## 13. Changes in Hormone Levels and Responsiveness During Aging Affect Behavior

# III

# TIME: HORMONAL EFFECTS ON BEHAVIOR DEPEND ON TEMPORAL PARAMETERS

## 14. Duration of Hormone Exposure Can Make a Big Difference: In Some Cases, Longer Is Better; in Other Cases, Brief Pulses Are Optimal for Behavioral Effects

# IV

# SPACE: SPATIAL ASPECTS OF HORMONE ADMINISTRATION AND IMPACT ARE IMPORTANT

# V

# MECHANISMS: MOLECULAR AND BIOPHYSICAL MECHANISMS OF HORMONE ACTIONS GIVE CLUES TO FUTURE THERAPEUTIC STRATEGIES

# VI

## ENVIRONMENT: ENVIRONMENTAL VARIABLES INFLUENCE HORMONE/BEHAVIOR RELATIONS

# VII

## EVOLUTION

# Preface

The first edition of this text was successful in showing that the science of hormone/behavior relations has matured to the point that we can state and prove principles.

The second edition adds three features, in addition to general updating.

First, our new coauthor Professor Jill Schneider (Lehigh University) has rendered the explanations more suitable for undergraduate reading. Second, we have paid more attention to the context in which hormones influence behaviors. This endeavor was supported by a grant to Dr. Schneider, IOS1257876, from the National Science Foundation. Third, we have added new material on epigenetics, material that was not known at the time of the first edition.

We emphasize that this text is not a review paper or an exhaustive reference source. Instead, the idea has been to illustrate each principle with a small number of well-chosen examples.

# Acknowledgments

I (Jill Schneider) would like to thank my coauthors for writing the first edition of *Principles of Hormone-Behavior Relations* and for the idea of organizing the chapters according to principles that can be applied to many behaviors. There is little scientific evidence for the idea that biologic mechanisms and principles should be divided according to discrete behaviors (sex, maternal, aggressive, ingestive, etc.); therefore the current organization will provide students with a more accurate view of the field. In addition, I thank Rae Silver for comments on a draft of Chapter 7. I thank all of my students and former students for their critiques of previous textbooks, especially those who gave vital feedback on drafts of this edition, including graduate students Noah Benton, Joe Brague, Jeremy Brozek, and Nunana Gamedoagbao, and undergraduates Zara Ahmad, Zekia Alhout, Matthew Asteak, Reem Azar, Allyson Briegel, Bhoomi Buptani, Jennifer Chen, James Copti, Trishna Dave, Taylor De Groot, Calla Dilli, David Ebhomielen, Minah Ebrahim, Cassandra Field, Sydney Gocial, Melanie Grajales, Devika Gupta, Christopher Hoke, Casey Hollawell, Maisara Huq, Brad Hutton, Robert Iulo, Kevin Jaramillo, Saran Kunaprayoon, Hannah Lahey, Haley Lombardo, Jessica Ludolph, Pooja Malhotra, Isha Markna, Natalie Martin, Molly McHugh, Marlee Milkis, Barrett Miller, Brianna Miller, Lauren Molina, Sean Nelson, Linda Nguyen, Christie Pai, Ashley Park, Jenna Pastorini, Jessica Pryor, Mackenzie Quinn, Christian Salcedo, Samantha Sarli, Danielle Schiraldi, Rachel Sternberg, Melanie Ting, Carlie Skellington, Jai Vaze, Amanda Walter, Alexis Wantanabe, Samantha Warner, Benjamin Warren, Montana Weitzel, Matthew Westenhaver, Kelsey Wieland, Ashleigh Williams, Grace Wong, Cynthia Xie, and Avery Young. I apologize if I did not remember to acknowledge other reviewers by name, and I want them to understand that their feedback has been important. Work on this project was made possible by a grant from the National Science Foundation IOS1257876.

# Introduction

**Dear Reader:** This book is written for both undergraduate and graduate students, some of whom may not have basic information about the field of neuroendocrinology and hormone -brain -behavior relations. The Introduction, therefore, is structured as a condensed primer of neuroendocrinology to provide a broad framework for the chapters that follow. Please review the Introduction fully before delving into the subsequent chapters. Our goal is to fully inform students at several educational levels about the exciting field of hormone -behavior relations and to inspire them to join us as colleagues in the future.

## WHAT IS NEUROENDOCRINOLOGY?

Modern living comes with many strange and disturbing consequences, as illustrated by the following three trends.

1. Chronic stress can lead to obesity, insulin resistance, type II diabetes, cardiovascular disease, ulcers, decreased immune function, and decreased resistance to disease (Kuo et al., 2008).
2. Men and women who work on rotating shifts, such as nurses, doctors, and airline pilots, have an increased risk of developing cancer, cardiovascular disease, diabetes, obesity, and cognitive problems (Conlon et al., 2007; Gan et al., 2015; Ha and Park, 2005; Hansen, 2006; Kivimaki et al., 2006; Marquie et al., 2014; Poole et al., 1992a, b).
3. Sexual identity, fertility, body fat distribution, and sex hormone action are changed by hormones and hormone-like compounds that enter our bodies from the air, water, and plastic vessels used to store foods and beverages (reviewed by Gore, 2008; Schneider et al., 2014).

These are just three of many unsolved mysteries in neuroendocrinology, the study of how chemical messengers bind to their receptors to regulate brain and behavior.

The nervous system is complex, and the number of chemical messengers in the endocrine system is staggering. How do we begin to understand the detailed mechanisms that underlie the chronic stress suffered by shift workers and airline pilots? How do we comprehend the altered memories and anxieties of patients with posttraumatic stress disorder

(PTSD)? How do we understand the behavioral effects of contraceptives, the cognitive and mood changes after menopause, the sexual side effects of treatments for Parkinson's disease and antidepressants, and the sexual-identity-altering effects of endocrine disruptors? To bring order to this complexity, neuroendocrinologists employ the general principles laid out in this book.

These principles have emerged from the study of hormonal effects on simple behaviors in healthy laboratory animals, and by observing anomalous behaviors in human patients and nonhuman animals with diseases of the endocrine system. Neuroendocrinologists have identified hormonal cascades: sequences made up of endocrine organs and their secretions. Each hormone binds to cells on the next organ in the sequence, thereby altering the secretion in the next hormone in the sequence. One such repeating pattern involves hormones secreted from the brain, which in turn stimulate secretion of hormones from the pituitary, which in turn stimulate secretion of hormones from a particular organ in the periphery. We can make inroads into understanding complex problems like chronic stress, obesity, jet lag, shift work, and PTSD by placing these problems within a known, generalized framework based on these recurring patterns of hormone action that involve brain, pituitary, and peripheral organs. This chapter provides an introduction to the brain, pituitary, and some of these peripheral endocrine organs.

The mechanisms by which hormones affect behaviors of all vertebrates, including humans, have achieved a primary spot in modern neuroscience. Terrific progress in this field is due to several factors. Stimuli and responses are relatively simple, so behaviors in their natural form can be evoked in the laboratory. Importantly, we can make use of the tremendous bodies of knowledge in hormone chemistry and pharmacology—especially for steroid hormones—to enrich and extend our ability to manipulate and measure neuronal events as they cause behavior. Steroid hormone receptors, discovered in the brain, turn out to be nuclear proteins, which are ligand-activated transcription factors, thus allowing us *pari passu* to do state-of-the-art molecular biology in the most complex organ in the body, the central nervous system (CNS).

The first edition of this book was an attempt to present our hard-won findings in an orderly fashion and make available to students some of the broadest generalizations that have emerged from the animal neurobiology literature. Included among these is the amazing development of proven causal linkages between specific molecular events and behavioral results.

The second edition attempts to bring the information up to date, which naturally brings more excitement to the prose. The original edition was restricted to mammals, but some of the principles are best illustrated using nonmammalian animal models, so we have included information in the new edition about birds, nonavian reptiles, amphibians, and fish. We have also attempted to share with the reader those aspects of behavioral neuroendocrinology that we found to be most profound and irresistible.

Hormone/behavior relations typically serve either the survival or the reproductive needs of the individual. To survive, individuals must maintain energy balance, fluid and osmotic balance, pH balance, and body temperature. When it comes to homeostasis, there are many physiological and metabolic responses, but those most crucial and immediate are the behavioral responses. Other behaviors do not function solely to protect the internal homeostasis, but rather promote reproduction through territorial competition, courtship, mating, and parental response sequences. The axioms proposed in this book include both types of hormone actions.

## ABOUT THIS BOOK

This book represents a first attempt to state most of the major truths of the general field of hormone/behavior relations and their mechanisms in a systematic fashion. We put forward and illustrate a number of simple statements, "principles" of hormone/behavior relationships supported by the literature to date. Each statement (i.e., principle) is the title of a chapter. That general principle of hormone/behavior relations is then exemplified in two ways—from basic scientific work in the laboratory and from clinical experience.

## WHAT THIS BOOK IS NOT

This is not meant to be a reference book or a collection of scientific review articles. In attempting to keep the book reasonably short and affordable, we have not tried to fill in background details from all the fields that border the integrative field of neuroendocrinology. Thus this text does not stand in for a steroid chemistry treatise, a neuroanatomy or neurophysiology primer, or a molecular biology handbook. A major bibliographic reference source has recently been published (*Hormones, Brain, and Behavior*, third edition, Academic Press, 2017), and compared to that source the examples in this text are neither exhaustive nor complete. They are intended simply to be physiologically clear and to represent a variety of species' (including human) endocrine systems and CNS or behavioral endpoints.

# A CONSISTENT PATTERN: PERIPHERAL ORGANS AND THE NERVOUS SYSTEM WORK TOGETHER

In most vertebrate species including our own, we see the pattern shown in Fig. 1.

In this generic endocrine cascade the following happen.

1. Changes in the environment are sensed, and the sensory signal is relayed to the CNS, consisting of the brain and the spinal cord.
2. The CNS alerts the peripheral organs using electrical communication via the peripheral nervous system (the PNS, or autonomic nervous system) and via chemical communication through the blood stream.
3. Chemical messengers from the brain travel in the blood stream and act as ligands by binding to specific receptors on the cells of endocrine organs. In response, the peripheral endocrine organs secrete new chemical messengers (hormones) that enter the blood stream.
4. These hormones, chemical messengers from peripheral organs, affect the brain and behavior (Fig. 1).

Depending on the type of stimulus and behavioral response, the pattern may be simple, involving one or two glands and their secretions, or complex, involving cascades of glands secreting chemicals that stimulate chains of other glands and other secretions.

In many of these cascades the chemical messengers fit the traditional definition of a hormone, i.e., they are chemical messengers secreted by endocrine glands and acting at a distance from where they are secreted. An endocrine gland is distinct from other types of glands in that it is ductless. Although hormones travel far and wide, their actions do not occur on every cell with which they come into contact.

Hormones mainly act on receptors that are specific to those hormones. Examples include leptin, secreted from fat cells (adipocytes), and insulin, secreted from the pancreas. Leptin and insulin travel in the blood stream and act on many organs that contain leptin and insulin receptors. But while many hormones are secreted from peripheral organs, the neurohormones are a special subclass of hormones secreted from neurons. Neurohormones fit the definition of a hormone because they act at a distance from where they are secreted. Gonadotropin-releasing hormone (GnRH), a neurohormone, is secreted from the terminals of neurons into the blood stream and travels outside the brain to the pituitary gland. Any chemical messenger that has "releasing hormone" in its name is a neurohormone, e.g., corticotropin-releasing hormone (CRH) and thyrotropin-releasing hormone (TRH). In contrast to neurohormones, neurotransmitters are secreted from the terminals of neurons in response to depolarization of a presynaptic cell, and travel only a very short distance across a synapse to affect specific receptors on a postsynaptic cell.

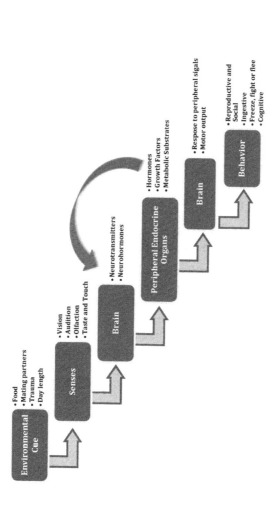

**FIGURE 1** A recurring pattern in behavioral endocrinology involves environmental changes communicated to the brain via the senses causing changes in the brain secretion of chemical messengers that bind to receptors in peripheral organs, such as the pituitary gland. These hormones bind to receptors on other peripheral organs, which in turn secrete other chemical messengers that alter brain and behavior.

Neurotransmitters are small molecules; examples include acetylcholine, the monoamines, including serotonin, dopamine, norepinephrine, and the amino acid neurotransmitters, including glutamate and gamma-aminobutyric acid (GABA). Neurotransmitters do not travel very far, and act rapidly. Neuropeptides, in contrast, tend to be larger than neurotransmitters (e.g., neuropeptides tend to be 3–44 amino acids in length), but smaller than proteins. Examples include neuropeptide Y, substance P, vasopressin, and oxytocin, all of which act in the brain relatively near their site of secretion. These definitions, however, are not absolute, as some neurotransmitters also act as neuropeptides and vice versa, and some neuropeptides are found in peripheral organs. Dopamine, for example, fits the definitions of both a neuropeptide and a neurotransmitter, and is found in both brain and stomach. Furthermore, some steroid hormones, such as estradiol, can be made in the brain and have more than one mode of action. At least one mode of neurosteroid action is thought to be almost as rapid as neurotransmitter action (Balthazart and Ball, 2006).

## Interim Summary i.i

Chemical messengers can be placed into categories, but some individual messengers can fit more than one of the following categories.

1. A hormone is secreted by a ductless gland and acts at a distance from the site of secretion. Protein hormones are composed of more than 50 amino acids, whereas peptide hormones are smaller.
2. A neurotransmitter is a small molecule (e.g., one amino acid) secreted upon depolarization of a neuron that acts rapidly at the synapse.
3. A neuropeptide is a molecule between 3 and 44 amino acids in length that is secreted in the brain and acts on nearby cells.
4. A neurohormone is a cross between a neuropeptide and a hormone; it is secreted by a neuron but acts at a distance.
5. A neurosteroid is a steroid hormone secreted in the brain and has rapid, neurotransmitter-like action.

One example of a classic neuroendocrine cascade is the hypothalamic–pituitary–adrenal (HPA) system (Fig. 2). This is the endocrine cascade most likely to be linked to the effects of chronic stress on obesity mentioned earlier. The HPA system involves the brain, the pituitary, and the adrenal gland.

In the HPA example, a stressor is detected by auditory, visual, or olfactory senses and relayed to a part of the brain known as the hypothalamus, which secretes a neurohormone, CRH, into the blood stream. CRH reaches the front lobe of the pituitary, the anterior pituitary, and stimulates the secretion of adrenocorticotropic hormone (ACTH), which in turn stimulates the secretion of glucocorticoids from the adrenal gland. The adrenal gland has an inner core, the medulla, and an outer layer, the cortex. The adrenal cortex

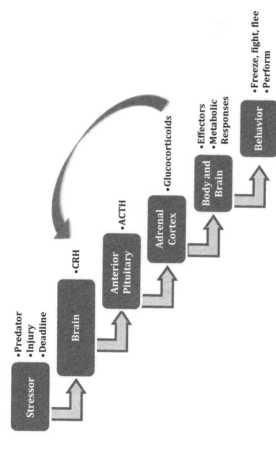

FIGURE 2   The hypothalamic-pituitary-adrenal (HPA) system, one arm of the stress response, is shown. In response to an environmental stressor, brain cells within the paraventricular area of the hypothalamus secrete corticotropin-releasing hormone (CRH) released from the median eminence into the pituitary portal plexus. CRH reaches receptors in the anterior pituitary and stimulates secretion of ACTH. ACTH stimulates secretion of glucocorticoids from the adrenal cortex. ACTH and glucocorticoids alter heart rate, respiration, and glucose levels, which allow the brain and body to respond to the stressor. Glucocorticoids also feed back on brain and pituitary to modulate the stress response so that the stress response does not escalate indefinitely.

is the site of glucocorticoid synthesis and secretion. One of these adrenal steroids is cortisol, the main glucocorticoid secreted in human beings. Cortisol facilitates the behavioral response, whether it be freezing, fighting, or fleeing from danger (Fig. 2). The HPA system and its role in obesity, jet lag, shift work, and PTSD are addressed later, but first it is important to understand the relation between the brain and the endocrine system.

One of the most critical principles of hormone/behavior relations is that the nervous and endocrine systems are not only linked, but in one respect they are almost one and the same. The brain has been referred to by some as the "largest gland in the body." While this may be hyperbole, it serves to highlight the shared function of cells in the nervous and endocrine systems. If one considers that a hormone acts at sites distant from the site of secretion, then the action of brain-derived neurohormones on distant cells in the periphery qualifies the brain as a *bona fide* endocrine organ. Thus many cells in the CNS not only interact with cells in the endocrine system, but they are endocrine cells. It can be argued that the brain is not only the largest gland in the body, but also the command center of the endocrine system.

The brain's signaling capacities are clearly crucial to the neuroendocrine system and its control of behavior by virtue of its connection to the senses. We cannot sense a threat, respond to a lover, or make appropriate decisions for our children without direction from the CNS. Our conscious thought process also initiates or attenuates neuroendocrine events. Humans can extend and exaggerate chronic stress through worry and grief, or can attenuate their reaction to stress by meditation or repeated practice (e.g., experience with public speaking and skydiving modulates the cortisol secretion that occurs during these activities). The brain's executive centers and connection to the five senses affords the hypothalamus its authority, and yet the hypothalamus is not a tyrannical dictator. The hypothalamus directs the secretions of the pituitary, and the pituitary directs hormone secretion in other peripheral cells, but the hypothalamus and pituitary also receive hormonal guidance from peripheral glands that modulate their neural output and behavior. Thus understanding hormone effects on behaviors, including obesity, jet lag, and PTSD, requires us to be not only endocrinologists but also expert neuroscientists.

## Interim Summary i.ii

The classic neuroendocrine cascade that controls behavior consists of chemical messengers going in the following sequence.

1. Brain.
2. To pituitary.
3. To a peripheral organ or organs, and then hormonal signaling back to the brain, which controls behavior.

# CENTRAL NERVOUS SYSTEM

This section begins with a brief overview of the organization of the brain. The brain is best understood by dividing it into components according to function, evolution, and geometric location in space. The most evolutionary ancient parts of the vertebrate brain, the brain stem, are those hidden deepest in the center and closer to the spinal cord. The brain stem includes the medulla oblongata, connected to the pons, which is connected to the midbrain. The brain receives sensory information from the skin, eyes, ears, tongue, and nose via the cranial nerves many of which enter at the base of the brain. In addition, the brain receives information from the peripheral organs about energy availability, salt balance, oxygenation of the blood, and hormonal state. Information from the peripheral organs (gonads, liver, pancreas, gastrointestinal tract, heart, and lungs) enters the brain via the vagus nerve to the dorsal vagal complex, which lies within the most posterior part of the brain, the medulla oblongata. The dorsal vagal complex is a nucleus with three reciprocally connected sections: the nucleus of the solitary tract (NTS), area postrema, and dorsal motor nucleus of the vagus. Information from sensory, gastrointestinal, and cardio-respiratory processes enters the dorsal vagal complex via the NTS (Fig. 3). Brain stem nuclei regulate automatic functions, i.e., those we do not need to control consciously, including respiration and heart rate.

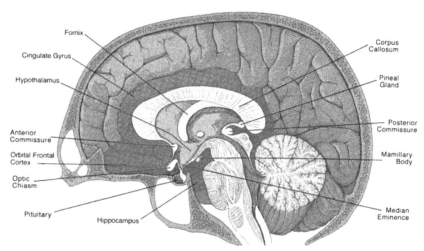

**FIGURE 3** Sagittal (midline) section of the human brain. The pituitary is at the base of the brain, connected to the median eminence of the hypothalamus by the pituitary stalk. The limbic structures (*darkly shaded*) lie around the corpus callosum and other central brain structures. *Modified from Krieger, D.T., 1980. In: Krieger, D.T., Hughes, J.C. (Eds.), Neuroendocrinology. HP Publishing, New York, p. 4; Kandel, E.R., Schwartz, J.H., Jessel, T.M., Kandel, E.R., Schwartz, J.H., Jessel, T.M., Kandel, E.R., Schwartz, J.H., Jessel, T.M., 2000. Principles of Neural Science, fourth ed. McGraw-Hill, New York, p. 987.*

Above the midbrain in primates, and in front of the midbrain in most other vertebrates, lies the limbic system, and above that the cerebral hemispheres. The limbic system is composed of relatively primitive structures, including the almond-shaped amygdala, located bilaterally on the outer edges of the brain, and the seahorse-shaped hippocampus, located centrally, which spans from hemisphere to hemisphere. The limbic system also encompasses the septal nuclei (medial olfactory nuclei), mammillary bodies, thalamus, hypothalamus, cingulate cortex, and orbitofrontal cortex. The system has extensions that reach into the cerebrum. It is the limbic system that comes into play when we deal with a stressful experience (a stressor), and when we suffer from the effects of chronic stress. It is involved in many aspects of motivated behavior, including courtship, copulatory, ingestive, and maternal behaviors. Connected and dorsal to (above) these deep brain structures lies the cerebrum with its different lobes: frontal, parietal, occipital, and temporal.

A brain area that appeared in the evolution of early vertebrates, the hypothalamus, lies at the base of the brain above (dorsal to) the pituitary gland. Just as all roads in Europe lead to Rome, many circuits in the mammalian brain lead to and from the hypothalamus. The hypothalamus contains other smaller sections, including the paraventricular nucleus of the hypothalamus (PVH), the location of cells that secrete the neurohormones known as the releasing hormones. These hormones include CRH, the neurohormone that initiates the HPA system. A great deal of attention in neuroendocrinology centers on the hypothalamus because cells in this area are essential for four patterns of basic physiological and behavioral functions: blood pressure and electrolyte composition (vasomotor tone, thirst, salt appetite); body temperature regulation (metabolic thermogenesis or shivering, seeking a different environment); energy metabolism (feeding, digestion, metabolic rate); and reproduction (hormonal control of mating, pregnancy, lactation). Within the hypothalamus lie clusters of cells, or nuclei. Specific cells within these nuclei are critical for the stress response and regulation of energy expenditure (PVH), reproduction and sexual libido (e.g., preoptic area (POA) in males and ventromedial hypothalamus (VMH) in females), hunger for food and thirst for water (e.g., arcuate nucleus, VMH, PVH, and lateral hypothalamus), and other so-called motivated states. In the anterior part of the hypothalamus lies the site of the biological clock or master circadian oscillator, the suprachiasmatic nucleus (SCN). It communicates with other parts of the hypothalamus as well as the pineal gland, the site of melatonin secretion. A great deal of this book concerns the hypothalamus and its many distinct and interconnected nuclei.

The thalamus lies above the hypothalamus, the hippocampus lies above the thalamus, and the amygdala lies lateral to (beside) the thalamus. The thalamus is a relay station between the cerebral cortex, where sensory perception and integration takes place, and the parts of the brain stem and spinal cord involved in movement. The thalamus brings information from sensory systems, integrates them with conscious thought, and relays the information to the brain areas that carry out motor actions. The cerebrum, enfolded in the cerebral cortex, surrounds these deep brain areas and plays an important role in sensory perception, generation of motor intention, spatial reasoning, conscious thought, language, and some types of learning and memory. Other types of memory, particularly emotional memory and fear learning, require connections to the hippocampus and amygdala, illustrating how the older and younger parts of the brain work together.

Throughout evolution, parts of the brain that were present in earlier forms of life have been overlaid with newer, more recently developed parts, culminating (at least at this stage of evolution) in the mammalian, primate, and uniquely human brains. When it comes to understanding human PTSD, chronic stress, compulsive eating, and obesity, we are called upon to understand how these higher brain areas involved in learning and memory interact with the limbic system and brain stem. While a great deal of neuroendocrinology involves the hypothalamus, understanding neural control of behavior requires attention to many other parts of the brain.

Structures in the forebrain and neocortex are sometimes perceived as the hallmark of the most recently evolved species and their higher cognitive abilities. Yet these same structures are under the influence of hormones and neuropeptide systems from more ancient parts of the brain. Cooperation among different brain areas is exemplified by the mesolimbic dopamine system. This system includes the ventral tegmental area, which is located in the midbrain, and its projections via the medial forebrain bundle to the forebrain areas, including the nucleus accumbens (NAcc). The NAcc receives other input from the hippocampus and amygdala. Cells of the mesolimbic system synthesize and secrete dopamine and other neurotransmitters. The system is central to learning, memory, and addiction. An excellent example of the cooperation between the brain and the periphery is the influence of the adrenal steroids, the glucocorticoids (cortisol and corticosterone) on neural architecture of the hippocampus and amygdala, and consequent effects on cognitive function, i.e., memory and learning. Optimal HPA activity in terms of glucocorticoid secretion is necessary for high levels of intellectual performance. Too little or too much HPA activity impairs learning and memory in a dose-dependent manner (Herman, 2013), and this might be key to understanding PTSD and anxiety disorders. When the hypothalamus influences behavior, it communicates with the autonomic nervous system, the endocrine system, and the immune system, all of which act together to produce coordinated physiological changes throughout the body.

Complex extant mammals evolved from simpler earlier species, some of which are now extinct, but some of which still exist in forms that have not changed since they first appeared on the Earth. The protochordates (animals that appeared on Earth before the evolution of vertebrates) managed to eat, drink, and reproduce without a hypothalamus. The ability of the brain stem to regulate behavior without the hypothalamus has been demonstrated experimentally. Laboratory rats that have had the hindbrain completely bilaterally transected from the forebrain show basic acceptance and rejection of food based on their energetic state and exogenous treatments with chemical messengers that increase or decrease food intake (e.g., increases in food acceptance after fasting or after treatment with orexigenic hormones, and increased food rejection after feeding or after treatment with anorectic hormones). Furthermore, food intake can be increased or decreased by application of hormones directly to the caudal brain stem (Grill, 2006; Grill and Hayes, 2009). These observations imply that many complex behaviors are influenced by the brain stem or an interaction between brain stem structures and forebrain structures. In addition, the brain cooperates with peripheral organs. Circadian oscillators (biological clocks) occur in organs other than the brain. For example, there is a circadian oscillator in the liver that influences the timing of hunger for food. There is no doubt that the SCN is the site of the master circadian oscillator; however, peripheral clocks contribute to many aspects of temporal synchronization of behavior. In general, it is a recurring theme that many different brain areas and peripheral organs cooperate to control individual motivated behaviors.

### Interim Summary i.iii

1. The brain can be divided into the brain stem, limbic system, and cerebral cortex, each of which is further subdivided.
2. The hypothalamus is closely related to the ancient limbic system and contains critical neuronal groups, such as those that secrete the neurohormones (the releasing hormones) and those that are critical for reproductive, parental, and ingestive behavior.
3. Brain areas are interconnected, and cooperate to regulate behavior.

## NEUROENDOCRINE CASCADES

Armed with some basic neuroanatomy, we can discuss in more detail the HPA system shown in Fig. 2.

1. When a blood-curdling scream reaches our auditory senses (the cochlear hair cells of the inner ear), the signal is processed by at least two routes: a fast subcortical route to the amygdala, and a slower route through the HPA system. The amygdala has connections to muscles in the body, and thus fast signaling via the amygdala might

account for the almost immediate reactions that occur prior to the time when we become conscious of the threat or emergency and the formation of conditioned fear responses that we cannot always recall (Campese et al., 2014). The HPA system is also initiated, but because it involves sending signals through the blood stream, it has a slower time scale. The sensory signal reaches the large cells of the PVH. Due to the size of these cells it is known as the magnocellular region.

2. Large cells in the magnocellular region of the PVH secrete the neurohormone CRH. CRH is secreted from axon terminals of the PVH cells into the capillary system that supplies blood to the anterior pituitary.
3. Cells in the anterior half of the pituitary contain receptors for CRH, and, in response to CRH binding to these receptors, the corticotrophs secrete the hormone ACTH into the blood stream.
4. ACTH reaches the adrenal cortex, binds to its receptors there, and stimulates the secretion of glucocorticoids. The main glucocorticoid in human beings is cortisol. Cortisol travels through the general circulation and binds to its receptors at many different sites in the brain and body to initiate the metabolic and behavioral aspects of the stress response.

This cascade from hypothalamus to pituitary to adrenals is essential for the high level of behavioral performance required to survive the various challenges that occur in a lifetime.

The stress response varies according to species, the individual, and the intensity of the stressor. In most cases, however, the behavioral response usually involves freezing, assessing the threat, and then either fleeing to safety or fighting to survive. In a less violent scenario the stress response might be initiated by awareness of an approaching deadline or the rising anticipation of a public speech or final examination. In all these scenarios, the glucocorticoids act in the body and brain to mobilize metabolic fuels, such as glucose and free fatty acids. These metabolic fuels are transformed from the storage molecules in muscle and adipose tissue into usable molecules that provide adenosine triphosphate (ATP), the energy currency of the cells. This cellular transformation is absolutely critical for our brains and body to have enough energy to react to the stressor (Fig. 2).

## Interim Summary i.iv

The HPA system that controls behavioral responses to stress consists of chemical messengers going in the following sequence.

1. Hypothalamic CRH from the magnocellular cells of the PVH.
2. CRH-stimulated secretion of ACTH from the anterior pituitary.
3. ACTH-stimulated secretion of glucocorticoids (e.g., cortisol) from the adrenal cortex.

# NEGATIVE AND POSITIVE FEEDBACK

Figs. 1 and 2 account for the initiation of a stress response, but when the threat or emergency is gone or resolved, what terminates the stress response? Does the hypothalamus run out of CRH? Does the pituitary exhaust its supply of ACTH? Can you wear out your adrenal cortex? The answer to all these questions is "no." Termination of the stress response is another critical function of peripheral hormones from the adrenal. After the stressful experience has passed, cortisol traveling through the blood stream reaches the brain and pituitary, where it binds to glucocorticoid receptors and inhibits production of hypothalamic CRH and pituitary ACTH. The consequent low levels of ACTH result in decreased glucocorticoid synthesis and secretion. This inhibitory process ensures that cortisol levels do not rise indefinitely, and is termed *negative feedback*. When hormones decrease the level of hormone secretion their action is referred to as *down regulation*. When hormones or other factors maintain levels of hormone secretion at a steady state their action is referred to as *clamped*. Negative feedback loops and their action to downregulate or clamp hormone secretion constitute a core principle of neuroendocrinology. Positive feedback is exactly the opposite, e.g., when a hormone action increases reactions that will, in turn, increase the amount of that very hormone.

Negative feedback has many important consequences. For example, a prolonged course of prescription glucocorticoids (e.g., prednisone or dexamethasone often prescribed for allergies or inflammation) can inhibit the patient's endogenous CRH and ACTH secretion for several weeks after the prescription ends. Abruptly ending an exogenous course of glucocorticoid treatment can results in dangerous hypocortisolemia and consequent low blood pressure, fatigue, and depression. For this reason the patient is often instructed to taper the dose gradually to release the hypothalamus and pituitary from the negative feedback effects of the prescription glucocorticoid. The prescribed gradual decrease in the dose is intended to allow the patient's own HPA system to reengage (escape the effects of negative feedback) so that the patient maintains the healthy effects of normal circulating levels of glucocorticoids. Negative feedback loops thereby regulate hormone secretion.

The HPA system exemplifies the inextricable link between CNS and the peripheral endocrine glands. The CNS is imperative for the reception of information from sensory cells, neurosecretory activity, motor and mental aspects of the behavioral stress response in response to stress hormones, and responding to the rising levels of cortisol by downregulating the stress response. The endocrine system influences how the brain processes sensory input, how the organism perceives and interprets environmental cues, and how the organism behaves in response to an environmental cue. The term *neuroendocrinology* reflects this reciprocal interaction between the endocrine and nervous systems.

The hypothalamic–pituitary–peripheral organ arrangement show in Figs. 1 and 2 recurs again and again in many different types of behaviors. The hypothalamic–pituitary–gonadal (HPG) system in male vertebrates provides a perfect example of a hormone cascade that involves negative feedback and downregulation by peripheral hormone action on the brain and pituitary (Fig. 4). The neurohormone GnRH is released from cells of the POA of the hypothalamus. GnRH stimulates secretion of two different anterior pituitary gonadotropins from pituitary cells known as gonadotrophs. These two gonadotropins are luteinizing hormone (LH), and follicle-stimulating hormone (FSH), which in turn stimulate testicular steroid production. In testicular Leydig cells, LH stimulates the synthesis of the progestagens from cholesterol and the synthesis of androgens from the progestagens. In the testicular Sertoli cell the androgen testosterone is reduced to dihydrotestosterone (DHT) and aromatized to estradiol. Androgens and estrogens play important roles in spermatogenesis and sexually dimorphic physical and behavioral characteristics. The testicular steroids are critical for sex and ingestive behavior, musculoskeletal development, male-typical genitalia, and complex sociosexual behaviors. Finally, the testicular steroids have negative feedback on the hypothalamic GnRH secretion and pituitary LH secretion, i.e., testosterone and estradiol downregulate GnRH and LH secretion.

## Interim Summary i.v

The HPG system controls steroid synthesis, gametogenesis, and reproductive behavior, and, like the other such cascades, involves negative feedback from the peripheral organs to the brain and pituitary.

1. Hypothalamic GnRH is synthesized in neurons of the POA.
2. GnRH stimulates secretion of LH and FSH from the anterior pituitary.
3. LH-stimulated and FSH-stimulated secretions of gonadal steroids from the gonads have negative feedback on the brain and pituitary, clamping secretion of GnRH, LH, and FSH.

The female HPG system is similar to that described in males, with the same hypothalamic and pituitary secretions and similar gonadal secretions. In both sexes, low circulating levels of gonadal steroids have acute negative feedback effects in brain and pituitary. There are some important differences between males and females.

1. In most species, the ovaries secrete more estrogens than androgens into general circulation, and the testes secrete more androgens than estrogens. Males of most species utilize more estrogens intracellularly after circulating testosterone enters the cell and is aromatized to estradiol.

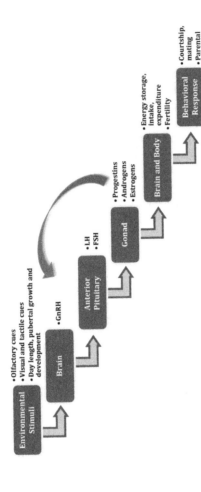

FIGURE 4   In the hypothalamic–pituitary–gonadal (HPG) system, signals from the environment interact with brain areas that secrete gonadotropin-releasing hormone (GnRH) from the brain. GnRH reaches the anterior pituitary where it stimulates secretion of luteinizing hormone (LH) and follicle stimulating hormone (FSH). LH and FSH stimulate the synthesis and secretion of gonadal steroids, which are important for energy metabolism and reproductive behaviors, including courtship, mating, and parental behavior. Gonadal steroids feed back on brain and pituitary so that steroid levels do not rise indefinitely.

2. In females, sustained high circulating concentrations of the ovarian steroid, estradiol, have positive feedback effects on hypothalamic GnRH, causing it to be secreted in 10-fold higher concentrations than in males. This upregulation of GnRH induces an enormous surge of LH from the pituitary, an event that induces ovulation. The HPG system of males produces a more or less continuous maturation of sperm, whereas the HPG system of females produces periodic ovulations. In some species these ovulatory cycles are spontaneous, and occur cyclicly on their own without sensory input. In other species ovulation only occurs when it is induced by stimuli from males.

### Interim Summary i.vi

In females of many species, just before the time of ovulation, the effects of prolonged high levels of estradiol have positive feedback on the brain and pituitary.

1. Hypothalamic GnRH is synthesized in the POA and mediobasal hypothalamus.
2. GnRH stimulates the secretion of LH and FSH from the anterior pituitary. In the early follicular phase of the estrous cycle, gonadal steroids have negative feedback on GnRH, LH, and FSH.
3. In the peri-ovulatory phase of the estrous cycle, LH-stimulated and FSH-stimulated secretions of gonadal steroids from the gonads have positive feedback on the brain and pituitary, resulting in a surge of GnRH and LH and an increase in FSH. These are necessary for ovulation.

## OTHER SIMILAR CASCADES

Another system with this familiar pattern is the hypothalamo–pituitary–thyroid (HPT) system (Fig. 5), which involves a negative feedback loop. Thyrotropin-releasing hormone (TRH) is a tripeptide (three amino acids) synthesized and secreted from cells in the PVH; it stimulates thyrotropin (thyroid-stimulating hormone or TSH) secretion from anterior pituitary cells known as thyrotrophs. TSH stimulates thyroxine and triiodothyronine secretion from the thyroid gland, and these thyroid hormones have widespread effects on metabolic rate.

Other anterior pituitary hormones are controlled in a similar way. Growth hormone (GH) is stimulated by growth-hormone-releasing hormone (GHRH) synthesized by cells in the arcuate nucleus of the hypothalamus. GH is inhibited by growth-hormone-inhibiting hormone (GHIH) synthesized in cells of the VMH. Axons from the arcuate and VMH carry the releasing and inhibiting hormones to the median eminence, where

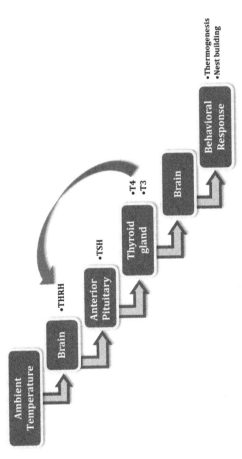

FIGURE 5   In the hypothalamic–pituitary–thyroid system, thyroid hormone-releasing hormone (THRH, also known as TRH) is synthesized in the brain and secreted into circulation in response to a fall in ambient temperature. THRH reaches the anterior pituitary and stimulates the secretion of thyroid stimulating hormone. TSH promotes the secretion of hormones of the thyroid gland, tetra-iodothyronine (T4) and triiodothyronine (T3). The thyroid gland is a small butterfly shaped gland located near the throat, and its secretions affect behavioral and physiological traits that involve cellular energy expenditure and the generation of heat. The thyroid hormones feed back on the brain and pituitary to ensure that the response to cold does not escalate indefinitely.

these neurohormones are released and travel to the anterior pituitary. There they reach the somatotrophs that secrete GH.

Similarly, anterior pituitary prolactin secretion is stimulated by many different chemical messengers, such as TRH, GnRH, and vasoactive intestinal polypeptide . Prolactin secretion is inhibited by dopaminergic neurons in the brain. Prolactin is secreted from lactotrophs in the anterior pituitary into the general circulation. Prolactin stimulates production of milk in the breast, and has many important effects on behavior. These are all additional examples of systems of hormones released from the hypothalamus acting on cells in the anterior pituitary, which affect other peripheral endocrine organs.

These hormonal cascade descriptions are oversimplified, in that other hormones can participate in the regulatory steps, as do the autonomic nervous system and immune system. For example, in the HPA system vasopressin, in addition to CRH, stimulates ACTH secretion from the anterior pituitary. The autonomic nervous system regulation of blood flow to the adrenal gland can modify adrenal response to ACTH stimulation. Furthermore, inflammatory cytokines such as interleukin-1 can be powerful stimulants of the HPA system. Some of these influences are considered in greater detail in later chapters.

As a general rule, cells in the hypothalamus secrete releasing hormones which travel to the anterior pituitary to stimulate trophic hormones. These trophic hormones affect peripheral organs in ways that influence growth, maturation, and development. Cells of the PVH synthesize CRH and TRH in their cell bodies and deliver these neurohormones via axonal projections to the median eminence. Similarly, cells in the POA and medial basal hypothalamus synthesize GnRH and deliver it via axonal projections to the median eminence. Cells of the VMH and arcuate synthesize GHRH, which also travels down axons to the median eminence. Thus releasing hormones tend to be synthesized in cells deep in the brain, and are then carried down long axons. They are released from terminal boutons at the base of the brain, close to blood vessels that carry them through the blood stream to their targets.

In contrast to the releasing hormones described in Table 1, other hypothalamic cells, rather than terminating in the median eminence at the base of the brain, send their axons beyond the base of the brain into the *posterior* pituitary. These axons stretch out of the brain and into the posterior pituitary from a structure known as the pituitary stalk, or infundibulum. The ancient Greek word for pituitary is *hypophysis,* and the posterior pituitary is also known as the *neuro*hypophysis because both the posterior pituitary and infundibulum are embryonically derived from the brain. In contrast, the anterior pituitary (hypophysis) does not derive from neural tissue and is known as the endocrine hypophysis or *adeno*hypophysis.

**TABLE 1**   Neuroendocrine Cascades Showing the Organs of Hormone Synthesis and Secretion (Left Column) and the Hormones (Rows)

| Organ (below)/ system (right) | Stress | Reproduction | Energy expenditure | Lactation | Growth |
|---|---|---|---|---|---|
| Brain (hypothalamus) | CRH | GnRH | TRH | VIP and DA | GHRH (stimulatory) and GHIH (inhibitory) |
| Pituitary | ACTH from anterior pituitary adrenocorticotrophs AVP from posterior pituitary | LH and FSH from gonadotrophs | TSH from thyrotrophs | Prolactin from luteotrophs | GH from somatotrophs |
| Peripheral Organ | Adrenal glucocorticoids | Gonadal steroids | Thyroid hormones T3 & T4 | Breast milk | IGF-1 |

*CRH*, corticotropin-releasing hormone; *GnRH*, gonadotropin-releasing hormone; *TRH*, thyrotropin-releasing hormone; *VIP*, vasoactive intestinal peptide; *DA*, dopamine; *GHRH*, growth hormone-releasing hormone (somatotropin); *GHIH*, growth hormone-inhibiting hormone (somatostatin); *ACTH*, adrenocorticotropic hormone; *LH*, luteinizing hormone; *FSH*, follicle stimulating hormone; *TSH*, thyroid-stimulating hormone; *GH*, growth hormone; *T3*, triiodothyronine; *T4*, tetraiodothyronine (thyroxine); *IGF-1*, insulin-like growth factor (somatomedin).

Fig. 6 portrays a sagittal (midline) section of the human brain showing the cells that project out of the brain into the posterior pituitary. The pituitary is shown at the base of the brain, connected to the hypothalamus by the pituitary stalk. As shown in the lower part of Fig. 6, arginine vasopressin (AVP) and oxytocin are produced in neurosecretory cells in the paraventricular (PVH) and supraoptic nuclei and carried into the posterior pituitary directly by the long axons of these cells. From the posterior pituitary, AVP and oxytocin are secreted into the general circulation. AVP stimulates the reuptake of water by the kidney and inhibits urination. Release of AVP during the stress response conserves bodily fluids and thereby maintains blood pressure. In cases of injury with blood loss, a massive release of AVP from the posterior pituitary is critical to maintain sufficient blood flow to the heart and brain, and the side effect is a feeling of urgent thirst. Oxytocin is synthesized in the brain, travels down axons, and is released from the posterior pituitary at the time of parturition to induce uterine contractions and in response to suckling stimulation to induce the milk let-down reflex. These two types of neuroendocrine cascades share in common the fact that the neuropeptides are synthesized in the brain and stimulate the pituitary. Where they diverge is that some cascades send hormonal ligands *through the blood to receptors on the anterior pituitary*, while in the other cascades hormonal ligands are released *from neurons in the posterior pituitary*.

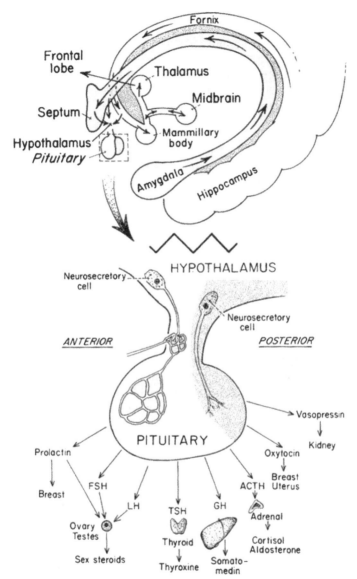

**FIGURE 6** The upper part of the figure illustrates particular limbic structures and their connections, including inputs into the hypothalamus. The lower part of the figure is an expanded portrayal of the pituitary and its connection via the pituitary stalk to the median eminence of the hypothalamus. Also illustrated are the neuroscretory cells that secrete releasing and inhibiting factors into the pituitary portal circulation, from where they influence the cells of the anterior pituitary; the neurosecretory cells that carry their hormone products directly into the posterior pituitary; and several anterior and posterior pituitary hormones and the peripheral glands and tissues that they in turn influence.

## Interim Summary i.vii

1. Many neuroendocrine cascades involve the hypothalamus sending hormones via the pituitary portal plexus to stimulate the secretion of other hormones secreted from the anterior pituitary.
2. Other neuroendocrine cascades involve the hypothalamus sending neural projections to the posterior pituitary.

# INTERACTION AMONG NEUROENDOCRINE CASCADES

These patterns of brain–endocrine cascades and the molecular mechanisms involved in each step of them can be applied to understand many complex problems in behavioral neuroendocrinology. Consider how knowledge of the HPA system can be applied to understanding the behavioral problems and obesity related to chronic stress. Food intake, energy expenditure, and energy storage as body fat and as glycogen in muscle are all related to metabolism and the processes whereby metabolic fuels are synthesized and broken down to be oxidized. Remember that the function of the so-called stress hormone, cortisol, is to increase the breakdown of stored molecules to provide metabolic fuels and energy for the stress response. The HPA system is therefore involved in energy metabolism. During the stress response, cortisol inhibits the secretion of insulin as well as cellular sensitivity to insulin action. Insulin is a hormone secreted by the pancreas that removes glucose and other metabolic fuels out of the blood stream into peripheral tissues such as muscle and adipose tissue (body fat).

The HPA system thus does not act in isolation, but rather it interacts with another hormone system: the system whereby pancreatic insulin moves fuels from circulation and into storage (Fig. 7). Insulin action can be seen as the opposite of cortisol action (Andrews and Walker, 1999). Insulin, typically secreted after meals, removes excess glucose from circulation and promotes the formation of stored glycogen in muscle and liver and stored triglycerides in adipose tissue. In fact, it is virtually impossible to gain body weight without insulin. Conversely, insulin resistance (a lack of glucose uptake in response to insulin) prevents storage of metabolic fuels and increases blood glucose levels. Sustained high levels of cortisol occur during chronic stress, and as a result insulin secretion is inhibited and responsiveness to insulin is decreased. Stress-induced insulin resistance occurs mainly in the skeletal muscles of the limbs (the arms and legs). It has been hypothesized that when the stress response is chronically triggered by persistent fears and imagined danger, cortisol remains elevated, and more and more glucose is released into circulation. The chronically stressed body reacts in unusual ways. In obesity-prone individuals,

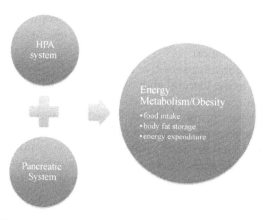

FIGURE 7 Different hormonal cascades interact to control energy metabolism (energy intake, storage, and expenditure): one of many hypotheses that might explain obesity resulting from chronic stress.

chronically elevated cortisol simulates the development of new adipocytes (fat cells) in the abdominal visceral region (Vicennati et al., 2014; Vienberg and Bjornholm, 2014). These new fat cells provide a storage area for excess glucose that must be removed from circulation (excess glucose is toxic to blood vessels and neurons). The problem is that visceral obesity exaggerates insulin resistance, inflammation, and cortisol secretion in a vicious cycle. Elevated cortisol, excess fuel storage, and peripheral insulin resistance can contribute to poor mental concentration, depression, increased hunger, and a preference for high-calorie (high-fat and high-sugar) foods (Andrews et al., 2002; Dallman, 2010; Epel et al., 2000; Paredes and Ribeiro, 2014). The mechanisms are not fully understood, but this example provides a glimpse of how neuroendocrinology is turning up new ideas that might help us understand complex biomedical problems.

Normal reproductive behavior is facilitated by the activity of the HPG system, the stress response is facilitated by the HPA system, and thermogenesis and activity are facilitated by the HPT system. These system do not act in isolation; rather, there is extensive cross-talk and feedback among them.

## Interim Summary i.viii

To explain the effects of stress on obesity, researchers from the fields of stress research and obesity research proposed a promising hypothesis.

1. The stress response and the HPA system are designed to get metabolic energy to tissues where it is needed to avoid danger, whereas the pancreatic insulin system is designed to promote glucose uptake after meals.

2. Chronic elevation of stress hormones inhibits insulin sensitivity, leading to a state of insulin resistance which prevents uptake of glucose from circulation.
3. In the absence of insulin-sensitive cells, the body generates new adipose tissue in the visceral abdominal area to accommodate the excess glucose.
4. Abdominal obesity promotes insulin resistance and a vicious cycle of overeating, body-weight gain, and further insulin resistance.
5. The mechanisms are not fully understood and are under study. They are presented here to exemplify the great importance of understanding the interactions among different hormone cascades.

# THE TIME COURSE OF HORMONE SECRETION AND SENSITIVITY TO HORMONES

If you want to know whether one child is heavier than another, you cannot answer the question by looking at a snapshot of the two children going up and down sequentially on a seesaw. Sometimes it will appear that child A is heavier and therefore pushing up child B, but at the next moment it appears the other way around. Similarly, a snapshot of a hormone's level at one time point may tell a different story than the systematic study of a pattern of hormones secreted over time. This can be seen at two levels of the HPG system. When portal blood is sampled and levels of GnRH are assessed every 8–10 min for several hours, it becomes clear that secretion of GnRH is pulsatile, with peaks in GnRH secretion roughly every hour depending on the species in question. Each hypothalamic pulse of GnRH leads to an immediate pulse of LH from the anterior pituitary, and not the other way around (LH does not stimulate GnRH secretion) (Figs. 4 and 6) (Clarke and Cummins, 1982). The pattern of GnRH and LH secretion is quite regular, and each species has its own characteristic pulse frequency. If GnRH is applied at a high and constant level over time, however, LH pulses will cease. Similarly, only pulsatile but not continuous application of LH stimulates gonadal steroid secretion and gametogenesis. In summary, gametogenesis, gonadal development, and gonadal steroid secretion are not stimulated by either a complete lack of LH secretion or a high and constant level of LH. The temporal pattern of pulsatile secretion is the critical variable. After years of study it turns out the most hormones are secreted episodically, in bursts, and the pulse frequency determines the functionality of the hormone.

The rhythms of hormone secretion that occur every 24 h are known as circadian rhythms, from the Latin *circa* (about) and *dia* (a day). Rhythms of cortisol occur on a 24-h cycle. Rhythms that occur at a higher frequency of more than once per day are known as ultradian rhythms (e.g., meal times), and rhythms that occur at a lower frequency than once a day are termed infradian rhythms. The 3–5-day estrous cycle of rats and the 28-day menstrual cycle of women are examples of infradian rhythms of hormone secretion, and these are also synchronized with circadian rhythms. The estrous cycle of the Syrian hamster, for example, occurs every 4 days, but the LH surge is strictly synchronized to occur exactly 2 h after the onset of the dark period of the photoperiod. Thus many neuroendocrine systems (the HPA, HPG, HPT, the somatic system, and those that utilize the hormones secreted by the anterior pituitary) are linked to the biological clock in the SCN. The time course of effects of hormones and behavior are explained in detail in later chapters.

## Interim Summary i.ix

Hormones are secreted in rhythmic bursts that fall into one of three categories.

1. Circadian: every 24 h.
2. Infradian: at a frequency of less than once every 24 h.
3. Ultradian: at a frequency of more than once every 24 h.

# HORMONES ACT VIA RECEPTORS

If hormones travel throughout the brain and body via the blood stream, how do they have effects on specific behavior and physiological processes? At least some specificity is conferred by hormone receptors that bind only one ligand or class of ligands. These receptors are located on the specific neurons or peripheral cells that determine the physiological process or behavior affected. Estradiol and other estrogens bind with high affinity to estrogen receptors (ERs) and not to androgen or LH receptors. Estradiol influences sex behavior via estrogen receptors in the VMH and other areas involved in control of sex behavior. Modern neuroendocrinology is concerned with hormone action on cognate receptors. Binding of hormones to their receptors triggers intracellular biochemical cascades that lead to gene transcription and translation and the synthesis of protein. There are primarily two types of hormone receptors: those spanning the cell membrane, and those within the cytoplasm.

## Membrane Receptors

There are three main types of membrane receptors, the G-protein-coupled receptors, tyrosine kinase receptors, and ligand-gated ion channels. Most of the hormones mentioned so far use G-protein-coupled receptors. GnRH, CRH, TRH, FSH, LH, AVP, and oxytocin as well as many other peptides bind to G-protein-coupled receptors. Members of the largest family of cell-membrane receptors are coupled by G-proteins to their intracellular second-messenger systems, usually adenylate cyclase or phosphoinositol. They are composed of loops of glycosylated amino acids that span the width of the cell membrane, with an extracellular binding domain and an intracellular tail. This type of receptor is very important in behavioral neuroendocrinology.

Knowing the type of receptor helps to create and synthesize specific antagonists for the receptor (molecules that mimic the binding of the ligand without activation of the signal transduction pathway) as well as agonists for it (molecules that mimic ligand action). We can say with certainty that LH binding to gonadal LH receptors is *necessary* for spermatogenesis because an antagonist to the LH receptors blocks spermatogenesis even in the presence of LH, FSH, and androgens. Many antagonists are available for peptides that act via G-protein-coupled receptors, and thus it has been relatively easy to determine whether or not those ligands are necessary for their suspected functions. In addition, we can say the LH binding to its receptors is *not sufficient* for spermatogenesis, because complete sperm maturation fails to occur in testicular cell cultures provided with only an LH agonist. On the contrary, complete spermatogenesis in testicular cell culture requires FSH and testosterone to be present in addition to an LH agonist.

On the other hand, some hormones are both necessary and sufficient for a specific function, others are necessary but not sufficient (as in the previous example), and still others are sufficient but not necessary for a particular function. Strategies for discovery include genetic "knock out" or "knock down" of the expression of genes that encode ligands and receptors. Knowledge of whether hormone action is necessary and/or sufficient guides biomedical treatments and drug discovery. Knowledge of hormone receptor mechanisms and the ability to synthesize specific agonists and antagonists are central to behavioral neuroendocrinology.

A second family of cell-membrane receptors has the effector activity (e.g., the enzyme tyrosine kinase) as part of the receptor structure. These receptors contain an extracellular ligand-binding domain, a single transmembrane-spanning amino acid sequence, and an intracellular kinase domain. Insulin, insulin growth factor I, and several other growth factors bind to this type of receptor.

A third family of cell-membrane receptors is comprised of ligand-gated ion channels. The effector mechanism for these receptors is a change in conformation of the channel which, depending on the receptor, allows the cations (positively charged electrically) sodium, potassium, or calcium to traverse the cell membrane. These receptors have four or six transmembrane-spanning domains. Their ligands are primarily neurotransmitters, such as acetylcholine (nicotinic cholinergic receptors) and GABA. Some hormones may act as allosteric (conformational) modulators of ligand-gated ion channels (e.g., progesterone and other steroid hormone modulation of the GABA receptor).

Other membrane receptors have a structure similar to those with intrinsic effector activity but do not contain such an effector; their mechanism of action is not yet understood. GH, prolactin, and several cytokines bind to this type of receptor.

## Intracellular Receptors

As mentioned earlier, positive feedback of estradiol on GnRH secretion requires sustained high concentrations of estradiol (over several hours). Why does this effect of estradiol take so long? Similarly, in females that have had the main source of estradiol removed by ovariectomy (removal of the ovaries), the restoration of mating behavior requires estradiol treatment for 24–48 h prior to behavioral testing with a male. One type of steroid action involves passive diffusion of the steroid hormone ligand through the lipid-based cell membrane into the cytoplasm of the cell. Once inside, the ligand binds to intracellular receptors. These liganded receptors pass through nuclear membranes into the nucleus of the cell, bind to specific DNA elements, and also can bind to other molecules serving as coactivators and corepressors. The unit then is known as a steroid-receptor complex. The response element for the estrogen receptor complex, for example, is known as the ER response element (ERE). The ER thus arranged acts as a transcription factor in that it binds to a specfic DNA sequence (AGGTCAnnnTGACCT) and promotes transcription of linked genes. This form of steroid action involving intracellular receptors acting as transcription factors is known as "classical genomic steroid action." Classical genomic action is characterized by relatively slow-acting and long-lived effects. For example, the classical steroid action whereby prolonged relatively high levels of brain estradiol stimulate the LH surge in females takes several hours.

In recent years it has been demonstrated that steroids, in addition to classical steroid action, can act via rapid effects on cell membranes (nonclassical, nongenomic steroid action), as mentioned above. Still more recently it has been demonstrated that steroid can act via membrane receptors to stimulate the transcription of genes other than those with steroid receptor response elements (nonclassical genomic steroid action).

## Interim Summary i.x

Hormones act in one of the following three ways.

1. Via membrane receptors. Examples include the action of protein and peptide hormones and neurosteroids that act on membrane receptors via signal transduction pathways or ion channels.

2. Via intracellular receptors. Examples include steroid receptors that are effective via classical genomic action.
3. Nonclassical genomic action.

## HORMONES HAVE THEIR OWN STEREOTYPES AND REPUTATIONS

To organize and bring meaning to their world, human beings tend to use short cuts, and these include character judgments based on stereotypes and reputations. Similarly, we tend to use short cuts to categorize hormones, which helps us to remember their main functions. This becomes problematic when we oversimplify, overgeneralize, and miss important new information about hormone function. Stereotypes and reputations are hard to shake, and they can lead to erroneous assumptions that blind us to the reality of hormone action.

For example, estradiol is often assumed to be a "female" hormone, whereas testosterone is assumed to be a "male" hormone. Testosterone, after all, was named from the male gonad, the testicle, where it is synthesized and secreted in copious amounts. The testicle was named after the Roman word for "testify," because in ancient Rome men swore their honesty and allegiance "upon their gonads." There is generally more testosterone circulating in the blood of men than in the blood of boys and women. In males of many species, circulating levels of testosterone are higher than circulating levels of estradiol. Testosterone is associated in our minds with masculinity, probably because there is a strong positive correlation between circulating levels of testosterone and the development of male secondary sex characteristics (facial and body hair, male-pattern baldness, masculine size, bone structure including the jaw, Adam's apple, deep voice, and body fat distribution) as well as levels of intermale aggression in many species of wild and laboratory animal. Estradiol, on the other hand, tends to be higher in the circulation of females than in males. The correlation holds in many different species. The problem occurs when the correlation does not hold or fails to predict a cause-and-effect relation. After long study of testosterone and estradiol in many different species, "testosterone as a male hormone" turns out to be an incomplete perspective inconsistent with many other facts.

First, both males and females synthesize both androgens and estrogens from cholesterol in many tissues, including the gonads and the brain. In males and females estradiol is the product of aromatization of testosterone, the chemical reaction catalyzed by the aromatase enzyme, aromatase. Other androgens can be converted to other estrogens. Furthermore, not all males have lower circulating concentrations of estrogens than females of the species. Stallions, for instance, have estradiol concentrations 100 times that in females (Kust, 1932; Zondek, 1934). Similarly, high amounts of estrogens are found in the urine of male donkeys, zebras, Grevy zebras, Kiang, and boars (Zondek, 1934). Furthermore, females of most species produce high levels of testosterone within the thecal cells of the ovary and many other bodily tissues, including the adrenal cortex, adipose tissue, breast, uterus, and brain. When local steroid synthesis is considered over the lifespan, the sex difference in androgens is not so great as we often assume. Female testosterone synthesis is much closer to male levels than previously thought (some estimates put female testosterone synthesis at 30%–40% of that of males). The important difference can be explained thus: in postpubertal, premenopausal women, estradiol is found in circulation and acts on ERs at a distance from where estradiol is synthesized; and testosterone is found mainly within specific tissues where it acts via androgen receptors. By contrast, in adult men circulating testosterone levels tend to be high and testosterone is transported to areas where it functions after being aromatized to estradiol (Simpson, 2003; Simpson et al., 2001, 2005; Simpson and Davis, 2001; Simpson and Jones, 2006).

It is almost silly to think of testosterone as masculinizing when this process requires aromatization of testosterone to estradiol and action on the ER. For example, in male birds territorial aggression is dependent upon testosterone from the testes, which reaches the brain through the circulation. In the brain, testosterone is aromatized to estradiol, which in turn binds to ER-alpha in the brain. In mice intermale aggression can be due to direct action of testosterone, estradiol, or the 5-alpha-reduced metabolite of testosterone, 5-alpha-DHT. In male laboratory mice, aggression consists of bites, lunges, and other offensive and defensive postures that are used to compete for resources and potential mating partners. Some neural controls over such aggression are turned on by estrogens and some by androgens, while some require a synergy between estrogens and androgens (reviewed by Simon, 2002). The relative importance of these three hormonal routes depends on the genetic strain and the species studied. In summary, in males and females the effects of estradiol on target tissue are direct, whereas the effects of testosterone might be indirect or direct. Testosterone can act directly via the androgen receptor, or indirectly as when testosterone is aromatized to estradiol and acts on ERs. The latter pattern occurs even in traits traditionally associated with masculinity.

Another stereotypical assumption is that progesterone, testosterone, and estradiol are "gonadal hormones," i.e., synthesized only in the gonads. In fact they are synthesized in many other tissues, including the brain, adipose tissue, adrenal cortex, breast, liver, and bone. When we think of them as only gonadal hormones, we limit our understanding of their important nonreproductive functions, which include vital roles in energy metabolism, ingestive behavior, bone and muscle growth and development, cardiovascular function, cognition, and mood (anxiety and depression).

Conversely, many hormones assumed to relate to "ingestive behavior" influence sex behavior. Leptin is a protein hormone secreted from adipose tissue cells (adipocytes), and the level of leptin secretion is positively correlated with adipocyte number and lipid content. The name leptin comes from the Greek *leptos*, meaning "thin," and this is consistent with data showing that short-term treatment with leptin (days to weeks) decreases food intake in many species. During starvation, animals respond by shutting down physiological processes that require high levels of energy expenditure, and this conserves energy to increase the chances of survival until food becomes available. Reproductive processes are energetically expensive, and are not necessary for immediate survival during starvation. During starvation, plasma leptin concentrations drop precipitously, leading to inhibition of the reproductive system as well as increased hunger and food intake. These starvation-induced effects on food intake and reproduction are prevented by leptin treatment in mice, rats, hamsters, and other species (reviewed by Ahima, 2004; Chehab, 2014; Schneider et al., 2013). Leptin is therefore effective at low levels when it acts a starvation signal to inhibit reproduction and promote food intake and body-fat storage. There are many such chemical messengers that promote overeating while they inhibit the reproductive system and sex behavior, and conversely many chemical messengers that inhibit food intake while they simultaneously stimulate the reproductive system and sex behavior (reviewed by Schneider et al., 2013). In no case do the names of these hormones reflect the multiplicity of functions.

Along these lines, hormones that were named for one function often turn out to have other functions, many of which have little or no relation to the original. Antidiuretic hormone (ADH), produced in the hypothalamus and transported to and released by the posterior pituitary gland, was so named because it helps the kidneys retain fluids, the exact opposite of a diuretic (a substance that promotes the production of urine). More recently, ADH was renamed AVP, in line with the fact that it causes vasoconstriction and consequent increases in arterial blood pressure. In response to reduced plasma volume, vasopressin secretion is stimulated by pressor receptors in the veins, chambers of the heart, and carotid sinus.

Behavioral endocrinologists, however, think of AVP in a different context. AVP is a neuropeptide critical for various sociosexual behaviors, including pair bonding in male prairie voles and marmosets, territorial aggression in Syrian hamsters, and parental behavior in rats (reviewed by Albers, 2015). Oxytocin, a hypothalamic hormone closely related to AVP, is well known as the hormone involved in the milk let-down reflex, and has multiple effects on social behaviors, including human bonding and parental behavior (reviewed by Carter, 2014).

Finally, hypothalamic GnRH-II, unlike GnRH-I, has little effect on pituitary gonadotropin release in well-fed animals. Rather, GnRH-II stimulates gonadotropin secretion only in food-deprived females after food availability has been restored (Temple et al., 2003). In other words, GnRH-II mediates the stimulatory effects of refeeding on pituitary gonadotropin secretion. Conversely, gonadotropin-inhibiting hormone inhibits pituitary gonadotropins in birds, but fails to live up to its name in some mammalian species; rather, in these mammals this RFamide-related peptide acts in the brain to inhibit GnRH secretion and reproductive behavior and facilitate territorial competition for resources and ingestive behaviors (Clarke et al., 2012; Klingerman et al., 2011; reviewed by Calisi, 2014; Kriegsfeld et al., 2010). Understanding the role of hormones in social behavior thus requires that we go beyond narrow-minded thinking about hormones in their original physiological roles. Subsequent chapters in this volume discuss the behavioral aspects of these hormones in greater detail.

# BEYOND HORMONES: THE COMMUNICOME

The interrelatedness of systems can be conceptualized as "omics," defined as "scientific disciplines comprising study of related sets of biological molecules" (Micheel et al., 2012). Classical examples are genomics, epigenomics, proteomics, and metabolomics. "Communicome" was first used to indicate "close coevolution of fungi with organisms present in their environment...the molecular mechanisms underlying microbial communication...giving insights into multispecies communication" (Cottier and Mühlschlegel, 2012). Communicome also is a useful term to conceptualize our current knowledge of the interrelatedness of hormones and other coevolved chemical messengers. As we learn more about the mechanisms of hormone regulation, we realize that hormones are not the only type of chemical messengers relevant to neuroendocrinology. Other chemical messengers with important effects on physiology, brain, and behavior may not meet the traditional definition of a hormone, yet they act via receptors and are essential to neuroendocrine function.

One such class of compounds, known as the eicosanoids, includes the prostaglandins, thromboxanes, leukotrienes, and a number of other similarly structured compounds. They are oxidized forms of 20-carbon fatty acids and are released by many different cell types (not only the ductless glands of the endocrine system). Prostaglandins are biologically active molecules that mediate the effects of hormones. For example, androgens and estrogens guide sexual differentiation, the masculinization or feminization of body, brain, and behavior, and the prostaglandin PGF2-alpha mediates the effects of androgens on the masculinization of brain and behavior.

At the frontier of neuroendocrinology lies the study of chemical messengers involved in cell damage, death, and rebirth. Cell death occurs by apoptosis, autophagy, and inflammation. Cell birth and death can be followed by cell rebirth, neural regeneration, and restructuring of neural circuitry. Neuroendocrinology must thus include the study of cytokines (small proteins released by cells of the immune system); chemokines (cytokines that induce chemotaxis, or movement toward a chemical attractant); growth factors (chemical messengers that stimulate cell growth and proliferation); neurotrophins (chemical messengers that promote the function, survival, and development of nerve cells); and acute-phase proteins that respond to injury and inflammation. Many of the molecules in these categories play central roles in processes related to neural plasticity in early development and adulthood, the physiological basis for learning and memory.

## OVERALL SUMMARY

The endocrine system consists of many ductless glands that secrete hormones which bind to receptors on other endocrine organs, including the brain, and these brain–hormone interactions control many behaviors that are important for survival and reproductive success. Hormone systems are organized into cascades, and many of these patterns involve a hypothalamic neurohormone, a protein hormone from the pituitary, and a hormone or hormone secreted from the peripheral endocrine organs. The nervous system contains *bona fide* endocrine cells (neurosecretory cells) that secrete chemical messengers (neurohormones) with all of the characteristics of hormones. Neurohormones stimulate the protein hormones secreted from the pituitary, which influence the secretion of hormones from peripheral organs. The hormones secreted by the peripheral organs naturally feedback negatively on the brain and pituitary, making sure that hormones levels do not rise indefinitely. In some cases, such as the female ovulatory cycle, peripheral hormones provide critical positive feedback. In other important neuroendocrine systems, neural cells extend into the posterior pituitary, the neurohypophysis, and secrete peptides that control many aspects of behavior and physiology. Disruption or overexpression of these hormone cascades can result in mental illness and disease. Understanding basic biology and solving biomedical problems can be facilitated by the principles of hormone behavior relations, but this requires having an open mind to the varied effects of hormones in different individuals, sexes, species, circumstances, and ages.

To build upon this brief overview, the reader may wish to consult Knobil and Neill's *The Physiology of Reproduction* (Academic Press, 1994/2004) and Jameson and DeGroot's *Endocrinology* (Elsevier/Saunders, 2016), and do targeted searches in Molecular Endocrinology and Endocrine Reviews.

# CHARACTERIZING THE PHENOMENA: HORMONE EFFECTS ON BEHAVIOR ARE STRONG AND RELIABLE

# 1

# Hormones Can Facilitate or Suppress Behaviors

The effects of hormones on behavior are powerful. Many behaviors simply do not occur without a requisite level of hormone secretion; therefore, hormones are central to our experience as human beings. For example, without the steroid hormone, testosterone, and its metabolite, estradiol, we would never blossom into full sexual awareness at puberty. Without the facilitatory effects of testosterone and its metabolites, we would live in a "silent spring," our lives devoid of bird song, the primary behavior whereby male songbirds attract mating partners. In most songbird species, male song is precluded by castration (removal of the testes) and restored by testosterone replacement. In a subset of bird species, the brain mechanisms for courtship song require stimulation by testosterone and/or its metabolite, estradiol. A similar hormonal effect is involved in hormonal control of agonistic behaviors, actions that harm or exclude other animals. Many of our favorite birdsongs, for example, are actually territorial displays used to compete with other males for female mating partners, and these behaviors are also dependent upon testosterone and its metabolites. The neurohormone, oxytocin is critical for the full expression of parental behaviors, actions that promote offspring survival, and affiliative behaviors, actions that bring animals in close contact. Treatments that block oxytocin binding to its receptors will prevent parental care and affiliative behaviors. Hormones have powerful effects on species-specific mating systems, i.e., whether couples practice monogamy, polygyny, and/or polyandry. Monogamy, the tendency to show preference for one mating partner, is facilitated by an array of hormones that includes both oxytocin and vasopressin. Without a fully functional vasopressin receptor, individuals tend toward promiscuous mating preferences, such as polygyny (each male mates with multiple females) and/or polyandry (each female mates with multiple males). The endocannabinoids are a class of endogenous (naturally occurring within the body) hormones essential for the pleasurable experience of eating food. Without the action of the

3

endocannabinoids binding to their receptors in the brain, users of marijuana would not experience "the munchies." Many other hormones have formidable effects on ingestive behavior (hunger for food, food intake, and foraging), and when we attempt to remove one hormone from the system, many other hormones step in to take its place. These are just a small sampling of behaviors that depend upon hormone action.

Conversely, hormones can decrease the chances that a specific type of behavior will occur, and in some cases, hormones bring behaviors to an immediate halt. Consider the stress response (Fig. 1.1). When individuals perceive an impending threat to their safety, specific activities are immediately suppressed, including foraging, ingestion, digestion, and copulation. When under attack, survival of a prey animal requires immediate evasive or defensive action and immediate cessation of all other activities. High levels of stress hormones can cause parenting birds to abandon their nests and cause parenting rodents to cannibalize their offspring. In the aftermath of an attack, cessation of energetically expensive behaviors

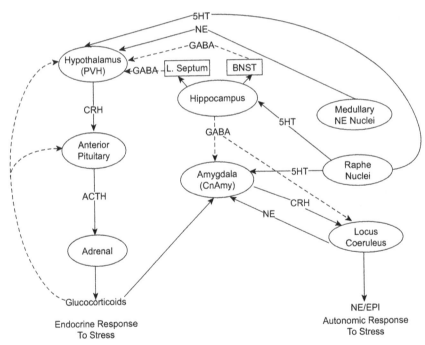

FIGURE 1.1   Neurochemical and neuroanatomical systems involved in stress responses. Key are the pathways involving corticotrophin-releasing hormone (CRH) originating in the paraventricular nucleus of the hypothalamus (PVH) and the central nucleus of the amygdala (CnAmy). Notice the two main outputs: endocrine responses and autonomic nervous system responses. *Modified from Kaufman, J., et al., 2000. Biol. Psychiatry 48, 778–790.* In addition, McEwen and colleagues reported strong glucocorticoid and mineralocorticoid inputs to the hippocampus, which we discuss further in Chapter 13.

must occur to save energy for healing and recovery. These behavioral priorities are orchestrated by hormones of the stress response, including adrenaline (a.k.a., epinephrine), corticotropin-releasing hormone (CRH), and glucocorticoids, among others.

Other disruptive hormones or hormone-like substances have their effects during development, and these effects result in deficiencies in adult behavior. For example, synthetic materials that are structurally similar to steroid molecules, such as those that leech from plastics, sewage effluent, pesticides, and herbicides, can interfere with steroids binding to their receptors, thereby disrupting developmental events that are normally controlled by steroids. These endocrine disruptors alter sexual differentiation, the process whereby males are masculinized and females are feminized. Endocrine disruptors have almost wiped out entire species by causing all-male populations of alligators, infertility in bald eagles, and an intersex condition in fish. Endocrine disruptors interfere with hormone action in all vertebrates: fish, amphibians, nonavian reptiles, birds, and mammals (including human beings), as well as in many invertebrate species. Normal, healthy hormone action is linked to reproductive success, whereas endocrine disruption is linked to cessation and alteration of ingestive and reproductive behaviors, obesity, insulin resistance, diabetes, alterations in body fat distribution, and reproductive failure (reviewed by Schneider et al., 2014).

Hormones have both "organizational" and "activational" effects on the brain and behavior. The term *activational* implies that increases in hormones affect neural circuits that are already developed and ready to be stimulated by hormones. When the circulating hormone levels fall or the hormone is prevented from binding to its receptor, the effects are removed on a relatively rapid time scale. The term was coined to describe the effects of hormones on adult behavior, which decrease or disappear when the hormone is removed from the system. For example, castration decreases aggressive behavior in many different species, and precastration levels of aggressive behavior are restored by treatment with androgens (Fig. 1.2). Activational effects are not permanent: when exogenous hormone treatments end, activational effects also end.

In contrast to activational effects, most *organizational* effects occur early in development when the foundation of brain structure is being built, and the organizational effects are permanent. After those initial organizational effects of hormones occur, continuous exposure to those hormones is not required to maintain those organizational effects. Hormones that bind to hormone receptors in the fetus, the infant, and in the juvenile set in motion a long series of developmental events that create permanent alterations in neural circuitry. These permanent alterations allow those circuits to be activated by key hormones that rise in adulthood. A main difference between organizational and activational effects is that organizational ones

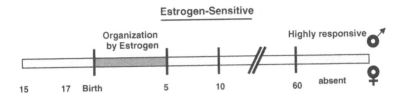

FIGURE 1.2  A summary of the major hormonal events and their timing in the establishment of androgen- and estrogen-sensitive regulatory pathways for offensive male-typical aggressive behavior in the mouse. Time is marked in days. The development of each pathway depends on exposure to specific testosterone metabolites during a restricted period shortly after birth. *From Simon, N., et al., Development and expression of hormonal systems regulating aggression. Ann. NY Acad. Sci. 794, 8–17. © 1996 New York Academy of Science. With permission.*

are permanent, whereas activational effects only occur in response to the hormone. Chapter 9 is concerned with organizational effects on behavior, whereas this chapter is concerned with activational effects.

## INTERIM SUMMARY 1.1

1. Some hormones have profound stimulatory effects on behavior, and these behaviors only rarely occur in the absence of hormones.
2. Some hormones inhibit or decrease the incidence of particular behaviors.
3. Hormone effects can be activational and reversible, whereas others are organizational and permanent.

### Examples From the History of Behavioral Endocrinology

The following historical account illustrates the fact that experimental behavioral endocrinology originated at the same moment as the field of endocrinology. A related point is that behavior is not only a legitimate topic in experimental neuroendocrinology; it is one of the best end points for any scientific experiment. Behaviors are visible to the naked eye, and quantification can often be accomplished without specialized equipment. For example, long before they could use radioimmunoassay to measure hormone levels, use a thermometer to take body temperature, or use a microscope to examine vaginal epithelial cells, ancient farmers and animal scientists could predict the time of ovulation in their female livestock by the unmistakable change in sex behaviors. Female livestock (e.g., cows) when they are approaching the time of ovulation (the periovulatory period) are said to be in "estrous" because they become increasingly agitated and restless, and the level of agitation increases further in the presence of an adult male of the species. The word *estrous* comes from a Greek word for gadfly, a biting, irritating insect that evokes frenzy in penned livestock (according to Freeman, 1994). Randy estrous behavior is a reliable marker for hormonal changes that underlie ovulatory cycles. Behaviors are useful in scientific experiments because behaviors are often stereotypical, remarkably robust, and easily quantifiable, and they change markedly according to the hormonal milieu. The following account shows that without the scientific study of behavior, we might not have the scientific study of endocrinology.

In 1849, many years before steroid hormones were discovered, the physician A. A. Berthold was moved by his curiosity to perform an experiment about the marked behavioral differences between roosters and capons. Both roosters and capons are adult male chickens (*Gallus gallus domesticus*), but capons have been castrated, whereas roosters have their testis intact. Berthold's independent variable (the variable he manipulated) was testicular secretion. He performed what is thought to be the first "remove-and-replace" experiment, i.e., removal of the gland (testes) and measurement of quantifiable traits (male-typical behaviors), followed by replacement of the glandular secretions and measurement the same traits. His dependent variables (those he observed after manipulating the independent variable) were the behaviors that are typical of a rooster but almost never seen in the hen: crowing, strutting, fighting other males, and mounting females. He noted that male-typical behaviors virtually disappear with castration and are fully restored by implantation of the testis. The restoration of rooster-like behavior occured after regeneration of the testicular circulation, but the behavioral restoration did not require any regeneration of the neural connections to the testis. Berthold therefore surmised that there is a circulating factor secreted by the testis that is necessary for male-typical behavior. It is very likely that male-typical sex and aggressive behaviors were the first trait of any kind to be experimentally linked to testicular secretions, and with this link, the scientific fields of endocrinology and behavioral endocrinology were born (Berthold, 1849). Though he had no immediate successor in this field of research, Berthold's methods and experimental design became the foundation for the study of endocrinology.

Many years later, after the techniques were invented to isolate and extract steroid hormones, it was discovered that rooster-like behavior could be restored in castrates by injection of testosterone. Since Berthold's time, remove-and-replace experiments using many different species demonstrate that intermale aggression, territoriality, courtship, and mating behaviors require sufficient testosterone and/or its metabolites (reviewed by Hull, 2017). For example, the incidence of male-typical mating behaviors in laboratory rodents such as rats and guinea pigs is significantly decreased by castration (removal of the testis) and fully restored by testosterone therapy, either by daily injections or by implants that supply a continuous infusion of testosterone. The powerful effect of testosterone on male-typical reproductive behavior is illustrated by the fact that the full repertoire of male behavior is restored even when the infusion supplies only 30% of the concentration of testosterone found in the normal, gonadally intact male (Damassa et al., 1977; Grunt and Young, 1952). In other words, the relation between testosterone concentrations and the level of sexual behavior is not linear, but rather, sexual behaviors are fully restored after a threshold concentration is reached, and beyond that concentration of testosterone, further augmentation does not lead to even higher levels of sexual behavior. Similarly, in gonadally intact, adult men, there is no significant correlation between circulating testosterone concentrations and the level of sexual function. It is, however, well-documented that treatment with physiologically relevant levels of testosterone effectively remedies the low sexual drive of hypogonadal men (e.g., men who do not synthesize testosterone due to pituitary abnormalities) (Wang et al., 2000; reviewed by Hull, 2017). As in rats and guinea pigs, in men, there is a threshold level of testosterone required for the full restoration of sexual motivation, and this threshold is well below the normal level of circulating testosterone. In summary, the effects of testosterone on behavior are powerful.

A more controversial topic at the time of this writing is whether changes in male sexual libido and erectile function can be accurately linked to falling levels of circulating testosterone that occur in some men as part of the natural aging process, and, if there is a true link, whether testosterone supplements are an appropriate healthy remedy. The answer is not as clear as some pharmaceutical companies would like us to believe (Snyder et al., 2016). Nevertheless, the early remove-and-replace experiments implicated testosterone in male reproductive behavior.

In recent years, behavioral neuroendocrinologists have refined their views on the role of testosterone in male sexual motivation and behavior. Whereas some effects of testosterone on reproductive behavior are due to this hormone binding to androgen receptors in the brain, other evidence reveals that testosterone effects involve other important downstream events. Testosterone can be metabolized to other steroids, and this is discussed in detail in Chapter 4. Furthermore, testosterone has important effects on many other peptides systems in the brain, not least of which is the dopamine system, and the combined effects of hormones are reviewed in Chapter 2 (reviewed by Hull, 2017). Other hormones effectively suppress male sexual behavior. In male rats, for example, sex behavior is inhibited by treatment with neuropeptide Y (Clark et al., 1985; Inaba et al., 2016; Poggioli et al., 1990). In summary, hormones have potent stimulatory and inhibitory effects on male reproductive behavior.

## INTERIM SUMMARY 1.2

1. Castration decreases male-typical sex behavior in many species, including men.
2. Male-typical sex behavior in some hypogonadal males can be restored by treatment with testosterone and can be inhibited in gonadally intact males by other hormones, including neuropeptide Y (NPY).
3. In some species, including human beings, there is a nonlinear relation between testosterone and male-typical behavior, such that, all other variables being equal, male-typical behavior in hypogonadal males can be restored by treatments that bring testosterone concentrations to a threshold level.

The facilitatory effects of sex steroid hormones on behavior are not restricted to males. Female courtship and mating behaviors depend on gonadal hormones, which fluctuate regularly over the estrous cycle of nonprimates and the menstrual cycle of primates.

To understand effects of hormones on female behavior, it is important to first understand the changes in hormones over the menstrual and estrous cycles. The ovulatory cycle can be divided into three distinct periods, each with a characteristic hormonal profile. The circulating concentrations of estradiol and progesterone in the blood stream are low in the beginning of the follicular phase, when the ovarian follicles are just beginning to mature. As the follicles mature and grow, the levels of estradiol begin to rise and eventually peak at the end of the follicular phase. In the next phase, the periovulatory phase, estradiol remains high, and luteinizing hormone surges, resulting in ovulation. After ovulation, the ruptured follicle transforms into the corpus luteum and begins to secrete both estradiol and progesterone. During the next phase, the luteal phase, progesterone levels rise, peak, and remain at a high plateau until just before menstruation. As the corpora lutea begin to regress, a sharp drop in circulating progesterone concentrations serves as a stimulus for menstruation in primates. In all spontaneously ovulating mammals, the cycle begins once again after the end of the luteal phase, when progesterone and estradiol fall, and the next round of follicles begin to develop and secrete estradiol (Fig. 1.3).

## Hormone Concentrations Over the Rat Estrous Cycle

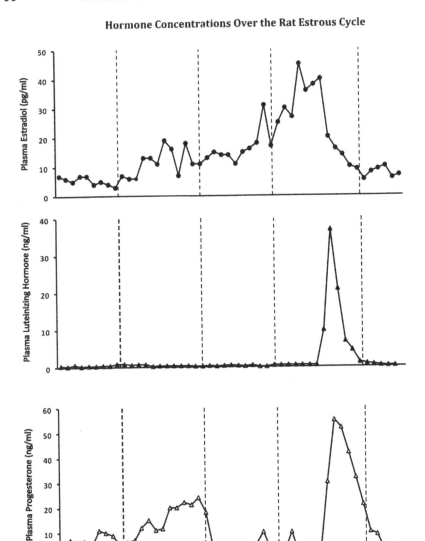

FIGURE 1.3   Plasma concentrations of estradiol (pg/mL), luteinizing hormone (ng/mL), and progesterone (ng/mL) over the phases of the estrous cycle (estrus, diestrus-1, diestrus-2, and proestrus). Blood samples were taken from 5–6 rats per group once every 2 h. *Adapted from Smith et al. (1975).*

The periovulatory phase in nonhuman mammals, including laboratory rats, is called the proestrous phase of the estrous cycle. During the proestrous period, when estradiol is at a high plateau and progesterone levels are rising, females reach their peak levels of sexual motivation. As they near the time of ovulation, female rodents will actively seek out adult, sexually experienced male conspecifics. In proestrous rats, females will respond to the close presence of an adult male by showing an array of courtship and mating behaviors, including lordosis behavior, the arched-back posture that allows males to obtain intromission (penile entry into the vagina). Within minutes of contact with an adult male, periovulatory females will freeze in place, and arch their backs, with the head and tail up and the midsection down. The number of times the female shows lordosis relative to the number of times the male mounts her is calculated to obtain the lordosis quotient. Removal of gonadal steroids from female rodents by ovariectomy decreases the lordosis quotient to almost zero. In ovari-ectomized rats, mice, and hamsters, lordosis is restored to levels shown by the periovulatory females by treatment with physiologically relevant doses of two ovarian hormones, estradiol and progesterone (reviewed by Blaustein, 2008). In Chapter 2, the importance of this combination of estradiol and progesterone will be discussed in detail, but the main point is that the ovarian steroids are implicated by classic remove-and-replace experiments.

The powerful effects of ovarian hormones go beyond lordosis behavior to include female motivation, also known as sexual desire or appetite. Whereas mating and copulation is categorized as a *consummatory* aspect of sex behavior, other behaviors are categorized as appetitive. Appetitive sex behaviors can be measured as the latency to approach an opposite-sex conspecific or to engage in courtship and mating behaviors (a short latency reflects a high level of sexual motivation). Other measures of sexual appetite include the amount of effort an animal will expend to work for a sexual reward or the level of courtship behaviors. For example, compared to ovariectomized females treated with a vehicle free of hor-mones, an ovariectomized female treated with estradiol and progesterone will run faster and work harder simply to gain access to an adult male (Cummings and Becker, 2012; Fig. 1.4).

Women engage in sex behavior in all phases of the menstrual cycle, which led many researchers to conclude that hormonal control of sex behavior in human differs from that of laboratory rodents and farm animals. When it comes to sexual motivation or sexual libido, how-ever, women do not differ markedly from other mammalian females (reviewed by Wallen, 2013). In cases in which women are reported to be vastly different from other animals in hormonal control of sexual motivation, it is often because hormone levels were not measured in the women studied, or because hormone samples were not frequent enough to capture the periovulatory increase in estradiol and progesterone. In some studies that report no significant changes in the sex behavior of women over the menstrual cycle, the results are in doubt because the measure of sex behavior was the frequency of intercourse or some other variable that is confounded by the motivation of the mating partner.

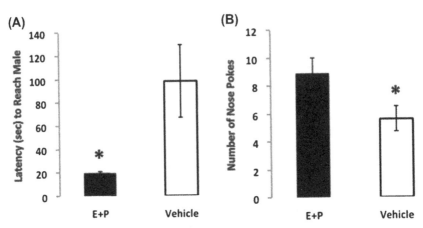

**FIGURE 1.4** In a remove-and-replace experiment, ovariectomized rats treated with either vehicle or estradiol (10 μg, 24 h prior to testing) plus progesterone (500 μg, 4–6 h prior to testing) were allowed to freely enter a compartment that contained an adult male tethered in place. (A) Ovariectomized females that received prior exposure to E plus P entered the male side of the apparatus significantly faster (at a shorter latency) than they did when they had not been hormonally primed (* = statistically significant differences where $P < .05$). (B) The number of nose pokes was used as a measure of female sexual motivation. Experimenters trained female rats to poke their noses into a hole to receive access to a tethered, adult male through an open door. The experimenters counted the number of nose pokes in ovariectomized females treated with either vehicle or E plus P. When the females were hormonally primed prior to the test, they worked significantly harder to obtain access to the male (i.e., the females showed significantly more nose pokes per door opening) than when not primed ($P < .0247$). *Adapted from Cummings, J.A., Becker, J.B., 2012. Quantitative assessment of emale sexualmotivation in the rat: hormonal control of motivation. J. Neurosci. Methods 204 (2), 227–233.*

The more relevant studies are those that assessed sexual behavior initiated by the female subjects and female sexual thoughts and fantasies. In studies of female monkeys and women in which estradiol, progesterone, and testosterone were measured daily, measures of female sexual motivation are significantly positively correlated with concentrations of estradiol, but not with progesterone or testosterone (Roney and Simmons, 2013; Wallen et al., 1984). Remove-and-replace experiments are perhaps most useful in settling the fact that estradiol facilitates sexual motivation in women. For example, in many women, sexual libido decreases after menopause, the cessation of ovarian steroid secretion, and sexual motivation is restored by treatment with estradiol alone at doses typical of the midfollicular peak plasma concentration of the menstrual cycle (Alexander et al., 2004).

Other hormones are inhibitory for sexual motivation and behavior in females. In rats, for example, lordosis is inhibited by stress-induced increases in glucocorticoids and by treatment with glucocorticoids in unstressed females (Madhuranath and Yajuryedi, 2011). In female rodents, the lordosis quotient is first increased by treatment with two daily injections of estradiol followed by a single injection with progesterone; however, prolonged treatment with progesterone has the opposite effect, preventing progesterone's effects on lordosis (Zucker, 1968; Meisel and Sterner, 1990). Similarly, when blood is sampled in women every day of the menstrual cycle, plasma progesterone concentrations are negatively correlated with sexual motivation (Roney and Simmons, 2013).

Recall that with regard to male sex behavior (discussed earlier in this chapter), the remove-and-replace experiments did not address the question as to whether hormone supplement can enhance sexual motivation when given in subjects with normal or slightly reduced circulating concentrations of gonadal steroids. Similarly, with regard to female sexual libido, remove-and-replace experiments cannot address the effect of supplementary steroids. In fact, there is no consistent effect of exogenous estrogen and progesterone treatment on sexual libido according to a large survey of the literature on women who took a combination of estrogens and progestagens as oral contraceptives. In this study, most women taking oral contraceptives (5358) reported no change in their sexual motivation; a small number of women reported an increase in sexual motivation (1826); and an even smaller number of women reported a decrease in sexual motivation (1238) (Pastor et al., 2013). These studies suggest that continuously elevated levels of estradiol and progesterone do not consistently enhance sexual motivation and, in some cases, might decrease sexual motivation. Furthermore, the discrepant behavioral effects of continuous vs. rhythmic or intermittent hormone treatment and the role of prior experience will be discussed in later chapters.

## INTERIM SUMMARY 1.3

1. Ovariectomy decreases female-typical consummatory sex behavior as well as appetitive sex behavior (sexual motivation) in many species, including women.
2. Female-typical reproductive behavior and motivation can be restored in some hypogonadal females by treatment with estradiol and/or progesterone depending on the species and prior experience.
3. Other hormones, including prolonged, high doses of progesterone, can inhibit female-typical reproductive behavior.
4. It is not clear that estradiol, progesterone, or testosterone supplements have significant effects on sexual motivation in women with intact ovaries and normal levels of steroid hormones.

## Ingestive Behavior

An intriguing example of the first principle of hormone-behavior relations is the control of ingestive behavior. Ingestive behavior includes eating and drinking and the underlying motivations or appetites. Food intake is categorized as a *consummatory* aspect of ingestive behavior and is typically defined as the amount of food eaten per unit time in laboratory animals housed at a constant room temperature (22°C). Hunger is the motivation to eat food, and it is categorized as an *appetitive* aspect of ingestive behavior. Hunger can be measured as the latency to eat food (a short latency reflects a high level of hunger) or the level of effort an animal will expend to reach food. In some species, including hamsters and human beings, a period of food deprivation is followed by increases in food hoarding or food shopping (appetitive behavior), but not by increases in food intake (consummatory behavior) (reviewed by Bartness et al., 2011). The appetitive and consummatory ingestive behaviors are increased by some hormones and decreased by others (Ammar et al., 2000 and reviewed by Schneider et al., 2013 and Keen-Rhinehart, 2013). Hunger and food intake increase in response to treatment with chemical messengers such as ghrelin, NPY, AgRP, or glucocorticoids. These chemical messengers are known as orexigens.

Conversely, anorexigenic chemical messengers decrease hunger and food intake. Anorexigenic chemical messengers include the adipocyte protein, leptin, and neuropeptides such as α-MSH, CART, and CRH. A partial list of chemical messengers that increase or decrease food intake are shown in Table 1.1 (reviewed by Schneider et al., 2013 and Keen-Rhinehart, 2013).

**TABLE 1.1**    Chemical Messengers That Influence Ingestive Behavior

| Increase Food Intake | Decrease Food Intake |
|---|---|
| Insulin (at levels that induce hypoglycemia) | Glucagon and glucagon-like peptide |
| Ghrelin | Leptin |
| Neuropeptide Y (NPY) | α-melanocyte stimulating hormone (α-MSH) |
| Agouti-related protein (AgRP) | Cocaine- and amphetamine-regulated transcript (CART) |
| Glucocorticoids | Corticotropin-releasing hormone (CRH) |
|  | Cholecystokinin (CCK) |
|  | Estradiol |
| Melanin-concentrating hormone (MCH) |  |
| Endocannabinoids |  |
| Hypocretin (orexin) |  |
| Gonadotropin-inhibiting hormone (GnIH) |  |

## Hormones That Facilitate Ingestive Behavior

### Ghrelin Facilitates Ingestive Behavior

Ghrelin is the most potent orexigenic hormone synthesized and secreted from the peripheral organs. Ghrelin is a peptide hormone secreted primarily from cells of the stomach, but also secreted in the hypothalamus. In most mammalian species studied, ghrelin tends to fall after the start of a meal and increases with increasing time after the end of a meal (Fig. 1.5). Peak levels of ghrelin tend to occur just before meal times. In nocturnal rodents that eat most of their food at the onset of the dark phase of the light–dark cycle, ghrelin secretion increases and plasma ghrelin levels rise sharply just before the onset of darkness (the timing of eating and rhythms of ghrelin will be discussed in more detail in Chapter 2). Food-deprived rodents have significantly higher circulating concentrations of ghrelin than rodents fed ad libitum (i.e., fed as much food as they want). The high levels of ghrelin in food-deprived rodents begin to subside soon after the same rodents begin to eat a meal. These results are consistent with a role for ghrelin in the initiation of circadian-timed meals and in the increases in food intake that occur after food deprivation.

Evidence for this idea is bolstered by the data showing that ghrelin treatment delivered by peripheral infusion, systemic injection, or by central infusion into the cerebral ventricles causes reliable and marked increases in food intake, even in animals that have been previously fed to satiation. Hormones dissolved in small amounts of liquid (usually a fluid that mimics the condition of the liquid normally found in the cerebral ventricles) infused into the ventricular system of the brain are carried to many brain

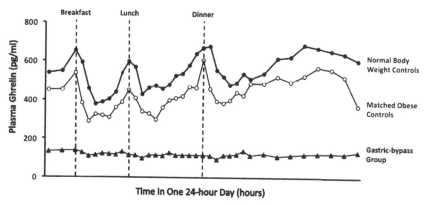

**FIGURE 1.5** Plasma ghrelin concentrations (pg/mL) are plotted over time in adult men of different categories: (1) normal body weight controls (*solid circles*), (2) obese controls (*open circles*), and (3) formerly obese patients after gastric bypass surgery (*solid triangles*). *Adapted from Cummings, D.E., Clement, K., Purnell, J.Q., Vaisse, C., Foster, K.E., Frayo, R.S., Schwartz, M.W., Basdevant, A., Weigle, D.S., 2002. Elevated plasma ghrelin levels in Prader-Willi syndrome. Nat. Med. 7, 643–644.*

areas from the lateral to the third and, finally, to the fourth ventricle in the brain stem. Thus, drugs delivered to the cerebral ventricles (i.e., drugs delivered by intracerebroventricular (ICV) injection) could be acting anywhere in the brain where there are ghrelin receptors. The ghrelin receptor, growth hormone secretagogue receptor (GHS-R), is found in abundance in the hypothalamic arcuate nucleus, as well as in the ventromedial hypothalamus (VMH), preoptic area (POA), suprachiasmatic nucleus, hippocampus, ventral tegmental area (VTA), substantia nigra, and dorsal raphe nucleus. GHS-R is also found in the brain stem, e.g., in the parabrachial nucleus, area postrema, and nucleus of the solitary tract (NTS). Both the name of the receptor as well as the term *secretagogue* refer to the fact that ghrelin binding to its receptor increases growth hormone (GH) secretion.

Which of these brain areas containing GHS-R are important for the effects of ghrelin on food intake? Does ghrelin increase food intake via increases in GH? The answers come from experiments that involve microinfusions of ghrelin that are directed at various brain nuclei. Those microinfusions directed at the arcuate nucleus of the hypothalamus are as effective in stimulating food intake as infusions into the cerebral ventricles. Furthermore, ghrelin has the same stimulatory effects on food intake when infused into the arcuate nucleus of GH-deficient mice, demonstrating that increased GH is not necessary for the effects of ghrelin on food intake (reviewed by Abizaid and Horvath, 2012; Andrews, 2011). Ghrelin is also effective in areas of the brain involved in learning and memory, making ghrelin an interesting candidate for mediating the habitual tendency to overeat highly palatable food (reviewed by Abizaid and Horvath, 2012; Andrews, 2011). ICV ghrelin infusion in Siberian hamsters increases both food hoarding and food intake, but the dose of ghrelin needed to increase food hoarding is lower than that for food intake (reviewed by Keen-Rhinehart et al., 2013).

The role of ghrelin in obesity is complex. In obese individuals, ghrelin concentrations are not only low; the expected surges around mealtimes and at nighttime are diminished (Fig. 1.5), and this has led some investigators to speculate that this leads to continuous eating throughout the day and night, rather than discrete circadian-timed meals. Given that ghrelin treatment increases food intake, it might be predicted that obese individuals would show higher levels of ghrelin than lean individuals, and/or they would be more sensitive to ghrelin's effects. The opposite, however, is the case, and this will be discussed in further detail in Chapter 2. The role of ghrelin in obesity is far from understood, but most evidence suggests that ghrelin is critical for the increases in hunger that promote animals to engage in ingestive behaviors necessary for survival, such as foraging, hoarding, and eating (reviewed by Abizaid and Horvath, 2012; Andrews, 2011).

Together, these experiments show that in sated animals, elevated arcuate nucleus concentrations of ghrelin are *sufficient* to stimulate the mechanisms that control food intake. In behavioral neuroendocrinology, the distinction between the word *sufficient* and the word *necessary* is very important.

Whereas the aforementioned experiments show that elevated levels of ghrelin in the arcuate are *sufficient* for elevated food intake, this does not show that ghrelin binding to GHS-R in the arcuate is *necessary* for increases in food intake. To determine whether ghrelin is necessary, experimenters use a drug to block ghrelin binding to its receptor (a specific antagonist) and then put the animal in a situation in which they would normally increase their food intake. One such experiment compared food deprivation–induced increases in food intake (known as hyperphagia) in subjects either ICV infused with a specific antagonist to the GHS-R or a vehicle. Indeed, food deprivation–induced hyperphagia is prevented by treatment with an antagonist to the GHS-R directed at the arcuate nucleus in rats (reviewed by Abizaid and Horvath, 2012; Andrews, 2011). Thus, it is concluded that binding of ghrelin to GHS-R is both necessary and sufficient for food deprivation–induced increases in consummatory ingestive behavior, consistent with its putative role in making sure animals are highly motivated to consume food for energy and nutrients necessary for survival.

One of the most interesting findings regarding ghrelin occurs in humans who lose impressive amounts of body weight after gastric bypass surgery. There are several different types of gastric bypass surgery, and in at least one type, the stomach is divided into a small upper pouch and a much larger lower "remnant" pouch. The small intestine is rearranged to bypass most of the stomach and receive digested food directly from the smaller pouch. The result is a very limited caloric intake, in general, and, specifically, a larger gastric load to the part of the gut that normally secretes ghrelin when the stomach is empty. The result of rapid gastric load in this section of the gut is that ghrelin secretion is inhibited, and this is hypothesized to be one factor that increases the effectiveness of weight loss after gastric bypass surgery (Cummings et al., 2002; Fig. 1.5).

## NPY Facilitates Ingestive Behavior

NPY was one of the first potent orexigens to be recognized, but unlike ghrelin, NPY is produced primarily in the brain. NPY secretion increases in response to food deprivation and decreases after eating a meal. Application of NPY to the cerebral ventricles or to specific brain areas, including the PVH, results in robust increases in food intake in sated animals. Increases in food intake in response to NPY are dose dependent, i.e., the higher the dose of NPY, the greater the increase in food intake (reviewed by Mercer et al., 2011). NPY has greater effects on appetitive than consummatory ingestive behavior. At low doses that do not affect the amount of food eaten (consummatory ingestive behavior), NPY has significant stimulatory effects on appetitive behaviors such as the latency to approach food and the level of food hoarding (Ammar et al., 2000 and reviewed by Keen-Rhinehart et al., 2013).

Most ingestive behavior researchers interested in NPY focus on a specific set of abundant arcuate nucleus NPY fibers that project to the PVH and POA and to a lesser extent the perifornical area (PFA), dorsomedial nucleus, and VMH. NPY binds to some of the many Y-type receptors in those areas. Other ingestive behavior researchers, however, point out that NPY and Y receptors reside all along the neuroaxis, and another abundant location of NPY cells is the brain stem, in particular the NTS. There are many different receptors for NPY, numbered Y1–Y7. Thus, experimenters interested in NPY action have used agonists and antagonists to the different Y receptors and asked, "Which of the Y receptors is necessary and sufficient for NPY-induced increases in food intake?" More specifically, to determine whether NPY action on a particular receptor in a particular brain area is *sufficient* for hyperphagia, an agonist to that receptor is applied in that brain area. To determine whether NPY action is *necessary* for hyperphagia, animals are put in a situation where they would normally increase their food intake, and in some of those animals, action of NPY is blocked with an antagonist to the receptor of interest. There are species differences, but in rodents, the action of NPY on the Y1, Y2, and Y5 receptors is necessary for NPY-induced increases in food intake. For example, a Y1 receptor *antagonist* infused into the PFA and PVH *prevents* increases in food intake, and a Y1 *agonist* infused into the same brain nuclei *stimulates* food intake in sated rats. Therefore, the Y1 receptor in particular appears to be important for food deprivation–induced increases in food intake (Mercer et al., 2011). In Siberian hamsters, the antagonist to the Y1 receptor prevents food deprivation–induced increases in appetitive ingestive behavior, i.e., food hoarding, when infused into the PFA, but not in the PVH, suggesting different brain areas are involved in appetitive vs. consummatory ingestive behavior (reviewed by Keen-Rhinehart et al., 2013).

### Other Hormones That Facilitate Ingestive Behavior

A similar story has emerged for the following hormones: agouti-related protein (AgRP), orexin (a.k.a., hypocretin), the endocannabinoids, and the glucocorticoids. AgRP is a protein found in abundance in the arcuate nucleus of the hypothalamus. This brain area has already been described with regard to NPY, and in fact, most cells in the arcuate that contain NPY also contain AgRP. In addition, AgRP and its receptors are found in the brain stem. AgRP acts primarily on the melanocortin-4 receptor (MCR4) to increase food intake. AgRP is higher in fasted than ad libitum–fed animals, and treatment with AgRP has potent, long-term stimulatory effects on food intake and food hoarding. AgRP effects on food intake and food hoarding tend to last far longer (days to weeks) than those of other orexigenic peptides. To determine whether AgRP is necessary for hunger and ingestive behavior, experimenters infuse an antagonist to the MCR4 receptor or create populations of mice that have a mutation that obviates the function of AgRP (termed genetic knockout mice). According to these experiments, the binding of AgRP to its receptor is necessary for food deprivation–induced increases in food intake, but it is not critical for overeating in response to a highly palatable (high fat and high sugar) diet (e.g., Denis et al., 2015). Thus, AgRP is likely a key process in survival after a period of food deprivation, but is not *the* factor (or not the only factor) necessary for the tendency to overeat when surrounded by tempting food treats.

Another endogenous group of ligands, the endocannabinoids, are linked to the pleasurable sensations associated with eating, especially the reinforcing properties of highly palatable foods. One potent endocannabinoid is arachidonoyl ethanolamide (a.k.a., anandamide (AEA)) found in brain areas of the mesolimbic dopamine reward system, including the VTA and nucleus accumbens shell. Anandamide binds to one of the cannabinoid receptors, CB1, and has strong effects on both the preference for highly palatable, high-fat foods and meal size (DiPatrizio et al., 2011). These behavioral responses are reversed by the administration of the selective CB1 cannabinoid antagonist, rimonabant. Rimonabant is a synthetic pharmaceutical product that was previously shown in clinical trials to be mildly effective in reducing body weight, and thus, it was marketed for a short time as an antiobesity drug. Subsequently, rimonabant was withdrawn from the market due to its alarming side effects on anxiety, depression, and suicidal thoughts. This outcome is not surprising when one considers that the CB1 receptors are the primary target of tetrahydrocannabinol (THC), the main ingredient in the psychoactive mood enhancer marijuana. Other hormones that increase food intake are orexin and the glucocorticoids, cortisol and corticosterone. The latter are secreted from the adrenal cortex in reaction to a stressor, and after a threat has passed, glucocorticoid-enhanced appetite might be important for recovery and healing. In chronic stress, glucocorticoids increase the preference for calorically dense foods and are thought to be the main hormone involved in stress-induced overeating, especially in women. Orexin is found in the lateral hypothalamus and increases food intake by preventing sleep and promoting food intake during the period when the individual is normally asleep. This chapter includes only a small fraction of the chemical messengers that increase hunger and food intake.

## INTERIM SUMMARY 1.4

1. Elevations in levels of ghrelin, secreted from the stomach, are necessary and sufficient to stimulate hunger and food intake.
2. Elevations in levels of NPY and AgRP secreted in the brain are necessary and sufficient to stimulate hunger and food intake in response to food deprivation, are not necessary for the drive to overeat a highly palatable diet.

3. Other hormones are critical for the pleasurable aspects of eating highly palatable food (e.g., the endocannabinoids), stress-induced overeating of calorically dense foods (glucocorticoids), and synchronizing food intake with the sleep–wake cycle (orexin).

## Hormones That Suppress Ingestive Behaviors

### Leptin Suppresses Ingestive Behavior

Leptin is one of a few peripheral protein hormones that decrease appetitive and consummatory ingestive behavior (reviewed by Flier, 1998). Leptin is secreted from a variety of tissues, but the bulk of peripheral leptin is secreted from adipocytes (fat cells) and is thus referred to as an adipokine. Leptin travels through the blood stream to its receptor (LepR) in the brain, in particular the arcuate nucleus of the hypothalamus, but also in the brain stem. Leptin is an example of a hormone that is not easily amenable to a "remove-and-replace" experiment. Leptin is produced in adipose tissue, which is distributed throughout the body, including under the skin, around the visceral organs, and within muscle and the liver. It has been challenging for biochemists to synthesize a leptin receptor antagonist. Another method for asking whether or not a hormone binding to its receptor is necessary for behavior is to measure behavior in animals that lack the gene that encodes the peptide. For example, ob/ob mice bear a homozygous recessive mutation in the gene that encodes leptin. These mice lack a functional leptin protein and develop obesity, diabetes, and hyperphagia. Treatment of ob/ob mice with leptin reverses every aspect of this phenotype, consistent with the idea that at least some leptin is necessary for normal food intake, body weight, and blood glucose homeostasis. More recently, an antagonist to the leptin receptor was made available, and it was demonstrated that treatment of rats with an antagonist to LepR increases food intake, consistent with the idea that leptin binding to its receptor decreases food intake. The most profound effect of leptin occurs in its absence or when leptin concentrations fall. Low levels of leptin are associated with voracious overeating, which can be attenuated by leptin injection; however, the effects of leptin plateau with increased doses, and continuous infusion of leptin can diminish effectiveness over time.

Problems in the secretion of leptin cannot account for widespread global obesity. Obesity researchers have searched the world over for obese patients with leptin deficiencies and have identified only a few people with mutation in the gene for leptin. Leptin therapy in morbidly obese individuals with mutations in the gene for leptin has proven moderately successful, at least initially, in decreasing body weight and food intake. Leptin treatment, both peripheral and ICV, decreases food intake in a variety of species, and it decreases food hoarding in Syrian and Siberian hamsters (reviewed by Schneider et al., 2013; Keen-Rhinehart, 2013), although most subjects rapidly develop a resistance to treatment with leptin alone (reviewed by Roujeau et al., 2014). After more than a decade of research on leptin and its receptor, there is no successful treatment for obesity that involves this hormone-receptor system. Most evidence suggests that a *fall* in leptin provides a critical signal to ensure that animals eat sufficient food to survive and reproduce (reviewed by Flier, 1998). Other functions of leptin are discussed in later chapters.

### α-MSH Suppresses Ingestive Behavior

The melanocortin agonists, cleavage products of the larger proopiomelanocortin (POMC) gene, have been known to be involved in control of food intake since the 1970s (Figs. 1.6 and 1.7). Treatment with melanocortin agonists reverses the voracious appetite of animals that previously have been food restricted or treated with other drugs that increased appetite. These agonists decreased food intake in a dose-dependent manner. Since then, hundreds of studies of rats, mice, hamsters, monkeys, and human beings confirm that melanocortin action can suppress the appetite for food, decrease food intake, and promote body weight loss. Not all POMC cleavage products affect food intake in the same way. The POMC-derived peptides include the pituitary hormone, adrenocorticotropic hormone (ACTH), the opioid peptide β-endorphin, and α-, β-, and γ-melanocyte-stimulating hormones (MSH).

An important group of POMC-containing cells resides in the arcuate nucleus of the hypothalamus, and these cells secrete the orexigenic peptide, α-MSH. ICV infusion of α-MSH decreases food intake in a variety of species. Studies that utilize an antagonist and agonist to particular melanocortin receptors (MCR) show that α-MSH action on the MCR-4s is necessary and sufficient for decreases in food intake. The endogenous agonist to the MCR-4 is α-MSH, and intracerebral treatment with this peptide decreases meal size and inhibits hunger (reviewed by Ellacott and Cone, 2006). You will recall from earlier in this chapter that AgRP is an endogenous antagonist (or a reverse agonist) to the same receptor, and treatment with this AgRP stimulates robust and long-lasting increases in meal size. In summary the arcuate nucleus contains cells that secrete POMC, and one of the POMC-derived peptides, α-MSH, as well as other cells that synthesize and secrete AgRP.

### Other Chemical Messengers That Suppress Ingestive Behavior

Other anorexigenic peptides include GLP-1, estradiol, and cholecystokinin. Glucagon-like peptide I (GLP-I) is secreted from mucosal endocrine L cells of the intestine in response to the influx of nutrients into the intestine. In response to a meal, there are two peaks in GLP-1 secretion;

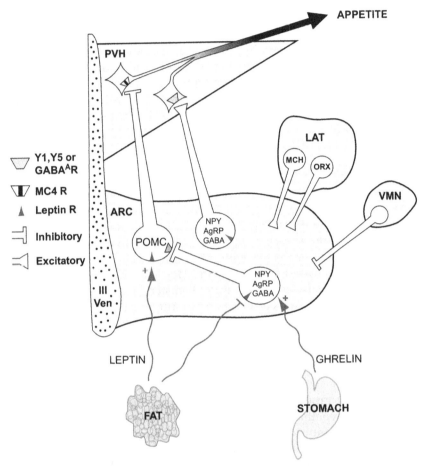

**FIGURE 1.6** Hormonal signals from stomach and fat impact neurons of specific chemical identities in the arcuate nucleus of the hypothalamus (ARC), lying next to the third ventricle (III Ven). These neurons, in turn, influence downstream neurons in the paraventricular nucleus (PVH) that facilitate food intake. A network of neuropeptide-expressing and neuropeptide-receiving neurons in the hypothalamus regulates appetitive behavior. The orexigenic peptides neuropeptide Y (NPY) and agouti-related peptide (AgRP) along with γ-aminobutyric acid (GABA) are coproduced by neurons in the arcuate nucleus (ARC) of the hypothalamus. Interspersed among these neurons are neurons that produce anorexigenic peptides, primarily α-melanocyte-stimulating hormone (α-MSH) derived from proopiomelanocortin (POMC). Both these neuronal populations project to the PVH, where α-MSH binds to MC-4 receptors to inhibit appetite. NPY and GABA released in the PVH activate the Y1/Y5 and GABA-A receptors, respectively, to stimulate appetite. Additionally, NPY released within the ARC nucleus binds to Y1 receptors on POMC neurons to reduce α-MSH synthesis. Thus, NPY stimulates appetite by direct action in the PVH and indirectly by suppressing POMC neuronal activity in the ARC. AgRP, coproduced with NPY, stimulates appetite by binding to MC-4 neurons in the PVH and preventing α-MSH action. Additional neuropeptides such as melanin-concentrating hormone (MCH) and orexins (ORX) from the lateral hypothalamus enhance appetite by upregulating the NPY network, and other unidentified signals from the ventromedial nucleus (VMN) inhibit this network. In sum, the appetite regulating axis in the

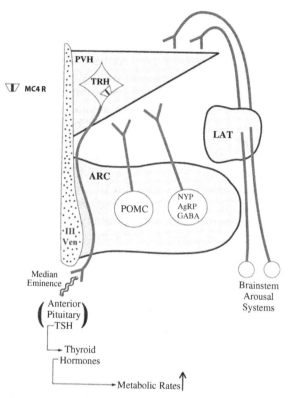

**FIGURE 1.7**   The activation of behavior, in general, as well as the influences over specific behaviors related to food depend not only on the input of metabolic energy but also on the expenditure of metabolic energy. Shown here are some of the neural inputs to the paraventricular nucleus (PVH) that influence nerve cells that secrete TRH (thyroid-stimulating thyrotropin-releasing hormone). In addition to interesting electrophysiological actions on other neurons (not shown here), the major action of TRH is in the pituitary, where it releases TSH and activates the thyroid gland. That is, TRH is released in the median eminence to travel down the portal vessels and cause the release of TSH. Throughout the body, thyroid hormones increase metabolic rates by both genomic (see Chapter 18) and nongenomic (see Fig. 16.4) mechanisms. In doing so, they prepare the body for the initiation of vigorous behaviors. *Modified from Barsh, G.S., Schwartz, M.W., 2002. Nat. Rev. Genet. 3 (8), 589–600.*

---

ARC and PVH is modulated by afferent hormonal signals that cross the blood–brain barrier and exert opposing effects. Food intake is suppressed by the adipocyte hormone leptin that conveys signals to the brain regarding the body's energy stores. Under conditions of positive energy balance, leptin binds to its receptors on NPY neurons to suppress NPY synthesis and release, and concomitantly, it activates receptors on POMC neurons to enhance α-MSH production. Conversely, hunger is signaled by ghrelin, a hormone secreted by oxyntic cells in the stomach. Blood ghrelin levels rise in conditions of negative energy balance and enhance appetite by activating NPY neuronal activity in the ARC. *Courtesy of P. and S. Kalra, University of Florida. See Morton, G.J., Schwartz, M.W., 2001. The NPY/AgrP neuron and energy homeostasis. Int. J. Obes. Relat. Metab. Disord. 5 (Suppl.), S56–S62 and Schwartz, M.W., 2001. Brain pathways controlling food intake and body weight. Exp. Biol. Med. 226 (11), 978–981.*

the first occurs within 15 min after meal initiation and is unrelated to entry of nutrients into the intestine. The second, larger peak occurs later when ingested nutrients contact intestinal L cells. GLP-I is of particular interest because gastric bypass surgery routes nutrients around the part of the intestine where nutrient infusion normally inhibits GLP-1 secretion, and thus, weight loss in gastric bypass patients might be at least partially related to elevations in this anorexigenic peptide (reviewed by Dailey and Moran, 2013; Masbad, 2014). Cholecystokinin (CCK) is another gut peptide with inhibitory effects of food intake that is also secreted in the brain. Both peripheral and central treatment with CCK receptor agonists decreases meal size, and CCK receptor antagonists increase meal size. Estradiol, in addition to promoting sexual motivation, decreases food intake, and this is discussed in detail in Chapter 3. Many other peptides in the brain and periphery decrease appetitive and consummatory aspects of ingestive behavior, and more are discovered each year.

## SUMMARY

This chapter has provided many examples of chemical messengers with profound effects on behavior. Here were just small samplings of the number of chemical messengers that stimulate and inhibit appetitive and consummatory aspects of behavior. There is no doubt that when the brain concentrations of these messengers are altered, there are changes in behavior.

The evidence presented in this chapter comes mainly from "remove-and-replace" experiments. This evidence legitimizes further research, but the results of a remove-and-replace experiment alone are not strong evidence that these hormones are involved in *natural* changes in behavior that occur over the ovulatory cycle, in the development of obesity, during fasting and dieting, or during the natural aging process. Remove-and-replace experiments include experiments in which organs are removed and hormones are infused, but they also include genetic knockout experiments in which the transcription of the gene that encodes the hormone has been rendered nonfunctional. Agonists and antagonists applied to specific brain nuclei have been used to determine whether hormones binding to particular receptors are sufficient (agonists) and/or necessary (antagonists) for a particular behavior. These techniques have laid the foundation for understanding brain mechanisms for the control of behavior by hormones.

As you probably noticed, more than one hormone can be sufficient for a particular behavior, depending on the behavior measured. You might wonder, for example, how both ghrelin and NPY can be sufficient for hyperphagia. In this case, there might be cascades of hormone action that involve two or more steps. For example, elevated ghrelin secretion from the stomach might reach cells in the arcuate that secrete NPY, and activation of the Y receptors might increase food intake and/or food hoarding. Thus, infusion of ghrelin appears to be sufficient in any animal in which the NPY/AgRP circuit is functional (Figs. 1.6 and 1.7). Similarly, you might have wondered why treatments with estradiol and progesterone are more effective than either hormone alone at facilitating the lordosis reflex in the presence of a male. Estradiol primes the brain for progesterone action. The ways that multiple steroid hormones work together to influence sex and ingestive behaviors are discussed in the next chapter.

Remove-and-replace experiments followed by experiments that employ agonists and antagonists continue to be very useful in behavioral endocrinology. These experiments alone, however, can provide little more than lists of inhibitory and stimulatory hormones. To move beyond simple lists toward wiring diagrams for neuroendocrine controls of food intake, for example, experimental hypotheses must be posed precisely, and experiments must be designed to test those precise hypotheses. If the hypothesis is "elevated levels of NPY in the arcuate constitute a sufficient stimulus to increase meal size," the resulting experiment will be different from one designed to test the hypothesis that "high-fat diet–induced obesity results from an oversecretion of NPY in response to elevated levels of fat in the diet."

To chase down such hypotheses, antagonists and agonists are combined with other techniques, including neuronal tract tracing and electrophysiological recording. Furthermore, the remove-and-replace experiments in this chapter do not address how hormones work together. Some hormones have additive effects, whereas others have synergistic effects (the effect of two hormones together is more than the sum of their individual effects). Chapter 2 describes experiments that elucidate "feeding-stimulatory" and "feeding-inhibitory" neural circuits that involve the hormones in this chapter. The next chapter begins with the circuitry involved in female reproductive behavior, the first neural circuit to be understood in detail from the sensory stimuli to the motor movements.

We have no doubt that the hormone/behavior relations you will read about in the rest of this text have clinical implications, from the management of specific behavioral responses to the activation of a wide range of behaviors. For example, consider the maladies caused by hypothyroidism (Fig. 1.8). The depression and dementia caused by an inadequate supply of thyroid hormones can take such extreme forms that the patient cannot really function in society. Many of the topics in this text will be of equal significance.

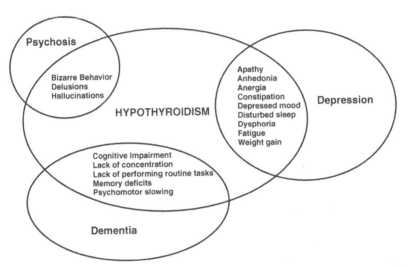

FIGURE 1.8   Most common psychiatric syndromes in hypothyroidism. *From Bauer, M., Whybrow, P.C., 2002. Thyroid Hormone, Brain, and Behavior. In: Pfaff, D.W., et al. (Eds.), Hormones, Brain, and Behavior, vol. 2. Elsevier, San Diego, p. 245 (Chapter 21). With permission.*

# 2

# Hormone Combinations Can Be Important for Behavior

Chapter 1 described the experiments that demonstrate a role of individual ovarian hormones in female-typical sex behavior. This might seem simple, but in reality, it was not easy to determine the exact hormonal "cocktail" necessary and sufficient to induce lordosis. This chapter illustrates that ovarian steroids must act in combination, in a particular order, over a particular time course to permit lordosis behavior in a particular species-specific context.

Chapter 1 listed individual hormones that stimulate or inhibit ingestive behavior, but this long list, in and of itself, does little to explain how ingestive behavior is controlled by combinations of hormones. This chapter illustrates (1) some of the neural circuits that are formed by cells that detect the presence of nutrients and/or available fuels and secrete combinations of hormones, (2) which *in turn* stimulate the secretion of hormones in other neurons, (3) which *in turn* stimulate the secretion of other hormones, and (4) which finally increase or decrease hunger for food. This chapter introduces a behavioral system that controls the intake of salt and water (dependent on combinations of hormones), and it concludes with some clinical examples.

## FEMALE REPRODUCTIVE BEHAVIOR

Even for the same pair of hormones, different temporal patterns of combinatorial action are important for different behaviors. Lordosis behavior, mentioned in Chapter 1, was the first mammalian behavioral circuit to be described in detail. The behavior endpoint was the easily quantifiable lordosis quotient, the number of times the female shows the stiff, arched-back posture known as lordosis, divided by the number of times the male mounts the female. The lordosis quotient of ovariectomized females is not restored to the level of the intact female by a single injection of estradiol or progesterone alone when given at doses that mimic natural circulating concentrations. Furthermore, the lordosis quotient is not improved by treating females with progesterone followed by estradiol. The most effective hormonal regimen to restore full lordosis in ovariectomized rats is estradiol given by subcutaneous (under the skin) injections followed at least 24 h by a subcutaneous injection of progesterone 2–5 h prior to exposure to an adult male conspecific (Table 2.1). The fact that estradiol must be given well in advance of progesterone provides a hint regarding the underlying mechanism that occurs in the estrous-cycling female as she nears the most fertile phase of the cycle. One way to describe it is to say that estradiol rising during the last half of the follicular phase primes the brain to respond to progesterone. According to endocrinologists of the late 20th century, progesterone was thought to originate from the ovarian follicle. More recently, it was discovered that "neuroprogesterone" is synthesized in the brain and is important for both the luteinizing hormone (LH) surge and lordosis. Regardless of the source of progesterone, these discoveries about estradiol and progesterone illustrate one way that hormones must work in combination to facilitate female sex behavior, and the mechanisms serve to synchronize lordosis behavior with fertility (ovulation) (Table 2.1).

To understand the exact mechanisms of estradiol priming, investigators ovariectomize females and give injections of estradiol, and then, at different time points, they examine molecular and electrophysiological

**TABLE 2.1**  Administration of Estradiol and Progesterone and Subsequent Observation of Plasma Luteinizing Hormone Levels and Behavior Have Revealed Many Parallels Between the Hormonal Conditions Sufficient for Ovulation and Those Sufficient for Female Sex Behavior Lordosis

|  | Ovulatory release of luteinizing hormone (LH) | Expression of lordosis behavior |
|---|---|---|
| Vehicle treated | O | O |
| Estrogens (E) only | ↑ | ↑ |
| Estrogens (E) followed by progesterone (P) (tested P + 2–5 h) | ↑↑↑ | ↑↑↑ |
| E followed by P administration longer (tested P + 20–25 h) | O | O |
| P (long administration) concomitant to E, followed by high E and declining P | O | O[a] |

[a] This hormonal pattern is optimal for parental behavior.

changes at the neural sites expressing estrogen and progestin receptors. These include the ventromedial hypothalamus (VMH), the medial preoptic area, and the medial amygdala. After one estradiol injection, it takes about 24 h to induce progestin receptors, which corresponds roughly with the time when progesterone is most effective in facilitation of lordosis. In intact females or in ovariectomized females treated with estradiol and progesterone, lordosis is prevented by treatments that prevent RNA transcription and protein synthesis and, hence, the synthesis of progestin receptors in the VMH. The appearance of progestin receptors after estradiol treatment and the absence of lordosis after blocking synthesis of progestin receptors support the idea that estradiol priming involves the induction of progestin receptors (e.g., Meisel and Pfaff, 1984, 1985).

To determine the important locations of estradiol priming, the action of estradiol is prevented by microinfusion with an antagonist specific to estrogen receptor alpha (ERα). In ovariectomized females, just before systemic treatment with estradiol, an ERα antagonist is delivered via a small cannula (tube) in amounts so small that they preclude diffusion outside the brain areas of interest. In different groups of females, the antagonist is delivered to specific brain areas just before estradiol treatments that are followed 24 h later by progesterone treatment. According to the data, lordosis behavior is prevented by pretreatment with an antagonist to the estrogen receptor (ERα) delivered into the VMH (Meisel et al., 1987, Table 2.2). Thus, even though the requisite level of estradiol is provided by the experimenter, estradiol is followed by the requisite level of progesterone, and a vigorous, sexually experienced male provides the requisite social cues to the female, her lordosis behavior is blocked if estradiol cannot bind to its receptor in one brain area known as the VMH. By contrast, when the antagonist is delivered to the medial preoptic area or the medial amygdala, or if the cannula aimed at the VMH misses its mark, lordosis is not prevented (the lordosis quotient is high in the presence of an adult male). An extra control group ensures that the antagonist to the ERα does not cause sickness or other ER-independent effects on lordosis. In this group (Control Group 2), the antagonist is delivered to the VMH 12 h *after* treatment with estradiol. In Control Group 2, the antagonist fails to prevent lordosis (the lordosis quotient is high) because the antagonist hits the VMH after estradiol has already initiated its effects. Other experiments in which robust lordosis is induced in ovariectomized females by microimplants of dilute estradiol into the VMH followed by systemic progesterone treatment show that estradiol in the VMH is sufficient to facilitate lordosis. Together, these results show that, in the presence of an appropriate male stimulus, estradiol binding to its receptors in the VMH is both necessary and sufficient for progesterone-induced lordosis.

If you understand the aforementioned experiment, it will not be difficult to design experiments that would determine whether progesterone

**TABLE 2.2**    The Experimental Design Used to Determine Whether Estradiol Binding to the Estrogen Receptor Alpha (ERα) in the Ventromedial Hypothalamus (VMH) is Necessary for the Full Lordosis Response in the Ovariectomized Female Rat Treated Systemically With Estradiol and Progesterone (Meisel et al., 1987)

| Group | Treatment 1 | Treatment 2 | Treatment 3 (systemic) | Lordosis quotient |
|-------|-------------|-------------|------------------------|-------------------|
| Experimental 1 | Antagonist to ERα in VMH | Systemic estradiol (48 h before testing) | Systemic progesterone (5 h before testing) | Low |
| Experimental 2 | Antagonist to ERα in the mPOA | Systemic estradiol (48 h before testing) | Systemic progesterone (5 h before testing) | High |
| Experimental 3 | Antagonist to ERα in the medial amygdala | Systemic estradiol (48 h before testing) | Systemic progesterone (5 h before testing) | High |
| Control 1 | Vehicle in VMH | Systemic estradiol (48 h before testing) | Progesterone (6 h before testing) | High |
| Control 2 | Systemic estradiol (48 h before testing) | Antagonist to ERα (48 h before testing) | Progesterone (6 h before testing) | Low |

binding to its receptor in the VMH is necessary and sufficient for lordosis in estradiol-primed, ovariectomized females. The caveat to these types of experiments is that the conditions under which a hormone binding to its receptor is sufficient must be specified. Whereas it was long thought that progesterone is necessary for lordosis facilitation, and that progesterone is part of the natural mechanism that controls the onset of lordosis in the estrous-cycling female, subsequent research showed that lordosis can be facilitated without progesterone if two estradiol injections are given several hours apart. A large body of data has accumulated showing that there is more than one mechanism whereby estradiol first inhibits lordosis prior to the estrous period and then facilitates lordosis during the estrous period. These mechanisms involve a neural circuit from the arcuate nucleus to the medial preoptic area to the VMH and plethora of hormones, including NPY in the arcuate and β-endorphin in the medial preoptic area. The time course necessary for this type of estradiol action on lordosis has been traced to changes in neuron morphology, such as the formation of specific types of neural spines that are critical for the formation of synapses. Thus, overall, the regulation of lordosis behavior is the result of hormones with combinatorial actions that create and strengthen neural connections in the lordosis circuit (reviewed by Rudolph et al., 2016).

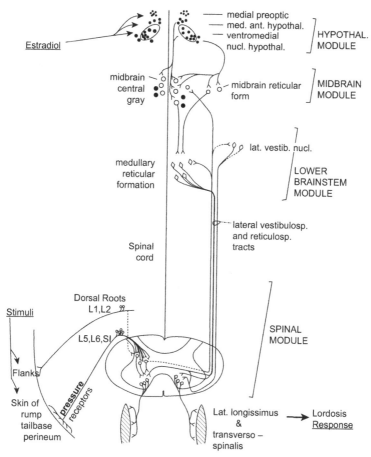

**FIGURE 2.1** Estradiol, followed by progesterone, acts at the top of the neural circuit for lordosis behavior in the ventromedial nucleus of the hypothalamus. The neural circuit is bilaterally symmetric and is plotted here on just one side for convenience of illustration. *From Pfaff, D.W., Drive, 1999. Neurobiological and Molecular Mechanisms of Sexual Motivation, MIT Press, Cambridge, MA, with permission.*

The neural circuitry and some of the genes supporting the mechanisms for this type of behavior have been worked out (Pfaff, 1999) (Fig. 2.1). Starting at the very end of the sequence, lordosis behavior is executed by muscles in the female's back, the lumbar epaxial muscles. How does the stimulation of gene transcription by estradiol in the VMH exert effects on those deep back muscles? Whereas microimplants with hormones and antagonists into particular brain areas have helped determine the hormones in the VMH necessary and sufficient for behavior, other electrophysiological and molecular biological methods elucidate the neural circuitry from brain to periphery and back (Pfaff, 1980; 1999).

In anterograde tract tracing, dye or stain is introduced into a very small brain area, where it is taken up by the nucleus of cells in that brain area. The stain is then transported forward (i.e., anterogradely) from the nucleus down the axon, thereby providing a picture of the projection sites of those cells that took up the stain. For example, a blue dye microinjected into the VMH leads to blue staining in the midbrain reticular formation, a projection site of neurons in VMH. A blue dye microinjected into the midbrain reticular formation leads to blue staining in the medullary reticular formation, a projection site of cells in the midbrain reticular formation, and so on, until the final step where the stain is found at the end of the circuit in the lumbar epaxial muscles that execute the lordosis response. The circuit can be confirmed using retrograde tract tracing, i.e., by injecting a stain into the lumbar epaxial muscles. A retrograde tracer is taken up by axon terminal boutons in the epaxial muscle and carried backward (retrogradely) to the nuclei of those cells to the medullary reticular formation.

Tract tracing reveals the existence of synapses in a circuit, but it cannot determine the functional significance of that synapse. To determine whether a particular collection of cell bodies, a nucleus, is necessary for behavior, the nucleus is lesioned prior to hormone treatment and behavioral testing. To determine whether a nucleus is sufficient, the area can be electrically stimulated. The combination of anterograde and retrograde tract tracing, lesioning, and electrophysiological recording was used to elucidate the circuit shown in Fig. 2.1 (Pfaff, 1999).

An alternative to step-by-step tract tracing is to use transneuronal viral tract tracing, a staining technique that employs a virus to carry the stain from synapse to synapse as the virus infects the nervous system. For example, when pseudorabies virus was injected into the lumbar epaxial muscle, and the brain was examined at a different time point after injection, the labeling was transported in the retrograde direction and appeared sequentially in the medullary reticular formation, then the periaqueductal gray, and finally in the VMH. This experiment confirms that there is an descending circuit from the muscles that control lordosis to the VMH (Daniels et al., 1999).

The experimenters were surprised, however, to see that the percentage of VMH cells labeled for oxytocin receptors (OTRs) was far higher than the percentage of VMH cells that were labeled for ERα (Daniels and Flanagan-Cato, 2000). These data suggest that oxytocin has an indirect facilitatory action on the lordosis behavior muscles. These data are also consistent with other experiments demonstrating that oxytocin binding to its receptor in the VMH is necessary for estradiol action on lordosis (e.g., McCarthy et al., 1994). In more recent experiments a number of other neuropeptides are implicated in the neural control of lordosis (reviewed by Rudolph et al., 2016).

You will recall from Chapter 1 that progesterone has biphasic effects on female sex behavior. It first stimulates and then inhibits lordosis. Stimulatory effects reach a maximum 4–6h after injection, but if the progesterone is infused for several more hours, the lordosis quotient decreases to zero. Finally, in a pregnant animal, estrogen levels are high and remain high during parturition. By contrast, progesterone levels start high but decline around the time of parturition, the time of birth. High estrogens and a decline in progestins make the optimal combination for maternal behavior.

## Interim Summary 2.1

1. Estradiol and progesterone combine to facilitate lordosis, the primary, essential female mating behavior.
2. Estradiol elevated for 24 h can prime the brain to respond to progesterone, specifically, by inducing progestin receptors.
3. There are multiple pathways by which estradiol can facilitate and inhibit lordosis that involve other hormones including oxytocin, NPY, and β-endorphin.

# AFFILIATIVE BEHAVIOR AND MONOGAMY

A different combination of hormone actions underlies social recognition, affiliation, and memory in many mammalian species, including primates. Several species of birds, rodents, and primates form monogamous pair bonds facilitated by multiple hormones. The biological definition of monogamy is a mating system of one male and one female forming an exclusive social pair bond for a period of time, which usually lasts at least until their offspring are weaned or fledged. Monogamy is associated with several quantifiable behaviors, such as partner preference, aggression toward novel opposite-sex conspecifics, and affiliative behaviors that include living and sleeping in close proximity and mutual grooming. Males of monogamous species often share in parental behavior. In prairie voles, for example, both males and females show a significant preference for mating and associating with their partner versus associating with a stranger, and they display more affiliative behaviors, and males attend the birth of, defend, carry, and groom their offspring. In polygyny, males mate with multiple females (although they might show a preference for one over the rest), whereas in polyandry, females mate with multiple males. In prairie voles, several days of cohabitation with an opposite-sex mating partner leads to a strong preference for mating and making physical contact with the familiar partner (Fig. 2.3). In prairie vole monogamy, strong partner preference is coupled with a violent rejection of strangers. By contrast, in a different species, the Montane vole, cohabitation with one partner has no effect on preference for spending time or mating with any particular partner (Fig. 2.3). This system has been very useful in showing how multiple hormones coordinate complex behavior patterns.

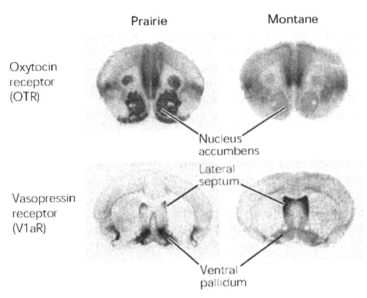

**FIGURE 2.2**    The distribution of oxytocin receptor (OTR) and the vasopressin 1a receptor (V1aR) differs markedly between species that are either monogamous or promiscuous. *Adapted from Nair, H.P., Young, L.J., 2006. Vasopressin and pair-bond formation: genes to brains to behavior. Physiology 21, 146–152.*

A monogamous pair bond requires at least three different conditions in prairie voles. First, requisite levels of oxytocin and vasopressin must be secreted in response to the continued presence of a mating partner. Second, the voles must have an abundance of OTRs and the vasopressin 1a receptor (V1aR) in specific brain areas, including the ventral pallidum and lateral septum. Polygamous species of prairie vole have a markedly different distribution of oxytocin and vasopressin receptors (Fig. 2.2). Third, increased levels of dopamine must bind to, first, the D2 receptor, and later, after the pair bond is formed, dopamine must bind to the D1 receptor to complete the bond and stimulate the rejection of novel adults. Promiscuous species of vole have many of the requisite neuropeptides, but they fail to respond to those peptides by showing partner preference or other affiliative behaviors, owing to the brain distribution of their peptide receptors.

Several teams of investigators have determined that in male prairie voles, pair bond formation requires both oxytocin and vasopressin binding to their cognate receptors in specific brain areas. In these experiments, subject males are allowed to cohabitate with an adult stimulus female for several days, and then, during a partner preference test, the subject males are provided with the option of spending time with either the familiar female or a novel female (a.k.a., a stranger). During the partner preference test, the male is free, but the strange and familiar stimulus females are tethered to their side of the apparatus. A partner preference or pair

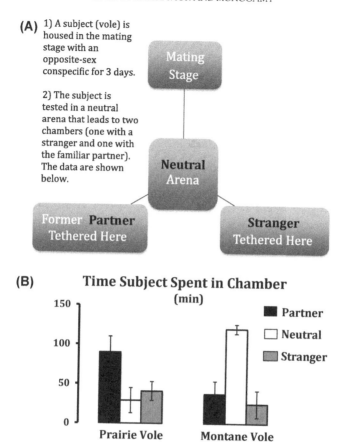

**(A)** 1) A subject (vole) is housed in the mating stage with an opposite-sex conspecific for 3 days.

2) The subject is tested in a neutral arena that leads to two chambers (one with a stranger and one with the familiar partner). The data are shown below.

Mating Stage

Neutral Arena

Former **Partner** Tethered Here

**Stranger** Tethered Here

**(B)** **Time Subject Spent in Chamber (min)**

■ Partner
□ Neutral
▨ Stranger

Prairie Vole          Montane Vole

FIGURE 2.3 (A) The experimental paradigm used to measure partner preference in voles (the amount of time a subject spends with a familiar mating partner, a stranger, or in a neutral arena). (B) Time spent (minutes) with a partner, stranger, or in an empty, neutral arena. *Adapted from Nair, H.P., Young, L.J., 2006. Vasopressin and pair-bond formation: genes to brains to behavior. Physiology 21, 146–152.*

bond is formed if the subject male spends significantly more time with the familiar female than with the stranger (Fig. 2.3). The formation of a partner preference requires several days of cohabitation, but the formation of a partner preference can be hastened by infusion of oxytocin or vasopressin into brain areas that contain receptors for those hormones. Similarly, pair bond formation can be prevented by infusion of antagonists to the oxytocin or vasopressin receptors during cohabitation (Cho et al., 1999; Johnson and Young, 2015; Lim and Young, 2004; Young et al., 2001). Follow-up experiments have used genetic manipulations to increase the transcription of genes for vasopressin receptors to determine whether this manipulation can facilitate the formation of partner preference and affiliative behavior in animals that would not otherwise show these behaviors. For "gene knockout" work, see Fig. 2.4.

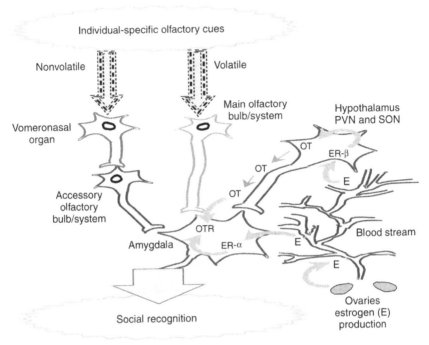

**FIGURE 2.4** A depiction of genomic and neuronal mechanisms for hormonal influences on social recognition. Estrogens produced in the ovaries circulate to the paraventricular nucleus of the hypothalamus (PVN, a.k.a., PVH) where, following binding to ER-β, they stimulate oxytocin (OT) transcription. OT is carried to the amygdala. Meanwhile, estrogens circulate to the amygdala where, following binding to ER-α, they stimulate OT receptor (OTR) transcription. OT, operating through OTRs in the amygdala, fosters social recognition. In mice, which are highly olfactory animals, olfactory signaling through both the main and the accessory olfactory systems provides the stimuli to the amygdala, to which the mice react using the estrogen-influenced OT/OTR system. Thus, a symphony of four genes— ER-α, ER-β, OT, and OTR—work in a coordinated manner. *From Choleris, E., et al., 2003. Proc. Natl. Acad. Sci. USA 100 (10), 6192–6197, with permission.*

Other investigators have focused on effects of other neurohormones and neurotransmitters that act downstream from the oxytocin and vasopressin targets.

Compared to changes in sexual motivation and territoriality, the formation of a pair bond reflects a higher level of complexity. To engage in a monogamous pair bond, brain systems must be engaged that recognize individual conspecifics, and then later recall those individuals and distinguish them from other conspecific strangers. Regardless of sexual motivation, their nervous system must add weight or salience to cues associated with the familiar, preferred mating partner and remove salience from cues associated with strangers. This sounds a lot like learning, so some researchers interested in monogamy and pair-bonding have turned their attention to areas of the brain that synthesize and secrete neurotransmitters involved in learning, memory, and addiction. Whereas oxytocin and vasopressin are released in a number of brain areas, including the ventral pallidum and lateral septum, these systems undoubtedly interact with the nucleus accumbens dopamine system, an area where dopamine is released when individuals engage in behaviors that are rewarding, e.g., eating highly palatable food, copulating, or pressing a bar that triggers a reward. Pair bond formation upregulates dopamine secretion and dopamine receptors, D1 and D2 in the nucleus accumbens shell, and pair bond formation and maintenance require the action of dopamine acting on D2 and D1 receptors (reviewed by Aragona and Wang, 2009). Whereas the D2 receptor is implicated in partner preference, dopamine action on the D1 receptor is necessary for selective aggression against the stranger (reviewed by Aragona and Wang, 2009). Similar mechanisms are at play when mothers recognize and bond to their newborn infants and when individuals learn behavioral sequences and patterns from their siblings and parents (reviewed by Insel, 2010). Many of these interactions involve oxytocin and vasopressin.

In addition, however, estrogens have an overriding effect in two ways (Fig. 2.4). Working through estrogen receptor-β, estrogens turn on the oxytocin gene; working through estrogen receptor-α, they turn on the OTR gene. Therefore, in this *combination* of hormones (i.e., estradiol and oxytocin), one has a superordinate relation to the other. Estrogens have to come first, to turn on the gene for the OTR, for instance, and oxytocin later.

These are prime examples of how hormones act in combination to produce changes in behavior.

## Interim Summary 2.2

1. Quantifiable behaviors associated with monogamy include partner preference and rejection of a strange member of the opposite sex, and these behaviors are facilitated by increases in the secretion of a number of chemical messengers, including oxytocin and vasopressin acting on their respective receptors.
2. Whereas both monogamous and promiscuous mammals secrete oxytocin and vasopressin during mating, only monogamous species show partner preference owing to a characteristic distribution of oxytocin and vasopressin receptors.
3. In addition to the aforementioned chemical messengers, the formation of partner preference and rejection of strangers is facilitated by dopamine. Partner preference is facilitated by dopamine acting on the D2 receptor, whereas rejection of strangers is facilitated by dopamine acting on the D1 receptor.

## INGESTIVE BEHAVIOR

As shown in Chapter 1, a long list of chemical messengers is involved in the control of ingestive behavior. Many of these chemical messengers are orexigenic, they increase food intake, and others are anorexigenic, they decreased food intake. The list alone provides only limited insight into the control of food intake. One part of the puzzle of ingestive behavior is how chemical messengers work together. Unfortunately, this is a very complicated puzzle, and pretending that it is simple exacerbates the growing problem of obesity. This chapter will explain three of many types of interactions among multiple chemical messengers that affect ingestive behavior: (1) multiple chemical messengers act in sequence to affect food intake, (2) hypothalamic feeding-stimulatory and feeding-inhibitory circuits interact with hormones in other brain areas and the periphery, and (3) short-term (satiation) hormones work with long-term (satiety) hormones. There are many other such interactions among hormones that affect food intake that are beyond the scope of this chapter.

### Chemical Messengers That Form Chemical Circuits

Peripheral hormones that increase food intake can act on receptors on cells in the brain that secrete orexigenic neuropeptides. These chains of neuroendocrine cells increase hunger and form what appears to be a "feeding-stimulatory" neural circuit. The feeding-stimulatory circuit has been elucidated by comparing the secretion of chemical messengers in animals that are predisposed toward eating (e.g., animals that have been previously food deprived or animals that are genetically predisposed toward hyperphagia) to the secretion of chemical messengers in animals that are not prone toward obesity and have had access to food ad libitum (at liberty). During prolonged fasting, sensations from an empty stomach (short-term signals) and low levels of metabolic fuels (long-term signals) lead to increases in the secretion of peripheral hormones such as ghrelin. As discussed in Chapter 1, ghrelin is a 28–amino acid peptide produced predominantly by the cells lining the stomach. Ghrelin was discovered in 1999 in relation to its stimulatory effect on growth hormone secretion. Its receptor is thus named the growth hormone secretagogue receptor. The 1a isoform of the growth hormone secretagogue receptor (GHS-R1a) is now called simply the ghrelin receptor (reviewed by Kirchner et al., 2012). Ghrelin levels begin to rise between meals and tend to peak just before meal times. During fasting, high levels of ghrelin result in increases in the secretion of at least three neuropeptides: neuropeptide Y (NPY), agouti-related protein (AgRP), and gaba-aminobutyric acid (GABA).

Particular neurons with cell bodies located in the arcuate nucleus of the hypothalamus synthesize and secrete both NPY and AgRP, and these NPY/AgRP cells contain the ghrelin receptor. The projections from these arcuate NPY/AgRP cells secrete the neurotransmitter GABA. There is mounting evidence that peripherally secreted hormones, such as ghrelin, and neuropeptides and neurotransmitters, such as NPY, AgRP, and GABA, form a feeding-stimulatory circuit. It has been hypothesized that the primary role of this circuit is to make sure that animals ingest sufficient food for survival and reproduction, and ghrelin responds in a way that is consistent with this hypothesis. Replenishing expended energy is an important and healthy response to stressors. Increased ghrelin secretion has been associated with psychosocial stress, mood, anxiety, and sleep. Ghrelin concentrations are elevated following metabolic stressors, including cold exposure, acute fasting, and caloric restriction, as well as psychological stressors, including chronic social defeat and unpredictable stress. The enhanced ghrelin plasma concentrations reported following stress are consistent with the idea that ghrelin acts as a defense against the consequences of stress. Ghrelin-deficient mice are anxious compared to normal mice, and this can be prevented by administration of ghrelin.

Whereas ghrelin is a "gas pedal" on the feeding-stimulatory circuit, leptin acts as a brake on this circuit. As discussed in Chapter 1, leptin is a protein composed of 167 amino acids and fits the definition of an adipokine, i.e., a hormone produced by adipocytes (fat cells). In contrast to ghrelin's stimulatory effects on NPY/AgRP cells, leptin inhibits these arcuate cells, so low concentrations of leptin that result from fasting stimulate the secretion of NPY and AgRP. Recall from Chapter 1 that leptin is synthesized primarily from adipocytes (body fat cells), and leptin is secreted in proportion to body fat content, and treatment with leptin decreases meal size and can prolong the intermeal interval. Furthermore, plasma leptin concentrations fall rapidly after fasting, even before there is a measurable loss of body fat. This fall in leptin is a potent stimulus for the activation of NPY/AgRP cells and overeating.

As mentioned in Chapter 1, treatment with any of these neuropeptides in the feeding-stimulatory circuit individually increases food intake in laboratory rodents. Furthermore, treatment with ghrelin in laboratory rodents mimics the effects of fasting, i.e., ghrelin treatment increases food intake, increases activation of cells that secrete NPY and AgRP, and increases NPY and AgRP secretion. Conversely, treatment with high doses of leptin inhibits NPY and AgRP secretion, in line with the idea that low levels of leptin induce NPY/AgRP-induced increases in food intake. Together, these results are consistent with the idea that during fasting, stimuli from an empty stomach and/or low levels of metabolic fuel oxidation stimulate the secretion of ghrelin and inhibit the secretion of leptin and insulin. It will be important to determine how deficits in fuel availability decrease the secretion of leptin and insulin and increase secretion of ghrelin. Furthermore, it will be critical to understand how these changes in peripheral hormones work together to promote a powerful drive to forage, hoard, and eat food. In summary, the metabolic and gastric consequences of fasting stimulate the synthesis and secretion of peripheral hormones and neuropeptides that stimulate food intake (reviewed by Sutton et al., 2016).

In contrast to food-deprived animals, those that have been allowed to eat ad libitum have high circulating concentrations of leptin and low circulating concentrations of ghrelin. As mentioned earlier, high levels of leptin inhibit the orexigens NPY and AgRP, and this is one way that leptin might inhibit ingestive behavior. In addition, other arcuate nucleus cells with leptin receptors synthesize proopiomelanocortin (POMC), the precursor to another peptide, α-MSH. Infusion of α-MSH into the third ventricle or microinfusion directly into the arcuate nucleus decreases food intake. The same cells synthesize another anorexigenic peptide, cocaine-and-amphetamine-regulated transcript (CART). The POMC/CART cells project to the lateral hypothalamus and other brain areas involved in hunger and food intake. Together, these and other results imply that, in well-fed individuals, high levels of leptin stimulate POMC/CART cells to secrete α-MSH and CART and thereby decrease food intake. The circuit by which leptin acts in the arcuate nucleus on POMC/CART cells is often called a "feeding-inhibitory" or "satiety" circuit (reviewed by Sutton et al., 2016).

The effects of leptin and ghrelin on these feeding-stimulatory and feeding-inhibitory circuits are not limited to changes in neuropeptide secretion; rather, these effects extend to energy expenditure, energy storage, and neural development of axonal projections and synaptogenesis (the formation of new synapses). For example, elevated plasma leptin concentrations increase synaptic input from feeding-inhibitory POMC/CART cells, and ghrelin increases synaptic input from feeding-stimulatory NPY/AgRP cells. In young animals exposed to starvation and low circulating levels of leptin during early development, altered development of hypothalamic circuits might predispose these animals toward obesity in adulthood. The obesity results from not only hormone-induced effects on food intake, but also hormone-induced decreases in energy expenditure, and it increases transformation of metabolic fuels into storage molecules and consequent body fat accumulation. The developmental effects of leptin and ghrelin on feeding-inhibitory and feeding-stimulatory circuits are implicated in the effects of high-calorie diets on the propensity to become obese (Horvath, 2006).

## Interaction of Hypothalamic Circuits With Other Brain Areas and Periphery

These feed-stimulatory and feeding-inhibitory circuits are obvious focal points for obesity research. Feedback circuits abound (Figs. 2.5 and 2.6). The simplest hypothesis to explain obesity might be that obese individuals oversecrete the orexigen (ghrelin) and undersecrete the anorexigen (leptin), but in fact, the opposite is true; obese individuals have the highest levels of leptin and lowest levels of ghrelin. The next obvious hypothesis is that in obese individuals the neurons that secrete orexigenic neuropeptides are more sensitive to ghrelin and less sensitive to leptin. This hypothesis is partially correct: for example, obese humans and some populations of laboratory rodents show (1) elevated plasma concentrations of leptin and insulin and decreased plasma concentrations of ghrelin and (2) leptin, insulin, and ghrelin resistance. For example, POMC cells

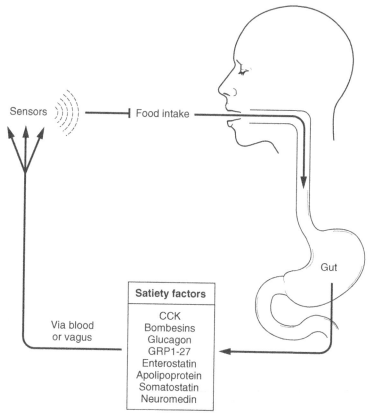

FIGURE 2.5  Several hormones secreted from the gastrointestinal tract, decrease food intake when injected peripherally or centrally, and thus, these factors are putative "satiety factors." Their relative importance and their mechanism of action is under investigation.

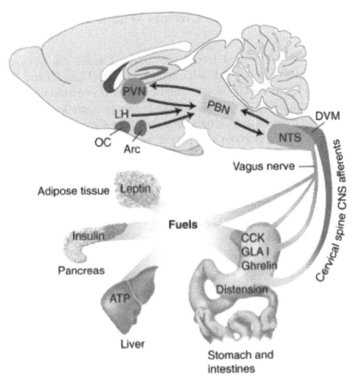

**FIGURE 2.6** A Diagram of the distributed neural control of ingestive behavior, which involves peripheral detection of signals that come from ingested food and overall disposition of metabolic fuels in the liver, muscle tissue, and adipose tissue. These signals are sent neutrally and hormonally via the caudal hindbrain to relay stations in the mid and forebrain. *Arc*, arcuate nucleus; *ATP*, adenosine triphosphate; *CCK*, cholesystokinin; *DVM*, dorsal motor nucleus of the vagus; *GLA1*, gamma-linolenic acid; *LH*, lateral hypothalamus; *NTS*, nucleus of the solitary tract; *OC*, optic chiasm; *PBN*, parabrachial nucleus; *PVN*, (a.k.a., PVH, paraventricular nucleus of the hypothalamus). *Modified from Schwartz, M.W., Woods, S.C., Porte, Jr., D., Seeley, R.J., Baskin, D.G., 2000. Central nervous system control of food intake. Nature 404, 661–671; Schneider, J.E., Wise, J.D., Benton, N.A., Brozek, J., Keen-Rhinehart, E., 2013. When do we eat? Ingestive behavior, survival, and reproductive success. Hormones Behav. 64 (4), 702–728.*

are said to be resistant to leptin because, in obesity, leptin fails to inhibit POMC gene transcription, protein synthesis, and/or secretion of the POMC cleavage product, α-MSH. NPY/AgRP cells are said to be resistant to ghrelin because in obesity ghrelin fails to increase NPY/AgRP secretion. Understanding obesity is complicated by the fact that obese individuals are resistant to both the anorexigenic leptin and the orexigenic ghrelin. Alternative hypotheses are being explored. Although ghrelin is decreased in the plasma of obese individuals and obese patients are resistant to ghrelin, plasma ghrelin concentrations are independently associated with type 2 diabetes, insulin resistance, and elevated blood pressure, suggesting that this hormone might play a role in the etiology of these complications.

As complicated as it is, the "hypothalamic perspective" described before actually provides an oversimplified model for control of food intake. Despite 100 years of research, we are facing a global rise in obesity. The incidence of global obesity has risen sharply, beginning in the 1970s from 13% to over 30%, and at the time of this writing, there is currently no reliable diet, exercise program, or pharmaceutical treatment known to prevent or reverse obesity. The rate of recidivism (subsequent body weight gain) is high for every type of intervention. The feeding-inhibitory and feeding-stimulatory circuits described earlier do not begin to encompass the neuroendocrinology of obesity, or even the neuroendocrinology of the initiation and termination of meals. They do, however, illustrate an important principle in behavioral endocrinology: hormones act in combination. The significance of this principle is that a "magic bullet" approach to obesity is unlikely to succeed.

First, as shown in Table 1.1 of Chapter 1, there are many other chemical messengers that influence food intake, and future research must unravel the relative importance of these different molecules. This redundancy has probably been the key to survival of extant populations of organisms, and it also creates one of the greatest challenges in neuroendocrinology. Second, receptors for leptin and ghrelin and cells that secrete NPY, AgRP, GABA, POMC, and CART are not confined to the arcuate nucleus or even to the hypothalamus, and these extrahypothalamic areas have been shown to participate in control of ingestive behavior. These brain areas include the caudal hindbrain, hippocampus, and ventral tegmental area. Brain areas involved in learning, memory, and reward interact with brain areas involved in hunger, satiation (the termination of meals), and satiety (the interval between meals). Furthermore, as discussed in later chapters, these hormones influence energy expenditure and body fat storage, and obesity often occurs without increases in food intake. Thus, although there is no doubt that various chemical messengers facilitate and inhibit food intake, future research must be aimed at understanding interactions among the sensory systems for the taste and smell of food, the caloric value of food, peripheral hormones, and how these sensory systems affect the neuropeptide systems involved in hunger, satiety, satiation, learning, reward, energy expenditure, and energy storage. Many hormones located in many brain areas and in the periphery work together to control food intake, energy storage, and energy expenditure (see Fig. 1.7 and reviews by Jovanovic and Yeo, 2010; Grill, 2006; Grill and Hayes, 2012).

## OTHER HORMONE COMBINATIONS

Some progress has been made in the study of hormone combinations with regard to the integration of short-term control of food intake (satiation) and long-term control of food intake (satiety). In the short term, in many mammalian species, eating and hunger are not continuous; rather, daily food intake is broken into meals. In the laboratory, meals are referred to as "eating episodes," defined as any amount of eating separated in time by other bouts of eating and from other activities (Kissileff, 2000). In experiments that use laboratory rodents, satiation is operationally defined as the end of a meal, the time at which the animal freely stops eating and does not eat again for a designated time period (Cummings and Overduin, 2007). In most laboratory species, each meal ends long before the food can be fully digested and metabolized and long before the gut reaches its maximum capacity. Termination of the meal and satiation is important to slow digestion to an optimal level to maximize absorption of nutrients and to prevent some of the adverse effects of rapid influx of glucose. Thus, there must be signals for meal termination other than maximum gut distension and a detection of the total caloric intake of the latest meal. For over 100 years, there has been an interest in the source of this satiation. A clue to the origins of satiation comes from a doctor's observation during the Civil War when an unfortunate soldier suffered a stomach wound that healed into a fistula, a tube that led from the inside of the stomach to the outside environment. Whenever the soldier ate, food passed through the oral cavity and esophagus, but it then left the stomach via the fistula; the patient was never able to satisfy his hunger. This ruled out esophageal and oral origins of satiation and suggested that satiation originated in the gut. Studies of many different species in the laboratory indicated that the termination of eating episodes is strongly influenced by signals from the small intestine. The stimuli that terminate eating episodes include submaximal, mechanical gastric distention and nutrient contact with the intestinal lumen, and the effect of these stimuli is to alter the secretion of a number of peripheral hormones, including ghrelin (decreased after meals), pancreatic polypeptide (PPY), glucagon-like peptide-1 (GLP-1), and cholecystokinin (CCK) (increased after meals) (Cummings and Overduin, 2007). The ovarian steroid, estradiol, can be included in this list because meal size increases with ovariectomy and is decreased by treatment with estradiol alone. At least one important hormone interaction is the potentiation of CCK-induced satiety by cyclic estradiol treatment. CCK is a peptide released from the small intestine when ingested food arrives, and treatment with CCK peripherally or centrally rapidly inhibits eating. The inhibition of eating is specific. CCK does not inhibit drinking but inhibits eating, even if the food is provided as a liquid diet so that the motor movements required for eating are identical to those for drinking. This effect is greater in ovariectomized females treated with estradiol injections every 4 days than in ovariectomized females treated with vehicle (reviewed by Geary et al., 2001). This is just one of many interactions among the chemical messengers that influences satiation.

In contrast to satiation, satiety refers to the interval between meals, or the long-term regulation of ingestive behavior. The reasoning behind discriminating between satiation and satiety is that signals from the full gut can terminate a meal, but the overall energetic status of the individual can shorten or lengthen the time between meals in addition to altering the size of the meal. A person who has just finished a month-long, low-calorie diet might eat smaller meals due to a diminished gut capacity but might find that they are hungrier sooner after the end of each meal (a shorter intermeal interval). Similarly, several years after gastric bypass surgery, meal size remains physically constrained. Patients must eat less than one cup of food per meal. Yet, some patients regain their lost weight by eating their one-cup food ration more frequently, with a very short intermeal interval. Another important point is that the gut hormones that terminate a meal may appear to decrease short-term food intake, but in an individual with low energy expenditure, the improved nutrient and calorie absorption caused by early meal termination might improve the efficiency of body fat storage. These considerations suggest that understanding body weight gain and loss must include attention to signals other than those that terminate meals. For example, signals that reflect body adiposity or the general availability of metabolic fuels might increase the secretion of the satiation hormones or increase sensitivity of the cells that secrete those hormones.

Interactions occur among the satiety factors. For example, an interesting and surprising interaction occurs between the POMC system and the endocannabinoid system. Recall from Chapter 1 that the cannabinoid system promotes hunger and ingestive behavior, and it is central to "the munchies" in marijuana users. Since POMC cells of the arcuate are implicated in satiety, it might be predicted that the endocannabinoids promote hunger by inhibition of POMC activity. To the contrary, activation of the endocannabinoid system, specifically the cannabinoid 1 receptor (CB1R), *increases* neuronal activity of POMC cells. Furthermore, the increase in POMC activity is essential for CB1-R-induced hyperphagia. The inhibition of POMC neuron activation blocks the ability of endocannabinoids to increase food intake. Recall from Chapter 1 that the POMC gene encodes the anorexigenic peptide α-melanocyte-stimulating hormone (α-MSH) and the opioid peptide β-endorphin. $CB_1R$ activation selectively increases β-endorphin but not α-MSH release in the hypothalamus, and systemic or hypothalamic administration of the opioid receptor antagonist naloxone blocks acute $CB_1R$-induced feeding. This mediation of endocannabinoid-induced eating via the endogenous opiate system is an example of some of the unexpected interactions that occur among chemical messengers that control ingestive behavior.

## Interim Summary 2.3

1. Ghrelin, NPY, and AgRP form a feeding-stimulatory circuit.
2. Leptin, POMC gene product, α-MSH, and CART form a feeding-inhibitory circuit.
3. Many other chemical messengers acting at hypothalamic and extrahypothalamic sites are involved in control of ingestive behavior, and more are discovered every year. There is currently no effective cure for obesity, and this is attributable in part to what seems like an endless array of these chemical messengers. More insights into obesity are provided in Chapters 3 and 14.

## HUNGER FOR SALT

One of the most powerful innate behaviors in humans and land animals, apart from thirst, is the hunger for salt (technically speaking, there are many types of salt, but here we are referring specifically to sodium). Sodium appetite is a deep-seated and fundamental urge ingrained through 30 million years of evolution; hence, Roman soldiers were paid in salt, and today, we still use the age-old aphorism for a valuable employee, "worth their weight in salt" (Denton, 1984). Through the ingestion and secretion of salt, animals maintain the volume of extracellular fluids, those fluids outside the cell, including the blood plasma. The correct volume of extracellular fluids is absolutely critical for healthy heart and circulatory function. Although a tight homeostasis (or equilibrium) in extracellular sodium is imperative, animals cannot manufacture their own endogenous sodium, so they must consume it from environmental sources. Animals will go to great lengths to ingest sodium and cannot live without it. The hormones involved in sodium appetite are surprisingly different from those involved in the appetite for food and water, although there is overlap and interaction among these ingestive behaviors. The two hormones that act on the kidney to conserve salt, angiotensin II and aldosterone, also drive salt-seeking behavior through their central actions. Angiotensin II comes from a precursor peptide, angiotensinogen. Angiotensinogen is manufactured in the liver and converted first to angiotensin I via renin synthesized in the kidney and then to angiotensin II via angiotensin-converting enzyme (the famous ACE, a target for many commonly used blood pressure medications). Angiotensin II is found in the brain and periphery. In laboratory rats, intracerebroventricular, but not systemic treatment with angiotensin II, stimulates the sodium appetite. Systemic or central treatment with aldosterone, a mineralocorticoid secreted by the adrenal cortex, increases sodium appetite. Blocking the actions of either angiotensin II or aldosterone in the central nervous system (CNS) reduces the voracious sodium appetite that is induced by sodium restriction, and blocking both hormones completely eliminates sodium appetite.

Angiotensin II acts at type 1 receptors on the anterior wall of the third ventricle of the brain. The actions of angiotensin on sodium appetite are distinct from angiotensin effects on thirst, which occur in the subfornical organ. In contrast to angiotensin II, aldosterone acts within the medial amygdala, as suggested by lesion studies and the presence of aldosterone (mineralocorticoid) receptors in that region. Furthermore, injection of mineralocorticoid receptor antisense antagonist into the medial amygdala abolishes sodium appetite (Sakai et al., 1996). The early work of Epstein and colleagues (Zhang et al., 1984) resulted in a synergy hypothesis, which implies that the effect of each hormone is enhanced by the actions of the other. The word *synergy* implies that the effects of the two hormones together is greater than the sum of the hormones given alone. For example, small doses of aldosterone and angiotensin II given alone have no effect on sodium appetite, but when they are given simultaneously, sodium appetite is markedly increased. The synergy was not only related to sodium appetite but also to motivating behaviors to seek salt in the environment, which has been observed in many species including birds. As well, this synergy for sodium appetite occurs very early in development, before the neonate is able to ingest sodium independently (Fitzsimons, 1998). The nature of this fascinating synergy is still under investigation.

The angiotensin II–aldosterone synergy and the involvement of other hormones has been studied at the cellular level. Whereas the stress hormone, corticosterone, increases angiotensinogen (the precursor to angiotensin I), mineralocorticoids, such as aldosterone, upregulate angiotensin receptors in the brain. There are two such receptors, the angiotensin 1 receptor, known for its action on blood pressure, and the type 2 receptor, active in the neonate and possibly involved in growth and development. Treatment with aldosterone increases angiotensin type 2 rather than type 1 receptors. There is a close association between the regulation of brain renin-angiotensin (renin being the kidney enzyme that converts angiotensinogen to angiotensin) and steroids, and this association allows animals to cope with stressful situations that involve repeated sodium depletion (e.g., through excessive sweating, exhausting attempts to escape from predators, or prolonged migrations when no salt ingestion is possible). The upregulation of central angiotensin by the genomic effects of mineralocorticoid and glucocorticoid receptors appears to be a major synergistic mechanism for survival in harsh conditions.

Just as hormones can act synergistically together to achieve an enhanced behavioral response, other hormones can act to diminish their effects. Atrial natriuretic peptide (ANP) opposes many of the actions of angiotensin II, including effects on thirst, drinking, blood pressure, urination, and sodium appetite (Fitzsimons, 1998). As its name indicates, ANP, which was discovered in the early 1980s as an extract from heart atria, has powerful diuretic properties. There are two other naturally occurring natriuretic peptides in mammals, brain natriuretic peptide, which was first isolated from pig brain and is actually synthesized mainly in the heart ventricles, and C-type natriuretic peptide. There are two main receptors, ANP-A and ANP-B, and a third receptor, ANP-C, which is thought to be a clearance receptor. All three peptides and their receptors occur in the brain, particularly associated with areas regulating fluid balance, such as the anteroventral third ventricle (AV3V) region, hypothalamus, nucleus of the solitary tract, and circumventricular organs. The mechanisms of these natriuretic peptides on sodium appetite is worth explaining in more detail.

Normally, ANP is in low concentrations in the circulation, but it can be released during atrial stretch, such as what occurs with volume overload. ANP increases urine volume, and ANP can act relatively quickly to alter fluid balance within the circulation, leading to hypotension (low blood pressure) (Woods, 2004). Fluid shifts from the circulation into the extracellular space, leading to increased hematocrit. At high doses, ANP is a vasodilator as well and has been marketed as a treatment for acute decompensating heart failure (ANP has a very short half-life and therefore must be infused intravenously). Some of the action of ANP is by inhibiting the release of renin and aldosterone and the synthesis of angiotensin. Thus, ANP injected centrally inhibits sodium appetite, likely through ANP-A, which increases cGMP (McCann et al., 2003; Blackburn et al., 1995). One of its sites may be the subfornical organ, because injection of natriuretic peptides in this region inhibits angiotensin-induced drinking.

In humans, ANP has been reported to depress osmotic thirst but not affect osmotically released vasopressin (Burrell et al., 1991). Thus, ANP stimulates natriuresis and diuresis in situations of volume expansion and/or increased plasma osmolality, as well as inhibition of the release and actions of components of the renin-angiotensin system, resulting in lowering of blood pressure. Other cardioprotective effects of ANP that are opposite to angiotensin II's effects include inhibiting the sympathetic nervous system and promoting vagal afferent reflexes to improve vagal tone (Thomas et al., 1998; Woods, 2004). It is interesting to note that the venom of snakes such as the green mamba and Australian inland python contain their own versions of natriuretic peptides that are potent and, unlike the mammalian forms, are long acting (Vink et al., 2012). These small peptides might have evolved in snakes to make escape by their prey less likely, by lowering cardiac output and preventing fluid-retaining responses to injury or hemorrhage.

## Interim Summary 2.4

1. Sodium appetite is increased by the synergistic effects of aldosterone from the adrenal cortex and angiotensin II in the brain.
2. Sodium appetite is decreased by ANP, which decreases secretion of angiotensin II and aldosterone
3. ANP opposes angiotensin II and aldosterone with regard to multiple aspects of sodium and water balance, blood pressure, and vagus nerve tone.

# CLINICALLY, THE SEQUENCE OF HORMONE TREATMENT CAN BE IMPORTANT

## Postmenopausal Hormone Replacement Therapy

Menopause denotes the cessation of menstrual cycles and occurs in women around the age of 50 (see also Chapter 11). It is due to a rapid decrease in the number of ovarian follicles. Circulating estradiol, produced during the follicular phase of the menstrual cycle, and progesterone, produced during the luteal phase, fall to very low values, and LH and follicle-stimulating hormone (FSH) production by the pituitary rises substantially. Many women experience a wide array of symptoms at menopause, many of them behavioral, including hot flashes, sweating, insomnia, headaches, dizziness, lack of energy, palpitations, digestive disturbances, difficulty concentrating, nervous tension, and decreased sex drive. Also, bone density decreases, thus increasing the risk of fractures, and the occurrence of cardiovascular disease increases with increasing age.

Estrogen replacement therapy can ameliorate many of these symptoms, but it carries certain risks. Initially, estrogen alone was used, but it became apparent that a major side effect was an approximately sevenfold increase in the occurrence of cancer of the uterine lining (endometrium). Thus, estrogen alone can be used in women after total hysterectomy (removal of both the uterus and the ovaries), but alternative hormone replacement strategies are necessary in anatomically intact postmenopausal women to guard against the development of endometrial carcinoma.

The next strategy shown to be effective in relieving menopausal symptoms was sequential hormone replacement, with estrogen followed by a progestin, which mimicked the normal monthly hormonal sequence and resulted in induced monthly shedding of the endometrium. This strategy reduced the risk of endometrial carcinoma while still ameliorating the symptoms of menopause and protecting against osteoporosis, but it did lead to monthly bleeding, an inconvenience for menopausal women. More recently, continuous administration of a combination of estrogen and a progestogen has been used, which confers a similar therapeutic effect and maintains an atrophic endometrium, thereby eliminating induced menses (although breakthrough bleeding can occur in some women).

A complication of combined estrogen and progestogen hormone replacement therapy (HRT) is increased risk of breast cancer. A number of epidemiological studies have concluded that a low daily dose of estrogen does not increase breast cancer incidence. An ongoing, large prospective study, however, in which women were randomized to estrogen alone, estrogen and progestogen, and placebo, discontinued the estrogen and progestogen arm after 5 years of follow-up because the adverse effects (significantly increased risk of breast cancer, pulmonary embolism, and stroke) outweighed the beneficial effects (significantly lowered risk of colorectal cancer and hip fracture). This finding has severely impacted the clinical practice of combined HRT for postmenopausal women; consequently, many women have reexperienced distressing menopausal symptoms because they have been concerned with the risks of hormone treatment. Alternative treatments directed at specific components of the menopausal state, such as decreased bone density, fortunately, are under active development.

## ASSISTED REPRODUCTION

Some women of reproductive age do not ovulate regularly during their menstrual cycles and therefore are candidates for assisted reproduction techniques. The first goal of such techniques is to promote mature follicles and, if possible, ovulation. Follicle growth and induction of ovulation can be accomplished hormonally by increasing circulating pituitary gonadotropins. Clomiphene citrate, a nonsteroidal estrogen that behaves as a competitive ER$\alpha$ antagonist, enhances release of pituitary LH and FSH, resulting in follicular growth, dominance of one follicle, and ovulation. For clomiphene to be effective, the patient must have an intact hypothalamic–pituitary–gonadal (HPG) axis. A similar result can be achieved by the direct administration of gonadotropins and is indicated when there is an inherent deficiency of LH or FSH or when clomiphene treatment has not been successful.

The goal of hormonally induced ovulation without assisted reproduction techniques is to produce one to two mature follicles, thereby minimizing multiple births. With assisted reproduction such as in vitro fertilization, however, the goal is to produce multiple mature follicles, so several eggs can be harvested, inseminated, and transferred into the patient's uterus, because the success rate with this technique is not high. With assisted reproduction, a sequential hormonal regimen is followed. Withdrawal bleeding may be initiated with progesterone or medroxyprogesterone administration for a week to cause shedding of the endometrial lining. A long-acting, gonadotropin-releasing hormone agonist such as leuprolide is given to downregulate the menstrual cycle, following which follicular induction is accomplished with gonadotropin administration. Eggs are harvested from several mature follicles for insemination and insertion into the uterus, and progesterone is then administered to support the endometrium and promote successful implantation. Thus, when assisted reproduction techniques are used, sequential administration of several hormones is necessary to suppress endogenous cycling, to develop mature follicles, and to prepare the endometrium for successful implantation of the inserted fertilized egg.

# MULTIPLE ENDOCRINE DEFICIENCY SYNDROMES

There are many endocrine deficiency syndromes (see Chapter 5), some of which involve several hormones. In partial hypopituitarism, deficiencies of gonadotropins and growth hormone often occur, and in panhypopituitarism, virtually all the anterior pituitary hormones are involved. Causes can be genetic, traumatic, tumors, and vascular insults to the pituitary. Other diseases involve the target endocrine glands (e.g., Schmidt syndrome, which is an autoimmune disease of the adrenal and thyroid glands that can result in complete exhaustion of basal glucocorticoid secretion and hypothyroidism). Replacement of the deficient hormones must be done sequentially in illnesses such as panhypopituitarism and Schmidt syndrome, when the adrenals and thyroid gland are both hypoactive. If thyroid hormone replacement is begun before the adrenal insufficiency is treated, the increase in overall metabolism produced by exogenous thyroid hormone can precipitate an Addisonian crisis with severe consequences. Therefore, the adrenal insufficiency is treated first, and then thyroid hormone administration is begun.

Thinking in terms of hormone combinations, while many of the examples illustrated in this chapter deal with synergies—one hormone helping another hormone to regulate a behavior—other examples show how one hormone can oppose another. See Fig. 2.7 for the ability of thyroid hormones to reduce the transcriptional effects of estrogens.

## Premenstrual Dysphoric Disorder

A troubling set of symptoms reported by some adult women is characterized by falling estrogen levels in concert with falling progesterone levels (Table 3.2). Negative (dysphoric) emotions and negative behaviors and social interactions can occur for 7–10 days prior to menses and

| Hormones administered | Estrogenic effects on: | |
|---|---|---|
| | Transcription through ER-α and ERE | Lordosis behavior |
| None | O | O |
| Estradiol (E) | ↑↑↑↑ | ↑↑↑↑ |
| Thyroxine (T) | O | O |
| E + T | ↑↑ | ↑↑ |

FIGURE 2.7   Co-administration of thyroid hormones can reduce the magnitude of an estrogen effect at the molecular level (transcription assays in neuroblastoma cells) and at the behavioral level (the female reproductive behavior lordosis). Considering combinations of hormones acting on behavioral mechanisms, therefore, the interactions can be subtractive as well as additive. *ER-α*, Estrogen Receptor-alpha; *ERE*, Estrogen Response Element on DNA.

TABLE 2.3   DSM-5 Criteria for Premenstrual Dysphoric Disorder

In most menstrual cycles during the past year, five or more of the following symptoms were present during the last week of the luteal phase (prior to menses), began to remit within a few days after onset of menses, were minimal or absent during the week postmenses, and at least one of the symptoms was 1 through 4:

1  Marked affective lability (e.g., feeling suddenly sad)
2  Marked irritability or anger or increased interpersonal conflicts
3  Markedly depressed mood, hopelessness, self-deprecating thoughts
4  Marked anxiety, tension, feeling "keyed up" or "on edge"
5  Decreased interest in usual activities
6  Difficulty concentrating
7  Lethargy, fatigability, or markedly decreased energy
8  Marked change in appetite, overeating, or specific food cravings
9  Hypersomnia or insomnia
10  Sense of being overwhelmed or out of control
11  Other physical symptoms (e.g., breast swelling or tenderness, headaches, joint or muscle pain, bloating, weight gain)

*Adapted from the Diagnostic and Statistical Manual of Mental Disorders, fifth ed., American Psychiatric Association, Washington, D.C., 2013, with permission.*

during menstruation itself. There also can be painful physical changes, including breast swelling, uterine cramping, and fluid retention. If this occurs with most menstrual cycles for at least a year and involves a variety of distressing symptoms including depression, anxiety, mood swings, irritability, fatigability, appetite and sleep changes, and physical changes, the condition can be formally diagnosed as premenstrual dysphoric disorder (Table 2.3), and treatment is recommended. The assumption has been that because the ovary undergoes profound changes in its production of estrogen and progesterone across the menstrual cycle, these hormones must be intimately involved in the etiology of this syndrome.

Many studies have examined basal and stimulated activity levels of the HPG axis in affected women. Studies of circulating estrogen and progesterone have been quite variable, in sum indicating that there is no excessive or deficient secretion of either hormone or testosterone (converted in women in small amounts from adrenal androgens) in patients compared to normal women. There may, however, be a positive correlation between severity of premenstrual symptoms and circulating female sex steroids, although they are still within the normal range.

Other possibilities also have been considered. The rate of change of gonadal steroids in the late luteal phase (e.g., progesterone decline) may be important—depressive symptoms become more frequent in girls than in boys at the time of puberty, when major sex hormone changes occur—and there is a predominance of cyclic mood disorders in women throughout adult life. Studies in which similar hormone changes have been experimentally produced in women with and without premenstrual dysphoria indicate that those without antecedent premenstrual complaints do not develop such complaints when their sex steroids are experimentally altered, whereas women who do have antecedent premenstrual dysphoria develop similar complaints with experimental alteration of their hormones. These findings suggest a heightened sensitivity of the CNS to changes in circulating hormones in women with premenstrual dysphoria.

Another consideration, and fitting with the differential CNS sensitivity hypothesis, is the role of neuroactive steroids, given that they have a predominantly inhibitory effect on CNS excitability, via GABA-A receptor modulation, as discussed previously. Studies of circulating concentrations of, for example, allopregnanolone, however, have been conflicting, with both higher and lower concentrations reported in premenstrually dysphoric versus normal women. Again, the rate of change may be an important factor. And, a potentially important influence on CNS sensitivity to gonadal hormone changes may be CNS-produced neurosteroids, which, as indicated earlier, may be concentrated differently in local CNS areas important in the mediation of affect and mood.

Finally, the role of sociocultural factors must be considered in premenstrual dysphoria (see Chapter 20). Some symptoms, such as sleep and appetite disturbances, may be relatively environmentally independent, whereas others, such as depression and irritability, can be provoked by interpersonal and other environmental stressors. Given that treatment of this major cause of discomfort in women remains empirical, attention to social as well as to biological factors during treatment is quite important.

# OUTSTANDING NEW BASIC OR CLINICAL QUESTIONS

1. Why do premenstrual physical and mental symptoms differ so markedly among women? There is very little information available to answer this question, except that, as noted previously, the rate of change of hormones appears to be a relevant factor, suggesting that differing receptor sensitivities among women predisposed to premenstrual dysphoria versus those without such symptoms play a role. The systems involved, however, remain obscure. Do gonadal steroid receptors within the CNS play a role? Do changes in neuroactive steroids, secondary to late luteal phase changes in gonadal steroids, play a role (e.g., falling allopregnanolone concentrations secondary to falling concentrations of its precursor, progesterone)? Do these hormone changes affect CNS amine neurotransmitter systems, such as serotonin?

2. Why do women with similar premenstrual mental symptoms respond so differently to similar hormone treatments? This question is a corollary of question 1, in that the vulnerabilities in hormonal and neurotransmitter systems leading to premenstrual dysphoria are the targets of hormonal and other replacement therapies. Because the hormonal and neuroendocrine links are not well understood, it is difficult to speculate on why symptomatic women respond differently to hormone treatment. Factors include the blood and tissue levels achieved with a specific dose of hormone, which may differ among women depending on differences in hormone pharmacokinetics across individuals, as well as differences in the innate sensitivity of hormone receptors and the metabolic cascades following receptor activation.

3. On the principle that the effects of hormones can be due to too much or too little available hormone (see Chapter 5), is it possible that the lack of counterregulatory hormones may be the principal cause of food intake and metabolic disorders?

# SUMMARY

Female reproductive behavior, ingestive behaviors (including food, salt, and water intake), and social recognition illustrate some ways that hormones work in combination. This is more the rule than the exception with regard to all behaviors.

# One Hormone Can Have Many Effects

Most of us are familiar with pharmaceutical advertising. The ads usually begin with leisurely scenes of smiling, satisfied users of a particular prescription drug and end with a rapid-fire list of alarming side effects of the same drug. Why are there so many "side effects," and how do we know they are not actually the main effects? Some answers lie in this chapter. First, many prescription drugs act on endogenous hormone receptors, and second, most if not all hormone receptor systems act on more than one behavior and/or physiological process. As you will see, scientists have imposed upon nature an organizational system, which includes categories of behaviors (e.g., sex, ingestive, and parental), categories of physiological systems (e.g., circulatory, nervous, digestive, and excretory), and categories of chemical messengers (e.g., sex steroids, stress hormones, and feeding peptides). The actual evolutionary forces that molded the traits we see in existing organisms know nothing of these categories. Evolutionary forces, such as natural selection, mutation, and genetic drift, are equally unconcerned with the claims of pharmaceutical corporations. Thus, real-life hormones have many functions that often cut across these artificial categories to optimize survival and reproductive success.

This phenomenon is illustrated by the dopaminergic drugs. The first pharmaceutical grade dopaminergic drugs, amphetamines, were prescribed during World War II to help depressed and exhausted soldiers perform their duties on the battlefield and to overcome their need for food or sleep. The mood-enhancing and appetite-suppressing effects of amphetamines became popular with the general public, but unfortunately, dopaminergic drugs have an additional affect. They cause dependence and addiction, defined as the physical need for the substance as the neurons adapt to the drug's cellular action, and therefore require increased amounts to maintain the desired effects produced by the original doses. By the end of World War II, over half a million United States citizens were using amphetamines, not only because it helped them feel confident, energetic, and less hungry, but also because it was difficult to stop taking amphetamines. The number of amphetamine users exploded to epidemic proportions by the 1960s (reviewed by Rasmussen, 2008), and in response to the consequences of addiction and withdrawal, the use of amphetamines was restricted, at least temporarily, only to explode again in the next decade (reviewed by Rasmussen, 2008). In more recent times, partially dopaminergic/noradrenergic/serotonergic "appetite suppressants" have been marketed under the names Phen-fen, Qnexa, and Qsyma, and most recently a similar compound has been introduced under the name, flibanserin, but for enhancing female sexual libido (Stahl et al., 2011). Dopaminergic effects are not restricted to appetite and energy metabolism, but they also include sexual libido and sexual performance. Ritalin and Adderall are dopaminergic compounds commonly prescribed at low doses, long term to children for attention deficit disorder, but the same drugs are used by adults (at higher doses) to enhance sexual motivation (Balon and Segraves, 2008; Csoka et al., 2008; Volkow et al., 2007). Methylphenidate (the main ingredient in Ritalin) and related dopaminergic compounds have long been prescribed to counteract loss of libido caused by antidepressant medications (Balon and Segraves, 2008; Csoka et al., 2008; Volkow et al., 2007). Well-known "side effects" of dopaminergic treatment for Parkinson disease include both sexual and gambling addiction. The sexual "side effects" of dopaminergic drugs come as no surprise to the behavioral endocrinologists who study reproductive behavior, because they know from their own studies that dopamine action is central to sexual motivation (e.g., Kleitz-Nelson et al., 2010; reviewed by Pfaus, 2011). Extreme doses of dopaminergic drugs induce hallucinations, psychosis, and schizophrenia, owing to the fact that dopamine systems are involved in learning the association between cause and effect. When brain dopamine levels are too high, associations are formed between unrelated events, one of the primary symptoms of psychosis. There is no doubt that dopaminergic prescription drugs have multiple behavioral and physiological effects because dopamine itself has many important functions. This chapter is divided into two main parts. The first part concerns chemical messengers that stimulate more than one behavior or physiological process, whereas the second concerns those that stimulate one process while inhibiting another. You will learn that the chemical messengers that control the stress response, eating, and drinking all have striking effects on sexuality.

# A SINGLE HORMONE CAN FACILITATE MORE THAN ONE BEHAVIOR

## The Hypothalamic–Pituitary–Adrenal System

The hormonal components of the stress response in the hypothalamic–pituitary–adrenal (HPA) system have multiple behavioral effects, and one way to make sense of this fact is that survival in a stressful environment often requires a wide variety of disparate responses. Consider the hypothalamic neurohormone, corticotropin-releasing hormone (CRH). When an animal senses a threat to its survival, CRH is secreted. CRH stimulates the release of adrenocorticotropic hormone (ACTH) from the pituitary gland, which, in turn, causes the release of adrenal glucocorticoids such as cortisol (figures that illustrate the HPA system are found in the Introduction to this book). The CRH that reaches the anterior pituitary is synthesized in magnocellular neurons of the paraventricular nucleus of the hypothalamus (PVH), and it is released by axon terminals of those magnocellular neurons into the median eminence of the ventral hypothalamus. CRH travels from the median eminence via the pituitary portal plexus to the corticotrophs of the anterior pituitary. Thus, a primary function of CRH is to promote the secretion of ACTH and glucocorticoids. In addition, CRH acts in other brain areas to potentiate anxiety, alertness, vigilance, increase blood pressure and heart rate, decrease gastric acid secretion, increase gastric motility, and inhibit food intake and sex behavior (reviewed by Valentino et al., 2013). Increased blood pressure and heart rate moves oxygen and metabolic fuels (e.g., glucose) to muscle and the brain to allow a rapid and strong response to a threat. Escape or strategic defense requires attention, focus, and vigilance toward the task at hand, and this is further facilitated by a strong inhibition of other motivations, such as the hunger for food and desire for sex. Thus, CRH decreases foraging, courtship, and mating, behaviors that are unlikely to increase survival during a life-threatening emergency. These effects are thought to be direct, central effects of CRH, because central treatment with CRH inhibits sex and ingestive behaviors. And central pretreatment with astressin, an antagonist to CRH receptors, facilitates sex and ingestive behaviors in animals in which stress was caused by either undernutrition or neuropeptide Y (NPY) treatment. Furthermore, the effects of CRH and their antagonists do not require changes in glucocorticoid secretion (Jones et al., 2004). Thus, CRH predisposes the individual toward a physical and emotional state that is most likely to increase survival in the face of danger. Our existence on our sometime volatile planet can be reasonably attributed to the multiple effects of CRH.

Some of CRH's actions on multiple traits occur via action on other neurotransmitters and neuropeptides. One suspected effect of CRH is increased and prolonged activation of the norepinephrine-releasing neurons of the locus ceruleus. Norepinephrine has been implicated in the production of anxiety states, a component of which may be inappropriately heightened arousal and vigilance in chronic stress or in posttraumatic stress disorder. CRH acts in other brain areas as well, so it is likely that the anxiogenic effect of CRH administration to experimental animals involves multiple sites of action. The relative specificity of CRH in anxiety production can be demonstrated in laboratory rats by examination of the anxiolytic effects of CRH receptor antagonists under several conditions: (1) when exogenous CRH is administered, (2) when rats are subjected to experimental stress, and (3) when rats are selectively bred to show innate heightened anxiety-like behaviors and to have a hyperactive HPA cortical system. Because CRH has anxiogenic effects in addition to its hormonal effects, because 30%–50% of psychiatric patients with major depression have increased HPA axis activity, and because about a third of such patients also have anxiety syndromes, CRH has been proposed as an etiological hormone in the pathogenesis of major depression. Consequently, drug development efforts for the treatment of depressed patients have focused on CRH antagonists. As with most new drugs, the effectiveness of CRH antagonists in the treatment of major depression so far has been modest, and as you might have predicted, some troubling side effects have led to the discontinuation of trials. These side effects are an example of this key principle of hormone-behavior relations (single hormones have multiple effects). Efforts toward the development of safer and more effective CRH antagonists for use in depressed patients are in progress.

At the level of the anterior pituitary, ACTH affects not only adrenal glucocorticoid secretion but also fear learning, i.e., the formation of learned associations during stressful experiences. ACTH and related peptides have consistent effects on learning in laboratory rats. For example, when different groups of rats are exposed to a red light just before receiving a mild electric shock, the rats eventually learn to escape or avoid the shock when they see the red light. In some experimental groups, rats are treated with ACTH just before pairing the light with the shock, whereas other groups receive a vehicle treatment by the same route. Both groups eventually learn that the red light predicts a shock, and they will reliably make moves to escape the shock at the appropriate moment. ACTH administration prior to testing improves (speeds up) the acquisition of an escape response. Subsequently, when the red light is presented repeatedly without the shock, the rats eventually stop trying to escape, and the response is said to be "extinguished." When ACTH is administered during the acquisition of an avoidance response, the response becomes more resistant to extinction. As well, ACTH-like peptides improve the acquisition and retention of passive avoidance behavior, likely by affecting memory-storage processes. One hypothesis under investigation is that ACTH triggers a state-dependent condition that enhances memory retrieval. In addition, ACTH may directly influence reward mechanisms (i.e., dopaminergic action). Rats will perform tasks repeatedly, such as pressing a lever or poking their noses in a small hole, if the task is paired with mild electrical stimulation of brain areas such as the mesolimbic dopamine system. Pretreatment with ACTH increases intracranial self-stimulation in rats. These effects of ACTH might be adaptive if they help the stressed individual remember where it had a stressful encounter and whether the ACTH-mediated responses led to a successful escape. Together, the effects of ACTH increase the chances that an animal will survive future encounters in the place and context in which they previously experienced the stress response.

Most amazing are the steroid hormones themselves. Adrenal glucocorticoid hormones such as cortisol and corticosterone have multiple effects on metabolism, the process by which cells obtain the energy they need to function. Response to stress requires extra energy to the cells in muscle and the brain. The primary energy currency of the cell is adenosine triphosphate (ATP), and ATP is synthesized via a chain of reactions that begins with the oxidation (addition of an oxygen molecule) of metabolic fuels. The metabolic fuels are glucose and free fatty acids. Glucocorticoids increase the processes that yield these metabolic fuels. First, glucocorticoids increase lipolysis, the breakdown of triglycerides released from the lipids in adipocytes. Triglycerides are broken down to free fatty acids and glycerol. Free fatty acids can be oxidized in the cellular organelle known as the mitochondria. Second, glucocorticoids increase glucose production in the liver by the process of gluconeogenesis, the making of new glucose from glycerol and amino acids. The availability of metabolic fuels enables the behaviors of the stress response: fighting and/or fleeing. When individuals are under chronic stress, however, the constantly elevated levels of metabolic fuels in circulation can increase the chances of developing abdominal obesity, insulin resistance, and Type II diabetes. These effects of glucocorticoids on metabolism are only a small part of their repertoire of effects on brain and behavior.

In addition to their effects on metabolism, glucocorticoids have many other effects on behavior, depending upon their levels and whether they are secreted acutely or chronically. High levels of glucocorticoids inhibit sexual motivation, and in the long term, they promote foraging, food hoarding, eating, and body weight gain (Table 3.1, Wingfield and Romero, 2001). The fact that CRH and ACTH tend to inhibit food intake, whereas glucocorticoids tend to increase food intake, is partially related to the time course of action and the difference between the immediate stress response and the effects of chronic stress. When an animal must respond to an immediate threat, such as a threatening predator or terrorist attack, there is a rapid sympathetic release of adrenaline (in the periphery)

**TABLE 3.1**   Effects of Glucocorticoid Hormones

| Short-term adaptation | Long-term disruption |
| --- | --- |
| Inhibition of sexual motivation | Inhibition of reproduction |
| Regulate immune system | Suppress immune system |
| Increase glucogenesis | Promote protein loss |
| Increase foraging behavior | Suppress growth |

*Adapted from Wingfield, J.C., Romero, L.M., 2001. Adrenocortical responses to stress and their modulation in free-living vertebrates. In: McEwew, B.S. (Ed.), Coping with the Environment. Neural and Endocrine Mechanisms. Oxford University Press, Oxford.*

and CRH (in the brain and pituitary). Later, after the immediate threat is over, the elevated levels of glucocorticoids can promote hunger and food intake, which can aid in healing wounds and recovery from the energy deficits incurred during the fight-or-flight reaction. Furthermore, the taste and metabolic consequences of high-calorie foods can attenuate the stress response and lower anxiety. In chronic stress and after long-term use of prescription glucocorticoids, high levels of glucocorticoid-induced consumption of high-calorie foods can promote obesity. The multiple effects of glucocorticoids on behavior and physiology are adaptive in the short term, but they can be maladaptive in the long term.

### Interim Summary 3.1

1. CRH initiates the cascade of hormone action of the HPA system and also has direct effects in the brain on anxiety, reproductive, and ingestive behavior.
2. ACTH stimulates secretion of adrenal glucocorticoids and also has effects on the acquisition and extinction of learned responses to stress.
3. The adrenal glucocorticoids influence many aspects of metabolism and behavior. Together the myriad effects of the HPA system coordinate the short-term responses to stress and account for many of the long-term, adverse effects of chronic stress, including obesity and sexual dysfunction.

## The Renin-Angiotensin System and Thirst

Another example of a single hormone having many effects is angiotensin II. Angiotensin II is an octapeptide: it has eight amino acids. Angiotensin II is found in the blood and in tissues, including the brain. When injected directly into the brain, angiotensin induces drinking behavior. This phenomenon, which was first discovered by Fitzsimons et al. in 1969, is one of the most compelling examples of a hormone-induced behavior. Within a few seconds of being injected with nanogram amounts of angiotensin into the brain, rats will immediately seek out the waterspout and begin drinking continuously for about 15 min. In addition to this dramatic form of behavior, the same dose of angiotensin II that induces drinking also stimulates an increase of sympathetic activity and the release of other hormones, including aldosterone, ACTH, and vasopressin (Fig. 2.1). Remember that vasopressin has at least two independent effects, increased reabsorption of water in the kidney and increased peripheral vascular resistance to raise blood pressure. Angiotensin II has well-known direct and indirect effects including but not limited to increased vasoconstriction, increased blood pressure, increased lipogenesis (creating body fat), decreased lipolysis (breaking down body fat), and increased sodium retention. The myriad effects are too widespread to be included here, and the discussion will be confined to effects on thirst and drinking.

Drinking behavior is quite complex, and the multiple effects of single hormones can be understood in terms of the many different organ systems involved in the behavior (Fig. 3.1). There are two different stimuli that induce thirst, and these are also part of the system that maintains fluid balance. Both the volume of fluids and the osmolarity (solute concentration) of the fluids are important, and both are regulated. First, a drop in blood volume (hypovolemia) is a decrease in the amount of fluid in the circulatory system. Hypovolemia is a potent stimulus for the release of renin from the kidney, which leads to the production of angiotensin II precursor, angiotensin I. In lung and kidney endothelium, the angiotensin-converting enzyme (ACE) catalyzes the reaction by which angiotensin I becomes angiotensin II. Concomitantly, vasopressin synthesis and secretion is stimulated. Angiotensin II-induced drinking and vasopressin-induced water retention attempt to restore blood volume.

In contrast to hypovolemic thirst, osmotic thirst results from cellular dehydration, a lack of water in the intracellular compartment relative to the extracellular compartment. A cell is essentially a bag of water, in which

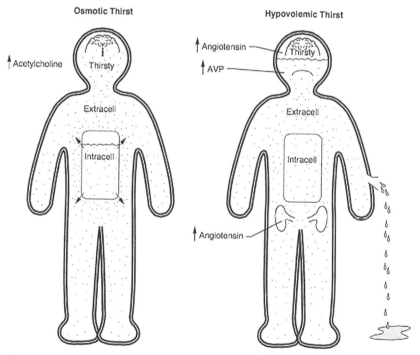

FIGURE 3.1   Two mechanisms for thirst: *Osmotic thirst* results from cellular dehydration without a change in volume. Excess salt in the extracellular compartment draws water from the intracellular compartment by osmosis. Acetylcholine in the brain stimulates thirst. *Hypovolemic thirst* results from reduced blood volume (e.g., hemorrhage) and activates renin release (see Fig. 2.2).

the bag is a semipermeable membrane. That is, water can passively diffuse in and out of the bag in the direction of the highest concentrations of solute (in this case, mostly salt). When the concentration of salt in the extracellular compartment is high, water from inside the cell will tend to flow out, resulting in cellular dehydration (Fig. 3.1).

In contrast to the control of hypovolemic thirst, osmotic thirst is mediated by the neurotransmitter acetylcholine, released in response to cellular dehydration. High levels of central acetylcholine stimulate drinking, and the increased water dilutes the fluid compartments to counter the cellular dehydration.

In addition, there are many aspects to hypovolemic thirst, and these help explain the multiple effects of angiotensin II. First, there is the increased drive or desire to drink; next, there is the coordination of motor movements required to drink, from locomotion and positioning of the body to the source of water to catching the water with the tongue and swallowing it. In addition, the behavior requires memory of a source of water. The initial action of angiotensin is to motivate the animal to drink (i.e., it makes the animal thirsty) to counteract hypovolemic thirst. Once the level of motivation is raised, information coming into the brain is divided into significant information (i.e., related to the water source) or nonsignificant information (i.e., unrelated to the water source). All of these aspects of thirst and drinking require coordinated hormone action.

The brain areas of angiotensin II action have been identified (Fig. 3.2). Angiotensin II stimulates angiotensin receptors in the hypothalamic region, and particularly the areas around the anterior ventral wall of the third ventricle (reviewed by Phillips, 1987). These receptors are connected by neurons to circuits within the brain that stimulate all of these complex activities leading to thirst and drinking behavior. The circuits have been elucidated using lesions (chemical destruction of a particular brain nucleus), electrical stimulation, microinjections of different doses of angiotensin and other chemical messengers, and even cream plugs to prevent access to sites on the ventricular wall. Lesioning has revealed two areas of significance: the subfornical organ (SFO) and the organum vasculosum of the laminae terminalis (OVLT). Lesioning either or both of these regions inhibits or abolishes the drinking response to central injections of angiotensin II. Electrical stimulation has shown that the SFO is connected neurally to the paraventricular nucleus (PVH) and the supraoptic nuclei (reviewed by Ferguson et al., 1990), and the lateral terminalis sends projections to the lateral hypothalamus. The exact mechanisms by which signals from the SFO and OVLT elicit drinking are under investigation. Both of these nuclei synthesize vasopressin. When stimulated by angiotensin II, these nuclei release vasopressin through neuronal axons coursing down axons to be released from terminals that lie in the posterior pituitary. From the posterior pituitary, vasopressin is released into the blood stream. Further studies have shown that the sites excited by angiotensin II are also connected to brainstem sites, for example, the rostral

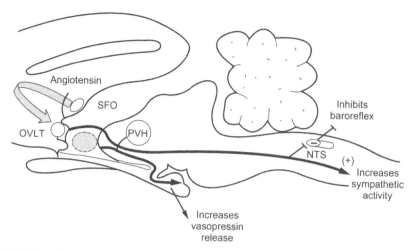

FIGURE 3.2  Angiotensin II has many effects. It acts on the subfornical organ (SFO), on the medial preoptic area (MPO), on the organum vasculosum of the lamina terminalis (OVLT), on the magnocellular nuclei of the anterior hypothalamus, in the paraventricular nucleus (PVH), and in the supraoptic nucleus (SO). As a result, vasopressin is released from the posterior pituitary gland, sympathetic autonomic nervous system activity is increased, and blood pressure rises because of inhibition of the baroreceptor reflex from the nucleus of the tractus solitarius. *From Phillips, M.I., 1987. Functions of angiotensin in the central nervous system. Ann. Rev. Physiol. 49, 413–435. With permission.*

ventral lateral medulla where sympathetic nerves are stimulated and affect the entire body. The effect of increased angiotensin II and vasopressin is to increase blood pressure by stimulation of norepinephrine secretion and norepinephrine-induced vasoconstriction of blood vessels. In summary, angiotensin II has direct effects on drinking by acting in the brain and indirect effects on blood pressure via stimulation of vasopressin.

In addition to central injections of angiotensin II, thirst can be stimulated by high doses of angiotensin II injections intravenously (into the blood stream). However, the concentrations necessary for this effect are so high that it has been difficult to determine whether the effects are from peripheral angiotensin II action or to opening up the blood–brain barrier by increasing blood pressure excessively and allowing angiotensin to enter the brain. In addition, angiotensin II can reach the circumventricular organs (CVOs), which have no blood–brain barrier. These include the SFO and OVLT. These examples caution that angiotensin might potentially work on multiple brain areas and in multiple peripheral organ systems involved in thirst and drinking.

These coordinated effects of angiotensin and vasopressin have lifesaving effects in response to injury. During hemorrhage, blood pressure and volume fall, often leaving the brain in dangerous deficit of oxygen and metabolic fuel. The fall in blood pressure increases the release of renin produced by the juxtaglomerular cells of kidney cortex. Renin released into circulation then acts on angiotensinogen in the liver to form angiotensin I.

In the lungs, ACE cleaves angiotensin I to angiotensin II, the octapeptide. Various aminopeptidases can break down angiotensin into other active metabolites, including angiotensin III and angiotensin IV. Vasopressin is stimulated, promoting water retention in the kidney, absolutely critical for maintenance of minimal blood pressure and survival during blood loss. One of the behavioral characteristics of hemorrhage is that eventually the patient becomes very thirsty. Whether this is due to high levels of angiotensin circulating in the blood entering the brain directly, to activating angiotensin receptors in the CVOs, or both has not been fully resolved. In the absence of injury, high levels of circulating angiotensin result in high blood pressure due to its vasoconstrictive effects on blood vessels. Imagine a large hose with water flowing at a particular rate. By constricting the hose diameter, the water volume is the same, but the rate of flow is faster due to the increased pressure. High blood pressure could force angiotensin II into the brain by opening the tight junctions in blood vessels that normally keep proteins out. Hormones that control drinking also control blood pressure, and both fluid balance and optimum blood pressure keep us alive.

Another health problem occurs when blood pressure is chronically high, which can result from a high-salt diet, obesity, and chronic stressors. If left unchecked, chronic high blood pressure can lead to atherosclerosis and heart disease among other problems. One common class of treatments for high blood pressure are the ACE inhibitors. ACE inhibitors decrease the general availability of angiotensin II, and thereby lower blood pressure.

Another interesting side effect of taking ACE inhibitors and other medications for high blood pressure is improved erectile function. Healthy erectile function involves efficient relaxation of smooth muscle and increased blood flow to the two erectile chambers, the corpora cavernosa and corporal sinusoids of the penis. Thus, the penis as a whole becomes firm when it is engorged with blood, and this is achieved only when smooth muscles of the penis become not firm, but *relaxed*, resulting in vaso*dilation* (the arteries of the penis open wider, allowing more blood flow). After ejaculation, erections subside due to vasoconstriction (i.e., increased blood pressure) of the arteries of the penis. Thus, chronically high blood pressure and chronic stress are linked to erectile dysfunction, which can sometimes improve while taking ACE inhibitors or adrenergic agonists commonly prescribed for high blood pressure. A urologist named Brindley famously injected himself with the adrenergic agonist phenoxybenzamine just before displaying his own "hard" evidence for these effects in front of the entire Urodynamics Society conference in Las Vegas, Nevada, in 1983. His work on this topic was subsequently published in the British Journal of Psychiatry (Brindley, 1983). Brindley's work could be viewed as an inspiration for the wildly "successful" drugs currently used for erectile dysfunction, such as Viagra and Cialis, which increase smooth muscle relaxation, decrease blood pressure, and promote erection by increasing availability of cyclic guanosine monophosphate. Note also that Viagra and Cialis, while they increase blood flow to the penis, can result in a dangerous plummet in blood pressure in men likely to be already taking blood pressure medication. Together, these sexual examples illustrate the wide array of effects of chemical messengers such as the lung peptide, angiotensin II, and noradrenergic alpha- and beta-blockers.

### Interim Summary 3.2

1. The protein angiotensinogen is synthesized in the liver, and when renin levels from the kidney reach the liver, this protein is converted to angiotensin I. In the presence of ACE, angiotensin I in the lung is converted to angiotensin II. Angiotensin II reaches the brain, where it has rapid, direct central effects on thirst and drinking.
2. In addition to effects on ingestive behavior (drinking), angiotensin II increases levels of vasopressin and norepinephrine and thereby increases vasoconstriction, increases blood pressure, lipogenesis, retention of sodium, and even erectile function.
3. Angiotensin II's myriad effects on vasoconstriction and concomitant effects on vasopressin and other chemical messengers can be lifesaving after injury that results in blood and other fluid loss.

## A SINGLE HORMONE CAN FACILITATE ONE BEHAVIOR WHILE INHIBITING ANOTHER

It has been noted in this and previous chapters that the traits we see in extant organisms are the result of evolutionary processes, including genetic drift, mutation, and natural selection. Traits are considered to be adaptive if they helped animals survive and increased reproductive success, which includes all of the processes involved in the conception, birth, weaning, or fledging of reproductively successful offspring. It is a curious fact that many of the orexigenic peptides discussed in Chapter 1 are strongly inhibitory for reproductive processes. Why do orexigenic molecules such as ghrelin, NPY, AgRP, MCH, and gonadotropin-inhibiting hormone (GnIH) all inhibit the hypothalamic–pituitary–gonadal (HPG) system and reproductive behavior? The inhibition of reproduction might, at first glance, seem counteradaptive. Furthermore, why, if both ingestive and sex behaviors are important, would hormones have the opposite effects on different behaviors?

One clue is that in the wild habitats in which animals evolved, resources were finite. Food is not abundant in all seasons; rather, food availability often fluctuates or is unpredictable. In some climates, food sources all but disappear in winter or after a storm, flood, or fire. Constraints on resources require animals to make tradeoffs. Eating food is the means by which animals acquire the energy necessary for all cellular activities. Of these different activities, reproductive processes are among the most energetically expensive. Thus, when food is scarce, investment in energetically costly reproductive processes can compromise the chances of survival. The individual must make a tradeoff. One individual might inhibit reproductive processes to gain the food necessary for survival. In terms of reproductive fitness, this individual essentially gambles that another chance to reproduce will become available in a future of energy abundance. If food resources eventually recover, this individual will have a high level of reproductive fitness. Another individual, however, might attempt to reproduce in the face of food scarcity, and in the worst-case scenario, the parent and offspring will die of starvation. In the best-case scenario, the conditions would improve rapidly, and this individual might end up with higher reproductive fitness than other members of the population that delayed reproduction. In female mammals, in which lactation is the most energetically costly process in the entire behavioral repertoire, mechanisms exist to "switch off" reproductive processes when food is scarce and switch them back on again when food becomes available. Not coincidentally, these same mechanisms promote vigilant foraging, food hoarding, and eating. Thus, a wide array of chemical messengers increases food intake and simultaneously inhibits reproductive processes.

In many animals the ingestive and sex behaviors are organized into a seasonal rhythm that ensures parents reproduce in an environment in which food is abundant. At nonequatorial latitudes, winters are cold and windy, and edible plants and prey animals become scarce. In equatorial latitudes, particular food sources become scarce during either the rainy or dry seasons (depending on the animal and its particular food source). To survive periods of drought, famine, harsh winters, or unpredictable natural disasters, most species developed the ability to overeat and store energy, either on the body as adipose tissue or in the home as a food hoard or cache. A sufficient store of food and/or body fat is a prerequisite for survival during food shortages or other energetic challenges and a prerequisite for successful reproduction. Animals that are able to compete for food and shelter will be able to compete in reproductive activities (courtship and mating, and for females, gestation and parental behavior). Thus, reproductive seasons are preceded by seasons of foraging, hoarding, and fattening, and there is a great deal of overlap in the mechanisms that control ingestive behavior, energy storage, and reproduction.

Chapters 1 and 2 contain descriptions of feeding-stimulatory hormonal cascades that are initiated during fasting. To review, food restriction or deprivation is accompanied by increases in the secretion of ghrelin from the stomach, orexin from the lateral hypothalamus, and NPY and AgRP from cells of the arcuate nucleus of the hypothalamus with reciprocal connections to the lateral hypothalamus and other projections to the PVH and other brain areas.

Treatment with all of these peptides increases food intake and food hoarding in well-fed animals, and blocking the action of these peptides prevents food deprivation–induced increases in food intake. Severe energetic challenges that increase food intake also inhibit the HPG system. This occurs whenever the level of energy expenditure exceeds the energy available. Energy availability is a sum of the energy stored on the body, mostly as adipose tissue, which can be broken down by lipolysis to release fatty acids, from glycogen in muscle, which can be broken down to glucose, and from food that is eaten, digested, and converted to glucose and free fatty acids. Very lean animals require constant food intake to fuel their daily activities and reproduction. Animals with a high body fat content have an energetic buffer to food deprivation. Thus, mild food restriction has little effect on the HPG system of fattened animals, but it is likely to inhibit the HPG system in lean animals. The effects of low energy availability on the HPG system are mediated, at least in part, by the same chemical messengers that increase food intake, i.e., NPY, AgRP, orexin, and ghrelin. Treatment with ghrelin, orexin, NPY, or AgRP inhibits the pulsatile secretion of luteinizing hormone (LH), the anterior pituitary glycoprotein that stimulates gonadal steroid secretion (e.g., Kiyokawa et al., 2011; Vulliemoz et al., 2004, 2005 and reviewed by Clarke, 2014; Schneider et al., 2013). The primary effects of energetic challenges are thought to occur at the level of the hypothalamus on gonadotropin-releasing hormone (GnRH) secretion, because in food deprived animals, treatment with GnRH can restore the full function of the HPG system and can even restore estrous cycle and ovulation in food-deprived females (reviewed by Schneider, 2004). Thus, the inhibition of LH by food deprivation is, in most cases, primarily the result of inhibition of GnRH secretion. GnRH, in turn, is stimulated by the peptide kisspeptin and inhibited by the peptide GnIH (reviewed by Dungan et al., 2006; Kriegsfeld, 2006). The latter is, like NPY, a potent orexigen (Clarke et al., 2012). The effects of food restriction on GnRH and LH secretion are mediated by decreases in kisspeptin and increases in GnIH (reviewed by Leon et al., 2014; Tena-Sempere, 2010). In summary, during severe energetic challenges in which energy expenditure exceeds the energy available from food intake and body fat stores, the HPG system is inhibited, and these effects are mediated by ghrelin, orexin, NPY, AgRP, and other orexigenic chemical messengers.

Severe energetic challenges inhibit the HPG system, but mild energetic challenges inhibit sex and courtship behaviors, even in cases where gonadal steroid levels are unaffected by the metabolic challenge. Central treatment with NPY has direct inhibitory effects on male and female sex behavior in rats and Syrian hamsters that are gonadectomized and treated with levels of gonadal steroids that normally induce sex behavior (e.g., Ammar et al., 2000; Clark et al., 1985; Corp et al., 2001; Jones et al., 2004; Kalra et al., 1988; Keene et al., 2003; Marin-Bivens et al., 1998 and reviewed by Clarke, 2014; Schneider et al., 2013). Ghrelin is also implicated in metabolic control of reproduction. The latency of male mice to mount a female is increased and the number of mounts is decreased by central treatment with ghrelin, and these effects are prevented by treatment with an antagonist to the ghrelin receptor (i.e., the growth hormone secretagogue receptor) (Babaei-Balderlou and Khazali, 2016). The RFamide-related peptide, GnIH, exerts inhibitory effects on reproductive behaviors, including courtship, sexual motivation, and the ability to defend resources necessary for reproductive success (reviewed by Calisi, 2014). When female Syrian hamsters are fed ad libitum (i.e., they are allowed to eat as much food as they want) and provided with the option to spend time with either food or adult male hamsters during their most active period after the onset of darkness, they invariably choose males. The levels of food hoarding and GnIH cell activation are low on every day of the estrous cycle. There is a slight increase in sexual motivation on the day of ovulation, but in general, levels of behavior and GnIH cell activation fluctuate very little over the estrous cycle in ad libitum–fed females. This all changes when females are mildly food restricted (75% of their ad libitum food intake) (Klingerman et al., 2010). In food-restricted females, food hoarding and GnIH cell activation are very high on all days of the estrous cycle, except for the day of ovulation, when food hoarding disappears, and the females spend all of their time with males (Klingerman et al., 2010, 2011). Their GnIH cell activation is also significantly lower on the day of high sexual motivation and ovulation (Klingerman et al., 2010, 2011). Central treatment with GnIH decreases the level of courtship solicitation in ovariectomized female Syrian hamsters treated with levels of ovarian steroids that normally increase courtship behaviors and mimics the effects of food restriction (Piekarski et al., 2012). In general, it appears that orexigenic hormones and neuropeptides are stimulatory for reproductive behaviors.

As you will recall, Chapters 1 and 2 describe a feeding-inhibitory circuit. After meals, a number of different hormones and neurotransmitters increase, including leptin, CCK, GLP-1, POMC-derived α-MSH, and CART. In many species, the chemical messengers that are secreted in well-fed animals, such as leptin, α-MSH, and CART, appear to be stimulatory for reproduction. Treatment with CCK-8 in monkeys inhibits pulsatile LH secretion (Schreihofer et al., 1993). GLP-1 application to an immortalized GnRH cell line stimulates the secretion of GnRH (Beak et al., 1998). Central treatment with the endogenous melanocortin antagonist α-MSH in rodents, sheep, and men stimulates pulsatile LH secretion (Alde and Celis, 1980; Backholer et al., 2009; Limone et al., 1997). The adipocyte hormone leptin is also stimulatory at many levels of the HPG system. The onset of puberty and adult estrous cycles and pulsatile LH secretion are inhibited by fasting in lean animals, and treatment with leptin during fasting prevents these effects (Ahima et al., 1997, 1996 and reviewed by Bellefontaine and Elias, 2014; Elias, 2012, 2014). A 48-h period of fasting inhibits estrous cycles in lean, but not fat, Syrian hamsters, and treatment with leptin during the fasting period preserves estrous cyclicity in lean Syrian hamsters (Schneider et al., 1998 and reviewed by Schneider, 2004). In summary, there is evidence that anorexigenic messengers can stimulate the HPG system.

Leptin is a potent stimulator of reproductive behavior. As discussed with regard to GnIH, mild food restriction in female Syrian hamsters inhibits sexual motivation and increases food hoarding. These effects on sex and ingestive behavior are fully prevented by treatment with leptin during food restriction (Schneider et al., 2007). When male rats are given the choice between drinking from a bottle of sucrose and mating with a sexually receptive female rat, those treated with a saline vehicle choose to drink from the bottle of sucrose, whereas those treated with leptin prefer to mate with the female and show increased ejaculatory frequency compared with the latency of the saline-treated male (Ammar et al., 2000).

As indicated in Table 3.2, there is a long list of hormones and neuropeptides with opposing effects on ingestive behavior and reproduction. This is only a partial list of chemical messengers that affect both ingestive and reproductive processes. There are some exceptions, but in most cases the chemical messengers that inhibit reproductive processes are stimulatory for ingestive behavior and vice versa. In the case of insulin, for example, high doses of insulin that result in hypoglycemia are inhibitory for reproduction and stimulatory for food intake. Other evidence, however, is consistent with

**TABLE 3.2** Chemical Messengers With Opposite Effects on Ingestive and Reproductive Processes (Schneider et al., 2013)

| Chemical messenger | Ingestive behavior | Reproductive processes |
|---|---|---|
| Insulin | −/+ | +/− |
| Ghrelin | + | − |
| NPY | + | −/+ |
| AgRP | + | − |
| Glucocorticoids | + | − |
| MCH | + | − |
| Endocannabinoids | + | − |
| Hypocretin (orexin) | + | −/+ |
| GnIH | + | − |
| Glucagon | + | − |
| GLP-1 | − | + |
| leptin | − | + |
| CART | − | + |
| CRH | − | + |
| CCK | − | + |
| Estradiol | − | + |

the idea that small increases in insulin after meals are inhibitory for food intake and permissive for normal HPG function. Leptin treatment enhances courtship and reproduction in mildly food-restricted animals but exaggerates the loss of sexual behavior in response to severe metabolic challenges. NPY has different effects on HPG function depending upon the steroid milieu. For the most part, those chemical messengers that increase hunger for food decrease the desire for sex and vice versa.

The overlap between the mechanisms that control ingestion and reproduction provides insight into the propensity to gain weight and become obese. Hormones that stimulate food intake also promote body fat storage and conservation of energy by decreasing energy expenditure. Orexigenic hormones ensure that animals acquire fuels for their immediate needs, but in addition, these hormones decrease energy expenditure and increase energy storage. Energy conservation and storage is critical for survival during potential future famine, but even more important in terms of evolution by natural selection, energy storage and conservation is critical for reproductive success. Conversely, hormones that inhibit food intake do so for at least two reasons. From our modern human perspective, we imagine that anorectic hormones should prevent obesity and its comorbidities. From an evolutionary perspective, however, mechanisms that inhibit food intake should do so only temporarily, just long enough to allow animals to engage in behaviors that increase reproductive success. In the energetically labile environments in which animals evolved, ephemeral inhibitory effects on ingestive were adaptive. Survival and reproductive success depended on high levels of energy expended on migration, foraging, and competition, and this required opportunistic overeating in anticipation of these energetically costly activities. These vigilant, anticipatory ingestive behaviors were interrupted only briefly just long enough to engage in courtship, mating, and offspring care. It would not be surprising if most anorexigenic messengers have only ephemeral inhibitory effects on food intake. In the present-day environment of ubiquitous food abundance, our predisposition for overeating and body fat storage is only rarely offset by high demands for energy expenditure or famine; therefore, this predisposition might be leading to obesity.

## Interim Summary 3.3

1. Reproduction, including sex behavior, courtship behavior, and the HPG system are inhibited during metabolic challenges, such as food deprivation or restriction in lean animals.
2. The effects of severe metabolic challenges are mediated, at least in part, by orexigenic hormones (such as ghrelin, GnIH, and NPY) that inhibit pulsatile GnRH secretion, which in turn inhibits the secretion of LH and the gonadal steroids.

3. Even mild energetic challenges decrease sexual motivation, even when the levels of gonadal steroids are not affected, and the effects of mild energetic challenges are associated with low levels of leptin and high levels of ghrelin, GnIH, and NPY.

## SUMMARY

This chapter illustrates the principle that individual hormones have myriad effects on a wide variety of behaviors. In some cases, more than one behavior or physiological process is stimulated, whereas in other cases, individual hormones increase one behavior while decreasing another.

# Hormone Metabolites Can Be the Behaviorally Active Compounds

The original "remove-and-replace" experiments showed, in many different species, that castration all but eliminates male-typical sex behaviors, and treatment of castrates with testosterone fully restores those behaviors. In addition, testosterone treatment increases sexual motivation, as measured by the eagerness of the male rats to be with estrous female rats and the short latency to perform copulatory behaviors. For example, compared to castrated male rats treated with vehicle, those treated with testosterone are quicker to begin mounting the female. In experiments designed to test their sexual motivation, testosterone-treated males run faster, work harder (e.g., they press a lever at a faster rate), and even endure more pain (mild electric shock) to reach a mature estrous female. Imagine the experimenters' surprise when the same facilitation of male copulatory behavior was achieved by treating castrated males with the so-called "female" hormone estradiol (Beach, 1942; Davidson and Bloch, 1969)!

This chapter explains that many hormones are metabolized to other hormones that are essential for the expression of behavior. The metabolized hormones are known as prohormones, and they include testosterone, progesterone, cortisol, proopiomelanocortin (POMC), and thyroxin. This chapter explains how behavioral neuroendocrinologists determine whether behaviors are influenced by prohormones and/or their metabolites.

In some chapters in this book, basic experimental findings are presented in the first half of the chapter, followed by clinical examples. This chapter, however, mixes the discussion of laboratory animal research with the discussion of clinical research. With regard to prohormones and their metabolites, a tremendous amount of experimental data is collected as a prelude to clinical trials, and conversely, clinical findings inspire many experiments using laboratory animal models. Thus, these two domains have not been separated in the following discussion.

*Principles of Hormone/Behavior Relations*
http://dx.doi.org/10.1016/B978-0-12-802629-8.00004-8

# TESTOSTERONE IS A PROHORMONE FOR OTHER ANDROGENS AND ESTROGENS

Testosterone, produced by the adrenals, testes, and ovaries, is a potent androgen that influences many cells in the brain and periphery. Testosterone binds to the androgen receptors located in both male and female reproductive organs, skeletal, cardiac, and smooth muscle, liver, kidney, adrenal cortex, pituitary, and many other organs. Metabolites of testosterone, however, also are important hormones, and these metabolites affect behavior via mechanisms that are different from the mechanisms by which testosterone affects behavior. Two important testosterone metabolites are dihydrotestosterone (DHT) and estradiol (E2). The chemical reaction whereby testosterone is converted to estradiol requires the action of the enzyme aromatase (a P450 enzyme), whereas the conversion of testosterone to DHT requires the enzyme, 5α-reductase.

Fig. 4.1 illustrates the conversion of testosterone to these two active hormonal metabolites. There are specific receptors for the two conversion products. Most of the androgen receptors for DHT lie in the peripheral tissues related to the development of secondary sex characteristics. DHT binds with higher affinity than testosterone to the androgen receptor. Estradiol acts via estrogen receptors located in both peripheral tissues and the CNS, and there is more than one type of estrogen receptor. Estrogen receptor α (ERα) and estrogen receptor β (ERβ) have received the most attention, but there are others (e.g., estrogen receptor X; Toran-Allerand et al., 2002). The two steroids in Fig. 4.1 are both comprised of four rings, labeled A–D. Each corner represents a carbon bond. Rings A–C are six-carbon rings, and ring D is a five-carbon ring. Aromatization involves the transformation of ring A of testosterone from an aliphatic ring to an aromatic ring (hence, the name of the enzyme for this reaction is aromatase). The aromatic ring, unlike the aliphatic ring, is a flat, cyclic, resonant conformation. Fig. 4.1, because it is static and two dimensional, does not convey (1) the dynamic resonance that confers stability to ring A nor (2) the major structural changes inherent in the enzymatic conversions that confer receptor specificity to these steroid compounds. In general, the structure of estradiol makes it compatible with estrogen receptors and not androgen receptors. Conversely, the structure of testosterone makes this steroid compatible with the androgen receptor.

## Estradiol Is Important for Adult Masculine Sex Behavior

The importance of metabolites of testosterone for species-specific behavior can be determined in a series of four experimental steps. Each step is designed to challenge the idea that one particular metabolite has direct effects on behavior and to further clarify the mechanism. The first

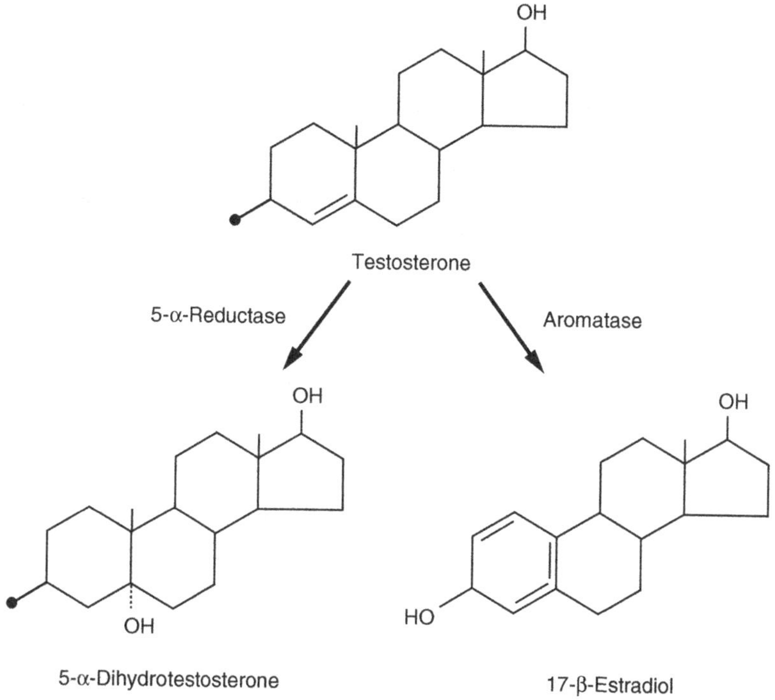

FIGURE 4.1    Testosterone is a prohormone for dihydrotestosterone (DHT) and estradiol. The conversions to DHT by 5-α-reductase and to estradiol by aromatase appear minor in this two-dimensional representation, but there are major conformational changes that confer receptor specificity to these steroid compounds. *From Gorski, R.A., 2000. Sexual differentiation of the nervous system. In: Kandel, E.R., Schwartz, J.H., Jessel, T.M. (Eds.), Principles of Neural Science, fourth ed. McGraw-Hill, New York, p. 1138. With permission.*

step is to determine whether a particular metabolite of testosterone is, in and of itself, sufficient for the behavior in question. If any variable X is *sufficient* for a process Y, then Y will always occur in the presence of variable X. For example, we can hypothesize that, in castrated males, estradiol alone is sufficient to restore sexual behavior to the level characteristic of the gonadally intact male. If the hypothesis is supported, castrated males will show significant decreases in levels of male-typical sex behaviors, and the behaviors will be fully restored by treatment with either testosterone alone or estradiol alone. If the hypothesis is refuted, treatment with a wide range of physiological doses of estradiol alone will fail to restore behavior to the level of the gonadally intact male. The same hypothesis can be tested with regard to the effects of DHT on male sex behavior.

If the data support a role for estradiol, the research does not end. This result raises more questions. Where does the estradiol come from? Is estradiol synthesized by the aromatization of testosterone? If estradiol alone is sufficient to restore behavior, a second step further clarifies the role of this metabolite. Castrated males are treated simultaneously with testosterone and a drug that prevents its aromatization to estradiol. Thus, we ask, "Is the aromatization of testosterone necessary for fully restored sex behavior?" If the hypothesis is supported, treatment with an aromatase inhibitor will block the expression of behavior in castrated males treated with testosterone. If the hypothesis is refuted, treatment with an aromatase inhibitor will have no effect on behavior in the testosterone-treated individual. The research would not end here.

Next, we can ask, "Is the binding of estradiol to ERα in a particular brain area necessary for restoration of behavior?" This hypothesis can be tested by experiments that contain the following groups of castrated males. Group 1 is given an antagonist to ERα implanted into a brain area of interest just before systemic treatment with a dose of testosterone known to increase male sex behavior. Group 2 is a control group treated with a central implant of a vehicle just before systemic treatment with testosterone. If the hypothesis is supported, treatment with the antagonist contained within the brain area will block the behavior even though the subjects are treated with behavior-inducing levels of testosterone. In these experiments, subjects that accidentally received antagonist implants that missed their targets can be used as controls. For example, an antagonist to ERα contained within the medial preoptic areas (mPOA) blocks the effects of systemic testosterone on behavior, but antagonist implants that missed the mPOA and landed in the nearby arcuate nucleus or diagonal band of Broca do not block testosterone-induced behavior. Depending on the species, these types of experiments have supported the role of ERα, but not ERβ, in control of some aspects of rodent sex behavior.

If specific strains of genetically engineered mice are available, they can be used to determine whether a particular receptor in a particular part of the brain is necessary for behavior. The procedure begins with insertion of an artificial piece of DNA into an embryonic stem cell. Stem cells have the ability to differentiate into nearly any type of adult cell; therefore, if a knockout occurs in a stem cell, the effects can be observed in any tissue in an adult. The piece of DNA that is inserted into the stem cell is chosen so that it has a nucleotide sequence homologous to the gene that is being knocked out. The inserted DNA, however, is inactive, and it contains a genetic tag, known as a "reporter gene," designed so that the inserted gene can be tracked. The nuclear machinery of the stem cell contains the gene to be knocked out, and it therefore "recognizes" the identical inserted sequence. After insertion and during the process of genetic recombination the cell exchanges all or a portion of its original copy of the gene with the inserted piece of DNA. This exchange results in a new recombinant gene with the wrong nucleotide sequence; therefore, it eliminates, or "knocks out," the function of the existing gene. For example, to determine whether ERα is *necessary* for reproductive behavior, the behavior is examined in animals in which ERα has been "knocked out" in a particular brain area of interest. Furthermore, to determine whether ERα in a particular brain area is *sufficient* for reproductive behavior, the behavior can be examined in animals in which ERα has been knocked out globally (i.e., in all cells of the brain and body) and "knocked in" only in the brain area of interest. Finally, we can use various molecular techniques, such as RNA interference, to decrease the expression of the genes that encode hormone receptors in particular brain areas to see whether this disrupts the behavior in question.

Many such experiments using laboratory rodents have confirmed the role of estradiol in masculinizing and defeminizing sex behavior (reviewed by Petrulis, 2013). In the process of studying prohormones and metabolites, we have learned a great deal about behavioral complexity. There are many different aspects of adult sex behavior, and it appears that part of this complexity is related to the differential effect of the various metabolites of testosterone. Some aspects of sex behavior reflect sexual motivation, whereas other aspects of sex behavior reflect physical ability. Examples of behaviors that reflect motivation include the latency to mount (the time it takes for the male to climb atop the female) and the latency to reach intromission (the time it takes for the male to make insertions of the penis into the vagina). These behaviors are measured in the presence of an adult female conspecific at the time when she is in the most sexually receptive part of her estrous cycle (known as the proestrous period). Another aspect of male sexuality measured in laboratory rodents is reflective of physical ability or performance and is analogous to erectile function in men. To measure erectile function in laboratory rodents, the experimenter restrains the male rat or mouse and applies a gentle tactile stimulus to the penis and measures the time it takes (the latency) to achieve an erection. The behavior is measured *ex copula*, i.e., in the absence of a potential female mating partner. Whereas mounting and intromission have a component of sexual motivation (i.e., they reflect in part the desire for sex), the latency to achieve an *ex copula* erection is a physical reflex related to the relaxation of muscles, vasodilation, and increased blood flow to the penis. The presence of a female is not required to achieve an erection. Castration increases the latency to *ex copula* erection as well as the latency to mount a female and to intromit the penis into the vagina. In sex behavior tests of castrated male rats treated with a vehicle, the latency to the first intromission is typically greater than 10 min, and the latency to achieve erection is often longer than 20 min (Fig. 4.2). By contrast, treatment with either testosterone or estradiol alone can reduce the latency to intromission to the short-latency characteristic of gonadally intact control rats (e.g., a latency of less than 3 min) (Fig. 4.2; Meisel et al., 1984). By contrast, short-latency penile erections are not restored by treatment with

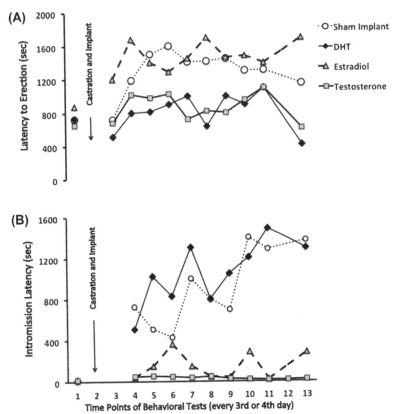

FIGURE 4.2   Dihydrotestosterone (DHT) alone, but not estradiol alone, is sufficient to restore penile erectile function; whereas, estradiol alone, but not DHT, is sufficient to restore sexual motivation. Male rats were tested for (A) the latency to achieve erection (*ex copula* penile reflexes in which the males were restrained on their backs and received artificial penile stimulation with no female rats present) and (B) intromission latency in response to an estrous female rat. The males were tested for baseline levels of responding at Time 1. At Time 2, they were castrated and implanted with either a sham implant or implants of crystalline testosterone, estradiol, or DHT. They were tested for the latency to erection beginning 4 days after castration/implant, again 3 days later, and again 4 days later, alternating testing on either the third or fourth day until the 39th day after castration/implant. They were tested for the latency to intromission beginning on the eighth day after castration and tested again 4 days later, 3 days later, 4 days later, alternating testing on either the third or fourth day until day 40 after castration/implant. *Adapted from Meisel et al. (1984).*

estradiol alone but, rather, by treatment with either testosterone or DHT. The results are consistent with the idea that DHT is necessary for penile reflexes, whereas estradiol is necessary for sexual motivation.

This hypothesis was explored further in many other experiments. For example, the restorative effects of testosterone on male copulatory behaviors were significantly decreased by pretreatment with an aromatase inhibitor (reviewed by Hull and Rodriguez-Manzo, 2017; Petrulis, 2013). Another line of evidence bolsters this conclusion. Genetically male mice (those with the XY sex chromosome complement) that lack a functional ERα show dysfunctional male-typical sex behavior. This result is consistent with the hypothesis that male sexual motivation and behavior require not only high levels of testosterone but also the aromatization of testosterone to estradiol and action of estradiol on ERα. It should be noted that there are exceptions, depending on the species, but in general, many such experiments have lent support for the "aromatase hypothesis" of control of male-typical sex behavior (reviewed by Petrulis, 2013).

## Estradiol Masculinizes the Brain During Development

In many vertebrate species, males differ from females in morphological (e.g., size), physiological (e.g., negative vs. positive feedback of estradiol on GnRH), and behavioral traits. Roosters crow and hens cluck. Lions have full, luxuriant manes, and lionesses do not. Big horn rams have enormous, curly horns, and ewes have smaller, straighter horns. Peacocks have a large, colorful plumage, and peahens have diminutive, dull feathers. Stallions mount estrous mares, and mares show lordosis in response to the stallion's mounting. These differences are ultimately related to the presence of the male sex chromosome complement, which determines the hormonal sex, which determines morphological and behavioral sex. Even in an individual with the female sex chromosome complement, prenatal exposure to male-typical levels of hormones can masculinize and defeminize morphology, brain, and behavior.

The process of masculinization and defeminization occurs in steps that are the same in most heterogametic species. In mammals, many male-typical behaviors can be attributed to the presence of the Y chromosome, which is the site of a gene that encodes testicular determination factor (TDF). TDF expression leads to the development of the male gonad, the testis, which secretes testosterone early in fetal development. Male-typical behaviors that occur in response to the increases in testosterone secretion at puberty can be attributed to the surge in testosterone that occurs during a critical time window of fetal development. In the normal male rodent fetus, exposure to high levels of testosterone during gestation or the first few days of neonatal life masculinizes the brain and behavior. Testosterone binding to androgen receptors during fetal development permanently alters the organization of the brain, so when the individual is exposed to testosterone in adulthood, male-typical behaviors occur. In fact, even in a genetic female (with an XX sex chromosome compliment), treatment with testosterone during fetal development masculinizes the brain and behavior, as first shown in guinea pigs (Phoenix et al., 1959) and later shown in many species including primates. Many years of pre- and neonatal testosterone manipulations in many species demonstrates that testosterone surges during a critical developmental period masculinize reproductive and nonreproductive behaviors. In normal XX individuals, the absence of the early testosterone surge sets in motion developmental processes that lead to female internal and external genitalia, as well as the feminine brain and behavior. In laboratory rodents, however, the high level of testosterone secreted from the fetal testes reaches the brain where it is aromatized to estradiol and then acts on estrogen receptors. Masculinization, therefore, results from estradiol binding to estradiol receptors during an early critical period.

In rodents, the conversion of testosterone to estradiol, a sequence of three enzymatic steps that go under the name, aromatase, within the CNS, is considered to be important for the early development of the "masculine brain." The aromatase enzyme is concentrated in areas of the brain related to sexual differentiation of the CNS, such as the hypothalamus. In the male rat or mouse, exposure of the developing brain to high concentrations of either testosterone or estradiol alone masculinizes the brain and sex behavior. In addition, the entire hypothalamic–pituitary–gonadal system is masculinized by early exposure to either testosterone or estradiol. Fetal treatment of XX rodents with estradiol leads to, for example, the tonic secretion of gonadotropins with negative feedback effects of gonadal steroids, a phenomenon that is characteristic of the adult male and is quite different from the cyclic secretion of these hormones in the adult female. Estradiol, however, does not have the same masculinizing effects on the genitalia. The genitalia include the spinal nucleus of the bulbocavernosi, the nerves that innervate the penis. These are masculinized by another metabolite of testosterone, 5-α-dihydrotestosterone (DHT). Exposure of the developing fetus to high concentrations of either testosterone or DHT prevents the death of cells in the spinal nucleus of the bulbocavernosi, and it promotes the elongation and fusion of the vaginal labia into the scrotum and stimulates the growth of the seminal vesicles, vas deferens, and phallus. In fact, DHT is not only sufficient, but it is also necessary for fully masculinized internal and external genitalia.

In addition to sex differences in reproductive behaviors, any sex differences in nonreproductive behavior also result from the differential exposure of the developing male and female CNS to testosterone, DHT, and estradiol. These include aggression and parental behaviors and many important human sex differences in response to pain, susceptibility to disease, reaction to drugs and endogenous hormones, propensity for obesity and eating disorders, and complex aspects of cognitive function, mood, and personality. The effects of hormones are often indirect (reviewed by McCarthy, 2016).

For example, a mutation in the gene that encodes the enzyme necessary for synthesis of DHT is responsible for a well-known condition in humans, 5-α-reductase deficiency. If 5-α-reductase deficiency occurs in a genetic male, a child with an XY sex chromosome complement, the genitalia of

Testosterone is a Prohormone for Both Male and Female Sex Steroids

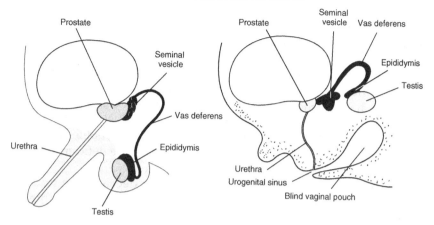

**FIGURE 4.3** Schematic representation of the genitalia of a normal man (left) and a patient with 5-α-reductase deficiency (right). Organs dependent on testosterone for normal development are in black, and those dependent on dihydrotestosterone (DHT) are stippled. *Adapted from Rey, R.A., Josso, N., 2016. In: Jameson, J.L., DeGroot, L.J. (Eds.), Endocrinology, seventh ed., Elsevier, Saunders, Philadelphia, PA, p. 2105. With permission.*

the newborn infant appear more like those of a newborn female or at least ambiguous. The phallus is diminutive or feminine (more like a clitoris than a penis), and the tissues that would form the vaginal labia of females and the scrotum of males remain unfused or only partially fused. Fig. 4.3 illustrates the genitalia of a normal XY man (left) and the genitalia of a 5-α-reductase-deficient prepubertal boy. The female-appearing external genitalia belie the presence of an XY sex chromosome complement, functioning testes (although undescended), and a masculinized brain secondary to fetal testosterone exposure. The behavioral consequences of this anatomical alteration illustrate how hormone metabolites can influence behaviors through indirect routes. Whenever a developmental event, such as androgen exposure, affects adult behavior, it might be a direct effect of the androgen on androgen receptors in the brain or an indirect effect through estradiol and subsequent signaling pathways. In addition, the effect of the hormone or metabolite might be indirect by way of an effect on the peripheral tissues. For example, hormonal stimulation of the phallus might lead to the development of male identity because having a phallus and the family knowing of the phallus might promote the development of male gender identity. A persistent theory in developmental psychology holds that gender identity is formed in the first 4 years of life when the child identifies with the gender of the spouse with matching genitalia. If this is true, it is very difficult to separate the direct effects of hormones on the brain areas involved in gender development and the indirect effects of hormones on the genitalia, which in turn affect gender development.

Before it was recognized that 5-$\alpha$-reductase-deficiency is inherited and therefore concentrated in certain families, affected infants were raised as girls (reviewed by Rubin et al., 1981). Upon reaching puberty, however, the increased secretion of testosterone from the pubertal testes led to some DHT being produced, so many prepubertal "girls" developed a male phallus with erections, scrotal testes, increased muscle mass, masculine body fat distribution (less subcutaneous adipose tissue), deepened voice, and male body habitus and psychological characteristics. Other male-typical attributes do not develop at puberty in these individuals, perhaps illustrating the differential effects of testosterone and its metabolites on different sexually dimorphic traits. Males with 5-$\alpha$-reductase deficiency fail to grow beards, male acne, or a male-typical prostate.

A great deal was learned about 5-$\alpha$-reductase deficiency in humans syndrome after it was recognized that the syndrome occurs at a higher frequency among affected families in several parts of the world (e.g., in the Dominican Republic). The result is that newborns with ambiguous genitalia in these families are less enigmatic, and parents and physicians can use their experience to counsel the families about the imminent development of the male genitalia. Some children with 5-$\alpha$-reductase-deficiency are raised as boys, some as intersex, and some as girls, and most are raised with the understanding that gender identity might change after puberty. Indeed, some adult individuals with 5-$\alpha$-reductase deficiency are raised as girls and, after puberty, enter into male heterosexual relationships and function physically and emotionally as men. Others have not entered into sexual relationships, whereas others have maintained female gender identity and engaged in sexual relationships with men.

It is noteworthy that, even after a lifetime of being raised as a girl, many adolescents with 5-$\alpha$-reductase-deficient individuals readily change their sexual orientation and gender identity from female to male at puberty. This was interpreted by some endocrinologists to be a demonstration of the primacy of nature over nurture or the direct effects of hormones on the early development of gender identity. They argued that individuals with 5-$\alpha$-reductase deficiency are masculinized by fetal testosterone, and though they have feminized external genitalia (due to their deficiency in circulating DHT), they "know" they are male throughout most of childhood, and thus, upon the development of the phallus, they easily transition from living as girls to living as men. In fact, men with this syndrome who were raised as girls report that they remember becoming aware of being a boy when they were only between the ages of 7 and 12. Thus, even though individuals with 5-$\alpha$-reductase deficiency are often socialized to see themselves as girls, the exposure of their fetal brain to prenatal testosterone has already masculinized their brains to be aware of their true gender identity and to become responsive to the rise in testosterone they will experience at puberty. Thus, it might be suggested that hormones are more influential than psychosocial factors imposed since infancy.

Another explanation for the smooth change from female gender identity to male gender identity at puberty, however, is that gender identity is influenced by the need for autonomy, independence, and financial success. In societies such as the Dominican Republic, girls enjoy far less independence, less encouragement to be involved in life outside the home, and fewer opportunities for careers and financial gain. It is not surprising therefore, that after being raised as a girl, they enthusiastically transition to living as a man. Some psychologists therefore note that it is premature to rule out environmental effects on gender identity (reviewed by Rubin et al., 1981). In summary, the final gender identity of children born with 5-α-reductase deficiency might be determined by prenatal testosterone and pubertal testosterone, psychosocial cues from parents, siblings, friends, and society, or some unknown combination of these factors. Whether the effects of hormone metabolites are direct or indirect, 5-α-reductase deficiency exemplifies the powerful effects of hormones and their metabolites on morphology and behavior.

There is more evidence that the discrepancies between physical sex and psychological gender in humans result from hormones, environment, and an interaction between the two. Table 4.1 summarizes the possible relationships among genital sex, sex of rearing, and gender identity in genetic (XY) men.

In experimental animals, the neurochemical and molecular bases of sexual differentiation of brain and behavior have been explored. The major findings add mechanistic detail to the clinical picture given before. Testosterone, having entered the brain of the neonatal rat or mouse, is converted to estradiol, and it is largely in the chemical form of estradiol that

**TABLE 4.1**  Sex/Gender Identity Relationships in XY Men

| Clinical description | Pathology | Genital sex | Sex of rearing | Gender identity |
|---|---|---|---|---|
| Normal male | None | Male | Male | Male |
| Transsexual | None | Male | Male | Female |
| Androgen insensitivity (complete) | Lack functional androgen receptors | Female | Female | Female |
| 5a-reductase deficiency | Lack dihydrotestosterone | Female (±) | Female (±) | Female to male |

*Adapted from Gorski, R.A., 2000. Sexual differentiation of the nervous system. In: Kandel, E.R., Schwartz, J.H., Jessel, T.M. (Eds.), Principles of Neural Science, fourth ed. McGraw-Hill, New York, pp. 1131–1148; Wallen, K., Baum, M.J., 2002. Masculinization and defeminization in altricial and precocial mammals: comparative aspects of steroid hormone action. In: Pfaff, D.W., Arnold, A.P., et al. (Eds.), Hormones, Brain and Behavior, vol. 4. Academic Press, San Diego, CA, pp. 385–423.*

the hormone exerts defeminizing activity. Three lines of evidence have been well established (reviewed by McCarthy, 1994):

1. Interruption of the function of the gene for the classical estrogen receptor, ERα, in the hypothalamus of the neonatal rat precludes the defeminizing actions of systemically administered testosterone. Normal females, with an XX chromosome complement, low circulating concentrations of testosterone, and low levels of brain aromatization during early development, will show robust female-typical sex behavior when treated with estradiol and progesterone as adults. They are *feminized* by the lack of testosterone and estradiol during gestation and the first days of life. Pre- or neonatal testosterone or estradiol treatment *defeminizes* the female's adult response to estradiol and progesterone. The defeminizing effects in animals with disruption of the gene for ERα are seen as a failure to show lordosis to adult treatment with estradiol and progesterone.

2. Blocking the action of the enzyme responsible for converting testosterone to estradiol, e.g., using an aromatase inhibitor, reduces the *masculinizing* effects of the androgen on adult male-typical sex behavior. The masculinizing effects of estradiol are seen as normal adult responsiveness to postpubertal treatment with testosterone. Pre- or neonatal testosterone or estradiol increases male-typical mounting and intromission, and these behaviors are blocked in those treated pre- or neonatally with an aromatase inhibitor. They are masculinized by estradiol and demasculinized by the aromatase inhibitor.

3. An estrogen receptor antagonist can reduce the amount of neuroendocrine masculinization and defeminization during the neonatal period. The physiological import of this androgen-metabolism step is manifold; the defeminized brain cannot manage to produce an ovulatory surge of luteinizing hormone, nor can it initiate feminine courtship and sexual behaviors in response to adult treatment with estradiol and progesterone.

### Interim Summary 4.1

1. Testosterone, synthesized and released from cells in the gonads and other peripheral tissues, can travel to other tissues, such as the brain, where it can be either aromatized to estradiol or reduced to DHT.

2. Pre- or neonatal aromatization to estradiol occurs in many tissues, including adipose tissue and brain, and it can organize the brain permanently to have the following:
   a. masculinizing effects on male-typical reproductive and agonistic behavior in response to adult treatment with testosterone;
   b. defeminizing effects on female-typical reproductive and agonistic behavior in response to adult treatment with estradiol and progesterone.

3. 5α-Reduction to DHT occurs primarily in peripheral tissues and is important for the following:
   a. penile reflexes and secondary sex characteristics (such a male pattern baldness) in adults;
   b. masculinization of internal and external genitalia, which in turn can influence the development of gender identity in humans.

# SIX ADDITIONAL EXAMPLES

## Proopiomelanocortin Gives Rise to Several Behaviorally Active Hormones

An important mechanism in endocrinology is the cleavage of protein hormones, yielding smaller peptides with different neuroendocrine activities. A prominent example is POMC, a prohormone for adrenocorticotropic hormone (ACTH) and several other hormones (reviewed by De Wied and Jolles, 1982). In humans, POMC is encoded by a single gene on the short arm of chromosome 2 and contains three exons and two introns. The structure is well conserved across species. In healthy humans, expression of the POMC gene occurs in abundance in anterior pituitary corticotrophs, to a limited extent in the hypothalamus, and to a minimal degree in some peripheral tissues. It also may be expressed in relatively high levels in certain tumors (e.g., in ACTH-secreting pulmonary carcinomas). POMC gene expression is stimulated by corticotropin-releasing hormone (CRH), arginine vasopressin, and the proinflammatory cytokine leukemia inhibitory factor. Conversely, POMC gene expression is inhibited by glucocorticoids (cortisol, corticosterone).

After synthesis of the peptide, POMC is first cleaved into two subunits, pro-ACTH and beta-lipotropin (β-LPH). Pro-ACTH is further cleaved into ACTH1–39 (the active, circulating corticotropin) and a second fragment that in turn is cleaved into N-proopiocortin (N-POC) and joining peptide (JP). ACTH1–39 itself is cleaved into alpha-melanocyte-stimulating hormone (α-MSH or ACTH1–13) and corticotropin-like intermediate lobe peptide (CLIP or ACTH18–39).

Cleavage of N-POC results in the production of γ-MSH. Finally, β-LPH, a product of the original prohormone POMC, is cleaved into β-LPH and β-endorphin. POMC, Pro-ACTH, ACTH, N-POC, JP, and β-LPH all can be detected in the human circulation. The other products do not circulate in detectable amounts but can have important hormonal effects (e.g., the opiate-like analgesic property of β-endorphin in the CNS). Some of these products are more abundant in nonprimate species: for example, α-MSH, which is produced in abundance in the

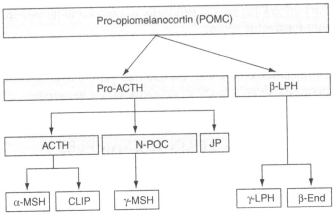

**FIGURE 4.4** *Proopiomelanocortin* (POMC) serves as a prohormone for smaller peptide hormones that have a variety of metabolic actions. *ACTH,* adrenocorticotropic hormone; *CLIP,* corticotropin-like intermediate lobe peptide; *End,* endorphin; *JP,* joining peptide; *LPH,* lipotropin; *MSH,* melanocyte stimulating hormone; *N-POC,* N-proopiocortin. *Modified from Harno, E., White, A., 2016. In: Jameson, J.L., DeGroot, L.J. (Eds.) Endocrinology, seventh ed. Elsevier, Saunders, Philadelphia, PA, p. 136. With permission.*

intermediate lobe of the pituitary in species such as the rat and mouse. Fig. 4.4 illustrates the peptide hormones and fragments produced from POMC.

POMC processing varies, depending on the tissue and species. In the human anterior pituitary, cleavage products are pro-ACTH (which is then cleaved to ACTH), N-POC, JP, and β-LPH. There is minimal production of gamma 3-MSH from N-POC or β-endorphin from β-LPH. In the rodent intermediate lobe, POMC present in melanotrophs is converted to smaller fragments including α-, β-, and γ-MSH, CLIP, and β-endorphin. α-MSH stimulates melanocytes (pigment cells) in skin and influences, for example, coat coloration in rodents and skin pigmentation in frogs. In the hypothalamus, ACTH is processed to CLIP and deacetyl α-MSH, and β-LPH is processed to β-endorphin. Thus, POMC serves as a prohormone for a number of smaller peptide hormones that have a variety of metabolic actions throughout the body.

## Dehydroepiandrosterone and DHEA-Sulfate Are Interconvertible and Have Different Potencies

The unconjugated steroid, dehydroepiandrosterone (DHEA), and its sulfate ester, DHEA-S, are present in the CNS and are both neurotrophic. They are neurosteroids; that is, they are synthesized *de novo* in the CNS (see Chapter 1) and are interconvertible. DHEA stimulates axonal growth, an effect that can be blocked by the glutamatergic N-methyl-D-aspartate (NMDA) receptor antagonist, dizocilpine (MK801). DHEA-S stimulates dendritic growth, but this effect is not blocked by NMDA receptor antagonism. In addition, DHEA and DHEA-S are excitatory neurosteroids that increase the neuronal firing rate. In this regard, DHEA-S is an allosteric γ-aminobutyric acid (GABA) type A receptor antagonist, decreasing the activity of inhibitory GABA systems in the brain. DHEA mimics this effect of DHEA-S, but with only about one-third its potency (Baulieu and Robel, 1996, 1998). DHEA plays an important role in territorial aggression, especially in cases where plasma testosterone concentrations are low (Soma et al., 2015).

## Angiotensin Metabolites

A theme encountered throughout the evolution of peptide hormones is the production of a large precursor that is cleaved and metabolized at specific amino acids to produce active end products and metabolites each with their own affinity for different but related families of receptors. As noted in previous chapters, angiotensin II is a well-established example of this principle, being the major effective peptide of the renin-angiotensin system. Angiotensin is well known for its actions to maintain fluid and electrolyte balance and blood volume through a range of actions in the kidney and vasculature as well as by stimulating $AT_1$ receptors in the circumventricular organs to induce thirst. The increased desire to drink to increase body fluid levels is important to protect the central nervous system from low blood pressure (McKinley and Johnson, 2004). The effects are accompanied by an increase in blood pressure through increased sympathetic vasomotor tone and the release of vasopressin, which is a potent stimulus of thirst. In response to renal nerve stimulation, low blood pressure or low salt intake, renin is released from the juxtaglomerular cells in the kidney and cleaves the 255–amino acid alpha 2 globulin precursor angiotensinogen to the 10–amino acid angiotensin I. The latter is converted to the eight–amino acid active angiotensin II by the carboxypeptidase angiotensin converting enzyme, which is also known as kininase II. Several different blood pressure medications target this enzyme.

The CNS contains all the elements of the peripheral renin-angiotensin system such as receptors, metabolizing enzymes, and precursor angiotensinogen. Angiotensinogen is mostly extracellular, and its mRNA is found in glial and in some neurons (Culman et al., 2002). The presence of renin in the brain has been controversial due to its very low levels, but there is now acceptance that both prorenin and renin exist in human and rodent CNS (Xu et al., 2011). A renin independent pathway has been suggested that produces angiotensin I from angiotensin 1-12, which has a two–amino acid c-terminal addition, compared to angiotensin I (Ahmad et al., 2014). The precise enzymatic pathway is not clear but may involve enzymes of the kallikrein-kinin system, cathepsins, and chymase (Carey, 2013). Angiotensin II is further metabolized by aminopeptidases to produce fragment peptides of different lengths including 1-7, 2-8, 3-8, and 3-7. The different angiotensin peptides emanating from this peptide cascade are involved in a wide range of actions throughout the brain, which can influence behavior, particularly those associated with regulation of fluid balance and blood pressure.

One of the main active peptide fragments is angiotensin 1-7, which is produced by angiotensin converting enzyme 2 from angiotensin II or from angiotensin I by neprilysin, which is also known as neutral endo-peptidase (Fig. 4.5). Ang 1-7 has its own specific G protein–coupled recep-tor called the Mas receptor through which it mediates vasodilatation at blood vessels and has antiproliferation and antihypertrophic effects (Xu et al., 2011). Within the CNS, Ang1-7 appears to be mainly involved in regulating baroreflex sensitivity but has been implicated in causing cogni-tive improvement (Sumners et al., 2013) by activation of the Mas receptor. Decarboxylation of aspartic acid residue of Ang 1-7 changes it to alanine, and the $Ala^1$-angiotensin 1-7 produced is called alamandine. The actions are largely similar to that of angiotensin 1-7. More recently, angiotensin 1-7 has been shown to prevent body weight gain via effects on energy expenditure (Morimoto et al., 2017).

Angiotensin 2-8, known as angiotensin III, is produced by aminopep-tidase A, which eliminates the first amino acid from angiotensin II. This modification does not change the peptide's potency at angiotensin type 1

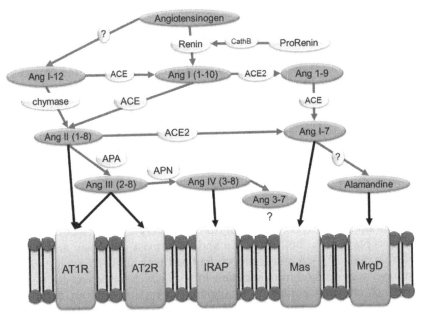

**FIGURE 4.5**   Metabolites of angiotensinogen. *NEP*, neutral endopeptidase.

receptors, and indeed, in the CNS, it has been suggested that angiotensin III might be the active peptide at this receptor. Blockade of aminopeptidase, which converts angiotensin II to angiotensin III, prevents the increase in blood pressure to angiotensin II (Reaux et al., 1999) and reduces thirst and sodium appetite, suggesting that both peptides are active in the CNS (Wilson et al., 2005). Centrally administered angiotensin III inhibits food intake in mice similarly to angiotensin II by acting on largely angiotensin type 2 receptors (Ohinata et al., 2008).

Angiotensin IV (3-8) and angiotensin 3-7 are products of angiotensin III metabolism by aminopeptidase A and aminopeptidase N, respectively. Angiotensin IV is a potent ligand at its own receptor, which is found in the brain, heart, kidney, adrenals, and blood vessels, and has been identified as membrane-bound insulin-regulated aminopeptidase (IRAP). In the brain the pattern of distribution of IRAP is distinct from angiotensin type 1 and 2 receptors and mediates the cognitive, sensory, and motor effects of angiotensin IV. Administration of angiotensin IV centrally enhances learning and memory and can reverse memory deficits in animal models of amnesia by inhibiting IRAP and possibly prolonging the actions of other peptides or changing glucose receptor trafficking. Angiotensin IV does not increase blood pressure or induce drinking. These effects are opposite to those of angiotensin II. Angiotensin 3-7, on the other hand, increases blood pressure when injected centrally but not via the $AT_1$ or Mas receptor (Ferreira et al., 2007). These studies illustrate the complexity of effects of different peptide fragments in a cascade that need to be considered when administering treatments that may alter the relative pools of each fragment in the system.

### *Interim Summary 4.2*

1. The POMC peptide is synthesized and cleaved to yield smaller peptides with important effects of behavior such as ACTH (anxiety and fear learning), α-MSH (ingestive and reproductive behavior), and β-endorphin (positive reinforcement, elevated mood, and reduction of pain perception).
2. DHEA and its sulfate ester, DHEA-S, are neurotrophic neurosteroids with important effects on aggression.
3. Angiotensinogen is converted to angiotensin I, which is cleaved to angiotensin II, the latter having myriad effects on thirst, drinking, blood pressure, and erectile function (more details in Chapter 3).

## Adrenal Steroids

Among steroid hormones produced in the adrenal cortex, one fascinating metabolic conversion is between two sets of steroids, each with important but different spectra of behavioral actions. Corticosterone is a classic glucocorticoid stress hormone, while aldosterone, a mineralo-corticoid, protects blood fluid volume by, among other actions, stimu-lating salt appetite. The enzyme steroid-18-hydroxylase accounts for the metabolic conversion from corticosterone to aldosterone. In molecular terms, this steroid conversion becomes even more interesting because of the different affinities of two nuclear receptors, both transcription fac-tors. Mineralocorticoid receptors (MRs) are high-affinity receptors with high selectivity for their ligands and a restricted neuroanatomical dis-tribution, notably, in the hippocampus. Glucocorticoid receptors (GRs) have lower affinity, readily bind both glucocorticoids and mineralocor-ticoids, and show a much broader neuroanatomical distribution. Most important, GRs mediate effects that are often opposite those of MRs. Within the field of stress-related behaviors themselves, liganded MRs seem to work through the rapidly responding CRH-1 receptor system. In contrast, GRs are more connected with the slower CRH-2 receptor system, through which they help to restore physiological balance fol-lowing stress.

How do we know about the relative effects of GR and MR agonists? At the intersection between stress physiology and learning, as well, adrenal steroids acting primarily through MRs or GRs have opposing roles. In the hippocampus, the electrophysiological phenomenon called *long-term potentiation* (LTP) is widely used as a tractable experimental model for some forms of learning. Pavlides and McEwan (1999) used LTP in adrenalectomized rats to study the effects of adrenal hormones on neurons in the dentate gyrus of the hippocampus in response to electrical stimulation of hippocampal inputs (Fig. 4.6). Whereas, the mineralocorticoid, aldosterone, produced a significant enhancement of LTP, a pure GR agonist produced the opposite result, a significant reduction of LTP.

Another enzyme for metabolic conversion among adrenal steroids, 11-beta-hydroxysteroid dehydrogenase-1 (11--HSD-1), reveals its medi-cal importance in an entirely different way. Glucocorticoids can be produced locally from inactive metabolites through the action of this enzyme. Because high levels of glucocorticoids are associated with vis-ceral fat, diabetes, and consequent mortality, the molecular controls over this enzyme and its consequence are of interest. When the gene for 11--HSD-1 is knocked out by homologous recombination, the resultant mice have reduced activation of GR-sensitive liver enzymes and, notably, have a diabetes-resistant phenotype. Masuzaki and Flier (2003) then created a mouse in which a transgene for 11--HSD-1 was overexpressed under the control of the strong aP2 promoter. These mice had pronounced diabe-tes, hyperlipidemia, and, despite high levels of the adipocyte hormone, leptin, they showed hyperphagia (increased food consumption). The rela-tive importance of a variety of behavioral changes to the complete health picture of these mice remains to be determined.

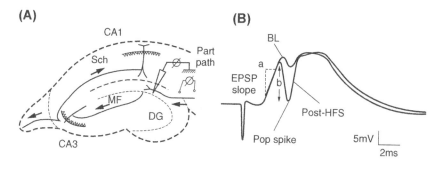

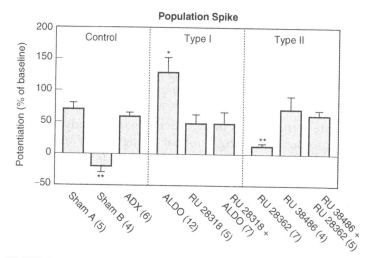

**FIGURE 4.6** Different steroid hormone metabolic patterns that yield agonists either for mineralocorticoid receptors (MRs) or glucocorticoid receptors (GRs) have opposite effects on long-term potentiation (LTP) during recordings from hippocampal neurons. Top panel: The electrophysiological setup; a stimulating electrode was inserted in the perforant pathway, an input to the hippocampus. A recording electrode was inserted into the dentate gyrus granule cell layer. *BL*, baseline; *DG*, dentate gyrus; *HFS*, high-frequency stimulation; *MF*, mossy fibers; *Sch*, Schaffer collaterals. The figure shows examples of electrical recording before potentiation (BL) and after potentiation (Post-HFS). Bottom panel: In the left box, the sham-operated animals were split into two groups. Sham A animals had low corticosterone levels, and showed normal LTP, while Sham B animals had high corticosterone levels and significantly suppressed LTP. The middle box illustrates MR receptor manipulations (also called type I receptors). A natural agonist, aldosterone (ALDO) stimulated increased LTP, whereas the receptor antagonist RU28318 did not. The antagonist blocked the ALDO effect. The righthand box illustrates GR receptor manipulations (also called type II receptors). The GR agonist RU28362 had the opposite effect of ALDO in that it reduced LTP, whereas the antagonist did not. Note that the GR antagonist blocked the agonist effect. *Modified from Pavlides, C., McEwen, B.S., 1999. Effects of mineralocorticoid and glucocorticoid receptors on long-term potentiation in the CA3 hippocampal field. Brain Res. 851 (1–2), 204–214; Pavlides, C., 1995. Hippocampal homosynaptic long-term depression/depotentiation induced by adrenal steroids. Neuroscience 68 (2), 379–385; Pavlides, C., Watanabe, Y., Magarinos, A.M., McEwen, B.S., 1995. Opposing roles of type I and type II adrenal steroid receptors in hippocampal long-term potentiation. Neuroscience 68 (2), 387–394.*

## Thyroid Hormones

The thyroid hormone predominantly secreted from the thyroid gland and circulating in the blood is thyroxine (T4, to reflect its four iodine atoms); however, the thyroid hormone predominantly found in the nuclei of cells in the brain and pituitary gland is T3. One iodine has been lost. The enzyme deiodinase type 2 (DIO2) is in fact present and active in the CNS and in the pituitary and accounts for the severance of one iodine atom. On nuclear thyroid hormone receptors, T3 is a more effective ligand than T4. As a consequence, the regulation of DIO2 synthesis and enzymatic activity powerfully controls the concentration of active thyroid hormone in nerve and glial cell nuclei and can buffer the brain's T3 defense against problems with thyroid gland production. For rapid actions of thyroid hormones (e.g., on $Na^+/H^+$ exchanger activity), the metabolite T3 is more effective than the prohormone T4. All of these details of thyroxine metabolism must bear on the ability of thyroid hormones to support actions requiring physical and mental energy and to affect mood.

A prime example is the control of ingestive behavior, body weight, adiposity, and reproduction during seasonal changes in photoperiod (day length) and ambient temperature. As described in Chapter 3, a combination of the annual change in photoperiod with internal biological rhythms synchronizes ingestive behavior and body fat accumulation to anticipate changes in food supply and energetic needs for mating, gestation, and lactation/brooding. For example, in response to the shortening day length of winter, Siberian hamsters cease reproduction, decrease their food intake, and lose body fat. These processes are spontaneously reversed in the spring. Whereas it might be predicted that these changes are mediated by the typical "feeding-inhibitory" peptides of the arcuate nucleus, to the contrary, these changes are most closely linked to effects of melatonin on the thyroid axis. The critical effect of long or short days occurs via nonneural cells, i.e., on a brain cell known as a tanycyte. Tanycytes are special glial cells found in the third and fourth ventricles of the brain with processes extending deep into the hypothalamus.

To look for clues to what genes and hormones are involved in the response to short days, researchers looked at expression in the Siberian hamster genome. They compared gene expression in hamsters housed either in long days (more than 14 h of light per day) or short days (less than 10 h of light per day). The most notable seasonal changes in gene expression in response to day length are in (1) the *TSHβ* (the gene for TSH) in the pars tuberalis and (2) deiodinase 2 (DIO2) and deiodinase 3 (DIO3) in the nearby ependymal tanycyte (nonneural) cells. This pattern of gene expression provides circumstantial evidence that the metabolism of thyroid hormones are central to the body weight response to short day lengths.

In support of this idea, tanycytes have receptors for TSH, and increases in TSH upregulate DIO2 gene expression and secretion. Whereas DIO2 is necessary for synthesis of T3, DIO3 converts T4 to inactive reverse T3 and deiodinates T3, with the contrasting effect of reducing local tissue concentrations of thyroid hormone. Across all seasonal species studied to date, there is a consistent pattern of high levels of DIO2 expression and very low levels of DIO3 expression in long days, whereas exposure of animals to short days tends to reduce DIO2 expression, but it markedly upregulates DIO3 expression. To test the idea that changes in body weight and ingestive behavior in short days are related to DIO2-mediated metabolism of thyroid hormones, Siberian hamsters were given either sham implants or implants with T3. Short day–housed hamsters with sham implants displayed the expected gonadal regression, hypophagia, and body weight loss response to short days, and these processes were completely prevented in hamsters with hypothalamic T3 implants. This provides evidence for control of body weight and food intake by the enzymes involved in metabolism of thyroid hormones in nonneural cells. Understanding how this cue from the environment (short day lengths) influence tanycyte gene expression of thyroid enzymes and understanding how the hormonal milieu might regulate neurogenesis and plasticity in the hypothalamus are likely to provide new insight into development of strategies to manage appetite and body weight (Ebling, 2014).

## Progestins

Progesterone (P) acts in the brain in the chemical form of progesterone itself and also acts as a prohormone. While P binding to the nuclear progestin receptor (PR) in hypothalamic neurons is proven to be a behaviorally important mechanism, P is also metabolized into reduced progestins such as 5-alpha-pregnane-3alpha-ol-20-one. The behaviorally important responses to these reduced metabolites, called *neurosteroids*, typically are not due to binding to nuclear PRs but instead are due to a potentiation of the action of the inhibitory neurotransmitter GABA through its GABA-A receptor. The effects of neurosteroids that are progesterone metabolites on behavior are rapid, not allowing time for new gene expression and protein synthesis (Fig. 4.7). Behavioral effects include not only the rapid facilitation of lordosis, a female sex behavior, but also the reduction of anxiety. The clinical importance of these developments is illustrated by the ability of high doses of progesterone metabolites, given to human patients, to act as anesthetics.

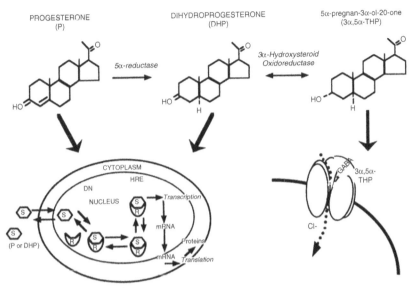

FIGURE 4.7   When progesterone is administered to an animal or a human, some of its behavioral effects are consequent to progesterone's metabolic products. The figure depicts CNS mechanisms of action of progesterone (P) and its metabolites, dihydroprogesterone (DHP) and 5-pregnan-3ol-20-one (3,5-THP). P and DHP have a high binding affinity for intracellular progestin receptors (PRs), whereas 3,5-THP is devoid of affinity for PRs. 3,5-THP acts as a positive modulator for GABA-A receptors to increase GABA's binding to its site on the receptor and thereby increase chloride ion influx. In the ventral tegmental area, progestins mediate sexual responses of rodents in part through actions of 3,5-THP at GABA-A receptors (Frye, 2001a,b).

## Interim Summary 4.3

1. Corticosterone (a stress glucocorticoid in rats and mice) is converted by the enzyme steroid-18-hydroxylase to aldosterone (a mineralocorticoid), and these two steroids have opposing effects on the neural processes that underlie learning and memory in rats and mice.

2. Thyroxine (tetraiodothyronine, T4) is converted to the more active compound triiodothyronine (T3) by the enzyme DIO2. Environmental regulation of this enzyme is important for control of seasonal changes in ingestive behavior in Siberian hamsters.

3. Progesterone is metabolized to other progestins (a.k.a., progestagens), such as 5-α-pregnane-3-α-ol-20-one in the brain, and this conversion is necessary for aspects of sex behavior and GnRH surges, among other behaviors and physiological processes.

# SOME OUTSTANDING NEW QUESTIONS

1. How might genetic conditions such as 5-α-reductase deficiency be detected early enough in gestation so gene therapy or hormone therapy might be instituted? Clinically, as well, knowing which men would be psychologically sensitive to 5-α-reductase inhibition would be useful for doctors treating prostate problems.

2. Females appear to be demasculinized and feminized by the absence of androgens and their metabolites during early development; but, what are the other genetic and hormone requirements for female development? Some evidence suggests that small amounts of various steroid metabolites are required for normal female development (not too much but not too little), and there are still many questions to be answered by including female subjects in experimental protocols.

3. It has been demonstrated that puberty is a second "critical period" for the permanent development of sexuality and social behaviors in response to later adult levels of hormones (reviewed by Sisk). What are the roles of androgens and their metabolites during this developmental window?

4. Should angiotensin II be the target for drug treatment of high blood pressure or should its metabolites?

# 5

# There Are Optimal Hormone Concentrations: Too Little or Too Much Can Be Damaging

The core principle of this chapter has important consequences for human health and will be illustrated using various examples of neuro-endocrine pathologies with adverse effects on behavior. These behaviors include the anxiety disorders (generalized, panic, or social anxiety disorders), depression, eating disorders (over- or undereating), excessive thirst, cognitive disorders (disorders of learning, memory, perception, problem solving, and attention), hyper- and hypoactivity, chronic fatigue, abnormal sexual motivation, erectile dysfunction, and aggressive disorders. Many of these problems are related to levels of hormones being too low, too high, or at optimal levels too early or too late.

Imagine the following scenario: A middle-aged woman, who formerly prided herself in her "hour-glass figure," is now plagued by periodic anxiety, irritability, and insomnia, which she believes are caused by her reaction to recent bouts of binge eating and body weight gain. Her physician notes that her weight gain is markedly concentrated in her abdominal, visceral area as well as in her face. Why, in this case, would her physician order blood samples for circulating hormones? In this chapter, you will learn about a disease known as Cushing syndrome, characterized by excessive secretion of the adrenocortical hormone, cortisol. In Cushing syndrome, elevated cortisol is likely a *direct* cause of abdominal fat accumulation *and* a direct cause of debilitating anxiety. Furthermore, as explained later in this chapter, appropriate treatment differs, depending on whether elevated cortisol levels are accompanied by elevated levels of the pituitary hormone adrenocorticotropic hormone (ACTH). Peripheral hormone secretion is often a result of a complicated cascade of hormones from pituitary and brain. Understanding these complicated cascades and the concentrations of hormones in healthy individuals has aided in the diagnosis and treatment of neuroendocrine diseases.

As illustrated in the previous example, the neuroendocrine mechanism has at least two components. First is the temporal envelope of hormone concentration; that is, the profile of circulating concentrations of hormone over different time periods. The second component involves the cellular response to hormone binding, related to the number of receptors and their binding affinity. In other words, the sensitivity of the cellular response to the hormone also helps determine whether a hormone concentration is "too little" or "too much." This sensitivity is mediated by specific receptors that detect and bind the hormone and by signaling cascades that are initiated by the hormone-receptor complex. If receptors are insensitive or reduced in number, then greater hormone concentrations are needed for an optimal effect. The following examples illustrate the importance of the timing of hormone secretion as well as the importance of the receptor sensitivity, receptor number, and the intracellular signaling mechanisms.

Another brief example illustrates the importance of the temporal pattern of hormone secretion. Blood concentrations of many hormones can be measured with accuracy, but two healthy patients who have their blood drawn at 0800h will have vastly different hormone concentrations if one of those patients regularly works the night shift. Cortisol, ACTH, thyroid hormone, melatonin, and testosterone are all examples of hormones that vary markedly in circulating concentrations over a 24-h period, i.e., the hormones are secreted in circadian rhythms. With regard to some of these hormones, the circadian rhythm is endogenous, i.e., it is controlled by a biological clock or circadian oscillator. If so, the rhythm will "free run" under constant conditions. Other circadian rhythms are secondary to environmental factors, such as daily meal times.

Rhythms in hormone secretion have consequences for behavior and mental health. For example, changes in the circadian rhythm of ACTH and cortisol can disrupt the circadian rhythm of behavior, causing insomnia, night panics, and midnight snacking, all of which tend to increase the accumulation of abdominal adipose tissue. Many endogenous hormones change with regularity over the day and night, and therefore both physicians and patients must be cognizant of these circadian cycles.

In addition to circadian hormonal cycles, there are hormonal cycles that span time periods longer than 24h (infradian rhythms). Consider the following scenario: A 55-year old female patient presents with mild body weight gain, memory loss, sudden mood changes, hot flashes, an increased incidence of bone fractures, and measurable bone loss. These are familiar symptoms of menopause, which is not a disease, but rather a developmental stage that occurs in women around age 50. The secretion of ovarian steroid hormones diminishes, fertility is lost, and other physiological changes occur (see also Chapter 11). After puberty and prior to menopause, ovarian steroid hormones fluctuate predictably over the roughly 28-day menstrual cycle, but as women near the time of meno-pause, the menstrual cycles become irregular until these cycles disappear. In some women, these natural hormonal changes are accompanied by many uncomfortable side effects.

The loss of menstrual regularity and low ovarian steroid concentra-tions provide a critical clue to diagnosis and treatment. Hormone replace-ment therapy (HRT) can provide relief for many of the postmenopausal symptoms. HRT for menopausal symptoms is controversial, however, and women and their physicians have questions. Should perimeno-pausal women be treated with a constant level of one or more ovarian hormones, treated with fluctuating hormone concentrations that mimic the natural fluctuations over the menstrual cycle, or allowed to undergo natural hormonal quiescence (no hormone treatment)? Can we find a dose and temporal schedule of ovarian hormones that alleviates symptoms of menopause without increasing the chances of heart disease and cancer? Menopause is a perfect example of an infradian rhythm with profound impact on health and disease.

In addition to timing, the receptor component is equally important, because receptor mechanisms determine the response to hormones and thereby alter the optimal concentration of hormones. For example, saving the lives of patients with Type 1 diabetes mellitus requires a treatment that differs from the treatment prescribed for Type 2 diabetes mellitus. This is because Type 1 diabetes mellitus results from low levels of insulin, whereas Type 2 diabetes mellitus results from low sensitivity to insulin accompanied by high levels of this hormone. When it comes to hormones, the optimal circulating concentration is dependent upon receptor mecha-nisms, as will be explained in detail in this chapter.

Finally, the temporal envelope of hormone concentrations can be disrupted not only by endogenous pathology of the endocrine system, but also by exogenous administration of the hormone. For example, obesity researchers held high hopes for leptin treatment as a cure for overeating and obesity. In initial trials, daily leptin treatment appeared to decrease food intake and body weight in some individuals, but over time, repeated exposure to high concentrations of leptin lowered the patient's ability to respond to leptin, and as a consequence, food intake and body weight rebounded, and sometimes even exceeded preleptin levels. Currently, a great deal of research is aimed at improving sensi-tivity to chronic treatment with leptin, insulin, and other hormones. Pharmaceutical agents of interest include various receptor-stimulating (agonist) and/or receptor-blocking (antagonist) drugs. The critical point is that for any given hormone, the optimal level depends upon changes in receptors, their signaling capabilities, and their previous exposure to the hormone.

## TOO LITTLE

### Insulin

Before the discovery of the pancreatic hormone insulin, many people suffered and died from a mysterious disease characterized by excessive hunger, thirst, and urination. In addition, they craved sweets, but consumption of sugary foods made the symptoms worse. Over time, these unfortunate patients developed excruciating pain in the extremities (known as peripheral neuropathy), body weight loss, muscle wasting, fatigue, depression, eating disorders, panic disorder, anxiety, infections sometimes requiring limb amputation, posttraumatic stress disorder, and premature death.

You might recognize this disease as diabetes mellitus, a.k.a. Type 1 diabetes (T1D). Late in the 19th century, physicians discovered that removing the pancreas, the body's main source of insulin, induces the diabetic state. The next obvious step was to produce an antidiabetic extract from the pancreas. This proved difficult, until it was discovered that insulin is produced in the *beta* cells of the pancreas by specialized structures known as the islets of Langerhans. The first successful treatments were islet extracts dubbed *pancreine, isletin*, and finally, insulin. In 1923, Banting and Macleod received the Nobel Prize in Medicine for development of injectable insulin for the treatment of diabetes in humans. In modern times, insulin is synthesized, based on the specific amino acid sequence for human insulin. In the treatment of T1D, accidental insulin overdose can be harmful, producing dangerously low blood glucose and shock-like symptoms. The dire effects of T1D and the reversal of its symptoms by careful insulin treatment exemplify the necessity of optimal circulating hormone concentrations.

To understand the importance of circulating insulin concentrations, it is useful to examine the function of insulin. Insulin's main function is to help maintain a steady level of metabolic fuels, especially glucose, in circulation. If blood glucose concentrations are too high, blood vessels and neurons are damaged, and if they are too low, cells will literally starve to death. This is critical in the brain, a tissue that primarily uses glucose for energy. Insulin promotes the transport of glucose and free fatty acids (metabolic fuels) from the blood stream to the inside of peripheral cells (cells in the body). Once the fuels (glucose and free fatty acids) are inside, they can be oxidized to serve the energetic needs of the cell. If no energy is required, the fuels can be shunted into adipocytes (fat cells) where they can be converted to storage molecules. Blood insulin concentrations must be high enough to shunt metabolic fuels out of the blood and into these peripheral tissues, and low enough to leave just the right concentration of glucose to be transported to the brain. In summary, in healthy individuals, the function of insulin is to promote the uptake of metabolic fuels into cells that can oxidize or store these fuels, to remove excess glucose from the blood stream, and to maintain a steady supply of glucose (i.e., to maintain glucose homeostasis).

Given this information, the symptoms of T1D make perfect sense. Without insulin, peripheral cells, including muscle, liver, and kidney, cannot acquire the metabolic fuels required for their function. Without insulin, it is impossible to accumulate body fat. Furthermore, without insulin, circulating glucose concentrations rise to dangerous levels (a condition known as hyperglycemia). Hyperglycemia is toxic to blood vessels and nerves, destroys peripheral circulation in the extremities, and decreases the ability of the circulatory system to deliver immune cells to heal skin or muscle injuries. Hypoglycemia acts as an osmotic stimulus to draw water out of cells, stimulating overproduction of urine and intense thirst. The original name for T1D, *diabetes mellitus*, stems from the Greek, *diabainein*, for "a siphon," referring to excessive urination, and *mellitus*, for "like honey," referring to the sweet smell and taste of the diabetic patient's syrup-like (glucose-rich) urine. T1D is often present from birth or early childhood, and healthy insulin levels differ depending on developmental maturity, body size, muscularity, body fat content, diet, and exercise. Matching insulin level to the needs of the changing individual is a challenge that lasts a lifetime.

### Interim Summary 5.1

1. Insulin is secreted from the pancreas in response to elevations in circulating concentrations of metabolic fuels (such as glucose).
2. Insulin functions to remove those fuels from circulation into tissue where they can be oxidized for cellular energy or stored in muscle and adipose tissue.
3. Lack of insulin, T1D, is a congenital condition that is usually discovered early in life. If untreated it leads to death from cellular starvation and damage to nerves muscles exposed to hyperglycemia.

## Arginine Vasopressin (Antidiuretic Hormone)

It is interesting to compare the symptoms of T1D, diabetes mellitus, to those of diabetes insipidus, a syndrome that is caused by a different mechanism. Patients with diabetes insipidus also have frequent and excessive urination and have been known to excrete more than 16 L of water a day. By contrast to those with diabetes mellitus, their glucose concentrations remain within normal limits. Patients with diabetes insipidus lack a different hormone, arginine vasopressin (AVP), a posterior pituitary peptide secreted in response to stressors such as injury and dehydration.

AVP is also known as simply vasopressin, but in many species, including humans, vasopressin contains the amino acid arginine. AVP is also known as antidiuretic hormone (ADH). A diuretic is a substance, such as alcohol or caffeine that promotes urination, whereas an antidiuretic prevents urination by promotion of water retention by the kidney. The main function of AVP is to decrease urine output. Diabetes insipidus can occur due to various mutations or abnormalities (e.g., head trauma, infection, or tumor) in cells of the posterior pituitary. These abnormalities result in little or no secretion of AVP.

To understand normal AVP function, it is helpful to remember that we are 75% water. Our circulatory system requires the maintenance of a healthy blood volume, and our cells require a particular concentration of salts (ions). Survival requires constant maintenance of homeostasis in blood volume and ion concentrations (osmolarity), and this is accomplished by altering behavior, specifically by drinking water. The amount of water ingested must be balanced by the quantity and concentration of urine produced, to keep osmolarity and blood volume within a healthy range. AVP is released from the posterior pituitary gland (the neurohypophysis), which is directly connected to the magnocellular cells of the paraventricular nucleus of the hypothalamus and supraoptic nucleus of the hypothalamus, where AVP is synthesized. AVP travels from the cell body where it is synthesized down axons to the posterior pituitary. From nerve terminals in the posterior pituitary, AVP is released into the general circulation and acts on the renal collecting duct to enhance water reabsorption. It does this by activating aquaporins, which are proteins with water channels that run from inside the cell to the apical membrane of the renal collecting duct cells. The aquaporins allow water to be reabsorbed across the membrane from the collecting duct back into cells and then into the blood. When blood volume and osmolarity fall out of balance, AVP release and its action on aquaporins are necessary to restore blood volume and osmolarity. Without AVP secretion, water is not reabsorbed and must be excreted. Excessive water drinking and urination is a hallmark of diabetes insipidus, another example of the need for optimal levels of hormones for health and well-being.

## Androgens

One live performance you may never have the chance to hear is the extraordinary voice of a castrato, a genetically male singer castrated prior to age nine. The last of the castrati died in the 1940s, but in the early 1700s, their voices were well known in the churches and operas of Europe. At the peak of their popularity, nearly 4000 boys were castrated and groomed for the choir of the Sistine Chapel in Rome and the Italian opera houses. The castrati voices were said to be higher in pitch and stronger than those of gonadally intact choir members, as a consequence of their lower circulating levels of testicular androgens. During normal male pubertal development, in boys with intact testes, the pubertal rise in secretion of testosterone and its metabolites increases the length and thickness of the larynx (vocal cords) and fuses the ends of the long bones of the arms, legs, and ribs. The onset of testosterone production produces a relatively low-pitched voice and terminates the pubertal growth spurt. Early-life castration, by contrast, preserves the childishly small larynx and extends the growth of the rib cage, producing a larger lung capacity than that of a gonadally intact male. Castrati were known for their loud, strong voices and exceptional vocal range. Fortunately, for the boys of Italy, this barbaric practice ended around the turn of the 19th century, but it provides a striking example of the effects of hormone concentrations that are too low.

Hypogonadism can be caused by pituitary tumors, endocrine disrupting compounds in the environment, or drug treatments that interfere with gonadotropin-releasing hormone (GnRH) or gonadotropin secretion. Hypogonadism is defined by low circulating testosterone concentrations (below 231 ng/dL), from blood sampled early in the morning. Testosterone therapy is recommended especially if low testosterone concentrations are accompanied by low sperm count and/or diminished sexual libido. This latter effect might have been on the minds of sultans in the Middle East and emperors in China who employed eunuchs (castrated men) to guard their harems. Similarly, chemical "castration" techniques (drug treatments that decrease testosterone secretion) are, in some geographic locations, used in an attempt to reduce sexual drive in male sex offenders. Aside from the ethical issues involved, the effectiveness of castration in harem guards and sex offenders is in doubt, because, in men with a great deal of sexual experience, castration fails to entirely eliminate sexual libido. Currently, castration is an important medical procedure used to lower the levels of androgenic hormones in patients with androgen-dependent cancer (e.g., androgen-dependent prostate cancer), but decreased sexual libido, decreased cognitive function, depression, and metabolic disturbances are problematic side effects. Whether hypogonadism is congenital or results from injury, drug treatment, or old age, it is associated with loss of muscle and bone mass, obesity, and insulin resistance. Whereas it is true that significant loss of circulating testosterone produces many negative effects, it is not clear whether these are direct effects of testosterone or effects of metabolites of testosterone.

### Interim Summary 5.2

1. Lack of AVP, secreted primarily by nerve terminals from the brain that release their contents from the posterior pituitary, results in diabetes insipidus, characterized by a deficit in the ability to retain water by the kidneys, and symptoms include excessive water drinking and urination.

2. Prepubertal deficits in the synthesis and secretion of androgens, primarily secreted by the testes and adrenals, lead to deficits in the development of secondary sex characteristics, including failure to develop a deep, masculine voice and failure to end the pubertal growth spurt. The symptoms of lack of adult androgens include infertility and low sexual libido as well as osteoporosis, the development of abdominal obesity, and insulin resistance.

## TOO MUCH

### Insulin

Insulin treatment is a lifesaver for patients with T1D, but insulin injections are no panacea. Diabetic patients must keep close track of their blood glucose concentrations throughout each day for their entire life. An accidental overdose of insulin can result in too much glucose being shunted out of the blood stream, causing hypoglycemia and an immediate deficit in metabolic fuels in the brain. Resulting symptoms include cold sweats, rapid heartbeat, blurred vision, trembling hands, nervousness, fatigue, headache, intense hunger and anxiety, and a general sense of confusion, and medical assistance may be needed. If hypoglycemia is prolonged, the patient can experience a dysregulation of body temperature accompanied by feelings of panic or confusion; if insulin levels continue to rise, the patient can lose consciousness due to insulin shock. Ingestion of sugar or treatment with another hormone, glucagon, can bring rapid relief. The main point is that too much insulin can be as fatal as too little, and it is worth repeating that one of the challenges of this century is to develop accurate methods of regulating circulating insulin concentrations.

It might seem confusing at first that T2D diabetes mellitus has symptoms in common with T1D (e.g., hyperglycemia, fatigue, and increased appetite); yet, T2D is characterized by high, rather than low plasma concentrations of insulin. This can occur when there is little or no cellular response to insulin, perhaps because the number of insulin receptors has been reduced. The number of insulin receptors on the surface of the cell is reduced (downregulated) or the signaling component of the receptor is defective, and thus the cell becomes resistant to insulin, thereby resulting in poor cellular glucose uptake and hyperglycemia. Other factors, such as elevated levels of fatty acids, adiponectin, and peroxisome proliferator-activated receptor (PPAR) alpha and gamma might play a role in insulin resistance. The latter receptor, PPAR is important because PPAR-gamma agonists increase insulin sensitivity and are used to treat T2DM. The adipocyte hormone adiponectin is of interest because in a particular population, Arizona's Pima Indians, there is a high incidence of obesity and

T2DM, closely associated with low circulating concentrations of adiponectin. This finding is compatible with the idea that high circulating concentrations of adiponectin protect against insulin resistance. This example illustrates that too much insulin can be the consequence of too little of a hormone necessary for healthy insulin sensitivity.

## Arginine Vasopressin (Antidiuretic Hormone)

In old war movies, mortally wounded soldiers often were portrayed as deliriously thirsty. It turns out that there is a grain of truth to this cinematic trope because massive blood loss (e.g., more than 15%–20% of a person's total blood volume) results in a release of the posterior pituitary hormone AVP (as well as angiotensin II). AVP is released in response to stressors that include injury, dehydration, and a sudden drop in blood volume. As noted earlier in this chapter, AVP stimulates thirst and water reabsorption by the kidneys as part of an adaptive suite of traits that restores blood pressure to equilibrium. It is therefore plausible that soldiers with massive blood loss would experience a surge in AVP secretion that produces intense thirst.

A related syndrome results in confusion, disorientation, delirium, seizures, and coma. The syndrome is caused by excess secretion of AVP, also known as ADH. Normally, AVP/ADH secretion is stimulated by a response to a drop in blood volume, but a dangerous pathology occurs when AVP/ADH rises in the face of normal or increased plasma fluid volume. The syndrome is therefore named syndrome of inappropriate antidiuretic hormone secretion (SIADH). In patients with SIADH, the kidneys retain water, resulting in abnormally low serum sodium relative to the amount of water (hyponatremia). Causes of SIADH include several classes of prescription drugs, postoperative stress, injury, central nervous system (CNS) disorders, some psychiatric disorders in which huge amounts of water may be consumed, and pulmonary disorders, particularly certain pulmonary cancers. In addition to confusion, disorientation, and delirium, other symptoms might include muscle weakness, tremor, and ataxia. Treatment of SIADH is aimed at correcting the low serum sodium, often by water restriction. Too much AVP thus can be as unhealthy as too little.

## Androgens

There has been a troubling syndrome among athletes and even among nonathletic teens who wish to lose body fat, gain muscle mass, and boost their self-esteem. The symptoms in males include paranoia, insomnia, irritable aggression, male-pattern baldness, acne, and increased sexual libido ironically accompanied by erectile dysfunction. In females, most of the symptoms are the same but also include irregular menstrual cycles and hirsutism. These are symptoms of anabolic androgenic steroid abuse. Longer-term effects might include permanent liver damage. Athletes may abuse testosterone and/or synthetic androgens that mimic testosterone's effects. The doses typically produce supraphysiological blood concentrations, which suppress secretion of GnRH from the hypothalamus and luteinizing hormone and follicle-stimulating hormone from the anterior pituitary.

How can an excess of this "male" hormone result in low sperm count? Androgens and their metabolites tend to downregulate testosterone receptors on the germinal cells of the testes, receptors that are necessary for sperm formation. Low sperm count and infertility thus are associated with anabolic androgenic steroid abuse. As well, excess testosterone has profound behavioral effects because testosterone, being a steroid, can easily cross the blood–brain barrier to enter the brain. Testosterone acts directly on androgen receptors or indirectly when it is metabolized by 5-alpha-reductase to 5-alpha-dihydrotestosterone (DHT), which activates receptors in the amygdala, hypothalamus, and limbic system. The receptors are inside the cell (cytosolic steroid receptors) and bind DHT, forming a receptor complex that migrates into the nucleus, binds to DNA, and initiates transcription. This process, transactivation, leads to the synthesis of proteins and the catecholamines (e.g., dopamine and norepinephrine). Norepinephrine stimulates the sympathetic nervous system for a "fight or flight" response. Unfortunately, the testosterone abuser too often chooses to fight. The secretion of another catecholamine, dopamine, is increased, and this neurohormone is stimulatory for male sexual motivation. Furthermore, chronically high levels of androgens increase the incidence of insulin resistance, cholesterol levels, atherosclerosis, liver disease, heart failure, and cancer. For these reasons, prolonged, high concentrations of androgens are unhealthy for both men and women.

### Interim Summary 5.3

1. Too much insulin can result from insensitivity to the effects of insulin (T2D, i.e., insulin resistance accompanied by hyperglycemia), a disease that typically occurs later in life, with eventual outcomes similar to those of T1D. Too much insulin can result from overdose of exogenous insulin, resulting in dangerous hypoglycemia.
2. Too much AVP, also known as ADH, can result in SIADH, in which the kidneys retain water, resulting in abnormally low serum sodium. Symptoms can include dizziness, delirium, and confusion, and in extreme cases, seizures and coma.
3. Excessively high levels of androgens in men and women can result in inappropriate aggression, irritability, paranoia, insomnia, male-pattern baldness, acne, infertility, and increased sexual libido ironically accompanied in men by erectile dysfunction. In the long term, high levels of androgens increase the incidence of cancer and liver and heart disease.

# CLINICAL CONDITIONS, ORGAN BY ORGAN

Clinical conditions in which target endocrine glands are underactive or overactive can be caused by pathology at several levels of the endocrine axis. The site of pathology can be in the target endocrine gland itself, the pituitary gland, the hypothalamus, or higher brain centers. Changes in gland size often result from under- or overstimulation; for example, the adrenal cortex atrophies (shrinks in size) and becomes hypoplastic (loses cells) without ACTH stimulation, and it hypertrophies (becomes overly large in size) and hyperplastic (increases in cell number) with excess ACTH stimulation. Different behavioral pathologies also result, depending on the hormone affected. Diagnosis is confirmed by measurement of circulating hormone concentrations under baseline conditions and following stimulation and suppression tests. Following are examples of clinical conditions related to adrenal, thyroid, and growth hormone (GH) function that illustrate these circumstances.

## Adrenal Gland Hypofunction

Whenever a patient appears with depression, fatigue, and emotional instability, it might be assumed that the first line of defense should be psychotherapy and a prescription for antidepressants. This is not necessarily the case because neuroendocrine disorders can cause similar symptoms. In the case of adrenal hypofunction, without hormone treatment the disease will worsen, eventually causing hallucinations, delirium, coma, and death. Treated with hormones, however, a high-quality life can be prolonged, as in the case of President John F. Kennedy, whose adrenal glands at the time of autopsy "were found to be almost completely gone." After Kennedy's death, it was revealed that he had undergone many years of treatment for Addison disease. The array of behavioral symptoms listed here are typical of adrenal insufficiency, which is classified into two types, depending on the source of the problem.

Patients who report weakness, fatigue, loss of appetite, dizziness, nausea, diarrhea, and abdominal, muscle, and joint pain are often tested for levels of adrenal hormones. Those who suffer from adrenal tumors or autoimmune diseases are likely to have primary adrenal insufficiency (Addison disease), and this will produce a recognizable pattern of hormone secretion. In primary adrenal insufficiency, both of the main adrenal cortical hormones, cortisol and aldosterone, are deficient. You will remember that cortisol secretion is stimulated by ACTH from the anterior pituitary, which is stimulated by corticotropin-releasing hormone (CRH) from the hypothalamus, and that increasing levels of cortisol normally have negative feedback action on the hypothalamus, inhibiting CRH and ACTH secretion. Thus, in cases of primary adrenal insufficiency (Addison disease), negative feedback to the hypothalamus and pituitary is reduced, and circulating ACTH levels are correspondingly elevated. Increased circulating ACTH can lead to overproduction of a related hormone, melanocyte stimulating hormone (MSH), which can decrease appetite and food intake and cause hyperpigmentation (dark blotches on the skin). To reiterate, in primary adrenal insufficiency, low circulating levels of cortisol and aldosterone are accompanied by high levels of ACTH.

In other cases in which circulating cortisol is deficient, ACTH levels also are decreased, and aldosterone levels are normal. This is a hint that something is wrong in the brain, where CRH is produced, or the anterior pituitary, where ACTH is produced. Secondary adrenal insufficiency, as this syndrome is called, results from pathology of the hypothalamus and/or pituitary gland; therefore, ACTH secretion is deficient, which then leads to reduced secretion of cortisol by the adrenal cortex. Because aldosterone secretion is regulated mainly by the renin–angiotensin system, its secretion is not impaired in secondary adrenal insufficiency. To recap, in secondary adrenal insufficiency, low circulating cortisol is caused by low circulating ACTH, and aldosterone concentrations are normal.

The actions of cortisol that are most affected in adrenal insufficiency are reductions in negative feedback regulation of ACTH secretion, modulation of the vasoconstrictor response to beta-adrenergic agonists, maintenance of cardiac muscle inotropy (systolic contraction strength), and antagonism of insulin secretion. In Addison disease, the actions of aldosterone that are most affected are reductions in the retention of sodium and excretion of potassium and hydrogen by the kidney (Fig. 5.1). Patients with adrenal glucocorticoid (cortisol) insufficiency show hypotension secondary to decreased peripheral vascular resistance and reduced cardiac output, increased heart rate (tachycardia), and, in some cases, low blood glucose (hypoglycemia). Patients with primary adrenal insufficiency also can have isosmotic dehydration secondary to reduced mineralocorticoid (aldosterone) secretion and hyperpigmentation secondary to increased ACTH secretion.

Adrenal insufficiency can be fatal; both glucocorticoid and, when indicated, mineralocorticoid HRT are mandatory. It is an excellent example of how similar behavioral symptoms can be caused by different arrays and patterns of hormone secretion. Our physicians' ability to help patients with Addison disease rests on scientific research that uncovered the hypothalamic–pituitary–gonadal cascade and its negative feedback loop. Fig. 5.1 illustrates some clinical features of a patient with primary adrenal insufficiency.

## Adrenal Gland Hyperfunction

Early in this chapter, we described a woman who reported massive weight gain concentrated in her abdominal area, i.e., she began sporting a "beer belly." The symptoms included weight gain in her facial area,

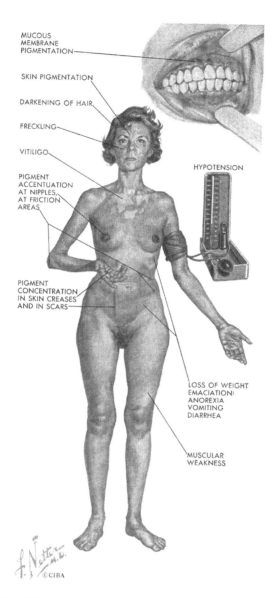

MUCOUS
MEMBRANE
PIGMENTATION

SKIN PIGMENTATION

DARKENING OF HAIR

FRECKLING

VITILIGO

HYPOTENSION

PIGMENT
ACCENTUATION
AT NIPPLES,
AT FRICTION
AREAS

PIGMENT
CONCENTRATION
IN SKIN CREASES
AND IN SCARS

LOSS OF WEIGHT
EMACIATION:
ANOREXIA
VOMITING
DIARRHEA

MUSCULAR
WEAKNESS

©CIBA

FIGURE 5.1   Clinical features of Addison disease. Patients would have behavioral symptoms of inadequate glucocorticoid levels. *From Netter, F.H., 1965. Icon Learning Systems, LLC. With permission.*

anxiety, irritability, insomnia, depression, attention deficits, memory loss, increased appetite, and binge eating. These are classic symptoms of adrenal hyperfunction, or Cushing syndrome, which results from excessive tissue exposure to cortisol. The excess cortisol production may be

ACTH-dependent or ACTH-independent. The former condition results from increased ACTH secretion from the pituitary, ectopic (abnormally placed) sources of ACTH secretion (for example, pulmonary cancer), and ectopic production of CRH, normally produced in the hypothalamus. Because many of the first Cushing patients had basophilic adenomas of the anterior pituitary (corticotroph tumors) that produced excessive ACTH, patients with this particular etiology are labeled as having Cushing disease. In contrast, ACTH-independent Cushing syndrome results from adrenal gland tumors and other pathologies and from exogenous glucocorticoid administration for therapeutic purposes. In these instances, negative feedback of cortisol or exogenous glucocorticoid treatments result in very low circulating concentrations of ACTH.

The secretion of ACTH and cortisol normally has a prominent circadian (24-h) rhythm that is related to the sleep/wake cycle. Little or no hormone secretion occurs in the early morning hours (midnight to 3:00 or 4:00 a.m.). Both hormones are then episodically secreted, with ACTH driving cortisol; both reach peak blood concentrations between 7:00 and 8:00 a.m., about the time of awakening. The amplitude of the circadian rhythm of circulating cortisol is about threefold, so over a 24-h period, tissues are exposed to widely varying cortisol concentrations. In contrast to this normal circadian fluctuation, in Cushing syndrome, cortisol is secreted not only excessively, but also consistently around the clock; that is, its circadian rhythm is blunted or lost. The high cortisol levels may fluctuate very little, or they may show episodic spikes. For example, tumors of the adrenal gland usually produce a high, steady output of cortisol, whereas stimulation of the adrenal gland by excess ACTH from a pituitary tumor often produces a high, fluctuating cortisol output. As normal sleep patterns require that low levels of circulating cortisol accompany high nighttime concentrations of melatonin, victims of Cushing disease often suffer from insomnia.

Irrespective of the cause of excess cortisol production, the clinical features of Cushing syndrome are similar. One of the earliest signs, increased fat deposition, occurs in the vast majority of patients, and difficulty in maintaining weight is a common first complaint. The fat deposition is primarily central (face and trunk) and leads to the characteristic "moon face," the "buffalo hump" on the upper back, and a collar of fat above the clavicles. Concomitant loss of subcutaneous tissue produces thinning of the skin, easy bruising, facial flushing, and reddish-purple striae over the lower trunk. Thinning of bones can result in fractures, typically of the feet, ribs, and vertebrae; and weakness of proximal muscles can lead to difficulties in mobility. Biochemical changes include retention of sodium, loss of potassium through the kidney, glucose intolerance or frank diabetes, reduced lymphocyte count, and suppressed thyroid and gonadal hormone secretion. As mentioned before, in ACTH-dependent Cushing syndrome, increased circulating ACTH and MSH also can cause hyperpigmentation, similar to that which occurs in primary adrenal insufficiency.

Prominent behavioral changes are characteristic of Cushing syndrome. Irritability occurs in most patients and is often the first behavioral symptom, beginning with the earliest manifestations of fat deposition and weight gain. Hence, it is plausible that an individual might confuse cause and effect. Irritability varies in intensity, with some patients simply being more sensitive to minor irritations and others feeling close to exploding emotionally. Depressed mood also occurs in most patients, varying in intensity from short spells of sadness to, rarely, feelings of helplessness and hopelessness. In contrast to primary major depression, the depressed mood of Cushing syndrome is usually episodic rather than sustained, often occurring suddenly and lasting 1–2 days, interspersed with nondepressed intervals. Autonomic activation, such as shaking, sweating, and palpitations, may be an accompanying feature. Also in contrast to patients with major depression, Cushing patients do not often experience social withdrawal, excessive guilt, or marked severity of depressive symptoms; they usually feel worse in the evening, not in the morning; and they often are cognitively impaired (see later). A few patients may experience episodes of elation or hyperactivity early in their illness, feeling more ambitious than usual and having increased activities and pressured speech. This phase often disappears as the illness progresses and other behavioral characteristics emerge.

Alterations in cognition are prominent. Memory impairment, especially for new information, is very common. Other symptoms include difficulty concentrating, shortened attention span, distractibility, slowed and scattered thinking, and, in more severe cases, thought blocking. Difficulty with mental subtraction and recall of common information occurs in many patients. As with affect and mood symptoms, cognitive impairments can range from minimal to severe. Of note, they are not accompanied by any disorientation or clouding of consciousness (delirium).

Alterations in biological functions comprise a third domain of disturbance. Fatigue occurs in all patients, and reduced sex drive occurs in most. Appetite disturbances also occur in most patients, increased appetite being more common than decreased appetite. Sleep disturbances occur in most patients as well and usually consist of insomnia in the middle and late parts of the sleep period, along with frequent, intense, bizarre, and vivid dreams.

Many otherwise healthy patients are treated with powerful adrenal steroid-like compounds, such as dexamethasone or prednisone. These drugs relieve joint pain and inflammation, but it is not surprising that prolonged use can produce symptoms similar to those with Cushing syndrome. These side effects include weight gain, elevated hunger and food intake, hyperactivity, and even mania. In contrast to most patients with endogenous Cushing syndrome, some patients who are receiving exogenous glucocorticoid treatment for medical conditions develop mental disturbances of a psychotic or confusional nature. These patients often receive high doses of potent synthetic steroids over a short period of time, in contrast to endogenous Cushing patients, in whom steroid increases can occur gradually over months to years.

Treatment of Cushing syndrome is, of course, directed toward the locus of pathology and often involves surgery for removal of a pituitary adenoma, an adrenal tumor, or an ectopic source of ACTH or CRH. If pituitary surgery fails to remove the entire tumor, irradiation of the remaining tissue can be undertaken. Drugs that interrupt the enzyme pathways for the synthesis of cortisol and that block cortisol receptors are effective in some cases. Fortunately, impairments in all behavioral domains improve with reduction of excess hormone secretion. However, neuroimaging studies suggest that some CNS structural and neurochemical abnormalities may not be completely reversed by treatment. In some patients with lingering depression, antidepressant medication can be helpful. Fig. 5.2 illustrates some clinical features of a patient with Cushing syndrome.

### Interim Summary 5.4

1. In primary adrenal insufficiency, e.g., Addison disease, low circulating levels of cortisol and aldosterone are accompanied by high levels of ACTH.
2. In secondary adrenal insufficiency, low circulating cortisol is caused by low circulating ACTH, and aldosterone concentrations are normal. In both primary and secondary adrenal insufficiency, symptoms include weakness, fatigue, loss of appetite, dizziness, nausea, diarrhea, and abdominal, muscle, and joint pain.
3. Excessive adrenal output, e.g., in Cushing syndrome, can result in body weight gain concentrated in the abdomen, shoulders, neck, and facial area, as well as anxiety, irritability, insomnia, depression, attention deficits, memory loss, increased appetite, and binge eating.

## Thyroid Gland Hypofunction

Physicians are faced with a dilemma when the symptoms of neuroendocrine disorders overlap. Characteristic symptoms of thyroid hypofunction include body weight gain, fatigue, and a decline in cognitive ability. In addition, patients with hypothyroidism might complain that they are always cold. Their skin is dry, and their hair is coarse, brittle, and thinning. Furthermore, their body weight gain appears to be caused by myxedema (abnormal deposition of mucopolysaccharides in the skin) as well as by water retention (edema). This condition can be confused with or caused by Addison disease (primary adrenal insufficiency), but in the case

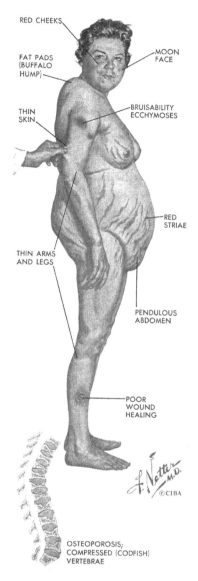

RED CHEEKS

MOON FACE

FAT PADS (BUFFALO HUMP)

THIN SKIN

BRUISABILITY ECCHYMOSES

RED STRIAE

THIN ARMS AND LEGS

PENDULOUS ABDOMEN

POOR WOUND HEALING

OSTEOPOROSIS; COMPRESSED (CODFISH) VERTEBRAE

**FIGURE 5.2** Clinical features of Cushing syndrome. Patients would have behavioral symptoms of a chronic excess of glucocorticoid hormones. *From Netter, F.H., 1965. Icon Learning Systems, LLC. With permission.*

of primary thyroid hypofunction the patient is suffering from a deficit in thyroid hormone secretion. The main function of thyroid hormone is to increase metabolic rate, thermogenesis (cellular heat production), and general activity, and therefore a deficit in circulating thyroid hormone will produce the opposite effects. Of importance, hypothyroidism is more common in the elderly, and it may be misdiagnosed in this population because the symptoms may be subtle. Geriatric physicians thus need to be aware that, in the older patient, mental complaints or changes without physical changes might be caused by hypothyroidism. Because of the shared symptoms between hypothyroidism and behavioral disorders, many psychiatric facilities routinely screen patients for thyroid status.

A hypothyroid patient is likely to show slowed organ function, including decreased heart rate. Other complaints include constipation and menstrual irregularities. Hypothyroid patients also are slow in movement and thought, and they have impaired concentration and memory. By contrast to those with Cushing syndrome, those with hypothyroidism will show excessive sleepiness along with classic depressive symptoms. Left untreated, patients may lapse into coma. Rarely, in addition to their other, severe mental impairments, patients may become extremely anxious and agitated, a condition termed myxedema madness.

The thyroid gland secretes two active hormones, triiodothyronine (T3) and thyroxine (T4). These are interconverted by enzymes in the thyroid and peripheral tissues. T4 is the major hormone secreted and serves primarily as a prohormone for T3, which is biologically more potent. T3 and T4 are rich in iodine, a necessary substrate for normal thyroid function that is supplied in the diet (e.g., iodized salt). Thyroid hormones have important effects in two broad areas: cellular differentiation during development, and maintenance of metabolic pathway activity in adulthood. Thyroid hormones regulate metabolic activity in all bodily tissues, including the brain.

As with several other endocrine deficiencies, hypothyroidism is classified into two types, depending on the site of pathology. Primary hypothyroidism, the most common type, is caused by hypofunction of the thyroid gland itself. Because negative feedback of thyroid hormones to the pituitary and hypothalamus is decreased, circulating thyroid-stimulating hormone (TSH) concentrations are correspondingly elevated. Secondary, or central, hypothyroidism results from pathology in the pituitary gland or hypothalamus; TSH secretion from the anterior pituitary gland is deficient, and the thyroid gland is inadequately stimulated.

The most characteristic finding of hypothyroidism is a general slowing of physical and mental activity involving many organs. Aberrant metabolic pathways lead to a build-up of hyaluronic acid and related compounds in interstitial tissues. Because these substances are hydrophilic, their presence leads to a mucinous edema of the skin and internal organs (myxedema).

During fetal and postnatal development, thyroid hormones in the CNS promote neuronal and glial cell proliferation and maturation, myelination, and synthesis of enzymes critical in neurotransmitter and neuromodulator pathways. Thyroid hypofunction during these early periods therefore can result in neurological abnormalities and general learning disability, in addition to delayed body growth (cretinism).

Treatment of hypothyroidism at all stages of life is straightforward: Thyroid hormone replacement is given to achieve a euthyroid (normal thyroid) state in all tissues. This is most often done with T4 and usually is done slowly, so as not to overtax the heart or produce additional mental disturbances such as mania. Fortunately, if patients are diagnosed and treated early in the course of their illness, most physical and mental symptoms are completely reversible. Fig. 5.3 illustrates some clinical features of an adult patient with myxedema.

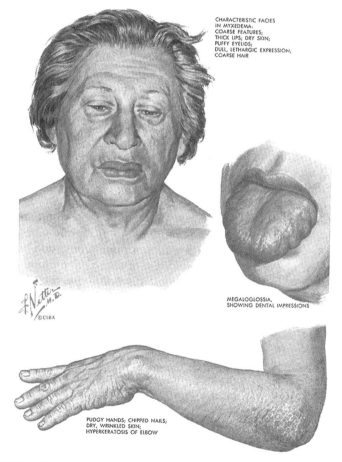

CHARACTERISTIC FACIES IN MYXEDEMA; COARSE FEATURES; THICK LIPS; DRY SKIN; PUFFY EYELIDS; DULL, LETHARGIC EXPRESSION; COARSE HAIR

MEGALOGLOSSIA, SHOWING DENTAL IMPRESSIONS

PUDGY HANDS; CHIPPED NAILS; DRY, WRINKLED SKIN; HYPERKERATOSIS OF ELBOW

**FIGURE 5.3** Clinical features of myxedema. Patients would have behavioral symptoms of inadequate thyroid hormone levels. *From Netter, F.H., 1965. Icon Learning Systems, LLC. With permission.*

# Thyroid Gland Hyperfunction

In the 1990s, President George H. W. Bush, his wife Barbara Bush, and their dog Millie all experienced nervous jitters, irritability, insomnia, and fatigue. These symptoms might sound familiar if you have been reading about other aforementioned neuroendocrine disorders. The Bush's and their dog, however, had a different disease. They were subsequently diagnosed with and treated for Graves disease, also known as *hyper*thyroidism.

As might be expected, hyperthyroidism is elevated concentrations of thyroid hormones, and it results in metabolic changes opposite those in hypothyroidism. Primary hyperthyroidism, the most common form, is caused by hyperactivity of the gland itself, frequently on the basis of an organ-specific autoimmune disease (Graves disease) resulting from circulating autoantibodies to the TSH receptor that mimic the effect of TSH on the gland. Psychological stress has been implicated as one precipitating factor, perhaps as an immune-response rebound following immunosuppression by stress-related hypothalamic–pituitary–adrenal cortical axis activation (see also Chapter 6). Prominent pathology includes diffuse enlargement of the thyroid gland and protruding eyes, secondary to an inflammatory increase in retro-orbital tissue volume (Graves ophthalmopathy). Hypermetabolic symptoms include heat intolerance, warm skin, sweating, increased heart rate and palpitations, fatigue, weight loss, and muscle weakness and wasting.

Common mental and behavioral complaints include nervousness, jitteriness, irritability, anger, anxiety, panic-like episodes, and emotional lability; rapid and disjointed speech; and insomnia and fatigue. Changes in cognition and memory are apparent on neuropsychological testing. About 10% of patients may have severe psychiatric illness including mania, psychotic states, and delirium; these conditions occur mainly when patients have a sudden, severe increase in thyroid hormone production (thyrotoxic storm). As with hypothyroidism, because of shared symptoms between hyperthyroidism and several behavioral disorders, psychiatric patients presenting with severe anxiety, panic attacks, and unexplained psychosis or delirium often are tested for thyroid status.

The primary treatment of hyperthyroidism is identifying the cause of excessive thyroid hormone production and eliminating it. Pharmacotherapy often is the immediate approach to controlling the hypermetabolic state. Treatments such as surgery to remove a hormone-secreting tumor and radioiodine therapy are more permanent approaches. Because the thyroid gland has a strong avidity for iodine, radioactive iodine will concentrate in and destroy glandular tissue. Following radioiodine therapy, the thyroid is usually underactive, so that T4 replacement must be given, generally for a lifetime. Fortunately, when patients are adequately treated and tissue metabolic rate returns to normal, most physical and mental symptoms are completely reversed. Fig. 5.4 illustrates some clinical features of a patient with thyroid hyperfunction.

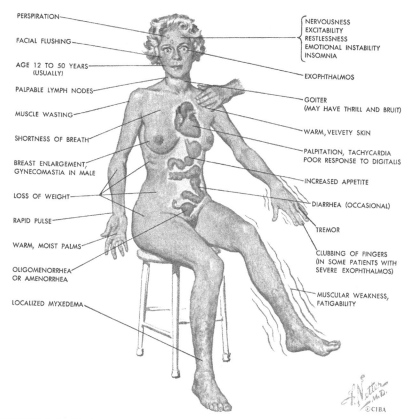

**FIGURE 5.4** Clinical features of Graves disease. Patients would have behavioral symptoms of a chronic excess of thyroid hormones. *From Netter, F.H., 1965. Icon Learning Systems, LLC. With permission.*

## *Interim Summary 5.5*

1. The symptoms of thyroid insufficiency resemble those of adrenal insufficiency; for example, there might be body weight gain, fatigue, and a decline in cognitive ability. In thyroid insufficiency, however, patients suffer from excessively dry skin and hair, feelings that suggest hypothermia, and their body weight gain is mainly attributable to abnormal deposition of mucopolysaccharides in the skin and water retention (edema).

2. Common symptoms of hyperthyroidism (e.g., which occurs in Graves disease) include heat intolerance, excessive sweating, elevated heart rate, bulging eyes, body weight loss, nervousness, jitteriness, irritability, anger, anxiety, panic-like episodes, and emotional lability, rapid and disjointed speech, insomnia, and fatigue.

## Growth Hormone Deficiency

GH secretion by the anterior pituitary is altered in response to stimulation by hypothalamic GH-releasing hormone and other secretagogues, such as ghrelin, and to inhibition by GH-inhibiting hormone (somatostatin), which themselves are regulated by higher CNS centers. GH acts directly to stimulate target stem cells, including neural stem cells. As these cells differentiate, they develop receptors for, and produce, insulin-like growth factor 1 (IGF-1; somatomedin-C). In response to GH stimulation, IGF-1 and its binding protein are produced in the liver and circulate in blood; IGF-1 also is locally produced in the kidneys, gastrointestinal tract, muscle, cartilage, and pituitary. Cells that express IGF-1 receptors are responsive to the growth-promoting effects of both circulating (endocrine) and locally produced (paracrine) IGF-1. GH acts through IGF-1 to enhance DNA, RNA, and protein synthesis and the growth of many tissues, including bone, muscle, cartilage, and CNS tissue (neural maturation and glial cell formation). Metabolic effects of GH/IGF-1 include antagonism of the action of insulin, stimulation of the breakdown of adipose tissue (lipolysis), and retention of nitrogen. IGF-1 also feeds back to the pituitary and hypothalamus to stimulate somatostatin secretion, thereby inhibiting GH release.

GH hyposecretion usually becomes apparent in early childhood, based on a failure to conform to established growth curves. The wide range of causes includes genetic deficiencies, birth trauma, CNS tumors, congenital abnormalities of the hypothalamus or pituitary, and psychosocial stress. Lack of normal amounts of GH and IGF-1 can result in general learning disability, defined as generalized neurodevelopmental disorder characterized by significantly impaired intellectual and adaptive functioning. In general, the longer the GH/IGF-1 deficiency persists, the more profound are the CNS deficits, so some individuals, even though treated with adequate amounts of GH or IGF-1, do not develop normal intelligence and may remain significantly intellectually impaired. This has occurred in conditions such as Laron syndrome (hereditary IGF-1 deficiency) when hormone treatment was not begun until adulthood. Most often, however, persons with GH deficiency are normal behaviorally and can be very accomplished during their lives.

Psychosocial failure of growth, or emotional deprivation failure of growth, has been reported in children who have suffered emotional and/or physical deprivation and abuse, including sexual abuse. Such children have reduced GH secretion, short stature, and failure to thrive. They may be hyperphagic as well, but despite a voracious appetite, they remain underdeveloped. Fortunately, most of these children improve their neuroendocrine and physical status when they are removed from the stressful environment (e.g., placed in a hospital for a period of weeks or in a foster home with a nurturing family). The syndrome of psychosocial dwarfism provides another example of the important interplay between stress-responsive, higher CNS centers and neuroendocrine function.

## Growth Hormone Excess

GH hypersecretion is usually caused by hyperplasia (excessive growth of normal cells) or a tumor of anterior pituitary somatotrophs. GH hypersecretion in early childhood results in overall excessive somatic growth (gigantism). Such individuals are tall, with a large trunk, long limbs, and proportionally large hands and feet. GH hypersecretion that begins in adulthood occurs after the epiphyses of long bones are fused and growth is complete, so overgrowth of only acral (peripheral) and soft tissues occurs (acromegaly). Prominent acromegalic features are enlargement of the jaw, nose, frontal bones of the skull, hands, and feet. The process is generally so slow that features other than cosmetic lead patients to seek treatment; these often are joint and back pains secondary to arthritis and joint degeneration.

Skin thickening, increased sweating, increased hair growth, voice deepening, and obstructive sleep apnea (owing to tongue, laryngeal, and pharyngeal soft tissue enlargement), and entrapment neuropathies such as carpal tunnel syndrome (owing to tissue overgrowth) also occur secondary to GH hypersecretion. Cardiac enlargement and diabetes are important causes of disability and death, and the occurrence of gastrointestinal cancer is increased. Excessive GH, therefore, shortens life expectancy, whether the onset is in childhood or adulthood. Behavioral effects of GH hypersecretion appear to be related primarily to the somatic difficulties patients develop. Gigantism produces an additional set of psychosocial problems related to peer-group relationships in school and other venues.

The treatment of GH excess includes surgery to remove a hormone-secreting tumor and pharmacotherapy with octreotide, a synthetic, eight–amino acid analog of somatostatin. Octreotide inhibits GH/IGF-1, glucagon, and insulin secretion. Clearly, attitudes of patients toward various kinds of social behaviors, including both romantic and aggressive behaviors, will be influenced secondarily by this kind of medical condition.

### Interim Summary 5.6

1. The symptoms of GH insufficiency during development often involve insufficiency in the secretion of IGF-1, and it results in small stature, general learning disability, and altered metabolism.
2. The symptoms of excessive secretion of GH during development are quite serious and include gigantism, acromegaly, obstructive sleep apnea, entrapment neuropathies, cardiac enlargement, diabetes, gastrointestinal cancer, and shortened life expectancy.

## Implications for Behavior

It goes without saying that the aforementioned endocrine maladies have consequences for the body type of the patients; thus, they can affect these patients' behavior in a number of ways, such as depriving them of the ability to enjoy some normal human interactions. Somatic changes raise issues of self-image and influence social perceptions by their families and associates. As we learn more about the mechanism of hormone action, we are finding that the hormonal changes can have direct effects on both body and brain. These endocrine pathologies illustrate the principle that hormones must be present at an optimal concentration, and this optimum level depends upon the time of day, developmental stage, and other environmental factors. They remind us of the indirect routes by which human behavior can be altered by these pathologies, such as altered social interactions.

## OUTSTANDING NEW BASIC OR CLINICAL QUESTIONS

The earlier examples indicate that hormone secretion outside the normal range can have far-reaching pathological consequences and can even be life threatening in some circumstances. Prompt diagnosis and treatment often returns the individual to a normal hormonal state, and the behavioral changes, as well as the physical changes, become normal. Nevertheless, there remain some intriguing questions about the antecedents of these abnormal hormonal states and their severity in different individuals, the answers to which will allow more precise diagnosis and treatment:

1. What is the interplay between genetic and environmental factors in the genesis of endocrine disorders? The answer, of course, depends on the disorder. Some are primarily constitutional, in that they have a prominent genetic or other physiological component (e.g., Addison disease), and others may develop after an unusual stress situation (e.g., psychosocial dwarfism). What is not understood is how environmental stress might trigger a pathological endocrine response in some individuals (characterized after the fact as "susceptible" individuals because they developed the illness) after an apparent environmental stress (e.g., Cushing disease, Graves disease).

2. What determines individual human CNS sensitivity to hormone changes; that is, why do some individuals develop severe mental symptoms with a certain hormone change, while others do not? For example, why can the same degree of hypercortisolemia in Cushing disease produce profound depressive symptoms in one patient but minimal symptoms in another? Why can the same degree of

hyperthyroidism in Graves disease produce severe anxiety symptoms in one patient but little or no anxiety in another? What factors determine the degree and type of disruption of neuronal function by abnormal hormone levels?

3. Regarding the neuroendocrine disorders described in this chapter, what is the metabolic significance of pulsatile versus constant hormone secretion? How much do disorders of the timing of hormone secretory episodes contribute to the behavioral disorders that can accompany endocrinopathies?

4. For endocrine pathology, what is the metabolic significance of the circadian rhythms of hormone secretion? When do rhythm disturbances lead to behavioral pathologies, such as the blunting of the prominent circadian ACTH/cortisol rhythm in Cushing disease?

5. There are other clinical conditions of too little or too much hormone secretion in addition to the examples given before. Can you think of some?

# Hormones Do Not "Cause" Behavior; They Alter Probabilities of Responses to Given Stimuli in the Appropriate Context

"His hormones made him do it!" Wrong. Hormones, in vacuo, do not simply spew out behavioral responses without regard to the incoming stimuli. Instead, to understand what hormones do, we must think in terms of whole systems beginning with the stimuli themselves and their interactions with hormones and receptors. In a specific set of circumstances, well-defined stimuli will have a quantitatively determined chance of increasing the probability of a measurable behavioral response. Even in the most robust hormonal effects, there is no guarantee of a response. We are studying living organisms, not robots. In the absence of the hormone in question, the number of occurrences, their latency, and their amplitude are recorded. Then the hormone in question is added, using a dose, route of administration, and time course deemed likely to be behaviorally effective. Again, the same well-defined stimuli are applied and the response measures taken. The experiment is repeated in a number of subjects sufficient for statistical analysis. If the hormone increases the incidence of the behavior of interest, the probability of occurrence goes up, latency goes down, and/or amplitude goes up. If the hormone represses that behavior, the opposite changes are seen. In all of the examples given in this chapter, the effectiveness of the hormone depends upon many other factors, including genetic predisposition (Chapter 8), experience, age (Chapter 13), developmental stage (Chapters 10–12), and the environmental context (this chapter and Chapter 7), including the time of day (Chapter 15), ambient temperature, food availability, day length, the relative safety from potential predators or other dangers, and the behavior of other individuals. Often, it is the contextual variable that is most critical to understanding the effects of hormones on behavior.

## BASIC SCIENTIFIC EXAMPLES

Laboratory animals popular for biomedical research include rodents because of their relatively low financial and space requirements, short life spans, easily quantified behaviors, and cellular and neuroendocrine similarities that they share with most mammalian species including humans. Most rodents live lives that are heavily dependent on olfactory stimuli, and thus, olfactory control of behavior has been well studied in rodents. Hormonal fluctuations during the estrous cycle significantly influence sensitivity to weak olfactory stimuli in rodents (reviewed by Moffatt, 2003). The peak in olfactory sensitivity to a variety of compounds occurs during the periovulatory period. Like female rodents, male rodents have behavioral repertoires that depend heavily on olfactory stimuli, particularly with regard to intermale aggression. A male under the influence of testicular androgenic hormones (a gonadally intact male or a castrate treated with testosterone) might well do a certain amount of territorial "flank marking" behavior in the absence of any stimulation; however, he certainly will respond to pheromones from a foreign male's scent mark by increasing his own territorial responses, including even covering up the other male's mark with his own flank gland odor. Similarly, with regard to sex behavior, a castrated male rodent might casually investigate sources of sexual pheromones from the estrus female of his species; however, the duration and avidity with which he approaches the estrus odor will be greatly increased if adequate levels of testosterone are circulating in his bloodstream. The species relevant olfactory stimuli are sensed via the olfactory and vomeronasal receptors, signaling through the main and the accessory olfactory bulbs, respectively, and they have profound impact via the cortical and medial amygdala and its connections to hypothalamus and other forebrain structures. Androgens and their metabolites act on steroid receptors in these latter brain areas to bring sexual salience to these olfactory cues (see Chapter 3, Fig. 3.5) (reviewed by Petrulis, 2013). The importance of olfactory context to the behavioral response to steroids is best illustrated by male Syrian hamsters in which copulatory behavior is stimulated by olfactory cues from female vaginal secretions. In castrated male hamsters, full-blown sex behavior in response to female vaginal secretions can be restored with a unilateral testosterone implant (on one side of the brain, not both) into the amygdala or medial preoptic area. If, however, the olfactory bulb is unilaterally lesioned *on the same side* of the brain as the testosterone implant, the smell of female vaginal secretions will not elicit male copulatory behavior. By contrast, copulatory behavior is fully restored by a unilateral testosterone implant if that implant occurs on the opposite side of the brain from the unilateral olfactory bulb lesion. Thus, in this latter group of males, even though one olfactory bulb is destroyed, input from the intact olfactory bulb can initiate the necessary synergistic interaction with the testosterone implanted on that same side of the brain. These clever experiments illustrate the physical interaction of olfactory input and testosterone binding to the androgen receptor in the same brain area to change the male's reaction to a social cue from the female (Wood and Newman, 1995). There are hundreds of examples of gonadal steroid effects on sensory thresholds in both male and female laboratory rodents as well as in other species.

The principle behind the discussion in this chapter is especially easy to illustrate with social behaviors because stimuli from social partners or contenders tend to synergize with hormones to influence behavior. Therefore, testing animals under circumstances where a full range of social interactions occur is becoming especially important in this field. A well-founded example is the role steroid sex hormones play in raising levels of aggression in humans as well as in other animals (see Chapter 1, Fig. 1.1; Chapter 12, Fig. 12.1).

## Aggression

In many species, males are more aggressive than females under a variety of circumstances. Both androgenic hormones (e.g., testosterone) and estrogenic hormones play a role in facilitating aggression. Among men, bodybuilders taking androgens of high potency to build muscles are storied, anecdotally, for occasional hostile acts known as "roid rage" (see Chapter 5). In the normal case, however, aggression in animals and humans does not occur in the absence of provocative stimuli from a conspecific. Among humans, this could be irritating communication, for example, from a person nearby. In the context in which most people show patience or mild annoyance, a man taking anabolic steroids is more likely to react with irritable aggression. This tendency can be curbed in men with extensive experience in nonviolent interactions or exaggerated in men with a long-standing habit of violence. In some contexts, aggression is entirely appropriate and evolutionarily adaptive. For example, another form of aggression, predation, occurs in carnivores such as the hungry lioness. Predatory aggression in the lioness is not spontaneous; rather, it is stimulated by a combination of low energy availability and stimuli from the presence of prey. Female aggression is common in mothers of newborn infants, but again, it does not occur spontaneously, but in response to a threat to her offspring. Intermale aggression and territoriality is common in laboratory rodents. Picture the classic encounter between two male animals, where both are impacted by testicular androgens (and estrogens, both directly and by metabolism of the androgens). The males notice each other. If and only if influenced by androgenic hormones, one goes into a defensive posture: in rodents, a "boxing" stance. As a result, the second individual mirrors the behavior of the first animal. Then, the first animal might assume gestures that signal an escalation, e.g., he might make a rapid head movement toward the other. As a result, the other individual defends against the attack and makes a biting movement back. Quickly, facilitated by androgens, the vicious, biting attack of one will lead to a response by the other that could lead to a rolling, biting, jumping fight. Blood will be drawn, and one might eventually be killed. What we are witnessing is a series of transitional probabilities, many raised by androgenic and estrogenic steroid hormones, from low-intensity, preliminary aggressive responses to vicious, high-intensity fighting. No response is independent of the social stimuli that preceded it. Androgens and estrogens do not cause the aggression; rather, they elevate aggressive response probabilities. The key to understanding the escalation of aggression lies in the next chapter on the reciprocal hormone-behavior relations.

The same is true for the well-studied lordosis behavior. When adequate schedules of estradiol and progesterone are given to female rats or mice to mimic the normal sex hormone conditions that lead to ovulation, the females do not simply spring into the swaybacked posture (lordosis) that allows the male to fertilize the newly released eggs, nor does the naturally estrus female that is in heat show lordosis "in a vacuum." Instead, the behavior is triggered by a combination of olfactory stimuli from the male combined with tactile stimuli on the flanks and rump consequent to mounting by the male. In the absence of estrogen and progesterone or tactile stimulation, lordosis will not reliably occur. Consistently, estrogens followed by progesterone raise the probability that such tactile stimuli will cause lordosis to occur in a sexual context (Fig. 6.1).

### Interim Summary 6.1

1. Hormonal fluctuations, such as those that occur over the ovulatory cycle, can alter the sensitivity to and processing of sensory signals (e.g., olfactory signals), changing the probability that those signals will elicit a behavioral response.
2. The probability of eliciting a particular agonistic behavior, such as intermale aggression, may be different for each animal, for each interaction one animal has with a particular individual, and for each interaction one animal has with subsequent individuals.
3. The probability of eliciting sex behavior, even the most reflexive of sex behaviors, depends upon adequate social stimuli from a potential mating partner.

The behavior of animals observed under strict laboratory conditions differs markedly from those observed under conditions more typical of the habitats in which those behaviors evolved. For example, in the early years of sex research using laboratory rodents and nonhuman primates,

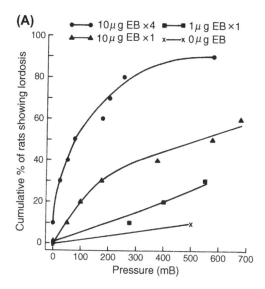

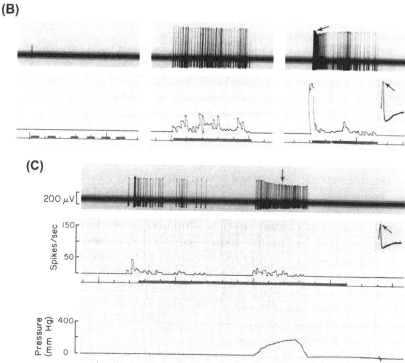

FIGURE 6.1    (A) Estrogenic hormones act to increase behavioral responses (due to electro-physiological responses) to specific stimuli. Increasing doses of estradiol benzoate (EB) amplify the ramps of behavioral responses to pressure on the skin of ovariectomized female rats sufficient for eliciting lordosis. (B) and (C) Electrophysiological responses of single primary sensory neurons in the dorsal root ganglion of an anesthetized female rat to pressure stimuli on the skin (pressure intensity shown quantitatively on bottom trace of each figure). Sustained pressure led to action potentials in a neuron that had no spontaneous discharge. A sudden peak of pressure, as would be caused by stimulation from a male rat during mounting, causes a high rate of firing. *From Kow, L.M., Montgomery, M.O., Pfaff, D.W., 1979. Triggering of lordosis reflex in female rats with somatosensory stimulation: quantitative determination of stimulus parameters. J. Neurophysiol. 42, 195–202; Kow, L.M., Pfaff, D.W., 1979. Responses of single units in sixth lumbar dorsal root ganglion of female rats to mechanostimulation relevant for lordosis reflex. J. Neurophysiol. 42, 203–213.*

sex behavior was most often studied in male–female pairs isolated from all other conspecific individuals in confined spaces from which they could not escape. In the real world, rats and monkeys live in large social groups and enjoy many behavioral options. Their sexual adventures are not confined to a small, enclosed area, and they are not inclined to engage in sexual activities with an arbitrary partner selected by an experimenter. When female sex behavior was observed in the laboratory, it first appeared that the male initiates contact with the female and then engages in a rapid series of mounts and intromissions that ultimately lead to ejaculation. It was assumed that the copulatory sequence was determined solely by the male, with the female serving as a passive recipient of sexual advances, so the female behavior was most often called sexual "receptivity." For example, it was assumed that the male determined the frequency of intromissions (the number of times the penis is inserted into the vagina in a given time period) as well as the interintromission interval (the time between intromissions). Later, other investigators designed experiments to determine whether females find copulation rewarding. In fact, under these conditions, copulation is not rewarding to female rats. Context is critical (reviewed by Yoest et al., 2014).

By sharp contrast, in more recent experiments in which female rats are tested in a seminatural environment, the female rats determine the rate and duration of sexual contact. Specifically, females control the intromission frequency, interintromission interval, and they tend to prolong the latency to ejaculation. Furthermore, the entire copulatory sequence is quite different than that originally observed in small, enclosed testing arenas. In the wild or in a seminatural environment, group-living females are the first to initiate a sexual interaction by increases in courtship behaviors, such as hopping (straight up, vertical jumps), darting (sudden forward movements terminated abruptly), and ear wiggles. These behaviors attract a number of adult males, and the female chooses her partners and decides the mating sequence. After receiving an intromission from any particular chosen male, the female makes a quick escape before she returns to either the same or a different male. This can be simulated in the laboratory by testing each female in a chamber from which the female can escape from the male rats. For example, in some experiments, female subjects that are smaller than their male stimulus partners are allowed to run away via a hole in the testing area. The hole is just large enough for the female but too small for the male. When tested in this special apparatus, after each intromission, the female escapes and then, after an average of 2 min, returns to receive another intromission. This female behavior is call *pacing*, because the female determines the rate at which she receives intromissions and the total duration of sexual contact. Hormonal control of pacing behavior could not be studied until behavioral endocrinologists discovered the appropriate context that revealed the existence of these behaviors.

Hopping, darting, ear wiggling, and pacing are precopulatory behaviors stimulated by rises in estradiol followed by the action of progesterone, but the preferred pace of intromissions is determined by the individual vaginal code of each female. The vaginal code is an innate preference, which is the intromission pattern that best facilitates the secretion of progesterone in that particular female. Increases in female levels of circulating progesterone achieved through pacing are important because high levels of progesterone stimulate the growth of the uterine lining. Sufficient thickening of the uterine lining facilitates the implantation of fertilized embryos. The higher the progesterone, the thicker is the lining of the uterus, and the more likely the embryos will develop into healthy offspring. In addition, paced mating is rewarding to the female, whereas unpaced mating is aversive. Females will perform operant tasks, such a pressing a bar, to gain access to the same male with which she experienced paced intromissions. Females will not press a lever to gain access to a male with which she has had unpaced mating. Furthermore, females will form a conditioned place preference to any place where paced mating has occurred. In a conditioned place preference test, animals receive a stimulus when they are in a chamber with specific characteristics, such as a rough gray floor and black walls. After exposure to a stimulus in that chamber, they are given a choice of two chambers that differ in appearance. If the female subjects have no preference, they spend an equal amount of time in both chambers. If, however, they have formed a conditioned place preference due to a rewarding stimulus, they will spend significantly more time in the chamber where they encountered that rewarding stimulus. When provided with a choice of chambers, female rats will chose the chamber in which they have previously received paced intromissions. Conversely, female rats will actively avoid chambers where they received unpaced intromissions. Reward is mediated in part by two chemical messengers, dopamine and β-endorphin. Dopamine is released into the nucleus accumbens, a brain area involved in reward, only when female rats are receiving paced intromissions, not when they receive unpaced intromissions, i.e., the same number of intromissions at a rate faster or slower than their preferred rate. In addition, the rewarding effects of paced mating are blocked by drugs that block the binding of β-endorphin to its receptors. By contrast, vaginocervical stimulation received at a pace determined by the male is aversive to the female, and the female will avoid any context in which she has received unpaced intromissions. The consequences of paced versus nonpaced mating in female rats underlines the importance of context in behavioral neuroendocrinology (reviewed by Yoest et al., 2014).

Male and female rats show differences in their mating strategies with regard to the pattern of intromission; yet the species is among the most successful on the planet. In a group setting, the preference for rapid intromission in the male is realized through mating with multiple females. Dominant males benefit by fertilization of multiple females. By contrast, the preference for paced mating in the female is realized through a series of "sampling" intromissions from many males followed by an escape from that male just prior to ejaculation. Sampling intromissions occur until she settles on the best male (from which she receives an ejaculation) and until she optimizes her own progesterone secretion. One benefit of pacing for the female is that by pacing, she (1) increases the chances that her eggs are fertilized by "the best" male (according to her assessment) and (2) increases the chances of successful implantation of the embryos fertilized by that male. Those fertilized embryos that are safely implanted in the uterine lining are more likely to develop into healthy offspring. Dopamine and endogenous opiates released during paced, fertile mating increase the chances that the female will mate and conceive additional litters of offspring by the same preferred father in the future. Thus, both sexes presumably benefit from the different, but complementary, mating strategies, behaviors that are controlled by hormones and behaviors that control hormone secretion (reviewed by Yoest et al., 2014).

These contextual effects are not limited to rodents. Hormonal control of primate sex behavior is masked in a laboratory setting in which male–female pairs are observed in enclosed spaces, and the effects of hormones depend upon social rank (e.g., Fig. 6.2). Long ago, when behavioral endocrinologists began to quantify aspects of primate sex behavior, they concluded that

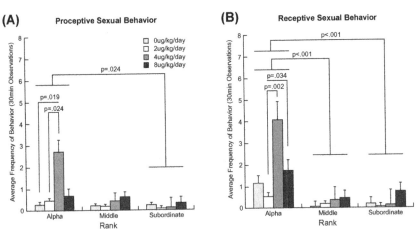

FIGURE 6.2 Interaction between female social rank (dominant [Alpha] or subordinate) and dose of estradiol on sexual behavior toward males. Both (A) proceptive and (B) receptive behavior were significantly affected by social rank, and there was a significant interaction between social rank and dose of estradiol, so estradiol had a stronger facilitatory effect on sex behavior in high-ranking females. Post hoc analysis showed activational effects of estradiol only in Alpha (dominant) females. Data are presented as average frequencies ± standard errors. *Adapted from Reding, K., Michopoulos, V., Wallen, K., Sanchez, M., Wilson, M.E., Toufexis, D., 2012. Social status modifies estradiol activation of sociosexual behavior of female rhesus monkeys. Horm. Behav. 62 (5), 612–620.*

hormones were not important. When male–female pairs of monkeys are tested for sex behavior in small cages from which the female cannot escape the male, copulation occurs every day of the female's menstrual cycle. This initial observation, along with the well-known fluctuations in estradiol and progesterone over the menstrual cycle, led to the erroneous idea that female primates are "emancipated" from hormonal determinism and are "receptive" to male advances every day of the cycle. By sharp contrast, for female primates living in large social groups, female-initiated copulations and approaches to adult males do not occur every day of the menstrual cycle. Rather, they are strictly confined to a small window of days around the time of ovulation (reviewed by Wallen, 1990). In other words, the laboratory testing environment used in the early days of behavioral endocrinology masked the effects of hormones on female sex behavior in primates.

Understanding the context of behavior is critical for understanding the adaptive significance of hormone-behavior relations in female primates. Whereas mating behaviors are necessary for getting genes into the next generation, they are not without risks to the survival of individuals living in large social groups that depend upon stable social hierarchies. Sex is risky. Courtship and copulatory behaviors expose wild animals to sexually transmitted diseases, parasites, predators, and to possible injury by rivals. Most important, survival of social primates in the wild depends upon stable dominant-subordinate relationships, and new sexual liaisons disrupt these hierarchies. For most of the menstrual cycle, adult females of nonhuman primate species do not casually initiate new sexual interactions. Their flirtations, solicitation, and displays of sexual motivation are concentrated around the periovulatory period of the menstrual cycle. This increased level of sexual enthusiasm is mimicked by treatment of ovariectomized females with estradiol alone. Wallen suggests that an important function of the periovulatory rise in circulating estradiol goes far beyond sexual motivation. He argues that the function of estradiol is to increase the incidence of risky sexual advances in social environments where these sexual advances are likely to have a substantial cost. Context is critical. For example, in some primate species, socially subordinate individuals experience more stress and typically have higher circulating levels of cortisol, whereas socially dominant individuals, the Alpha individuals, are less stressed. In the same species, treatment with estradiol dose-dependently increases sexual motivation and male affiliation in Alpha females, whereas subordinate females showed no positive effects of estradiol, even at higher doses (Fig. 6.2; Reding et al., 2012). In sum, the effects of hormone differ according to social context, including the levels of stress and dominance status that alter hormone sensitivity.

These important effects of hormones on female risk-taking are masked in small laboratory cages where only one male is present. In more general terms, these experiments illustrate the importance of context in our ability as scientists to (1) assess the effects of hormones on behavior and (2) to interpret their adaptive significance.

## Interim Summary 6.2

1. Behavioral endocrinologists acquire only a limited understanding of hormonal control of behavior by studying male–female pairs in small, enclosed spaces.
2. Sex behavior is rewarding to female rats only in certain contexts. Whether female sex behavior initiates the neuropeptides of the reward system depends upon the ability of the female to control the mating sequence.
3. The importance of hormones on female primate behavior is masked by studying male–female pairs, and it is revealed by studying females in social groups in seminatural environments.
4. Females are differentially responsive to estradiol depending on their social rank and level of stress.

## Energy

The energetic context is also important for understanding hormone effects on behavior. Energetic context is related to the amount and caloric value of available food, the energy requirements for thermogenesis and locomotion, and the amount of energy stored as body fat or stored in the home as a food cache or hoard. Individuals that become preoccupied with mate searching, mate guarding, courtship, and mating cannot simultaneously remain vigilant about foraging for food, eating food, provisioning their territories, and hoarding food. The individual must choose between ingestive and reproductive behavior. This is critical because food stored as body fat and in the home will be needed for survival during future food shortages and for the energetically costly processes related to rearing offspring. Thus, hormones control behavioral priorities. For example, during most of the estrous cycle, hormone levels predispose females toward foraging, hoarding, and eating. For the small time window of highest fertility, hormones predispose females toward courtship and mating. This obvious fact is obscured in animals housed in typical laboratory conditions, i.e., in isolation with unlimited food. The effects of hormones on ingestive behavior are brought into sharp relief in animals that are energetically challenged (Fig. 6.3; Klingerman et al., 2010; Abdulhay et al., 2014; Schneider et al., 2013).

Most research on ingestive behavior has been conducted in male (not female) rodents housed in small, enclosed spaces, under constant conditions of ambient temperature, day length, with unlimited food and water (reviewed by Schneider et al., 2013). The benefits of these traditions are related to the need for experimental control, but the benefits come with inevitable liabilities.

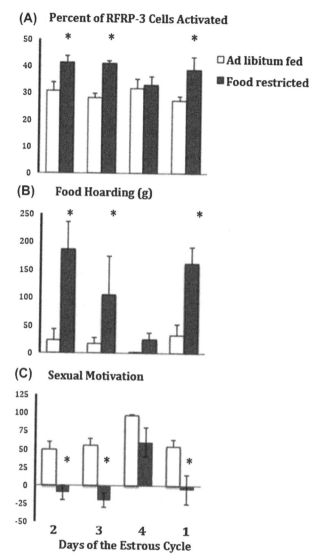

**(A) Percent of RFRP-3 Cells Activated**

Ad libitum fed
Food restricted

**(B)    Food Hoarding (g)**

**(C)    Sexual Motivation**

Days of the Estrous Cycle

FIGURE 6.3   The effects of the hormones of the estrous cycle on ingestive behavior depend upon the energetic context. (A) The number of brain cells containing RFamide-related peptide-3 (RFRP-3) that were activated in females either fed ad libitum (white bars) or previously food restricted (gray bars), (B) amount of food hoarded in 90 min, (C) preference for visiting and adult, sexually-experienced male versus visiting a food source. In mildly food-restricted female Syrian hamsters (energy limited to 75% of ad libitum intake), food hoarding (gray bars, middle graph) fluctuates markedly over the days of the estrous cycle. The effects of ovarian steroids are masked in females fed ad libitum (energy abundant). *From Klingerman, C.M., Krishnamoorthy, K., Patel, K., Spiro, A.B., Struby, C., Patel, A., Schneider, J.E., 2010. Energetic challenges unmask the role of ovarian hormones in orchestrating ingestive and sex behaviors. Horm. Behav. 58, 563–574; Schneider, J.E., Benton, N.A., Russo, K.A., Klingerman, C.M., Williams 3rd, W.P., Simberlund, J., Abdulhay, A., Brozek, J.M., Kriegsfeld, L.J., 2017. RFamide-related peptide-3 and the trade-off between reproduction and ingestive behavior. Integr. Comp. Biol. (in press).*

Standardized, artificial conditions hide the reality of hormone–behavior relations as they exist in the real world. Contrary to the notion that ad libitum–fed animals are the healthy control group, rodents housed with unlimited access to standard laboratory chow are overweight, insulin resistant, hypertensive, hyperglycemic, hyperinsulinemic, and hyperleptinemic (reviewed by Schneider et al., 2013). In nature, however, animals do not typically have access to unlimited food with no energy expenditure requirements. To the contrary, they must work for their food by searching and foraging, hunting, gathering, and hoarding (reviewed by Bronson, 1989).

Thus, to examine the effects of the hormones of the ovulatory cycle on ingestive and sex behavior, female hamsters fed ad libitum are compared to those that are mildly food restricted, i.e., they are fed 75% of their ad libitum diet (Fig. 6.3). In fed or mildly food-restricted females, ovarian steroid levels are normal (they are not compromised by mild food restriction). Experimenters measure the preference for visiting food or visiting a potential male mating partner. They also measure food hoarding, food intake, and lordosis duration. Female hamsters housed with ad libitum food intake (as much food as they can eat) consistently prefer to visit males rather than hoard food on every day of the estrous cycle. On the day of ovulation, they visit males and show lordosis, whereas on the other 3 days of the estrous cycle, they visit the male and make many vaginal scent marks all around his wire mesh restraining cage. Ad libitum–fed females make almost no visits to the food, and they hoard very little food on every day of the estrous cycle (Fig. 6.3). By contrast, food-restricted females show marked fluctuations in food hoarding that are in the opposite direction of their circulation levels of estradiol. They hoard great amounts of food on 3 infertile days of the estrous cycle, and they hoard almost no food on the ovulatory day, when they stay in close proximity to the male. As noted in Chapter 3, the vigilant food hoarding on the nonestrous days is linked to elevation in the activation of hypothalamic cells that secrete RFamide-related peptide-3 (RFRP-3), the mammalian ortholog of gonadotropin-inhibiting hormone (GnIH). Food hoarding and RFRP-3 cell activation are highest on the nonfertile days and lowest at the time of ovulation, whereas preference for the male and levels of estradiol are highest at the time of ovulation and lowest on the other 3 days of the cycle (Fig. 6.3; Klingerman et al., 2011). These results illustrate that ingestive and sex behaviors are coordinated by interactions among RFRP-3 and the hormones of the estrous cycle. These interactions make no sense on their own, but in the context of the environments in which the animals evolved, these interactions optimize reproductive success in response to fluctuating energy availability. These same results are seen in cold-housed females fed ad libitum, i.e., in female hamsters housed at cold ambient temperatures; large fluctuations in food hoarding occur that are in the opposite direction of fluctuations in sexual motivation (when food hoarding is high, sexual motivation is low). Similar results are seen in females housed with running wheels (housed in warm temperatures and fed ad libitum). Together, it appears that the effects of hormones on sexual motivation are unmasked by (1) increasing the energy that must be spent on heat production, (2) increasing the energy expended on exercise, and (3) decreasing the availability of energy as ingested food. These results suggest that it is not the presence or absence of food *per se* that is the critical variable, but rather the availability of energy as determined by the balance of energy intake, storage, and expenditure, i.e., the energetic context (Abdulhay et al., 2014 and reviewed by Schneider et al., 2017).

## Affiliative Behavior

Oxytocin (OT) priming for affiliative behavior makes another fine example. Oxytocin is associated with behaviors such as living in close proximity to conspecifics, partner preference, trust, and maternal attachment. The very nature of social behavior dictates that the hormone (in this case, oxytocin) cannot "cause trust and cooperation" all by itself. It can only influence the probability of positive and friendly responses to social stimuli. In the case of monogamous voles, the formation of a preference for one female sex partner takes several days to develop, and it does not occur after only 1 day of contact. The appropriate environmental context for development of affiliation and partner preference includes mutual physical, tactile, and olfactory stimulation continuously without a break. In rodents, sheep, and human mothers, the nonapeptide, oxytocin, acting as both a hormone and a neuropeptide/transmitter, relies on the olfactory stimulus coming from the potential mate, offspring, or a potential opponent for its behavioral effect, which is revealed as an increased response probability, appropriate to the particular conspecific. During the time course of the developing social bond, the strength and rate of bonding becomes something that lies partially outside each individual. The affiliative behavior under study is thus more than the hormonal effects on one brain because the bond is mutually co-created and is not permanent once it is set in motion. These are the particular challenges and fascinations of studying social behaviors, and they become very important when thinking about how hormonal agents might be used clinically for issues such as social anxiety disorder and sexual dysfunction. In the case of sexual dysfunction, the individual gets the prescription, but it is the couple and their dynamic affiliation that will be affected.

*Interim Summary 6.3*

1. The effects of hormones on ingestive behavior are masked by studying only males of the species housed in isolation, with unlimited food, in enclosed spaces in which the animals have no options, and the effects of hormones are unmasked by studying animals in a seminatural environment.
2. The hormones of the estrous cycle, via neuropeptides such as RFRP-3, predispose animals toward vigilant foraging, hoarding, and eating during the nonfertile phase of the cycle, and toward courtship and mating during the fertile phase of the cycle, but these effects are masked in females housed with unlimited food and no mating partners.
3. Oxytocin effects on affiliative behavior will not occur in the absence of appropriate social cues that must be continuously present in a specific time frame. Furthermore, different affiliative behaviors, such as pair-bonding, develop at different speeds and with variable magnitudes of response dependent upon mutually co-created interactions between pairs of individuals.

## CLINICAL EXAMPLES

### Hypercortisolism as a Factor in Major Depression

Major depression is a psychiatric syndrome defined by a number of signs and symptoms, as indicated in Table 6.1. Crucial is the point, according to this chapter's principle, that a terrible mood, in and of itself, is not sufficient to diagnosis clinical depression. It is the altered probability of negative emotional responses in the face of environmental stimuli (that would not evoke such gloom in the normal person) that makes a major depressive state. With this in mind, consider the role of glucocorticoid hormones.

Many studies have shown that 30%–50% of patients with major depression have increased hypothalamo–pituitary–adrenal cortical (HPA) activity, as indicated by increased circulating adrenocorticotropic hormone and cortisol concentrations, as well as resistance of the HPA axis to suppression by the synthetic glucocorticoid dexamethasone. Ancillary indicators of HPA axis hyperactivity include elevated cerebrospinal fluid corticotropin-releasing hormone (CRH) content, downregulation of pituitary corticotroph CRH receptors, and adrenal gland enlargement. The more severe the depression, the more likely there will be increased activity of this endocrine axis. Successful treatment results in return of HPA activity to normal levels, as well as amelioration of depressive symptoms. If

**TABLE 6.1**  DSM-5 Diagnostic Criteria for a Major Depressive Episode

A. Five or more of the following nine symptoms (one of which is either depressed mood or loss of interest or pleasure) have been present for at least 2 weeks and represent a change from previous functioning:
1. Depressed mood (e.g., feels sad) or affect (e.g., appears tearful) most of the day nearly every day (in children and adolescents, can be irritable mood)
2. Markedly diminished interest in all or almost all activities most of the day nearly every day
3. Significant weight loss or gain or decrease or increase in appetite nearly every day (in children, can be failure to make expected weight gains)
4. Insomnia or hypersomnia nearly every day
5. Psychomotor agitation or retardation nearly every day as observed by others
6. Fatigue or loss of energy nearly every day
7. Feelings of worthlessness or excessive or inappropriate guilt nearly every day
8. Diminished ability to think or concentrate or indecisiveness nearly every day
9. Recurrent thoughts of death, suicidal ideation, a specific plan for committing suicide, or a suicide attempt
B. The symptoms cause clinically significant distress or impairment in social, occupational, or other important areas of functioning.
C. The symptoms are not due to the direct physiological effects of a substance (drug or alcohol) or a medical condition.

*Adapted from the 2013. Diagnostic and Statistical Manual of Mental Disorders, fifth ed. American Psychiatric Association, Arlington, VA. With permission.*

clinical improvement occurs with treatment but HPA axis activity remains elevated, there is a greater likelihood that relapse will occur when treatment is discontinued.

There has been some debate as to whether the increased HPA axis activity is an epiphenomenon resulting from altered CNS neurotransmission and has no influence on the depressive illness or whether it compounds the severity of, or even is causally related to, the depression. Some investigators believe CNS glucocorticoid receptors are subsensitive in depressed patients, leading to reduced hippocampal negative feedback on the HPA axis and toxic influences of the increased cortisol on hippocampal neurons, resulting clinically in the memory disturbance seen in some depressed patients (depressive pseudodementia). A proposed treatment for major depression, therefore, has been reduction of circulating cortisol with glucocorticoid synthesis inhibitors such as ketoconazole or blockade of glucocorticoid receptors with receptor antagonists such as mifepristone (used as an "abortion pill" because it also blocks progestin receptors). These drugs, however, have not proven to be effective stand-alone treatments for depression; at best, they may provide some adjunctive help to primary antidepressant therapy. Perhaps the most important observation in the relationship between the HPA axis and major depression, as noted earlier, is that one-half to two-thirds of patients meeting diagnostic criteria for major depression have normal HPA axis function.

## Hyperthyroidism as Predisposing to Anxiety and Irritability

In Chapter 5, we introduced the subject of thyroid gland hyperfunction; here, we expand upon that treatment. It is well understood that abnormally high thyroid hormone levels are associated with irritability, but this is not a temperamental change divorced from the patient's environment. Rather, the syndrome refers to a tendency to express annoyance (or worse) under circumstances that might be irritating, but which a euthyroid person would suffer gracefully. Clearly, there are many situations in which no people, including hyperthyroid patients, are angry. Across a range of increasingly difficult problems, more and more people become fed up, until circumstances are reached in which everyone has had enough. The irritability curve for the hyperthyroid person is shifted significantly to the left.

## Psychosis

What about patients for whom the principle discussed in this chapter is, in fact, not true? In such a case, certain hormone levels (be they sex hormones, stress hormones, or thyroid hormones) would allow a behavior to be set off *in vacuo*. The objective external situation would not affect the behavioral outcome. Such a patient is, in fact, psychotic. He or she is operating independent of reality, and the hormone is simply allowing the psychotic behavior to be expressed.

## SOME OUTSTANDING QUESTIONS

1. The vast majority of research has been conducted with animals housed in small cages with unlimited food. Is this the appropriate control group, given that they are essentially living in an all-you-can-eat buffet with very little opportunity to engage in species-specific activities?
2. In each case, basic or clinical, what are the loci for hormone effects? Do hormone actions at several levels of a neural circuit interact with each other? Multiplicatively? When are the hormone actions directly on the sensory apparatus? Are hormone-caused changes in responses to stimuli sometimes physiologically on the motor side, with a strong alteration in response readiness?
3. Among human subjects, why do some individuals' thresholds for responsiveness to sensory stimulation, secondary to hormonal changes, vary so widely?
4. Multiple chemical sensitivities—abnormal responses to a wide variety of odors including headache, dizziness, and nausea—are seen much more frequently in women than in men. Why?

# Hormone-Behavior Relations Are Reciprocal

Changes in hormone secretions can predispose an individual toward behaviors, but this idea is incomplete at best. For example, "testosterone-poisoned dictator," "testosterone-fueled heavy metal bands," and "testosterone-driven violence" are contemporary clichés that stem from the assumption that high levels of aggression, competition, and machismo are caused by elevated levels of testosterone. After all, men tend to have, on the average, higher circulating concentrations of testosterone than women, and more men than women are involved in military coups, heavy metal music, violent crimes, and contact sports. Should violent criminals be acquitted based on the "testosterone defense; " and is castration a just and logical sentence? Correlation is not causation, and, in this case, cause and effect are confounded. Sharply elevated levels of plasma testosterone can result from engaging in conflict and competition, and if the effect accumulates over time, it is very difficult to argue that innate, baseline levels of circulating testosterone account for all the differences in aggressive behavior.

As discussed in this chapter, the idea that men are more aggressive because of their innate levels of testosterone is oversimplified for many reasons. Most important, levels of testosterone and other hormones are affected by environmental factors, including social experience. For example, sharp spikes in circulating testosterone concentrations occur in response to participation in competitive behaviors, i.e., attempts to acquire and defend real or perceived resources such as status, territory, mating partners, food, and power. Effects of competition (both winning and losing) on hormone secretion have been observed in many different species, including *Homo sapiens* (reviewed by Gleason et al., 2009; Hamilton et al., 2015; Hirschenhauser and Oliveira, 2006; Soma, 2006; van Anders and Watson, 2006; Wingfield et al., 1990).

Moreover, in human beings, the spikes in circulating testosterone are higher in (1) the winners of competitions relative to those of the losers, (2) both male and female winners, (3) sports fans after they watch their favorite team win, and (4) the winners of the many different types of competition, whether those competitions are gladiatorial, such as football, or virtual, such as the video game Tetris (Carre' and Olmstead, 2015; Zilioli and Watson, 2014; Wood and Stanton, 2012; and reviewed by van Anders and Watson, 2006; Hamilton et al., 2015). This chapter reviews the mechanisms whereby behaviors can change hormone synthesis and secretion in the brain and periphery. We explore the idea that our own level of hormone secretion and our sensitivity to hormones can change as a result of watching the behavior of others. In addition, we explore the idea that neuroendocrine mechanisms are sensitive to our own actions, which then predispose us toward subsequent behavioral choices. Previous chapters explained that changes in hormone secretion can account for initial changes in behavioral tendencies, but this chapter reveals that reciprocal hormone-behavior interactions can explain certain elaborate behavioral sequences. Understanding the detailed molecular mechanisms that underlie reciprocal hormone-behavior interactions will have far-reaching implications for understanding the flexibility and plasticity of behavior in general.

## DANIEL LEHRMAN'S RESEARCH ON RING DOVES

It is difficult to understand reciprocal hormone-behavior interactions using human beings as experimental subjects, but great strides have been made using other well-chosen species as animal models. Some of the most inspiring research in this area was conducted by Daniel Lehrman et al. at Rutgers University on the complex sequence of interactions displayed by male and female ring doves during courtship, mating, nest building, incubation, and parenting. The ring dove and the closely related species, the common pigeon, are common birds that are ever present in our environment. It is important to realize that almost any one of us might have made the same discoveries, if we had Lehrman's curiosity and patience for careful, thorough observation combined with clever experimental manipulation. The distinctive behaviors of paired ring doves are easily recognizable and quantifiable, occur in a predictable sequence, and the entire cycle from courtship to fledging the offspring is completed in only a few weeks. Just as this model system is near perfect for scientific discovery, it is near perfect for illustration of the reciprocal effects of hormones on behavior.

First, the male ring dove begins an aggressive display of bowing and cooing, which initially causes the female to retreat for cover. The male responds to the female's coy performance by bringing nesting materials, such as straw or hay, to a suitable nesting site. As he stands near the nest site, his call changes from an aggressive cackling coo to a soothing, one-syllable lullaby, the nest coo, until the female responds in kind.

Next, both members of the pair team up to build a nest, and the female begins to reveal her commitment to their joint effort by remaining in or very near to the nest and by cooing in her own feminine voice. If the experimenter were to try and dislodge the female, she would cling desperately to the nest and drag it along with her, a sign that copulation is imminent. Courtship and nest-building behaviors last somewhere between 7 and 11 days, after which the male commences with copulation, and the female lays two fertilized eggs.

After the two eggs are laid, the expectant mother and father share in incubation duties, i.e., alternating sitting on the eggs, for 2 weeks until the eggs hatch. During incubation, the parents develop a crop, a glandular sack in the gullet that produces a nutritious substance known as crop milk. Both male and female share in brooding and the feeding of the offspring, called squabs, by regurgitating crop milk into their greedy little mouths until they fledge at the age of 15–20 days. Finally, in the days after the offspring have fledged, the "empty nesters" bow and coo, and their courtship begins anew. The reproductive cycle of the ring dove is 6–7 weeks long (reviewed by Lehrman, 1964, 1965).

Lehrman carefully documented this reproductive sequence and noted that the behavioral sequence never occurs in ring doves housed in isolation. Moreover, the latter stages of the sequence cannot be induced before the previous stages have occurred. A male ring dove commits to building a nest only after his courtship performance elicits a nest coo from the object of his affection. He does not copulate until the female demonstrates her motivation and commitment; the former is signified by her nest coo, whereas the latter is signified by her steadfast attachment to the nest. She will emit vocalizations that initiate mating only after several days of cooperative nest building with her potential mating partner. Furthermore, if you were to present a premade nest filled with dove eggs to a healthy, adult dove living in isolation, the lone dove, whether male or female, would fail to incubate those eggs, no matter how long you waited. If, however, you present a new, premade nest with eggs to a male after he has courted a female and helped her to build a nest and mated with her, he will incubate those eggs within a few hours as if they were his own eggs. Similarly, when presented with strange, hungry squabs, adult doves neither brood nor feed the strange squabs unless those adults have previously courted, built a nest, laid eggs, and incubated eggs. Something about the process of courtship, nest building, and incubation awakens the birds' motivation to engage in parental behaviors (reviewed by Lehrman, 1964). How can we explain this chain of events? How are females transformed from birds with zero interest in nest building to birds obsessed with nest building?

I. CHARACTERIZING THE PHENOMENA

# THE FEMALE'S NEST BUILDING IS STIMULATED BY GONADAL SECRETIONS PLUS SIGNALS EMITTED DURING COURTSHIP

Of course, hormone levels change predictably over the course of this behavioral sequence, but changes in hormones alone are not sufficient to orchestrate the complex interactions. Lehrman et al. made substantial progress in understanding ring dove behavior, and what is more remarkable, they did so many years before techniques were developed to measure rapid changes in brain hormone secretion and synthesis. Lehrman knew that female pigeons (a species closely related to ring doves) ignore nesting materials, such as straw or grass, if they are housed in sensory isolation, whereas female pigeons enthusiastically collect straw and incorporate it into nests if the females are able to peer through a window at the courtship displays of an adult male pigeon (Matthews, 1939). Lehrman therefore hypothesized that, like pigeons, female ring doves' sense cues from the male's behavior, which in turn transform her from a nonnest builder to a nest builder. Furthermore, he hypothesized that the female's transformation is mediated when visual cues stimulate female hypothalamic hormone secretion.

In support of this idea, the growth of the ovary is significantly, positively correlated with the female's mounting level of enthusiasm for nest building, which develops in response to courtship. The female's nest building behavior is eliminated if she is ovariectomized prior to pairing with a male. In addition, female nest building in ovary-intact females is precluded by removing all visual and auditory cues that arise from the male. Thus, both intact ovaries and visual/auditory cues from mating partner are required for the transformation to nest builder (reviewed by Lehrman, 1964, 1965).

What is the male doing to stimulate the female? Is it his mere presence or something that he does that stimulates the development of the ovary and the female's behavior? Ovarian development and nest building are *not* stimulated by the sight of a female, a juvenile male, or a *castrated* male, but only by the sight of a gonadally intact male. Only the gonadally intact male shows the requisite bowing and cooing that stimulates the female reproductive tract and behavior. Ovarian development is therefore not stimulated by the sight of any ring dove, but by visual stimulus from an adult, hormonally intact, male ring dove. At first, it was concluded that the sight and sound of the male's coo stimulated the auditory receptors of the female, thereby promoting her hypothalamic cells to secrete gonadotropin-releasing hormone (GnRH) (Fig. 7.1). GnRH was assumed to flow via the pituitary portal vessels to the anterior pituitary to stimulate luteinizing hormone (LH) secretion, which was assumed to stimulate ovarian steroid secretion, leading eventually to ovulation. Furthermore, it was hypothesized that high levels of ovarian steroid secretion are permissive for increased female nest-building behavior in response to the courtship behaviors of the adult male.

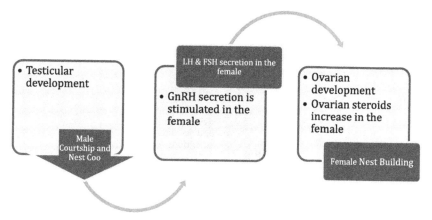

FIGURE 7.1  Lehrman's incomplete hypothesis to explain the effect of the male ring dove's behavior on ovarian development, which in turn results in increased secretion of ovarian hormones, which in turn increase levels of nest building in the female. It was later found that the female's own cooing is the sufficient stimulus for ovarian development (Fig. 7.2).

The same conclusion was reached by Lehrman's British contemporary, R. A. Hinde, who studied the effect of male courtship song on female canaries (Hinde, 1965). This was a bold conclusion, but confirmation of effects of male behavior on female GnRH, LH, and ovarian steroids came more than 40 years later. In many ways, Lehrman and Hinde were correct, with some important caveats.

## THE FEMALE'S OWN BEHAVIOR INCREASES HER OWN GONADAL STEROID LEVELS

Even though this was an important insight, Lehrman's conclusion was only partially correct. In later experiments, ovarian development was measured in muted female doves that could hear the male coo but could not emit their own nest coo. Half of the muted females were exposed to recordings of the nest coo of an adult male, whereas the others were exposed to the recordings of the nest coo of a female. In the unmuted female, ovarian development was accelerated by the male's recorded nest coo, whereas in the muted female, ovarian development was only accelerated by the female's recorded nest coo. Thus, it appears that the *female's own cooing* is the most proximate stimulus for her own ovarian development (Fig. 7.2, Cheng, 1986). Neural activity in single hypothalamic cells and levels of LH were measured in females that were exposed to recorded songs of either male or female doves. Both types of calls increased single-unit hypothalamic activity and LH levels, but the LH increment was three times greater for birds hearing female nest coos than for birds hearing male nest coos (reviewed by Cheng et al., 1998). To summarize, the

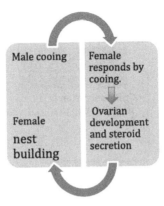

FIGURE 7.2    Cheng's hypothesis to explain the effect of the male ring dove's behavior on the female's behavior, which in turn stimulates ovarian development, which in turn increases levels of nest building in the female.

male's coo elicits the female's coo, and hearing the sounds of her own coo promotes the development of her ova, gametes, and subsequent behavior. The idea that one's own behavior can alter the hormones that control subsequent behavior has profound implications for understanding many different kinds of animal behavior, including learning and addiction.

    The aforementioned data reveal mechanisms whereby courtship and nest building can influence hormone levels, but this leaves many unanswered questions. What starts the male's courtship to begin with? The male's initial aggressive courtship does not occur in juveniles or castrated males, and it is associated with increases of circulating testosterone, but not estradiol. Termination of the aggressive male-like display and the switch to nest solicitation and nest cooing requires a responsive female combined with elevated levels of progesterone. Whenever testosterone is implicated in control of behavior, modern behavioral neuroendocrinologists ask whether testosterone is acting directly via androgen receptors or indirectly after being aromatized to estradiol (and then acting on estrogen receptors) (reviewed by Cornil et al., 2013; Soma, 2006). In fact, courtship in male and female ring doves is prevented by treatments with drugs that block the aromatization of testosterone to estradiol (reviewed by Belle et al., 2003). Thus, estradiol might be the most critical hormone for courtship behavior in both males and females. In addition, estradiol induces progestin receptors, and cross-talk among steroid receptors might be important for the switch that occurs when progesterone terminates the aggressive coo and facilitates nest solicitation (Belle et al., 2005; reviewed by Belle et al., 2003). These experiments provide a deeper understanding of so-called "testosterone-fueled" male behaviors because even though plasma concentrations of estradiol do not fluctuate over the reproductive cycle of male ring doves, both male and female nest cooing are dependent upon estradiol action on estrogen receptors in the hypothalamus and augmentation by progesterone. Furthermore, these experiments opened the doors to the possibility that one effect of social interaction might be to alter brain steroid receptors and the activity of enzymes involved in steroid synthesis (reviewed by Belle et al., 2003; Cornil et al., 2013; Soma, 2006).

## Interim Summary 7.1

1. In birds such as ring doves and canaries, ovarian development in the female is stimulated in the presence of a male of the species who sings a species-specific courtship song.
2. An individual's own behavioral responses in a relationship can alter their own endogenous hormone secretion. For example the male ring dove's courtship vocalization leads to changes in the female's hormone level, but this change is actually due to the fact that his vocalizations elicit the female's own courtship vocalizations. Hearing the sounds of her own coo promotes the development of her ova, gametes, and subsequent behavior.

## NEST BUILDING AND VISUAL CUES FROM THE PARTNER'S NEST BUILDING STIMULATE PROGESTERONE SECRETION

What changes the birds from nest builders to incubators? Building a nest is a frenetic, energy-consuming behavior, whereas incubation is the opposite. It requires calm patience and a slowed metabolism.

The hormonal profile is also different. In both male and female ring doves, the process of engaging in nest-building behavior is followed by increases in circulating concentrations of progesterone that peak just prior to the onset of incubation. This type of correlation provides circumstantial evidence that progesterone plays a role in the onset of incubation.

A good scientific strategy to test such an idea is to see whether the complete behavior occurs even when you take away selected components of the system. In other words, you ask, "is progesterone necessary and/or sufficient for incubation?" Incubation behavior does not occur in birds with low circulating levels of progesterone (e.g., birds that have been gonadectomized and adrenalectomized). This is consistent with the idea that progesterone is necessary, but it is also consistent with a role for all other ovarian hormones, of which there are many. The role of progesterone was tested by giving daily hormone injections to gonadectomized birds housed in isolation for 1 week before the birds were paired and presented with nest-building material. The onset of incubation behavior in gonadectomized males and females (housed with opposite-sex mating partners and nest-building material) was not affected by treatment with estradiol, testosterone, or prolactin alone; rather, the onset of incubation was accelerated in males and females treated with progesterone alone (reviewed by Lehrman, 1965), and in later studies, incubation was most effectively elicited by a combination of estradiol plus progesterone (reviewed by Cheng, 1979).

What is the stimulus that increases progesterone secretion? If you have been paying attention, you might guess that there are three obvious possibilities: the sight of a mating partner diligently building the nest, the opportunity to build a nest, or both. When the nesting materials are present, the rise in ovarian steroid levels and consequent incubation behavior fails to occur unless the birds are with their mating partner, or at least allowed to watch their mating partners through a clear glass plate. The sight of the mating partner alone, while necessary, is not sufficient for changes in nest building. Levels of circulating progesterone and incubation behavior will not rise in paired ring doves unless nest building materials are available. The effects of the mating partner and the nest on incubation are additive, and both stimuli must be maintained for 7 days for the nest builders to transform into incubators; remove either the mating partner or the nest, and the onset of incubation will be delayed. The conclusion is illustrated in Fig. 7.3.

## Interim Summary 7.2

The entire chain of ring dove behaviors is linked by the reciprocal hormone-behavior contingencies.

1. Levels of testosterone must be elevated and testosterone must be aromatized to estradiol prior to and during male courtship.
2. Male and then female nest cooing is required for elevated ovarian estradiol (in females) and must occur prior to nest building.
3. Nest building and cues from mating partners must be sustained prior to elevated levels of progesterone in both sexes.
4. Levels of progesterone in both male and female must rise before the start of incubation.

## INCUBATION AND WATCHING THE PARTNER INCUBATE STIMULATE PROLACTIN SECRETION

What transforms the birds from incubators into brooders (birds that care for newborn offspring)? Clues come from the physiological events that are critical for the survival of the offspring. For example, there is a significant, positive, cause-and-effect association between incubation behavior and the concurrent development of the crop, a pouch in the gullet

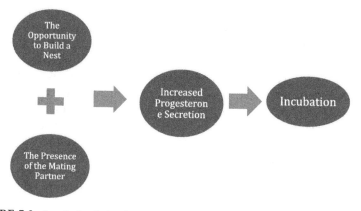

**FIGURE 7.3**  Incubation behavior in male and female ring doves requires elevated levels of progesterone, the secretion of which is stimulated by engaging in nest building and by the presence of the mating partner engaging in nest building.

of both males and females that produces crop milk. Crop milk is a calorie-dense, nutrient-rich food that is essential for offspring survival and healthy maturation. The parents' crops develop in response to increases in the pituitary hormone prolactin. As the parents take turns incubating their eggs, their anterior pituitaries secrete higher and higher levels of prolactin (Goldsmith et al., 1981), which reaches their gullets and causes the lining of the crop to thicken and the crop cells to differentiate into glands that produce milk. After their crops develop and their eggs hatch, both parents regurgitate the crop milk into the mouths of the newborn squabs. One logical hypothesis for control of brooding and feeding is that incubation stimulates prolactin secretion, high levels of prolactin increase crop development, and crop development stimulates behavior, but this idea was not supported.

An alternative idea is that high levels of prolactin stimulate behavior independently of crop development. If so, it should be possible to use prolactin treatment to stimulate parents to feed squabs prior to crop development. Indeed, exogenous prolactin treatment early in incubation accelerates the onset of brooding and feeding squabs, even before the crop milk is synthesized. Even though the parents have no crop milk at this premature stage, prolactin-treated parents feed squabs with a mixture of partially digested seeds. In summary, just as nest cooing leads to nest building, which increases progesterone-induced incubation, it appears that incubation leads to increased secretion of prolactin, which is necessary for parental physiology, motivation, and behavior.

What is it about the incubation period that stimulates crop growth and prolactin secretion? The first hint came in the 1930s when it was noticed that male pigeons fail to develop their crop when housed in isolation, but males develop a crop if (1) they remain paired with their female partner or (2) if they are allowed to observe her through a clear window sitting on the nest (Patel, 1936). The knowledge that this occurs in pigeons makes it no surprise that a similar phenomenon is at work in ring doves. When an isolated male ring dove is allowed to watch his partner through a clear glass plate as she incubates their eggs, crop development and circulating prolactin levels increase along with the motivation to feed squabs. When circulating prolactin was measured in birds that were housed with (1) nests with eggs and a mating partner, (2) nests with eggs but no mating partner, and (3) a mating partner but no nest or eggs, it was discovered that a nest with eggs is necessary but not sufficient to maintain prolactin secretion. The presence of the mating partner was critical (Ramsey et al., 1985).

This is not the complete story because the male does not automatically invest in the energetically costly process of crop development simply because he can see a female incubating. It is true that when crop growth and incubation behavior are measured at different times after the ring dove lays her eggs, there is a significant positive correlation between the time spent incubating eggs and growth of the crop. The site of the female, however, is not sufficient in all males. When members of the pair are separated using the clear, glass plate at varying times after the first two eggs were laid, not all males develop high levels of prolactin, crop milk, and parental behavior. One essential parameter is the duration of close contact with the female, her nest, and her eggs before separation. First, the males must make close contact with the nest and eggs prior to separation by the glass plate, for at least 72h after the second egg is laid. If the male is removed from the female and nest too early, the male's crop will not develop and levels of prolactin do not rise, even if he can see her sitting on the nest through a glass window. This suggests that some initial tactile stimulation is important in addition to the visual cues from the partner. Later work showed that prolactin secretion, consequent crop development, and the readiness to feed offspring is accelerated by applying tactile stimulation to the bird's ventrum (underbelly). This suggests that the act of sitting on eggs creates a stimulus to the bird's ventrum that *initiates* prolactin secretion, whereas the visual cues from the partner's incubation behavior *maintain* prolactin secretion (reviewed by Lehrman, 1965). This pattern of hormone-female interaction might be adaptive for a male ring dove living in nature, since development of the gullet and crop milk is energetically expensive, and would be a wasted investment if his female partner were lazy, lost, or killed during the early stages of incubation. Thus, after incubation commences, without the site of the female sharing in the incubation process, male crop development ceases, leaving him "hormonally free" to return to courtship with a new female. In the last stages of incubation, however, after the crop is developed, the presence of the mate and even high levels of prolactin are not so critical for the maintenance of incubation behavior (Ramsey et al., 1985) (Fig. 7.4).

## Interim Summary 7.3

1. In ring doves, testosterone, estradiol, and participation in courtship interact to produce the conditions for nest building.
2. Nest building enhances secretion of progesterone, and high levels of progesterone interact with the sight of the mating partner building a nest to produce the conditions for incubation.
3. Tactile stimuli during incubation and the sight of the mating partner incubating interact to initiate and maintain the secretion of prolactin.

## VOCAL STIMULI FROM SQUABS STIMULATE PROLACTIN SECRETION

The chain of behavior-hormone relations cannot end with incubation because natural selection depends not only upon fertility but offspring survival, and the newly hatched squabs need food to survive. In the next part of the sequence, stimulation from the squawking squabs leads to sharp spikes in prolactin secretion, crop growth, and milk production/secretion, and it encourages the parents to regurgitate into the offspring's hungry mouths. Removal of the squabs and replacing them with unhatched eggs prevents the normal posthatching increase in crop growth and crop milk production. Together, these results demonstrate that hormones are not autonomous triggers of behavior; rather, hormones respond to behavioral context, and changes in hormone levels increase the incidence of behaviors appropriate for that context.

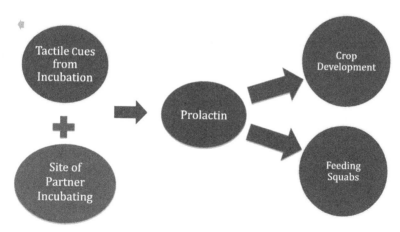

**FIGURE 7.4**  Parental behaviors, such as feeding squabs, in male and female ring doves require elevated levels of prolactin, the secretion of which is stimulated by engaging in incubation and by watching the mating partner engage in incubation. In parents with artificially elevated prolactin, the onset of feeding behavior, including regurgitation into the mouths of squabs, can be accelerated even before crop development.

## LEHRMAN'S CONCLUSION AND ITS LATER CONFIRMATION

In summary, early work, primarily by Lehrman and his students, identified a double set of reciprocal hormone-behavior relations: (1) hormones stimulate behavior, and when an animal engages in behavior, its own hormonal secretions are altered, and (2) the behavior of one animal stimulates hormone secretion in a second animal, thereby changing the second animal's behavior. The second animal's behavior stimulates hormone secretion in the first animal, and again, it changes the first animal's behavior (Fig. 7.5). These reciprocal interactions can create the long and elaborate sequences involved in reproductive cycles.

Long before techniques were invented for measuring GnRH secretion and gene expression, Lehrman, Hinde, and others formulated the following prescient hypothesis, "The physiological explanation for these phenomena lies partly in the fact that the activity of the pituitary gland, which secretes prolactin and gonad-stimulating hormones, is largely controlled by the nervous system through the hypothalamus" (Lehrman, 1964, 1965; Hinde, 1965).

## RECIPROCAL HORMONE-BEHAVIOR RELATIONS HAVE IMPORTANT CONSEQUENCES FOR SURVIVAL AND REPRODUCTIVE SUCCESS

The study of song birds has been particularly useful for understanding reciprocal hormone-behavior interactions. Whereas ring doves and pigeons make relatively simple one-syllable coos, song birds sing complex songs that are learned from adults during an early critical period of development. Species-specific songs cause neuroendocrine changes in

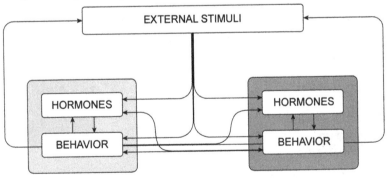

FIGURE 7.5   Interactions that appear to govern the reproductive behavior cycle are suggested here. Hormones regulate behavior and are themselves affected by behavioral and other stimuli, and the behavior of each bird affects the hormones and behavior of its mate (Lehrman, 1964).

the brains of the listener, whether the listener is an adult, reproductively active female, an adult, male competitor, or a juvenile male who is trying to learn his own species-specific song. The ability to hear and respond to bird song can determine survival and reproductive success in the listener. For example, in female canaries and budgerigars, egg laying and courtship are stimulated by the sound of male courtship song (Brockway, 1965), and the longer and more complex the song, the more eggs laid at a shorter latency (Kroodsma, 1976). Female courtship and nesting behaviors are dependent on estrogens, and estrogen secretion is augment by long, spring-like days combined with hearing the song of a robust, adult male (Hinde and Steele, 1978). Together, these mechanisms increase the likelihood that offspring are born to highest ranking males in a time of food abundance, and the study of these phenomena has elucidated important principles in hormone-behavior relations.

The effects of social cues on hormones extend to mammals and have effects on other types of behavior, some of them more complex than sexual motivation. For example, social cues and recognition affect the hormones that influence affiliation, the social behaviors that bring animals together and maintain social bonds. In mammalian species, including prairie voles and marmosets, several days of social contact with an opposite-sex conspecific results in the development of a monogamous pair bond. In a monogamous pair bond, members of the pair show a significant preference for mating and associating with their partner versus a stranger, and they display affiliative behaviors, such as mutual grooming and sleeping while maintaining physical contact. In polygyny, males mate with multiple females, whereas in polyandry, females mate with multiple males. Many such species are promiscuous; however, in prairie voles, several days of cohabitation with an opposite-sex mating partner leads to a strong preference for mating and making physical contact with the familiar partner. In prairie vole monogamy, strong partner preference is coupled with a violent rejection of strangers. Extensive research on prairie voles has revealed an array of chemical messengers that rise in specific brain areas during the formation of a partner preference, including corticotropin-releasing factor, oxytocin, vasopressin, and dopamine. The mechanisms are discussed in detail in Chapter 2.

Similar mechanisms are at play when mothers recognized and bond to their newborn infants, and when individuals learn behavioral sequences and patterns from their siblings and parents (reviewed by Insel, 2010). Monogamy is closely linked with maternal bonding and offspring care in females and males. In species with monogamous mating systems, the offspring are altricial, i.e., they require the constant care from both mother and father (whereas in promiscuous species, the offspring are precocial, and they require no care or only the care from the mother for a short period of time). Maternal bonding, maternal care, and cues from offspring have profound effects on the neuroendocrine system of the parents, and these,

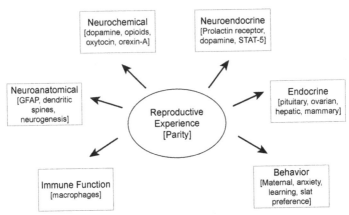

FIGURE 7.6  Constellation of systems affected by the prior reproductive experiences of pregnancy, parturition, and lactation. *Adapted from Bridges, R.S., 2016. Long-term alterations in neural and endocrine processes induced by motherhood in mammals. Horm. Behav. 77, 193–203.*

in turn, have short-term and long-term effects on behavior and ability of the parents. These effects involve the neurotransmitters (especially dopamine), the neuropeptides, oxytocin and orexin A, and the pituitary hormone, prolactin, as well as ovarian steroids (Fig. 7.6; reviewed by Bridges, 2016). Giving birth and caring for offspring also has important effects on immune function, neurogenesis, and neuromodeling. These reciprocal hormone-behavior interactions have adaptive value; they promote the survival and reproductive success of the species.

In summary, behavior of one individual alters neuropeptides and neurohormones in another individual to influence reproductive maturation, seasonal reproductive development, pair bonding, maternal behavior, learning, and memory. You might predict then that the behavior of one individual can induce a massive reorganization of brain areas in another individual, and the behavior of the second individual induces the same plastic restructuring in the first individual. What is even more fascinating is that some changes in behavior occur rapidly, but effects on the hypothalamic–pituitary–gonadal (HPG) system are relatively slow.

## Interim Summary 7.4

The effects of one animal's behavior on another animal's hormones increase reproductive fitness.

1. More complex male bird song leads to higher fertility and eggs laid by female birds that hear those songs.

2. Signals that monogamous species of voles and monkeys receive from their mating partners initiate brain mechanisms involved in pair bonding, and pair-bonded parents are more successful at rearing offspring.

## MECHANISMS

As more and more species have been studied in different social contexts, it appears that there is often a mismatch between the onset of behaviors and the rise of circulating gonadal hormones (reviewed by Atkins-Regan, 2005). One important piece of the puzzle is the knowledge that neuropeptides and neurotransmitters, including gamma aminobutyric acid (GABA), glutamate, norepinephrine, and dopamine, change rapidly in the brain and have direct effects on brain mechanisms that control behavior. In addition, GnRH can affect behavior directly by acting in the brain, prior to effects on the cascade of hormone secretion involving the pituitary and gonad. The rapid, facilitatory effect of GnRH on reproductive motivation occurs in many vertebrate taxa (reviewed by Gore, 2013). In addition, rapid effects of external cues on behavior can be mediated by steroid hormones that are synthesized within synapses of specific circuits (reviewed by Saldanha et al., 2011). Steroids synthesized in brains cells or in synapses are called neurosteroids (Baulieu, 1991). Neurosteroid synthesis can be stimulated by activation of steroidogenic enzymes, such as aromatase, and the time course of effects on behavior is similar to the time course of neurotransmitter action (seconds to minutes) (e.g., Hayden-Hixson and Ferris, 1991; reviewed by Balthazart and Ball, 2006; Forlano et al., 2006; Heimovics et al., 2015). Neurosteroids sometimes act too rapidly to be mediated by intracellular receptors and subsequent initiation of gene transcription (traditional genomic steroid action). Rather, neurosteroids act by signal transduction pathways activated when they interact with the cell membrane (reviewed by Vasudevan and Pfaff, 2008). The rest of the chapter examines what is known about slow versus rapid effects of social cues on the synthesis and secretion of neurohormones, such as GnRH, neurosteroids, such as estradiol, and neuropeptides, such as gonadotropin-inhibiting hormone (GnIH) and kisspeptin.

## SOCIAL CUES CAN RAPIDLY INFLUENCE GNRH

Certainly, Lehrman and Hinde were correct in their observation that social cues in ring doves and canaries promote the cascade of hormones of the HPG system, but the effects of male song on the brain and pituitary can be rapid; they occur within minutes to hours. For example, when female white-throated sparrows are exposed to the complex court-ship song of a male, circulating levels of LH increase significantly within 1 h. The increase in LH is presumed to be stimulated by increased GnRH secretion, although it is difficult to measure levels of neurohormones and neuropeptides secreted from hypothalamic cells. Researchers study-ing birds, fish, and small mammals have used indirect estimates of neu-ropeptide secretion. One commonly used method is to determine the percentage of identified cells that show detectible immediate-early gene expression. Transcription of immediate-early genes, such as *c-Fos, c-Jun, and Zenk,* are correlated with cellular activation, and in identified cells, the percentage of those cells that also express immediate-early genes is taken as proxy for activation. In female white-throated sparrows, record-ings of male song played for a duration of only 42 min induces significant increases of cellular activity of cells in (1) the auditory centers involved in the production and detection of bird song and (2) the brain areas where GnRH-1 cells are located (medial basal hypothalamus). There is no detectible activation of GnRH-1 gene transcript or immediate-early gene expression in GnRH-1 cells *per se.* We know, however, that LH is stimulated by GnRH-1, and therefore the rapid effect of male song on female LH and on activation of cells in the auditory centers and medial basal hypothalamus implies that hearing male song results in rapid synaptic transmission from auditory neurons to neurons that stimulate GnRH-1 secretion (Maney et al., 2007).

Song-GnRH interactions fit the pattern envisioned by Lehrman and Hinde in the 1960s (reciprocal hormone-behavior interactions) because, just as the sound of male song affects hypothalamic cells in females, the presence of the female affects GnRH cells in males. In male listeners, how-ever, the mechanism of action on GnRH cells differs from that in female listeners. When male European starlings are housed in close proximity to females, they show significant increases in GnRH gene transcription compared to males that are housed in isolation from females (Stevenson and Ball, 2009). These effects of female behavior on male GnRH can be even more rapid in some species. Male ring doves allowed to engage in courtship interactions with females show a 40% increase in GnRH gene expression and a 60% increase in cells that contain the GnRH peptide, and the increase was apparent within 2 h of the social interaction (Mantei et al., 2008).

There are at least three major ways GnRH can be affected by social cues. First, social cues can alter the number, size, and sensitivity of neural cir-cuits that impinge on GnRH cells. Second, social cues can affect transcrip-tion of the gene for GnRH and/or synthesis of the peptide. Alternatively, social cues can increase secretion of GnRH that has already been synthe-sized. In some cases, the level of GnRH in the median eminence or in the pituitary portal plexus has been measured.

In species from a wide array of vertebrate taxa, all of the aforementioned mechanisms are represented (Bakker et al., 2001; Cornil et al., 2013; Fabre-Nys et al., 2015; Fernald, 2015; McEwen, 1999; Rissman, 1997; Wilczynski and Lynch, 2011).

In birds, GnRH exists in more than one isoform, but the form most closely linked to the HPG system and fertility is GnRH-1. GnRH-1 cell bodies are located in more than one brain area, but those in the preoptic area are most closely linked to control of the HPG system. In male ring doves, the sound of female nest cooing increases the expression of the gene for GnRH-1 in the hypothalamus (increases the levels of GnRH-1 mRNA) as well as increases in circulating LH and testosterone (Feder et al., 1977; Goldsmith et al., 1981; Mantei et al., 2008; Silver et al., 1980). Together, these results are consistent with Lehrman's original idea that auditory cues reach the hypothalamus to stimulate GnRH secretion, which stimulates pituitary gonadotropin secretion, which stimulates ovarian hormone secretion, gametogenesis, and courtship behavior.

In female musk shrews, the HPG system is activated by cues from the male. Interaction with adult males promotes rapid changes in GnRH-II cells (Dellovade et al., 1995), consistent with the idea that social interactions promote reproductive development and behavior via GnRH, although less information is available regarding the effects of the female on the male HPG system. Vaginocervical stimulation increases the activation of GnRH cells in female ferrets (Bakker et al., 2001). A similar effect of male cues stimulates the HPG system in sheep. In adult, female sheep (ewes), reproduction is seasonal, and mating occurs in decreasing day length of autumn, so the lambs are born in spring, a time of food abundance. During the nonbreeding season (when GnRH, LH, and steroid secretion are low), LH surges and ovulation can be induced by close contact with adult, male sheep (rams), and the phenomenon is known as "the ram effect." Direct contact with a ram or exposure to ram odor increases the percentage of GnRH cells expressing Fos, the protein product of the immediate-early gene *c-fos* in the preoptic area, and these effects are more pronounced in sexually experienced compared to sexually naïve ewes (reviewed by Fabre-Nys et al., 2015). The ram effect is clearly an effect of social cues on the HPG system much as Lehrman and Hinde envisioned because the rapid effects on GnRH do not immediately induce female sex behavior; rather the cues from males induce a silent LH surge (no sex behavior) followed approximately one estrous cycle later with another LH surge, this time accompanied by sexual motivation and behavior. This scenario is consistent with a stimulatory effect of social cues on GnRH, which stimulates the gonadotropins, which stimulates ovarian steroid secretion, and consequent female-typical sex behavior.

Male rodents depend largely upon olfactory cues from females for maximal reproductive stimulation. In male mice and hamsters, indirect evidence suggests that GnRH secretion is affected by social interactions. In adult male mice and hamsters, exposure to adult females increases LH secretion without concomitant effects on GnRH mRNA. This is reminiscent of the mechanism in female white-crowned sparrows, and the interpretation is that cues from a potential mating partner stimulate secretion of GnRH in cells that previously have synthesized and stored sufficient GnRH peptide. The increase in GnRH secretion presumably stimulates LH secretion, which in turn increases gonadal androgen secretion, thought to boost the male's sexual maturation, motivation, and fertility. A different mechanism is in play when a male mouse encounters another adult male. In this case, GnRH mRNA in male mice is significantly increased within 90 min of exposure to another male, and this hormonal response to male cues might facilitate LH and testosterone secretion as well as dominance and territory defense (Gore, 2013; Richardson et al., 2004).

Understanding how sensory cues are transmitted to GnRH cells has been a slow and difficult process in mammalian systems. Mammals have two systems for detecting and reacting to olfactory cues: the main olfactory system, in which odors are detected by the olfactory bulb, and the accessory olfactory system, in which pheromones are detected by the vomeronasal organ. Early work in Syrian hamsters was consistent with the idea that olfactory cues from female hamsters are detected by receptors in the male vomeronasal organ and transmitted to the hypothalamic areas involved in sexual motivation via the accessory olfactory system and medial nucleus of the amygdala (reviewed by Swann et al., 2009). Other work using transgenic mice implicates a direct connection from the main olfactory bulb to GnRH cells, but no neural connections from the accessory olfactory system (Yoon et al., 2005). Similarly, the main olfactory system mediates the effects of social olfactory cues in ewes, pigs, rabbits, and ferrets (reviewed by Fabre-Nys et al., 2015). Other cues, tactile, visual, and auditory, most certainly influence neuroendocrinology and behavior in mammals, and they are likely to act via multiple neural routes from sensory receptors to GnRH and other effector cells (Petersen and Hurley, 2017).

HPG responses to social cues occur even in the vertebrate class that was first to evolve. For example, in male cichlid fish, *Haplochromis burtoni*, males are remarkably responsive to social cues that affect the HPG system and behavior. These fish live and breed in groups with a strict social hierarchy, and cues from conspecifics cause neuroendocrine and behavioral transformations in rank. Only a small percentage of males are dominant, territorial, and reproductively active at any one time (reviewed by Fernald, 2015). The territorial males are easily distinguished by their larger size, bright coloration, and aggressive behavior. Nonterritorial males display none of these characteristics and are reproductively suppressed with little chance of mating. Fortunes change, however, when a dominant male dies of natural causes, is killed by a competitor, or disappears after a natural disaster. In such cases, survival and reproductive success depend upon the ability to either climb up or step down in rank in response to environmental changes. The loss of the dominant, territorial male is sensed by the nonterritiorial male, and sensory cues from the loss alter his neuroendocrine status, i.e., the suppressive effects of the dominant male are removed. In the laboratory, switching a nonterritorial male from his all-male group to an all-female group results in significant increases in GnRH-1 gene expression and size of GnRH-1 cell bodies in the preoptic area, the location of the brain where the GnRH cells necessary for stimulation of LH and gonadal hormones. The increases begin within 3 days of the switch to an all-female group and are significantly higher by 7 days after the switch, and these changes precede the behavioral change from a nonterritorial to a territorial, reproductively active male (White and Fernald, 2002). These and many other rapid changes occur in the ascending male's hormonal profile and precede his change in territorial behavior. Slower changes in the HPG system are more closely correlated with changes in reproductive behavior and fertility (Fernald and Maruska, 2012).

As in fish, reciprocal behavior-hormone interactions occur in amphibians, including anuran species (frogs and toads) (reviewed by Wilczynski and Lynch, 2011). The most important sensory modality in these species is audition. Male frogs, for example, emit loud courtship calls that increase in volume over time, and as more and more males join in the chorus, the sound builds to a frantic crescendo. The nightly performance depends upon a fascinating interaction between testicular steroids, the presence of adult, fertile females, and perhaps most important, the sound of other males in the vicinity. The interaction fits the classic pattern predicted by Lehrman/Hinde. Testosterone secretion in spring is a necessary requirement for male calling, and the sound of other males calling and the act of calling itself stimulates the secretions of the HPG system, with increases in GnRH-1 mRNA and testosterone in recipient frogs, which in turn stimulates calling in those frogs with elevated testosterone. The system is a successful adaptation, since females are very unlikely to ambulate toward the call of a lone male, but they hop with enthusiasm toward a chorus of males. The neuroanatomical and sensory details of the system are beyond the scope of this book, and the reader is referred to the original work and reviews of Wilczynski et al. (reviewed by Wilczynski and Lynch, 2011).

## Interim Summary 7.5

1. In some species, one animal's courtship behavior influences the entire HPG system, so sensory cues (auditory, olfactory, or visual) stimulate secretion of GnRH, which stimulates secretion of LH, which stimulates secretion of gonadal steroids, which have long-term effects on behavior.
2. In other species, one animal's courtship behavior influences the secretion of GnRH (or GnRH-II), which can have direct, rapid effects on behavior.

# SOCIAL CUES RAPIDLY AFFECT NEUROPEPTIDES AND NEUROTRANSMITTERS

Activation of GnRH cells, and concomitant stimulation of GnRH secretion, is often accomplished by modulatory neurons that contact GnRH cells near the axon terminals where GnRH is secreted. The modulatory neurons either inhibit or stimulate secretion GnRH, and there are approximately 5 million such cells in most vertebrate species (Herbison, 2006). Some of these modulators are classical neurotransmitters, such as norepinephrine, dopamine, GABA, and glutamate. Sensory cues that occur during social interactions affect dopaminergic and noradrenergic activity in a variety of species, and this system has been highly conserved during the evolution of vertebrates (Korzan and Summers, 2007). Effects of experience on these monoaminergic systems and their subsequent effects on future behavioral tendencies go beyond reproductive and territorial behaviors to more general activities such as the stress response, learning, memory, and resilience in the face of trauma (reviewed by McEwen, 1999; Sapolsky, 2003).

Two more recently discovered examples of peptides that are affected by social experience and modulate GnRH and behavior are the RF-amides, including GnIH. GnIH inhibits GnRH secretion and kisspeptin secretion, and both of these events stimulate GnRH secretion. Kisspeptin and GnIH may be controlled by noradrenergic cells that receive input from the auditory, visual, and olfactory sensory system. GnRH is subject to modulation by gonadal steroids, but GnRH cells lack the requisite steroid receptors. The cells that release kisspeptin and GnIH, however, are enriched with receptors for estradiol and androgens, and they are thought to be important in steroid modulation of GnRH in response to social cues. Kisspeptin likely mediates "the ram effect" on the HPG system and other such phenomena in mammals (reviewed by Fabre-Nys et al., 2015). In birds, GnIH, first identified in Japanese quail, is rapidly affected by both reproductive hormones and by behavioral interactions that determine social status (Amorin and Calisi, 2015; Calisi et al., 2011; Tobari et al., 2014; Ubuka et al., 2005).

# SOCIAL CUES RAPIDLY AFFECT LOCAL LEVELS OF NEUROSTEROIDS

In this millennium, investigators have perfected technology that allows measurement of very low concentrations of neurosteroids. One example is estradiol, which is made in the brain when testosterone is aromatized via activation of the enzyme aromatase. Local estradiol concentrations have been measured using in vivo microdialysis, a technique that uses a very small probe, a tiny tube made of a semipermeable membrane, to collect small volumes of neurosteroids from the surrounding extracellular space in the brain area of interest. The microdialysis probe can be fixed to the brain area of interest, so brain fluid samples can be collected from the brains of free-living animals before and after exposure to social cues. Prior to the development of the in vivo microdialysis of neurosteroids, experimenters conducted experiments in which different groups of animals would receive different social experience prior to the histological preparation of the brain and blood collection. With in vivo microdialysis, however, neurosteroid levels in specific brain areas are assessed in the behaving animal before and after the introduction of social cues (either cues from a living conspecific or from a recording of their song). Extracellular estradiol concentrations increased within 30 min of the start of interactions with adult females in the brain area known as the auditory caudomedial nidopallium (Remage-Healey et al., 2008). The auditory caudomedial nidopallium is analogous to the mammalian auditory cortex, the brain area where input from the auditory receptors is processed, especially during language and song learning. This area is known to show high levels of expression and activity of the enzyme aromatase (reviewed by Saldanha et al., 2011), and aromatase activity is stimulated rapidly by social cues from environmental stimuli as well as by pharmacological treatments (reviewed by Balthazart and Ball, 2006; Cornil et al., 2006; Forlano et al., 2006; Soma, 2006). In a subsequent experiment, when male zebra finches were exposed to conspecific males, concentrations of estradiol rose while concentrations of testosterone fell in this same brain area within 30 min (the caudomedial nidopallium) (Remage-Healey et al., 2008). Together, these studies are consistent with the idea that social cues increase aromatase expression and activity allowing the conversion of testosterone to estradiol. As described in previous chapters, estradiol acts by nongenomic mechanisms to facilitate rapid changes in behavior (reviewed by Vasudevan and Pfaff, 2008). Activation of aromatase, estradiol synthesis, and rapid changes in these particular brain areas are thought to play a role in the social interactions that involve singing and listening to complex song, and therefore they have implications for understanding social transmission of language. As we develop a deeper understanding of rapid effects of neurosteroid synthesis and action, we will gain a better understanding of reciprocal hormone-behavior interactions.

## Interim Summary 7.6

The behavior of one animal can influence the brain and behavior of a potential mating partner, at least, by affecting the following:

1. Norepinephrine, dopamine, GABA, and glutamate,
2. Kisspeptin and GnIH,
3. Activation of the enzymes of steroid synthesis, such as aromatase.

# SUMMARY

The chapter began with an examination of the notion that human sex differences in aggression are caused by innate sex differences in circulating testosterone. This idea, at first glance, might appear to be in direct conflict with the barebones behaviorist interpretation: "Aggressive men are the product of a lifetime of societal pressures that encourage and train men to be aggressive and women to be cooperative and nurturing." The *reciprocal interactions* among behaviors and hormones renders this conflict moot. The idea that sex differences in aggression are due purely to innate levels of circulating testosterone and similar notions (higher oxytocin makes females more maternal, fluctuating hormones over the menstrual cycle make female behavior more variable, or dopamine makes all stimuli more pleasurable) are also oversimplified, given the *reciprocal* effects of hormones on behavior. Learning and endocrinology are not two different things. Learning involves hormones and other chemical messengers, and individuals learn from each other and from their own experience.

The idea that high levels of aggression result from elevated concentrations of testosterone requires a closer examination. First, even in males of countless species, testosterone might be acting after aromatization to estradiol or reduction to dihydrotestosterone. Alternatively, high levels of testosterone in males might be a fortuitous correlate of aggression, and aggression might be related to any number of neurosteroids, neurotransmitters, or neuropeptide levels. Most important, levels of testosterone and other hormones are affected by environmental factors, including experience (reviewed by Gleason et al., 2009; Hamilton et al., 2015; Hirschenhauser and Oliveria, 2006; Soma, 2006; van Anders and Watson, 2006; Wingfield et al., 1990). Some evidence consistent with this idea has come to light. In one experiment, hormone levels were measured before and after women participated in two kinds of staged social encounters. In one type of encounter, they wielded power over another individual, whereas in another type of encounter they engaged in neutral interactions in which neither person was dominant or subordinate. Women who wielded power increased their mean circulating testosterone concentrations by an average of 10%–15%, whereas those who engaged in neutral encounters decreased mean circulating testosterone concentrations, and the differences were statistically significant (van Anders et al., 2015).

The astounding implication is that behavioral habits can increase levels of hormone secretion, and so-called "innate sex differences" in hormone secretion might be augmented or reversed by years of socially enforced behaviors. In other words, higher levels of circulating testosterone in men could be, in part, a consequence of socialization that encourages men to wield power and women to cooperate and/or nurture. In this case, it is inaccurate to say that low circulating concentrations of testosterone in women are innate, since engaging in behaviors of a dominant individual can significantly raise testosterone concentrations in women. It is equally erroneous to say that testosterone is not involved in aggression, since practicing dominant behavior leads to elevated testosterone concentrations. This chapter alludes to some of the possible mechanisms and suggests that they are likely more complex than simple effects of social cues on circulating concentrations of testosterone. Circulating hormones can be metabolized to other hormones in the brain, and testosterone can have complex interactions with other steroids (e.g., the glucocorticoids), neuropeptides, and neurotransmitters.

In humans, it is difficult to mechanistically analyze reciprocal hormone-brain interactions at the level of neurosteroid synthesis and neural remodeling. This topic will be of primary importance in humans, however, because it has become clear that brain circuits are remodeled by the stress hormones affected by stressors from other humans, such as early-life trauma and abuse, competition for limited resources, and health-damaging habits (drug addiction and diets high in sugar and fat). Constant exposure to these hormones, including norepinephrine, epinephrine, and glucocorticoids, have many dire consequences on neural circuitry that controls human behaviors, particularly overeating, learning, memory, attention, mood (depression), and impulse control (McEwen, 2010). Unless the world becomes significantly more peaceful in the future, research in reciprocal behavior-stress hormone interactions will be of critical importance (McEwen and Tucker, 2011; McEwen, 2016).

# HISTORY: HORMONE EFFECTS CAN DEPEND ON FAMILY, GENDER, AND DEVELOPMENT

# Familial/Genetic Dispositions to Hormone Responsiveness Can Influence Behavior

On the battlefield, after a terrorist attack, or during a family tragedy, some individuals always seem to act more courageously than others. The neuroendocrine stress response is almost universal; that is, in virtually every species studied, the secretion of epinephrine and glucocorticoids increases in response to various stressful stimuli. Stressors are many and varied, and include those stimuli that predict a coming threat, as well as those that are aversive, novel, unpredictable, uncontrollable, and ambiguous. Individual differences in hormone secretion in response to identical stressors have been documented in many species, including human and nonhuman primates. For example, in response to aggressive threats and competition, individuals vary widely in their secretion of adrenal cortisol and catecholamines. This chapter is concerned with the sources of these individual differences.

Even before the discovery of genes as the vehicle of inheritance, it was clear that certain behaviors "run in families." An individual's family background can predispose him or her toward more or less sensitivity to a given hormone in a given situation. An individual from one family may react to a particular hormone secreted at a particular level, whereas an individual from a different family may not react at all. Family background can also affect other aspects of neuroendocrine function, including the level of hormone secreted in response to a stimulus, the timing of hormone secretion and response, and the strength of motor and behavioral responses to a hormone. The influence of the individual's family can be genetic and/or environmental, because offspring and parents often share a portion of their genome and much of their environment (i.e., they share experiences that are common to both parent and offspring, such as diet, socioeconomic status, trauma, and education). One important feature of the environment is the treatment of the individual by family members.

This includes the distinct possibility that the individual will copy behavioral tendencies exhibited by his or her parents. For example, an abused child is more likely to abuse his or her own children, and childhood trauma is often a better predictor of adult behavior than genetic makeup.

Genetic effects stem from the DNA sequence passed from parent to offspring. The modern synthetic theory of evolution states that genes are the unit of inheritance and that traits acquired during a lifetime do not alter the DNA in the gamete, and therefore acquired traits are not passed on to the next generation. There is an important caveat to this fact. When an individual experiences certain stimuli in the environment (e.g., diet, climate, stress), especially during early development, chemical tags or labels are sometimes added to certain nucleotides in the DNA sequence. Some chemical tags mark the nucleotide sequence so that, when transcription of the gene is initiated, there will be more transcription, whereas other tags mark other nucleotides for less transcription. These tags can be passed from generation to generation. Environmental effects on the level of gene transcription are known as *epigenetic effects*, and although epigenetics does not involve evolution (defined as changes in allele frequency), some epigenetic effects on transcription can be passed from generation to generation. Thus, phenotypic traits in an individual can result from environmental influences on gene transcription that occurred in the individual's parents or grandparents (see Chapter 10). These can be manifest as differential DNA methylation. Methylation is a process by which methyl groups (one carbon atom bonded to three hydrogen atoms ($CH_3$)) are added to DNA. When located in a gene promoter, DNA methylation typically acts to repress or "turn off" gene transcription. Epigenetic effects can also occur via transcriptional barriers composed of DNA-binding proteins and other nuclear proteins, among other mechanisms. These involve genes on both the autosomes and the sex chromosomes.

The SRY gene, which encodes a protein that leads to the development of testes, determines whether the fetus will be exposed to high levels of androgens, and this in turn has strong influence on the behavioral responsiveness to gonadal hormones in adulthood. In addition to effects of SRY and other genes on the Y chromosome, some sex differences arise from differences in the "dose" of genes on the X chromosome or their parent imprint. For genes located on the X chromosome, females have two alleles, whereas males have only one. Many experiments since the 1950s have revealed effects of pre- and neonatal hormones on behavior, but more recent studies have revealed sex differences that are not mediated by early gonadal hormones. These include differences between males and females in food intake, locomotion, the propensity to store body fat, aggression, parental behavior, habit formation, nociception, and social interactions. These are influenced by the "gene dose" of the X chromosome. For example, the X chromosome encodes proteins that influence adult responsiveness to gonadal hormones and glucocorticoids, and having two X chromosomes, one X chromosome, or no X chromosome influences a variety of behavioral and physiological traits independent of effects on sexual differentiation of the gonad (Arnold and Chen, 2009). The complete absence of one X chromosome or complete monosomy (a karyotype of XO; one cause of Turner syndrome) leads to intellectual disability and susceptibility to heart disease. In addition, there are known genetic predispositions for depression, alcoholism, drug dependence, addiction, and schizophrenia. As described in Chapters 1–3, there are many genes that encode hormones that increase food intake, including the gene for ghrelin, neuropeptide Y, certain NPY receptors, and the melanocortin receptor agonists. There also are genes that encode hormones that decrease food intake, including leptin, the leptin receptor, and the POMC gene. Mutations in these genes lead to hyperphagia, obesity, and diabetes. Many of these mutations alter the response to gonadal and adrenal steroid hormones.

Not all genetic differences arrive from single mutations. Most of the variation that we see in body weight and adiposity is, in fact, polygenic; that is, the variance is attributable to many genes. It is thought that more than 70 genes contribute to variation in these energy balancing traits. Traits in which the variance in populations is influenced by multiple genes are known as *quantitative traits*. When natural selection favors certain levels of behavior, such as a high propensity to store fat or a high propensity to build a nest, the mean level of these traits tends to increase slowly over many generations. This can be mimicked in the laboratory or on the farm, where artificial breeding programs increase the mean value of morphological traits over successive generations. The same basic process is involved in the domestication of dogs and commercial breeding of dogs on the basis of various physical and behavioral traits. Domestication, whether it occurs in dogs, cattle, rats, or chickens, involves changes in gene frequencies over many generations that favor a dampening of the hypothalamo–pituitary–adrenal cortical (HPA) system in response to stress. The examples in this chapter are so obviously relevant to questions of human behavior, and the effects of family background and early environment are so important for later human behavioral pathology, that this chapter is not divided into separate experimental and clinical discussions.

# GENETIC DIFFERENCES IN THE NEUROENDOCRINE RESPONSE TO STRESSORS

The stress response invariably involves secretion of hypothalamic corticotropin-releasing hormone (CRH), adrenocorticotropic hormone (ACTH), catecholamines (epinephrine, norepinephrine, and dopamine), and glucocorticoids, and this occurs in every species studied. Yet, there are differences among individuals of the same species in the neuroendocrine response to stressors. Individuals with the same baseline cortisol, for example, differ in their ACTH and cortisol increases in response to a stressor. Often, the individuals with the lower stress-induced increases in circulating cortisol are those that display the most physically aggressive behavior. Furthermore, in such individuals, high levels of physical aggression are coupled with lower levels of performance on cognitive tasks. What is the source of this variation? Glucocorticoids and catecholamines interact during the stress response, and these interactions differ according to genetic predisposition. In human males, for example, high levels of aggression are associated with variation in the gene that encodes catechol-o-methyltransferase, COMT, an enzyme that degrades the catecholamines norepinephrine and dopamine. Natural genetic variation at any particular genetic locus is known as a polymorphism. The polymorphic gene that encodes COMT, for example, has more than one genetic variant, or allele, at the COMT locus. One such polymorphism in the COMT gene results in decreased degradation of dopamine, with resulting increased cortisol secretion in response to stress. Furthermore, this COMT variant leads to increased anxiety, low levels of competition, altered central processing of fear, exaggerated pain perception, and increased cognitive function. A different polymorphism in the COMT gene has the opposite effects, including decreased cortisol response to stress, stress resilience, social dominance, and decreased cognitive function (Qayyum et al., 2015). This is but one of many examples of genetic differences that lead to changes in the neuroendocrine systems that control behavior.

Genetic differences in this chapter include natural allelic variation, such as the polymorphisms just described for the COMT gene, as well as mutations, rare occurrences of a nucleotide change, i.e., a deletion, insertion, or rearrangement of large sections of gene or chromosome that can lead to a change in the function of the encoded protein. Similar studies have examined why some individuals who experience life stressors develop major depression, whereas others who experience the exact same life stressors have little or no depressive symptoms. A polymorphism in a region of the gene that encodes the serotonin transporter influences individual differences in response to stressful life events. This region of the gene is known as the promoter because, when stimulated, it initiates transcription of the gene. Humans with one or two copies of a particular allele of the serotonin transporter promoter polymorphism display more depressive symptoms, diagnosable depression, and suicidality in response to stressful life events compared to individuals homozygous for a different allele (Caspi et al., 2003). The COMT and serotonin transporter polymorphisms are prime examples of how genes and environment interact to determine behavioral traits such as depression, anxiety, and aggression.

# EARLY LIFE REACTION TO STRESSORS

Early childhood stress has profound effects on adult behavior, and these can be passed to future generations. In humans, childhood neglect and abuse can have permanent consequences on adult behaviors, including resilience in the face of stress, propensity for substance abuse and addiction, depression, anxiety, violence, and abusive parental behaviors (reviewed by McGowan et al., 2009). Early experimentalists who studied primate behavior learned that, although baby monkeys are precocial in their ability to ambulate and feed themselves without parental assistance, their social development is severely stunted by maternal separation. In some studies, if infants are separated from their mothers soon after birth, they show stages of protest, despair, and withdrawal. Upon reunion with their mothers, the infants show considerable mother-directed behavior, with increased clinging and contact and altered social development that lasts throughout life. Fig. 8.1 shows social withdrawal in young monkeys that had been separated from their mothers. If young monkeys are fully socially isolated for an extended period (e.g., 6 months), for months thereafter, they continue to show disturbed behavior, including self-mouthing, self-clasping, and huddling, with almost no locomotive or exploratory behavior, or social behavior when group-housed. Understanding the neuroendocrine mechanisms of this phenomenon is critical for a medical response to orphans, refugees, and victims of child abuse and neglect. It is not yet known whether the mechanisms involved translate to human depressive behavior. Work on primate models is controversial, but many investigators argue that the disturbed-rearing models in rodents and primates may serve as test beds for pharmaceutical interventions that may prove useful in the treatment of human depressive disorders. For example, clinically useful antidepressants have been shown to ameliorate some of the behavioral disturbances in maternally deprived animals.

Thus, there are established connections among early adverse experiences, life-long neurohormonal and other central nervous system (CNS) disruptions, and behavioral disturbances. The long-term physiological and behavioral changes depend on both the genetic makeup of the animal (species and strain differences) and the type of stressor. In some situations, the long-term changes can be modified by corrective parenting after the stressful period.

It might be premature to extrapolate from neonatal maternal separation studies in rats and monkeys to specific psychiatric syndromes such as major depression, but there is little doubt that the basic experimental findings in laboratory animals have intriguing implications for human behavior and psychopathology. For example, epigenetic regulation of the gene that encodes the glucocorticoid receptor (GR) has been linked to

FIGURE 8.1  Monkeys separated at very young ages from maternal and social supports, even while adequately nourished, developed abnormal emotional and social behaviors. Seen here, they appeared sad and withdrawn. *From McKinney, W.T., Suomi, S.J., Harlow, H.F., 1971. Depression in primates. Am. J. Psychiatry 127, 1313–1320. With permission.*

variation in human behavioral abnormalities in response to child abuse. The conservatively stated implications are the following:

1. Maternal stresses during pregnancy may have important consequences in the fetus because of the relatively lengthy period of intrauterine development in primates versus rodents.
2. As well, early childhood experience can shape one's perception of the world throughout adult life, particularly one's physiological and psychological responses to stressful life situations.
3. Both physical and psychological childhood abuse, depending on its severity and extent, can predispose an individual to, or even directly result in, pathological physical conditions as well as psychiatric disturbances.

4. The specific type of behavioral/psychiatric disturbance depends on many factors, including one's genetic predisposition to developing depression, panic disorder, chronic anxiety, schizophrenia, etc.
5. Some hormonal interventions (e.g., reducing excessive glucocorticoid activity with synthesis blockers or receptor antagonists in those patients with major depression who exhibit excessive HPA axis activity) may aid in treatment, but evidence to date indicates such hormonal interventions are clearly not useful as sole therapeutic agents.

Using laboratory rodents, we have learned a great deal about the neuroendocrine mechanisms that underlie early effects of maternal care and neglect. Long-term behavioral effects on many behaviors occur in murine offspring that have been deprived of their mother's care (*maternal deprivation*). Other long-term effects occur when newborn offspring are handled by experimenters before being returned to their own mothers or foster mothers (*effects of neonatal handling*). Both maternal deprivation with neonatal handling and maternal deprivation without neonatal handling alter the HPA system in the offspring, and these effects last into adulthood (Table 8.1). Furthermore, the effects can persist into future generations.

In the most commonly studied paradigm, mouse pups are removed from their mothers for a few minutes and handled by the experimenter. These pups emit distress calls, and after the return to their mother, they receive elevated levels of maternal care, including increased licking and grooming. Mouse pups that are maternally deprived show high HPA responses to stress, whereas those that are maternally deprived and receive neonatal handling show lower HPA responses than those that experienced maternal separation with no handling. The effects in the group that receive neonatal handling are traced to the extra care the pups receive from their mother after handling. These effects on the offspring differ according to whether the mother is attentive (high licking and

**TABLE 8.1** Treatment of Baby Rats by Mother or Experimenter Can Have Long-Lasting Consequences for the Animal's Temperament, Especially Its Responses to Stress

| | CRH | Hyperresponse to Stress, Depression, Anxiety? |
| --- | --- | --- |
| Separation from mother | ↑ | Yes |
| Neonatal handling and contact | → | No |

Separation from the mother—which surely deprives the baby of the licking and handling it would receive in the nest and may also allow body temperature to fall—is associated with permanent changes in the release of corticotrophin releasing hormone levels (CRH; see neural circuitry in Fig. 1.1) and with a disposition toward hyperresponsivity to stressful situations, depression, and anxiety. In contrast, excess handling and contact can actually render the adult response to stress more adaptive than the normal case.

grooming) or inattentive (low licking and grooming) toward her pups. Pups of high-licking-and-grooming mothers have reduced plasma ACTH and corticosterone responses to acute stress, increased hippocampal GR messenger RNA expression, enhanced glucocorticoid feedback sensitivity, and decreased levels of hypothalamic CRH messenger RNA, compared to pups of low-licking-and-grooming mothers. Perhaps most remarkably, high-licking-and-grooming mothers raise pups that are also high licking and grooming, suggesting that early life experience has significant effects on adult maternal behavior (e.g., Liu et al., 1997).

As a stress-responsive endocrine system, the HPA axis is important in regulating metabolic activity when animals are confronted with environmental stressors, and these effects are different at different times in development. The widespread effects of glucocorticoids have been discussed elsewhere in this volume (e.g., the Introduction and Chapters 1–3). In the fetal and neonatal rat, components of the HPA axis, as well as other endocrine axes, mature at different rates and reach peak activity at different times. For example, plasma ACTH and corticosterone reach their peaks during late embryogenesis, whereas GR binding does not peak until more than 3 weeks after birth. Some components of the HPA axis peak perinatally and then decline to adult levels in the neonatal period. Thus, various aspects of the HPA system can be affected in different ways by environmental stresses, depending on when those stresses occur. Early experimental work applying stressors to neonatal rats suggested a stress-nonresponsive period in the first 2 weeks of life, but subsequent data (e.g., the fact that corticosterone secretion is blunted but ACTH secretion is not significantly altered following a variety of stressors) have led to the newer concept of a stress-hyporesponsive period, during which adrenal cortical responses to stressors are blunted but not absent. During the hyporesponsive period, however, a specific stressor (separation of pups from their mother) still leads to enhanced adrenal cortical sensitivity to ACTH and, consequently, exaggerated corticosterone responses compared to nonseparated pups.

The effects of maternal separation and subsequent reuniting with the mother are complex and involve many aspects of the pup's physiology. First, there are effects of separation on body temperature in the poikilothermic pups (i.e., they have not yet developed the ability to control their own body temperature). Second, isolation from the mother means that the pups are not nursing, and therefore they are not receiving the same nutrition as unseparated pups. Furthermore, some separation paradigms might cause adrenal hypersensitivity to ACTH, which might be expected to decrease nursing, which would be expected to result in food deprivation. Third, there is natural variation in the maternal grooming of the pups, which, as mentioned in a previous paragraph, appears to be a powerful antidote to maternal separation. The separation-induced changes in the HPA axis are highly correlated with the level of licking and grooming by the mother. Of interest, this effect of maternal care can be mimicked by stroking the anogenital region of separated pups with a wet brush in the same frequency and intensity of maternal licking. This illustrates that excellent maternal behavior is a complex behavior that provides many critical offspring requirements, and the scientific study of this phenomenon requires the control of many physiological variables.

When experiments are designed to control for these variables (temperature, suckling, and maternal licking and grooming), it is clear that the HPA axis can become hyper- or hyporesponsive to stressors, and this effect persists into adulthood. In separated pups, basal ACTH and corticosterone secretion appear to be normal. The ACTH/corticosterone stress response is exaggerated and resistant to negative feedback, e.g., the inhibitory effects of dexamethasone suppression. Maternal separation increases offspring CRH content in the hypothalamic paraventricular nucleus (PVN), where neurosecretory cells for CRH are located, and in the median eminence, where CRH is released. Of importance, there also appears to be reduced GR density in several CNS areas, including the hippocampus. These areas are important for negative feedback regulation of the HPA axis. In addition, there is reduced GR density in the prefrontal cortex, an area implicated in several psychiatric syndromes. In contrast to the effects of maternal deprivation, neonatal handling of rats has the opposite (i.e., "calming") effect on the HPA axis: lessening the ACTH/corticosterone stress response; reducing CRH and arginine vasopressin content of the PVN, median eminence, and pituitary stalk; and increasing GR density in the hippocampus and prefrontal cortex.

There are clear, long-term neuroanatomic and behavioral consequences of the interrupted mothering of neonatal rats, among them, the loss of one of the histological regions of the hippocampus known as the *cornu Ammonis* 3 (a Latin term for ram's horn, CA3) neurons. In addition, interrupted maternal care leads to altered mossy fiber morphology and upregulation of CRH expression in the hippocampus, along with impairment of long-term memory in adult rats. This may be a direct effect of CRH, in that elevated corticosterone levels are not necessary for these effects to occur, and they can be mimicked by CRH injections during the neonatal period. As mentioned, impaired negative feedback regulation of the HPA axis secondary to reduced GRs in the hippocampus may be the key element in the excessive and prolonged HPA axis responses in stressed adult rats exposed to neonatal maternal deprivation. This, however, is likely not the only mechanism, as maternal separation also causes decreased CNS expression of neurotrophins, particularly brain-derived neurotrophic factor (BDNF), and decreased GABAergic neurotransmission, gamma-aminobutyric acid (GABA) being a widespread CNS inhibitory neurotransmitter that inhibits the HPA axis. In summary, neonatal neglect affects the development of neurogenesis and the transcription of genes for and synthesis of CRH and other neurotransmitters.

Increased CRH levels secondary to maternal deprivation also may affect CNS circuits underlying the anxiety response. Several pathways are putatively affected by decreased BDNF and GABAergic input, a major one being from the central nucleus of the amygdala to the locus ceruleus. CRH produced in the amygdala stimulates the locus ceruleus to release norepinephrine, a neurotransmitter that has activating and anxiogenic effects. The locus ceruleus has 50%–70% of the norepinephrine content of the CNS. Thus, effects of stress on offspring CRH can alter adult anxiety in response to a stressor.

Genetic influences modulate the familial (i.e., parenting) aspect of neonatal stress, in that there are differences among inbred strains of mice in the resiliency of exposed pups. Inbred strains of mice are for all practical purposes genetically homogeneous due to many generations of purposeful brother–sister mating. For example, BALB/cByJ mice exhibit more stress-related HPA axis activity and behavioral disturbances as neonates than do C57BL/6ByJ mice. Early-life handling of BALB/cByJ mice and cross-fostering of BALB/cByJ neonates to C57BL/6ByJ dams reduce the HPA axis stress response and prevent associated behavioral disturbances. In contrast, cross-fostering C57BL/6ByJ neonates to BALB/cByJ dams does not induce stress hyperreactivity in the C57BL/6ByJ pups. These findings indicate that qualitatively different familial interactions can have over-riding effects in certain animal strains that are genetically predisposed to high stress reactivity, but not in other strains that are more stress-resistant.

As well, there are major species differences in physiological responses to neonatal handling. In contrast to rats and mice, neonatally handled boars had greater corticosteroid-binding globulin and lower circulating free and total cortisol concentrations at age 7 months than did nonhandled control boars. Also, in contrast to rodents, 7-month-old handled and control boars had similar ACTH and cortisol stress responses and similar GR densities in the frontal cortex, hippocampus, hypothalamus, and pituitary gland (Fig. 8.2).

Thus, there are established connections among early adverse experiences, life-long neurohormonal and other CNS disruptions, and behavioral disturbances. The nature of the long-term physiological and behavioral changes depends on both the genetic makeup of the animal (species and strain differences) and the type of stressor. In some situations, the long-term changes can be modified by corrective parenting after the stressful period.

These basic experimental findings have intriguing implications for human behavior and psychopathology, even though some investigators have drawn tenuous extrapolations between the neonatal maternal separation studies in rats to specific psychiatric syndromes such as major depression. The conservatively stated implications are summarized here.

| Target | Postnatal Handling | Maternal Separation |
|---|---|---|
| CRH mRNA (PVH, CnAmy) CRH receptor binding | ↓ | ↑ |
| GR mRNA GC negative feedback on CRH | ↑ | ↓ |

FIGURE 8.2 This figure shows the effects of early life stress on aspects of the adult hypo-thalamo–pituitary–adrenal system in rats. Maternal separation exaggerates the hypotha-lamic–pituitary–adrenal response to stress, whereas postnatal handling and subsequent increases in maternal behavior ameliorate these effects of maternal separation. These effects include changes in corticotropin-releasing hormone (CRH) messenger RNA (mRNA) in the paraventricular nucleus of the hypothalamus and central nucleus of the amygdala (CnAmy), and glucocorticoid receptor (GR) mRNA in the hippocampus and glucocorticoid (GC) nega-tive feedback effects on CRH. Treatment of neonatal rat pups by mother or experimenter can have long-lasting consequences for adult temperament of the offspring, especially their responses to stress. Separation from the mother, which deprives the pups of the licking and handling they would receive in the nest and may also allow the fall of body temperature and caloric intake, is associated with permanent changes in the release of CRH levels. These, in turn, lead to a disposition toward hyperresponsivity to stressful situations, depression, and anxiety. In contrast, after a period of separation, excess handling and contact can actually increase the amount of maternal behavior received by the pups and program the offspring toward adult resilience in the face of stress. *Adapted from Francis, D.D., Meaney, M.J., 1999. Maternal care and development of stress responses. Curr. Opin. Neurobiol. 9, 128–134.*

## Interim Summary 8.1

1. Early neonatal experience can shape one's perception of the world throughout adult life, particularly one's physiological and psychological responses to stressful life situations.
2. Both maternal separation and the maternal response to neonatal handling can predispose an individual to, or even directly result in, alterations in the HPA system.
3. Mouse pups that receive extra maternal care after separation from their mother show reduced plasma ACTH and corticosterone responses to acute stress, increased hippocampal GR messenger RNA expression, enhanced glucocorticoid feedback sensitivity, and decreased levels of hypothalamic CRH messenger RNA. There are additional effects on other stress-related neurotransmitters.
4. The specific type of behavioral/psychiatric disturbance depends on many factors, including one's genetic predisposition, nutrition, and other types of stress.

# SEX CHROMOSOMES AND HORMONE SENSITIVITY

Complete androgen insensitivity in XY males is relevant to the topics of several chapters in this text. An example of the disconnect between genotypic and phenotypic sex is complete androgen insensitivity in boys born with the XY sex chromosome complement. Androgen insensitivity results from a mutation in a gene on the X chromosome, and thus XY individuals are always androgen insensitive, i.e., they synthesize a nonfunctional androgen receptor protein. Regardless of the tissue, the lack of androgen receptors renders the tissue unresponsive to testosterone and/or dihydrotestosterone (DHT). Because the patients are XY, they develop normal testes capable of normal levels of testosterone and Mullerian regression factor secretion, and they therefore undergo normal fetal regression of Mullerian duct derivatives (fallopian tubes, uterus, and upper part of the vagina). However, because of the absence of androgen receptor function, all tissues that would normally respond to the action of testosterone and DHT remain unmasculinized, including the Wolffian ducts, which cannot develop into seminal vesicles and vas deferens without androgenic action. This demasculinization includes the internal and external genitalia, as well as body fat distribution and breast development.

Individuals with androgen insensitivity are typically born and raised as girls. These genetic males come to medical attention in infancy because of undescended, inguinal testes or at puberty because, as phenotypic girls, they do not develop menstrual cycles. As adults, they have a paucity of male-typical facial and axial hair, normal female breast development secondary to unopposed estrogen action, and generally tall stature. They are most always raised as girls and can have satisfactory vaginal intercourse without surgery; lacking ovaries, however, they are sterile. Fig. 8.3 illustrates an XY male with complete androgen insensitivity.

As far as can be discerned from interviews, patients with androgen insensitivity have female gender identity and sexual attraction toward males. Thus, it has been inferred that sensitivity to androgens is necessary for masculine gender identity, because without androgen sensitivity, gender identity is feminine. It is important to point out that if androgen action is necessary, it could be either a direct effect of androgens on the brain areas involved in gender identity or an indirect effect of androgens that occurs via masculinization of the genitalia, which is thought by some to influence the development of gender identity as the individual experiences life with a masculinized body. Androgen insensitivity syndrome is an example of a genetic change with far-reaching consequences on the ability of hormones to influence physiology and behavior.

Repeating earlier material in a different context, a related example comes from 5-alpha reductase deficiency in XY males. While testosterone has many androgenic effects throughout the body and is responsible for

**(A)**    **(B)**

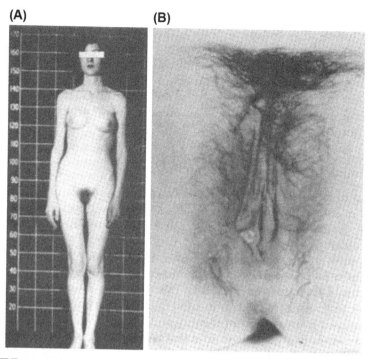

FIGURE 8.3   XY male with complete androgen insensitivity. The subject is tall, with normal breast development, a paucity of pubic hair, and female external genitalia. *From Rey, R.A., Josso, N., 2016. In: Jameson, J.L., DeGroot, L.J. (Eds.), Endocrinology, seventh ed. Elsevier, Saunders, Philadelphia, PA, p. 2108. With permission.*

virilization of internal structures (e.g., testicular development), DHT is required for virilization of the external genitalia (i.e., penile growth and scrotal development). If the enzyme that converts testosterone to DHT, 5-alpha reductase, is genetically deficient, the external genitalia of the newborn infant appear female or at least ambiguous (male pseudohermaphroditism). Fig. 4.2 (Chapter 4) schematically illustrates the genitalia of a normal man (left) and the genitalia of a 5-alpha-reductase-deficient prepubertal boy. The female-appearing external genitalia belie the presence of an XY sex chromosome complement, functioning testes (although undescended), and a masculinized brain secondary to fetal testosterone exposure.

Before it was recognized that this syndrome is inherited and therefore is concentrated in certain families, affected infants were raised as girls. Upon reaching puberty, however, the increased secretion of testosterone from the pubertal testes led to some androgens being produced that promoted development of a male-typical phallus, complete with erections, scrotal testes, male hair distribution, deepened voice, and male body habitus and psychological characteristics, to the initial consternation of parents and family members. However, the syndrome was recognized to occur in affected families in several parts of the world (e.g., in the Dominican Republic), so the appearance of ambiguous genitalia was no longer a mystery. Families were forewarned that the children might become more masculinized at the time of puberty (at about 12 years of age), and therefore they were psychologically prepared for the change in gender identity. The syndrome even received a nickname, "hueva doces," which translates to "eggs (testes) at 12."

## Resistance to Thyroid Hormone

Indeed, some individuals with this syndrome have entered into heterosexual relationships and have been able to function physically and emotionally as men. Others have led isolated lives or retained some female gender identity.

The outward switch of gender identity from female to male at puberty in 5-alpha-reductase-deficient individuals was initially interpreted as the primacy of nature over nurture; that is, it was interpreted to mean that sex hormones are more influential than the psychosocial signals that had imposed a female gender identity on these individuals since infancy. Another interpretation, however, is that environment is a major determinant of the transition from female to male gender identity in peripubertal individuals with 5-alpha-reductase deficiency. In societies where women enjoy less personal agency and power, it seems reasonable that peripubertal individuals offered a choice of future gender identities would choose the gender identity with more agency and power (in the Latin American countries, where this syndrome is more common, that gender identity is male). Despite the early recognition that newborns from affected families and with ambiguous genitalia might have masculine pubertal development, these infants have been raised either as boys or ambiguously as girls. The nature-versus-nurture dichotomy, therefore, is not as straightforward as some would believe. In general, the 5-alpha-reductase syndrome is a clear example of genetic factors that alter the action of hormones.

## PRECOCIOUS PUBERTY

Puberty, another critical period of sexual differentiation, is a process of sexual maturation that extends over several years and, in the United States, usually begins at 10.6 years of age in Caucasian girls, 8.9 years in African-American girls, and 11 years in boys of both races. A century ago, the average age of a girl's first period in the United States and Europe was about 16 years of age. In recent years, that age has dropped to 13 (Euling, *Pediatrics*, 2008). Precocious, or abnormally early, puberty occurs more often in girls and usually results in sexual maturation between ages 6 and 8, although it can occur earlier. Such children exhibit signs of physical sexual maturity and body awareness, but their emotional and heterosexual behavior more closely matches their chronological age. Onset of sexual activity may be earlier than average, but it usually occurs around the normal age for such activity. Because of the disparity between physical appearance and chronological and emotional age, such children are susceptible to social isolation and depression. Because they might appear to be older than their chronological age, they might receive unwanted sexual advances or find that trusted parental figures keep them at a distance. They are also subjected to cruel comments from peers and might gravitate toward older, more mature children. As their age cohort normally matures into adolescence, children with precocious puberty become less of an anomaly, and their social adjustment becomes less difficult. They may, however, remain of short stature because their long-bone growth has been prematurely halted by their accelerated hormonal development. There is a genetic predisposition for precocious puberty: in 15 families, four mutations in the *MKRN3* gene have been linked to precocious puberty. The *MKRN3* gene is responsible for coding a protein called makorin ring finger protein 3, which is thought to help tag other proteins for degradation. The genetic mutations result in truncated *MKRN3* proteins and disruption of *MKRN3* protein function. The symptoms of precocious puberty only manifest themselves when the father is the carrier of the mutated gene. *MKRN3* deficiency, due to a loss-of-function mutation, leads to the withdrawal of hypothalamic inhibition and prompts pulsatile gonadotropin-releasing hormone secretion, resulting in precocious puberty. Precocious puberty also has been linked to epigenetic effects related to endocrine-disrupting compounds in the environment, which mimic the action of endogenous hormone ligands and act as agonists or antagonists to endogenous hormone receptors.

## Interim Summary 8.2

1. Genetic mutations that alter particular proteins; e.g., steroid receptors or enzymes involved in steroid synthesis, can disrupt normal endocrine function.
2. A mutation in the gene that encodes the androgen receptor results in complete lack of masculinization in genetically male individuals (individuals with an XY chromosome complement).
3. A mutation in the gene that encodes the enzyme necessary for conversion of testosterone to DHT results in XY individuals with ambiguous genitalia, but pubertal increases in androgen synthesis usually results in the development of male genitalia as adults.

# RESISTANCE TO THYROID HORMONE

An important inherited hormonal syndrome is that of resistance to thyroid hormone (RTH). The hallmark of RTH is a significant decrease in tissue responsiveness to thyroxine. Refetoff et al. (Refetoff, 2000) pioneered both the elucidation of this syndrome and the discovery of its genetic mechanisms. The first two mutations they reported for the explanation of RTH were in the ligand-binding domain of the thyroid hormone receptor-beta (TR-beta) gene. Now, a large number of mutations has been identified, including some in the so-called "hinge" region of TR-beta adjacent to the ligand-binding domain (Fig. 8.4). On the one hand, John Baxter at the University of California, San Francisco have used X-ray crystallography

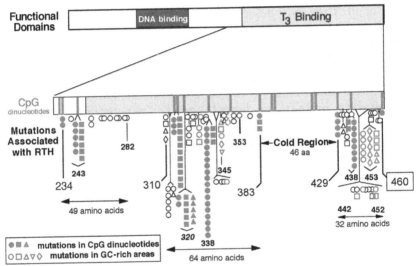

FIGURE 8.4  Familial lack of response to thyroid hormone has been associated with particular mutations and has obvious behavioral implications. Shown here is a rendition of the TR-gene and certain genetic alterations. (Top) Location of natural mutations in the TR-molecule associated with RTH; schematic representation of the TR-gene and its functional domains for interaction with TREs (DNA binding) and with hormone (T3 binding). (Bottom) The T3-binding domain and distal end of the hinge region that contain the three mutation clusters are expanded and show the positions of CpG dinucleotide mutational hot spots in the corresponding TR-gene. The locations of the 99 different mutations detected in 158 unrelated families are indicated by various symbols. Identical mutations in members of unrelated families are represented by the same color and pattern of vertically placed symbols. Cold regions are areas devoid of mutations associated with RTH. Amino acids are numbered consecutively starting at the amino terminus of the TR-1 molecule according to the consensus statement of the First International Workshop on RTH. TR-2 has 15 additional residues at the aminoterminus. *From Refetoff, S., 2000. Resistance to thyroid hormone. In: Braverman, L.E., Utiger, R.E. (Eds.), Werner & Ingbar's the Thyroid: A Fundamental and Clinical Text, eighth ed. Lippincott, Williams & Wilkins, Philadelphia, PA, pp. 1028–1043. With permission.*

to determine the three-dimensional structure of this region of TR-beta and then to explore the consequences of the clinically important mutations. The structural consequences of these ligand-binding domain mutations allow increased flexibility and disorganization of this region of TR-beta. Importantly, thyroid hormone binding is reduced, and release of corepressors is impaired. On the other hand, Refetoff (2000) has taken the lead in identifying patients with RTH who do not have mutations in either TR-alpha or TR-beta, the only two known thyroid hormone receptors. The search for the mechanisms of disease in these patients devolves upon the analysis of other, related nuclear proteins (see Chapter 18). All of these familial-resistance-to-thyroid-hormone syndromes have obvious implications for patients' hormonal controls over mental energy and mood, impacting both their intellectual and their social capacities.

## PHENYLKETONURIA, DOWN SYNDROME, HYPOTHYROIDISM, AND INTELLECTUAL DISABILITY

Phenylketonuria (PKU) is an autosomal recessive disease in which phenylalanine hydroxylase in the liver is absent, leading to a lack of conversion of phenylalanine to tyrosine, and thus to hyperphenylalaninemia and PKU. This is caused by mutations of the genes coding phenylalanine hydroxylase. Until recently, it was supposed that a lack of tyrosine production was the cause of intellectual disability; however, a new theory proposes that hyperphenylalaninemia specifically impairs the N-methyl-D-aspartate (NMDA) receptor (of glutamate) function. The NMDA receptor appears to play a role in the formation of neural networks during development.

Pharmacological blockade of NMDA receptors disrupts memory and learning in animals. In PKU, wherein L-phenylalanine levels in the blood reach 1200 mM versus 55–60 mM in normal subjects, intellectual disability and postnatal brain damage are characteristic. Electrophysiological studies on the hippocampus of the brain showed that L-phenylalanine selectively depresses glutamate receptor activation, indicating that it is the excess phenylalanine and not the lack of tyrosine that leads to intellectual disability (Glushakov et al., 2002). Children are normal at birth but steadily develop severe intellectual disability. The effects of PKU are primarily on the brain. The lack of glutamate action is believed to reduce the excitation of cells needed for normal synaptic development.

Another genetic example is Down syndrome, chromosome 21 trisomy, in which there is an extra chromosome 21 in addition to the two normally present. Children with Down syndrome develop intellectual disability, and hypothyroidism is very prevalent. Thyroid hormone acts powerfully on brain development, particularly to increase cell metabolism and the growth of neurite processes; presumably, the lack of adequate neuronal growth and metabolism contributes to the intellectual disability.

# OBSESSIVE–COMPULSIVE DISORDER

In the movie, *As Good As It Gets*, Jack Nicholson memorably portrayed an obsessive–compulsive man who unwrapped new bars of soap while washing his hands, avoided stepping on cracks, and repeated various rituals prior to performance of common, everyday habits. This is example of obsessive–compulsive disorder (OCD). OCD includes hoarding behavior, Tourette syndrome, compulsive grooming, and other behaviors that appear to have a both a hormonal basis and a genetic basis. In many cases, too little serotonin (5HT) is a factor. So, too, is dopamine, through its action on the D1 receptor in the brain. The exploration of specific brain mechanisms in humans is difficult; however, people displaying excessive handwashing and those diagnosed with trichotillomania (pulling out their hair) have parallels in the behavior of laboratory animals. Mice with a mutation in the homeobox (or Hox) gene groomed themselves (and their cage mates) with such compulsive aggressiveness that they developed bald patches and skin lesions. There seemed to be no secondary cause for this behavior, such as pain or irritation of the skin (Greer and Capecchi, 2002). Although the proteins and possible hormones associated with this ancient gene are not known, the extraordinary behavior is a potential example of how a gene mutation can illuminate a behavior of importance in psychiatry. There also are clear contributions of environmental stressors to OCD-like behaviors in animals, such as acral lick in dogs and stereotyped, repetitive pacing and other behaviors in caged animals.

## Mechanisms

Some of the fastest progress in analyzing detailed mechanisms of familial/genetic contributions to hormone responsiveness is coming from experiments delving into genetic determinations of simple animal behaviors. This is a massive field, extending well beyond the boundaries of laboratory science. The idea of genetically engineering animals is not new and dates back to domestication of animals by the ancient Greeks. For generations, cattle have been bred for docility, high body fat content, tender meat, and high dairy milk output. Dogs have been bred to achieve various desirable behavioral tendencies (trainability, herding, retrieving, fighting, and affiliation with humans) and to avoid undesirable tendencies. Undesirable traits also can result from too much inbreeding; that is, mating of close relatives can increase the probability of offspring that are homozygous for deleterious alleles. Other undesirable traits that emerge after selection can result from epistasis (when genes affect more than one trait, only one of which may be desirable) and from genetic correlations (genes occurring in linkage groups, resulting in traits being inherited in adaptive complexes that segregate together over generations). In some horse breeds, selection is based on strength, e.g., draft horses, whereas in others, selection is based on speed, e.g., racing thoroughbreds. Thoroughbreds are selected for speed that occurs in bursts over relatively short distances, whereas other long-distance racehorses are selected for both speed and endurance. The fastest thoroughbreds tend to have a very low bone mass and very high distal bone strength. Speed is correlated with bones with an exceptionally high strength-to-mass ratio. There is a fear, however, that thoroughbreds will be at risk for fatal fractures if the limits of strength are exceeded when bone mass falls to critically low levels. These examples illustrate the fact that changes in phenotypes through genetic engineering can be limited by genetic linkages. With such caveats in mind, artificial selection has produced a number of useful laboratory animals, such as those selected for diet-induced obesity and those selected for resistance to diet-induced obesity, i.e., the tendency to gain body weight or resist gaining body weight in response to exposure to high-calorie diets. By comparing obesity-prone and obesity-resistant lines, experimenters hope to unlock insights into the obesity epidemic and its reversal.

A more frequently used type of laboratory population is the mutant strain, in which members of the population carry a genetic anomaly at one allele. At the end of the 20th century, there was an explosion in the use of transgenic animals, most commonly mice, that were purposefully engineered with specific genes inserted into their genome. Transgenic lines are created by three basic methods: DNA microinjection, retrovirus-mediated gene transfer, and embryonic stem cell gene transfer. In the earliest of these techniques, DNA of a mouse is injected into the pronucleus of another mouse's fertilized egg, which is then allowed to develop to an appropriate embryonic stage and then implanted into the oviduct of another female. This female gestates the embryo to term and gives birth to transgenic offspring. The second method uses a retrovirus, a microorganism that carries its genetic material as RNA instead of DNA. The genetic sequence of interest is inserted into the retrovirus, and then the retrovirus is used to infect the host organism (often a mouse). In one of the early retroviral transfer experiments, a primate sequence of interest was inserted into a virus, which then was used as a vector to infect a mouse. The resulting mouse was a chimera, an individual with a mix of primate and mouse cells, some cells containing the primate sequence and other cells containing the original mouse sequence at the same locus. After many generations of inbreeding of chimeric animals, a transgenic line is created, in which all individuals of that line are homozygous for the inserted gene. In

II. HISTORY: HORMONE EFFECTS CAN DEPEND ON FAMILY

embryonic stem cell–mediated gene transfer, genes of interest are inserted into embryonic totipotent stem cells, i.e., cells at initial stages of development that retain the capacity to develop into any type of tissue in the body. As in the retrovirus-mediated gene transfer, this technique results in a chimeric individual that is used to breed transgenic lines.

The transgenic animals thus created can bear an altered gene sequence that renders the encoded protein nonfunctional. These are called transgenic knockout lines. Alternatively, the transgenic animals can carry a unique promoter sequence attached to a gene of interest that causes over- or underexpression of the gene of interest. Examples of the use of transgenic knockouts include estrogen receptor $\alpha$ (ER$\alpha$), ER$\beta$, and aromatase gene knockouts. Knockout of ER$\alpha$ receptor produces a female mouse that will not respond to estrogen plus progesterone treatment with increased courtship and locomotor behaviors or with normal female sex behavior. In male ER$\alpha$ receptor knockout mice, Sonoko Ogawa, at Rockefeller University, tried to elicit aggressive behaviors but was unable to do so, in spite of the fact that normal wild-type littermate controls were very aggressive in response to androgenic hormones. Other knockouts show that synthesis of estradiol by aromatization of testosterone is dependent on ER$\alpha$ receptors for masculinization of behavior, whereas binding of estradiol to ER$\beta$ receptors is important for defeminization of some behavioral traits. A functional ER$\alpha$ receptor is essential for normal fertility, body weight, and adiposity, and knockout of this receptor results in lack of pubertal development and obesity.

In studies of melanocortin receptors, melanocortin receptor 4 (MC4R) knockout animals develop obesity, whereas MC3R knockout animals were no different from wild-type with regard to body weight. Among rat studies carried out in a traditional neurobiological laboratory setting, the high attack latency and low attack latency rats produced by Leiden University's Ron deKloet et al. are of special interest for further neuroendocrine experimentation. Of course, in light of 21st-century functional genomics, studies with genetically altered mice swamp all of these examples. For example, the ability of Randy Nelson, now at Ohio State, to produce hyperaggressive mice by altering genes affecting nitric oxide provides the opportunity to obtain chemical and cellular detail with respect to this behavior pattern, which is so sensitive to steroid sex hormones (see previous discussion). Hypothalamic overexpression of vasopressin owing to a single nucleotide polymorphism in the promoter area of the vasopressin gene has been associated with high anxiety behavior in a rat model (Landgraf and Wigger, 2002). In these and many other examples with gene knockout and transgenic mice, a detailed, functional genomic understanding of familial influences on hormone/behavior relations is being realized.

An important limitation to these earlier transgenic studies is that the altered gene sequence is present in all cells throughout the brain and periphery. This makes it difficult to determine the mechanism of the gene product. In addition, when the gene is knocked out from an early embryonic stage, other developmental events can compensate for the loss of the gene, masking its potential importance in adult function. In the worst-case scenario, embryonic gene knockouts are lethal, and no live offspring survive for the experiment. Recently developed techniques overcome these limitations by permitting the knockout of the gene or knockdown of gene expression in particular cells of interest and in animals that are already developed. This is accomplished by more recent technology that allows cutting and pasting of the genome in adults. Thus, scientists can determine whether the encoded protein is necessary and sufficient in specific cells or tissues to produce acute behavioral changes. For example, with regard to the adipocyte protein leptin, mutations in the gene for leptin or the gene for the leptin receptor prevent pubertal development. Leptin receptors are present in many areas of the body and brain. To determine where the leptin receptors critical for puberty are located, experimenters use conditional knockout animals. One common method for creating conditional knockout animals is the Cre-lox recombination system. An enzyme that breaks DNA, known as the Cre recombinase protein, recognizes specific sequences of DNA, the loxP sites, to excise target pieces of the genome. When the Cre recombinase enzyme specifically recognizes two lox (loci of recombination) sites within DNA, genetic recombination occurs; that is, strands of DNA exchange information, causing a deletion or inversion of the genes between the two lox sites. An entire gene can be removed to inactivate it. A useful aspect of this system is that it is "inducible," i.e., the gene is not affected during baseline testing of behavior, but then a chemical can be added to the animal's food or water that induces the knockout just before experimental testing of behavior. The experimenter thus can compare the behavior before and after conditional knockout. Two of the most commonly used inducers are tetracycline, which activates transcription of the Cre recombinase gene, and tamoxifen, which activates transport of the Cre recombinase protein to the nucleus.

To determine where leptin receptor is needed for normal pubertal development, experimenters produce conditional knockout of the gene for the leptin receptor in the brains of adult rodents. In this way, it was determined that the leptin receptors critical for normal pubertal development are located in the brain. More sophisticated versions of this technique have been used to knock out the leptin receptor in the brain and then to selectively reactivate the leptin receptor in particular brain nuclei at any time point during development. In rodents that were otherwise null for leptin receptors, reexpression of leptin receptors solely in the ventral premammillary nucleus induced puberty and improved fertility without affecting body weight and adiposity (reviewed by Cavalcante et al., 2014). This result overturned the idea that critical leptin receptors reside in the arcuate nucleus or ventromedial nucleus of the hypothalamus and refocused attention to the ventral premammillary nucleus, which is an underappreciated neural hub for information coming into the hypothalamus from olfactory and hormonal inputs. This is an excellent example of experimenters' ability to dissect the genetic controls of aspects of behavior and physiology.

The genetic dissection of traits has been made commonplace by the use of RNA interference (RNAi) and by clustered regularly interspaced short palindromic repeats interference (CRISPR) and CRISPR-associated proteins (Cas), together known as the CRISPR/Cas system. These techniques take advantage of naturally occurring pieces of RNA that bind to newly synthesized RNA to block or downregulate gene expression. For DNA to result in the synthesis of its encoded protein, it must first be transcribed to RNA and then translated to an amino acid sequence. The process of RNA transcription can be prevented by the use of specific small bits of interfering RNA (siRNA). To use siRNA knockdown, double-stranded RNA is synthesized with a sequence complementary to a gene of interest and introduced into a cell or organism, where it is recognized as exogenous genetic material and activates the RNA interference pathway. Using this mechanism, researchers can cause a drastic decrease in the expression of a targeted gene. Studying the effects of this decrease can show the physiological and behavioral roles of the gene product. Since RNAi may not totally abolish expression of the gene, this technique is called knock*down*, rather than knockout. The CRISPR/Cas system makes gene editing simpler and less expensive than ever before and has vastly increased the number of knockdown experiments in behavioral endocrinology. By delivering the Cas9 nuclease complexed with a synthetic guide RNA into a cell, the cell's genome can be cut at a desired location, allowing existing genes to be removed and/or new ones added. Whereas traditional transgenic animals were created by sequential recombination in embryonic stem cells and/or time-consuming intercrossing of mice with a single mutation, the CRISPR/Cas system can be used for editing multiple genes simultaneously. Furthermore, the CRISPR/Cas system can be induced at will by using an external stimulus such as light or small molecules. We now can expect an explosion of new studies using gene editing in behavioral endocrinology.

Of considerable importance, gene editing in humans likely will soon be underway in clinical trials. What will be the technically feasible and ethically acceptable use of pharmacogenomics to predict the metabolism of hormones, the activation of particular hormone receptors, etc., for the purpose of ameliorating behavioral disorders? The traditional approach to both endocrine and behavioral studies is a linear one, tracing the causal relationship of a change in a single hormone or neurotransmitter to its effects on metabolism and behavior. In this current genomic/proteomic era, massive amounts of data on gene activation and suppression, transcriptional changes, protein formation and conformation, etc., are being produced. These mega-data sets will require shifts in conceptual approach to their analysis, including consideration of interlocking and interactive endocrine, neurotransmitter, and receptor networks that influence multifaceted behavioral changes. The clinical application of genomic informatics, via large datasets collected with gene-chip technologies, is beginning to foster the prediction of exogenous hormone and drug effects in individuals. Gene activation and suppression profiles following hormone and drug administration may someday be of predictive value as to who will respond therapeutically, what dosage schedule will be required, and who will have unacceptable side effects, based on each person's pharmacokinetic and pharmacodynamic handling of the hormone or drug in question. Pharmacogenomics will aid in dissecting the multiple molecular/metabolic steps that result in the final, physiological outcome.

A series of methodological issues, however, must be confronted before this clinical application can be accepted with confidence. These include achieving sufficient sensitivity, reliability, and reproducibility of large-scale genomic screens in individuals; ensuring consistency of methodologies across laboratories; developing mathematical algorithms for handling extremely large multivariate datasets, including setting thresholds for gene activation and suppression compared to reference standards; amalgamating these large datasets across laboratories into megasets that can be accessed and analyzed by investigators in many laboratories; and, of course, securing financial support for all these expensive activities.

## SOME OUTSTANDING QUESTIONS

1. How do we interpret contradictory results when strong evidence using pharmacological antagonists and agonists do not agree with the results from conditional gene knockout experiments?
2. Can we assume that concepts elucidated by using genetic dissection in laboratory strains of mice can be generalized to mice living in natural habitats or to other species, including human beings?
3. What are some of the remedial interventions for children who suffer neglect and abuse, and should we emphasize pharmacological treatments, a behavioral modification approach, or both?
4. To avoid the development of intellectual disability in a newborn, what should you look out for?
5. Low birth weight has a number of causes. What is the difference between low birth weight and low gestational weight? What do you think the consequences might be of one versus the other?
6. How do the hormonal consequences of differential early environments translate at the biochemical level into the neural differences that will affect adult behavior?

# Sex Differences Can Influence Behavioral Responses

In 2014 the US government passed legislation requiring all federally funded research to include both male and female laboratory animals, cells, and tissue. In 1993 similar legislation was passed requiring the inclusion of women and girls as subjects in all clinical trials and federally funded research on human subjects. The reasons could not be clearer: first, all vertebrate animals, tissues, and cells have a sex; and second, sex has a profound impact on function. The human relevance is unequivocal. Women differ from men, and girls differ from boys, in many clinically relevant traits. For example, there are significant sex differences in the incidence of many diseases. Women have a higher incidence than men of Alzheimer's disease and many autoimmune diseases. Men have a higher incidence than women in autism spectrum disorder. Men show a slightly higher incidence of schizophrenia than women, but women with schizophrenia tend to have a later age at onset, less severity of symptoms, a higher level of function, and different cognitive deficits and neural abnormalities than men. Other important sex differences in humans are in the main effects and side effects of many prescription drugs (e.g., antipsychotics and antidepressants). Furthermore, men and women have different thresholds of pain perception and differ in their responses to analgesics, pain medications, and anesthetics. Beyond the clinical world, sex differences are equally relevant in veterinary, farm, zoo, wildlife, and laboratory animal care. Prior to 2011 the vast majority of basic and biomedical research excluded female subjects (Beery and Zucker, 2011), and to correct this bias will take a concerted effort and injection of funding. In both animal experimentation and clinical medicine the impact of sex differences on hormone/behavior relations is so pervasive that only some of the best examples can be recounted here.

Sex differences are influenced by many biological and environmental factors, and in behavioral endocrinology, as in all scientific disciplines, it is important to work from agreed-upon terminology and definitions. The word *gender* is used in biology and many fields of academia to refer to societal prescriptions for masculine and feminine traits. Thus, by definition, *gender* refers to human subjects. Gender is flexible, population-specific, time-specific, and can be learned. Gender can be influenced by the ways in which we are treated by others and the ways in which we see ourselves, e.g., whether we have male or female external genitalia (penis and scrotum versus clitoris and vaginal labia), and whether our own genitalia match the genitalia of our mother or our father. Gender can also be influenced by the ways in which we are taught to behave. Anthropologists report that two genders exist in most cultures, but many cultures differ in their definition of gender. For example, in the Mwila tribe of Angola, femininity is expressed as a hairstyle that resembles a large octopus with long, thick tentacles. The hairstyle is made by mixing hair with mud, butter, and cow dung. Men in the same society do not wear this hairstyle, and their gender is expressed by a far less elaborate hairstyle. This gender difference is not found in most other societies, and is neither genetically programmed nor hormonally dictated. In Western societies of the 1900s–1950s period it was considered feminine to wear curly tresses, whereas it was considered masculine to have shaved hair (e.g., a "flat top" or "short back and sides"). In the same geographic area, but in more recent times (the 1960s), it was considered feminine to wear long straight hair with bangs, but it was also considered masculine for men to have similar hairstyles. Long-haired men in the 1960s did not result from mutations on the Y chromosome or a deficit in prenatal hormone exposure.

These changing male–female differences are part of the flexible gender prescriptions that are particular to different cultures in different time periods. Gender prescriptions, including ideas about appropriate dress, hairstyles, careers, and personality traits, are largely learned by imitation, instruction, and subtle or overt messages from parents, siblings, schools, the media, and the rest of society. The word *gender*, therefore, does not apply to nonhuman animals. There is a large and rich literature in sociology, anthropology, and psychology that explores the environmental influences on gender. For example, the Oxford English Dictionary cites the book *Sex, Gender, and Society* as the defining text for modern usage of the word *gender* (Oakley, 2015). The distinction between gender and biological sex is important, because each society has its own definitions of gendered behavior and there is often a wide assumption that sex and gender will match. However, an individual can have male biological sex but take on a wide array of gender roles, and individuals with a consistently male gender can be of intermediate biological sex (i.e., intersex).

In contrast to gender, *sex* is biological. The word applies to most animal species and has many components: chromosomal sex (e.g., whether a mammal has two X or X and Y chromosomes), gonadal sex (whether one has ovaries or testes), hormonal sex (whether the brain and body are exposed to high or low levels of gonadal steroids during early development), internal genital sex (whether an animal has seminal vesicles and vas deferens or fallopian tubes), genital sex (whether a mammal has a penis and scrotum or clitoris and vaginal labia), and behavioral sex. Behavioral sex is relevant to any species in which there are sex differences in behavior. Many behavioral sex differences are in reproductive behaviors, but these differences can affect other types of behaviors, including parental, agonistic, and cognitive. In any particular individual all these components of sex may be male, female, or mixed, so in one individual the sex chromosome complement might be XY male but hormonal sex may be female (e.g., no functional androgen receptors (ARs)). Behavioral sex also has many components.

In human subjects, sex and gender are difficult to untangle. Behavioral sex/gender differences appear in childhood play behavior (males tend to engage in significantly more rough-and-tumble play and prefer different toys), preferred gender roles (females tend to prefer the role of mother or princess, whereas males tend to prefer the role of father or king), and gender identity (whether one thinks of oneself as male or female). Sexual orientation can be seen as partially related to sex differences but also unrelated to them, and therefore there are four or more categories of sexual orientation. For example, individuals can have a female gender identity and heterosexual orientation, a female gender identity and homosexual orientation (lesbian), a male gender identity and heterosexual orientation, or a male gender identity and homosexual orientation (gay male). The genetic, hormonal, and physiological influences on homosexuality differ depending upon whether the subject is a lesbian (homosexual woman) or a gay male (homosexual man). Some lesbians classify themselves as "butch" if they have masculine traits or "fem" if they have feminine traits. Furthermore, within males or females these are not simple, binary categories, and thus different degrees of masculinized gender identity and sexual orientation are possible, and these are not fixed in time. Individuals change orientation, and some change their sex and/or gender identity with surgery and/or hormone treatment (e.g., this may be true of some transsexuals). Some individuals, for example, have a partially masculinized female gender identity and bisexual orientation, whereas others have a feminine gender identity and bisexual orientation. As is indicated by the term "gender identity," these human behaviors are influenced to an unknown degree by genes, hormones, and the environment. Nevertheless, the preponderance of evidence shows that they have a strong biological component. It has been challenging to tease apart these influences, and comprehensive treatments are found in books and review articles by prominent behavioral neuroendocrinologists (Balthazart, 2011a, b; Berenbaum and Meyer-Bahlburg, 2015; Hines et al., 2015).

II. HISTORY: HORMONE EFFECTS CAN DEPEND ON FAMILY

The biological mechanisms that underlie behavioral sex differences are only partially understood, but there is no doubt that one important influence is the steroid hormone milieu during early development. Most research has centered on the mechanisms by which androgens secreted from the testes during prenatal and neonatal development initiate processes that masculinize and defeminize the brain and behavior. It can be argued that we have made the most progress in studying sex differences in songbirds, in which there is a clear sex difference in the production of complex territorial/courtship songs. With regard to at least a few songbird species, we know that testosterone is aromatized to estradiol in specific brain areas involved in male-typical song. Estrogens masculinize aspects of these brain areas by their early developmental effects on neurogenesis and cell death. The masculinized brain areas comprise neural circuitry necessary and sufficient to produce male-typical songs. In addition, with regard to these birds we know that changes in the early development of these brain areas, via changes in cell death, neurogenesis, and altered cytoarchitecture, lead to sex differences in behavior (male-typical song). In contrast, for mammals we have little information about how masculinized or feminized brain structures produce behaviors: we have documented the masculinizing effects of early hormone action on some brain areas, and we have documented the effects of the same hormone action on some behaviors, but we have not documented the link between the specific brain areas and behavior. Thus for laboratory rodents we do not know how the sex differences in the brain produce the sex differences in behavior, but we have a great deal of knowledge about how hormones act during development to alter the brain.

# BASIC EXPERIMENTAL EXAMPLES

## Reproductive Behaviors

To summarize briefly the effects of early androgens and estrogens on the brain in laboratory rodents, these include differences in the size (volume) of identified nuclei that result from differential cell death, neural and glial genesis, dendritic branching, and synaptic patterning. Some of these brain nuclei include the sexually dimorphic region of the medial preoptic area, which has a significantly larger volume in males, and the anteroventral periventricular nucleus (AVPV), which has a significantly larger volume in females. The AVPV is important for the ability of gonadotropin-releasing hormone (GnRH) cells to produce a surge of GnRH in response to positive feedback from estradiol, but the AVPV has no direct effect on a sexually dimorphic behavior. In addition to these two areas, there are sex differences in the hormonal modulation of dendritic spine synapses in the hippocampal pyramidal neurons, but again we do not know how these sex differences relate to sex differences in adult behavior. Other examples in this chapter include sex differences in brain areas involved in food intake and the response to pain.

Most work on the mechanism of neonatal hormone action in rodents is concerned with sex differences in reproductive behavior. As discussed in other chapters, the adult sex difference that has been studied most often is the consummatory performance of copulatory behaviors in response to adult levels of gonadal hormones. Male rodents show significantly more male-typical sex behavior in response to adult treatment with androgens than do females treated with the same adult androgens. As noted in previous chapters, prenatal and neonatal androgens and estrogens initiate developmental events (organizational effects) that lead to the masculinization and defemininization of the neural substrates thought to influence sex behavior. Masculinizing effects of testosterone aromatized to estradiol in the brain lead to male adults that show mounting, intromission, and ejaculation in response to adult levels of androgens in the appropriate sexual context, i.e., the presence of a sexually receptive estrous (fertile) female. Defeminization by early androgen exposure leads to female adults that fail to show typical courtship behavior and lordosis in adulthood in response to high levels of estradiol and progesterone and the presence of a sexually experienced adult male conspecific. A large body of literature concerns the effects of early estradiol on the neurotransmitter systems in these different sexually dimorphic brain areas (medial preoptic area, AVPV, and the hippocampal pyramidal neurons).

Another interesting sexual dimorphism in the brain is in the medial preoptic area. This is not a consistent volume difference, but rather a difference in the synaptic architecture. You will recall that each neuron has an axon and one or more smaller processes, the dendrites, upon which synapses form. Dendrites have smaller processes known as spines, and the more spines per dendrite, the more synapses are possible on that dendrite. Dendritic spine density (number of spines per unit of dendrite) is twice as great in male as in female rodents. The male-typical density is stable throughout life and significantly correlated with aspects of male-typical sex behavior, although whether this spine density and stability are necessary and sufficient for masculinized sex behavior is unknown. There are sex differences in the density of peptidergic inputs from certain brain areas to the preoptic area, including galanin, substance P, and cholecystokinin (CCK). These inputs are organized by early estrogens and not changed by adult hormonal treatment. Furthermore, there are regional sex differences that might influence sex differences in behavior. Male rodents have more CCK-positive cells in the medial preoptic nucleus, whereas female rodents have more CCK-positive cells in the periventricular preoptic nucleus. Neurotransmitters are logical downstream mediators of early organizational effects of steroids, but there are likely to be many redundant peptide systems that are engaged by early androgens. When we remove any one peptide from the system, or upregulate the activity of any one receptor, other peptide systems are likely to show compensatory changes.

If neurotransmitters are not the primary targets of prenatal hormones, what are those targets? Prostaglandins are the primary target in at least one system, the sexually differentiated spine density of the medial preoptic area. The masculinization of synaptic patterning in this brain area of the rodent hypothalamus can be fully mimicked by neonatal intracerebral treatment with the prostaglandin E2 (PGE2). Prostaglandins are not hormones *per se*, but rather they are fatty-acid-derived molecules that are present in many different cell types, including endocrine and nonendocrine cells. Prostaglandins are well known for their association with the inflammatory response and fever. One of the main effects of nonsteroidal antiinflammatory analgesics is to inhibit prostaglandin action. PGE2 in particular is a membrane-derived, rapidly acting signaling molecule that acts on G-protein-coupled receptors. PGE2 in dendritic spines activates a classic signaling pathway that involves adenylate cyclase, production of cyclic adenosine monophosphate, activation of protein kinase A, and phosphorylation of the glutamate receptor 2 subunit of glutamate AMPA receptors. There is a link between AMPA receptors and masculine spine density. The movement of AMPA receptors to the membrane induces the formation and stabilization of spine synapses, which in turn leads to the masculinized medial preoptic area of the hypothalamus. It is astounding that injection of PGE2 into the brains of female (XX) rodent pups during the early critical period of neonatal development fully masculinizes the synaptic spine density and adult sex behavior! This neonatal treatment with PGE2 has no masculinizing effect on other sexually dimorphic brain areas. Thus PGE2 can mimic effects of estradiol on the organization of a very specific, sexually dimorphic characteristic of the brain (reviewed by McCarthy et al., 2015).

Another surprising aspect of this sexually dimorphic system in the medial preoptic area is that early treatment with PGE2 also masculinizes nonneural cells, including astrocytes and microglia. Astrocytes are a type of glial cell, while microglia are immune cells with multiple functions that include the secretion of prostaglandins, promotion of cell survival, and phagocytosis of dead cellular debris. The secretion of prostaglandins from microglia is necessary for the complete masculinization of dendritic spine density in the medial preoptic area by early estradiol treatment (reviewed by McCarthy et al., 2015). This is interesting, because many neuropsychiatric disorders that are gender biased in incidence and/or severity have been hypothesized to have altered immune function, e.g., autism, schizophrenia, and depression. Thus understanding how hormones and the immune system interact during development should facilitate an understanding of susceptibility to neurological disorders. The link between prenatal steroid action and the immune system and the role of this link in development and sexual differentiation represent a new frontier in behavioral neuroendocrinology.

It is important to understand that masculinization via estradiol and prostaglandins is only one of many possible mechanisms whereby different parts of the brain become sexually differentiated. For example, some aspects of the brain are masculinized via estradiol, whereas others are masculinized via testosterone. In some brain areas estradiol acts directly on mechanisms that affect dendritic branching, while in others the estradiol receptors upregulate transcription of genes for prostaglandins (which in turn can affect dendritic branching). These mechanisms are involved in masculinization of brain areas, but other mechanisms govern defeminization. In at least some mammalian systems androgens, estrogens, and progestagins masculinize behavior during one critical period and defeminize behavior during a different critical period (reviewed by McCarthy et al., 2015). With regard to some aspects of sexual differentiation, estradiol acts via classical estrogen receptor action (the estrogen receptor response element on the gene), and with regard to other aspects of sexual differentiation, estradiol acts rapidly via membrane receptors. Furthermore, some aspects of the neural architecture are sexually differentiated during puberty and adolescence via differential cell death, neurogenesis, and gliogenesis. As is examined later in this chapter, peripubertal architectural remodeling underlies sex differences in vulnerability to addiction and psychiatric disorders that emerge during puberty (reviewed by Juraska et al., 2013). Still other brain mechanisms are masculinized by protein products of genes on the X chromosome. Sexual differentiation arises according to X chromosome dose, i.e., whether the individual has one or two X chromosomes (Arnold and Burgoyne, 2004). The consequences of having multiple mechanisms for sexual differentiation is that there is no such thing as a masculinized or feminized brain; rather, each individual has a mosaic of neural characteristics that are likely masculinized or feminized to different degrees (reviewed by McCarthy et al., 2015). The following examples are other nonsexual behaviors that are organized by gonadal steroids during an early critical period.

## Parental Behaviors

An assumption across a wide range of studies of mammalian behavior is that females will display effective parental behaviors and males will not. In most species females show immediate maternal care when offspring are born, and the lack of experience or learning required for the full suite of behaviors suggests biological determination. Females that have never been pregnant or given birth, however, can learn maternal behavior, and even juvenile males and females, with repeated experience, can show high levels of maternal care. In some species, such as prairie voles, California mice, Mongolian gerbils, and Djungarian hamsters, males, like females, show immediate-onset parental care of their own offspring, including all behaviors shown by females except lactation. Males of these species have adult levels of circulating androgens that are no different from those of other species, and their parental behaviors persist in the face of exogenous androgen treatment. The neuroendocrine bases for paternal behavior in these species are not fully understood (Gonzalez-Mariscal, 2002). However, in most species, including laboratory rodents, there is a sex difference in immediate-onset care of offspring. This might be relevant to our own species, because in a large number of human societies, for social and economic reasons documented by Chodorow (1999), the larger burden of parental care has been assumed by the mother. For species that show sex differences in parental care, prenatal and neonatal hormones influence the development of the brain, particularly the preoptic area of the hypothalamus, which might explain the sex difference in adult levels of parental care. In species where immediate-onset paternal care is not shown, its lack is likely due to neonatal actions of androgenic hormones on the brain. Castration of the male rodent on the day after birth increases the likelihood of maternal-like behavior. Conversely, exposure of the genetic neonatal female to androgens reduces maternal-like behavior and favors infanticide of newborn offspring. Later, in adulthood, high levels of testosterone weigh against short-latency maternal-like behaviors in those neonatally androgenized individuals. Estrogens combined with progesterone and prolactin (Bridges, 2016) favor parental behaviors in individuals that were not exposed to prenatal or neonatal androgens.

## Sex Differences in Stress, Fear, and Agonistic Responses

Sex differences in stress and fear responses comprise an important part of animal behavior studies. There are also sex differences in the severity of specific symptoms in depressed human subjects. The incidence of depressive episodes is twofold to threefold greater in women; women tend to respond more robustly to stressful events than do men; and women exhibit more severe symptoms when depressed. These epidemiological findings parallel the sex difference in hypothalamo–pituitary–adrenal (HPA) axis function in response to stress, whereby females show a more robust increase in circulating cortisol than do males (Fig. 9.1). In laboratory animals peripheral administration of estradiol raises stress reactivity, as measured by the circulating stress hormone corticosterone, in females much more than in males. Of all the experimental groups, females with estradiol implanted directly into the medial preoptic area had the highest circulating corticosterone levels after stress (McCormick et al., 2002).

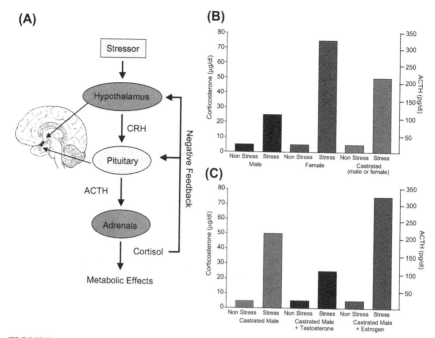

**FIGURE 9.1** The hypothalamo–pituitary–adrenal (HPA) axis is a well-described neuro-endocrine circuit that begins when a cluster of cells in the hypothalamus receives signals that convey stress-related information from the brain (A). Such cells, located in the paraventricular nucleus of the hypothalamus, produce corticotrophin-releasing hormone (CRH). The secretion of CRH into the hypothalamo–hypophyseal portal vasculature occurs in response to both physical and psychological stressors. As well, arginine vasopressin (AVP) is carried by neuronal axons in the paraventricular and supraoptic nuclei down the pituitary stalk to the posterior pituitary. In turn, CRH and AVP bind to specific receptors on pituitary corti-cotroph cells, which causes them to release adrenocorticotropic hormone (ACTH). ACTH is then transported through the general circulation to its target organ, the adrenal glands. The adrenal glands, located on top of the kidneys, respond to ACTH by increasing the secretion of the steroid hormone cortisol. (In rats the primary adrenal steroid is corticosterone.) The release of cortisol initiates a series of metabolic responses aimed at alleviating the harmful effects of stress. Additionally, a negative feedback mechanism is targeted to both the hypo-thalamus and the anterior pituitary as well as to other brain sites such as the hippocampus. This reduces the concentration of ACTH and cortisol in the blood once the state of stress subsides. A sex difference in the HPA axis response to stress has been well characterized in the rat model and is depicted graphically in (B). In general, rats (whether male, female, or castrated) have similar levels of stress hormones when in a nonstressed state, although reduction of baseline stress by environmental enrichment reduces corticosterone more in males than in females (Belz et al., 2003). Following stress, both male and female rats respond with a rapid increase in circulating ACTH and corticosterone. This response is much more pronounced in females than in males. Neutering of male or female rats extinguishes this difference, suggesting that the sex difference is at least partly due to circulating estrogen in females. Results from such a study are shown graphically in (C). Castrated male rats, devoid of circulating testicular hormones, are compared with castrated male rats treated with either testosterone or estrogen. Castrated males treated with testosterone and then stressed display a reduced ACTH and corticosterone response, typical of intact males. On the other hand, estrogen-treated males display a greater ACTH and corticosterone response to the stressor, more characteristic of females. It thus appears that testosterone can act to inhibit HPA func-tion, whereas estrogen can enhance HPA function. Understanding of such hormonal effects allows us to appreciate more fully how estrogen and androgen interact in the development of pathologies such as depression, potentially leading to improved preventive and/or phar-macological approaches to mental health (Rubin and Carroll, 2009).

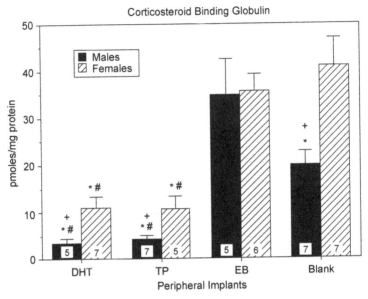

**FIGURE 9.2** Some of the sex differences in hormonal responses to stress could be due to circulating hormone binding proteins. McCormick et al. (2002) gonadectomized male and female rats and studied the effects of specific hormones. As expected, even without hormone treatment there was a sex difference (see Blank). Note also the sex differences following administration of dihydrotestosterone (DHT), a reduced testosterone metabolite, and testosterone propionate. *From McCormick, C.M., Linkroum, W., et al., 2002. Peripheral and central sex steroids have differential effects on the HPA axis of male and female rats. Stress 5 (4), 235–247. With permission.*

For neurobiologists, the simplest way to think about sex differences is usually to consider mechanisms in the forebrain or the anterior pituitary gland. In addition, there are sex differences in a protein in the blood that binds corticosterone, corticosteroid-binding globulin (CBG). CBG can regulate the amount of free, and therefore available, circulating corticosterone (Fig. 9.2). Such differences in blood proteins do not always bear directly on hormone/behavior relations, but they should always be considered.

The magnitude and direction of sex differences in response to stress may depend on the species studied and the type of stress imposed. Siberian hamsters form monogamous male/female bonds, so long-term separation from a mate is a form of social stress. During chronic separation male hamsters show behavioral changes reminiscent of depression, including elevated circulating concentrations of the stress hormone corticosterone. As you might expect, the effects are mediated by the HPA system, especially by CRH (Bosch et al., 2009). In contrast, females showed decreased circulating corticosterone concentrations. In rats, females have greater corticosterone responses to restraint stress than do males (Aloisi et al., 1994; Babb et al., 2013).

Differences between male and female animals in aggressive responses are widespread across mammalian species. Depending on the species, males are usually more aggressive than females when they encounter an unfamiliar male conspecific (although females of some species are more aggressive than males when defending the nest and offspring). In those species in which males show high levels of intermale aggression, castration reduces aggression, which is restored by treatment with androgens. In contrast, ovariectomized females do not increase levels of aggression when they receive the same adult treatment with testosterone. The male tendency to respond to adult treatment with testosterone by showing intermale aggression can be masculinized in XX females by treatment with early androgens, estrogens, or combinations of steroids. Some of the hormonal underpinnings of aggressive responses are illustrated in Chapter 3.

In some species females are very aggressive by nature, and these females are exposed to high levels of androgens during development. Female spotted hyenas are more aggressive and dominate over the males. The sex differences in aggressive behavior appear very early in juvenile development, with female cubs engaging in violent scratching, biting, and wrestling, compared to less intense aggression in male cubs. As adults, females fight and engage in ritual aggression that determines who eats first and who mates with the best males. In contrast to female rank, the male dominance rank is determined solely by the rank of the individual male's mother. Aggressive behavior in adult female hyenas is correlated with the prenatal levels of androgens during the latter half of gestation, and prenatal treatment with antiandrogens leads to female cubs that show reduced levels of fighting with siblings. In addition to marked aggression and dominance over males, the females possess highly masculinized genitalia that are indistinguishable from those of adult males. Females are born without a vaginal birth canal, but rather posses a phallus and urethra; as adults, female hyenas give birth to a fetus through the urethra, which is a laborious process associated with high neonatal mortality.

Both aggression and genital development are at least partially linked to exposure to high levels of androgens during fetal development. Some aspects of the female genitalia become masculinized prior to a surge in testosterone that occurs in both male and female fetuses, suggesting that some of the masculinization of female hyenas is independent of early hormone action. The aspects of the female genitals that make birth difficult are, however, determined by gonadal steroid action during a critical period of development. Prenatal treatment with antiandrogens alters the genitalia of female offspring; when these females grow up and become pregnant, they give birth to healthy offspring more easily than those not treated prenatally with antiandrogens. However, this same prenatal antiandrogen treatment alters penile development of the male (XY) siblings; when these males grow up, they develop a penis that is too short and has the wrong shape for insertion into the female's genital opening. Thus early gonadal steroid action is critical for healthy formation of the male genitalia to allow successful mating with females, and the same gonadal steroid action leads to female genital development associated with a difficult birth process for both mother and offspring. Even though females of this species are genitally masculinized and more aggressive than males, the masculinization occurs by a process similar to the process of masculinization in other mammals.

It is also interesting to note that the position of a developing XX female in the uterus (e.g., whether she is located between other females, between two males, or between one male and one female) can affect her anatomical development, her neuroendocrine status, and her eventual behavior. Female mice that have been positioned between two males *in utero* grow into adults that are less attractive to adult males, more aggressive than females generally, more likely to defend food sites, and show deficits in lordosis behavior and significantly more mounting of other adult female conspecifics. This set of effects is thought to be due to the females' prenatal exposure to androgenic steroid hormones from their neighbors (see also Chapter 10). These effects of the intrauterine environment are examples of effects that are not programmed by genes on the Y chromosome (and concomitant development of the testes with high levels of androgens).

## Sex Differences in Responses to Pain

As well as the sex differences in aggression and in responses to stress and fear, there are male/female differences in responses to pain. The clinical implications are obvious. Morphine is the most widely prescribed opiate for the alleviation of persistent pain, but it is significantly less potent in women than in men (reviewed by Loyd and Murphy, 2014). Mechanisms in the brain that suppress responses to pain are called antinociceptive responses. For example, males are much more sensitive to pain suppression exerted at the level of the midbrain by opioid peptides that work through μ-opioid receptors. These results in experimental animals have application in clinical medicine (see below).

A different set of pain control mechanisms contributes to male/female differences. Mogil et al. (2003) used a forced cold-water swim to cause a reduction in the stress-induced analgesia pain response. Females were much less sensitive to this analgesia being blocked by the neurotransmitter N-methyl-D-aspartate, implying that females had a special pain control system not possessed by the male experimental animals. The female-specific system is amplified by estrogens. This pain control mechanism in females appears to depend both on sex differences in k-opioid receptors and on a female-specific action of a different system, the melanocortin type 1 receptor. Thus not only do multiple neurochemical systems contribute to controlling the sensation of pain, but the sex differences in such responses also have multiple origins.

## Sex Differences in Ingestive Behavior, Body Weight, and Body Fat Distribution

For many behaviors, sex differences result from a two-step process. In step 1 androgens or their metabolites, acting during an early critical period, set developmental events in motion that organize neural circuits (organizational effects). These neural structures endure into adulthood. In step 2 these circuits respond to postpubertal levels of gonadal hormones (activational effects). For example, neonatally masculinized neural circuits that control intermale aggression in mice remain stable into adulthood and are activated by the postpubertal rise in testosterone. Masculinized circuits are activated by exogenous testosterone treatment in castrated male subjects, but not in ovariectomized females or in neonatally castrated males. Similarly, this two-part process underlies the development of sex differences in food intake and body weight. Just as male-typical sex behavior is more prevalent in male than in female rats, body weight and food intake are greater in male rats. In rats, male-typical sex behavior, food intake, and body weight are all decreased by adult castration, and the effect is reversed by testosterone replacement. Furthermore, adult gonadectomy significantly decreases the incidence of sex behavior and increases body weight and food intake in females, and has opposite effects in males. Analogous to male-typical sex behavior, adult food intake is more responsive to the stimulatory effects of adult testosterone treatment in males and neonatally androgenized females than in nonandrogenized females. As well, adult food intake is less responsive to the suppressive effects of estradiol in males and neonatally androgenized females than in nonandrogenized females (reviewed by Schneider et al., 2014).

Male mice have a significantly higher daily food intake than female mice throughout life. In addition, they show higher levels of postfast hyperphagia: after a period of food deprivation, when they are given food their daily food intake is higher than before the start of food deprivation. Males are also more sensitive to the inhibitory effects of the adipocyte protein, leptin, on food intake. These sex differences in ingestive behavior have been linked to sexual dimorphisms in the arcuate nucleus of the hypothalamus (ARH) in cells that contain neuropeptide Y, agouti-related protein, and proopiomelanocortin (POMC). Male rats show significantly lower POMC gene expression and neuronal projections from POMC cells to the ARH compared to females. Perinatal androgens masculinize the sexually dimorphic circuits of the ARH. Daily food intake,

| | Altered behaviors of Estrogen Receptor Knockout (ERKO) mice | |
| --- | --- | --- |
| | In the female | In the male |
| Aggressive behaviors | ↑ | ↓ |
| Sex behaviors | ↓ | ↓ |

FIGURE 9.3    Two points can be made from the analysis of mice whose classical estrogen receptor (ER) gene has been knocked out (ERKO). (1) The effect of an individual gene on a specific behavior can depend on the gender in which the gene is being expressed. ERKO females are more aggressive than wild-type female littermate controls, whereas ERKO males are less aggressive than male controls. (2) Normal sex behaviors in both genders depend on the integrity of the ER gene (see Ogawa et al., 1996, 1998a,b).

postfast hyperphagia, and POMC gene expression and projections are fully masculinized in females treated with either testosterone or dihydrotestosterone (DHT), both of which act via ARs. This is consistent with the organizational hypothesis: the idea that testosterone masculinizes the energy-balancing system during early development via action on ARs in the ARH (Nohara et al., 2011; reviewed by Schneider et al., 2014).

## Genes and Sex Differences in Behavior

The contributions of specific genes to sex differences in behavior are being explored. A dramatic example is the phenotypes of female and male mice in which the gene for the classical estrogen receptor has been knocked out (ERKO) (Fig. 9.3). With respect to certain social behaviors, the sex roles of these animals were reversed. Females were very aggressive; they behaved like males in agonistic encounters, showing vicious offensive attacks with biting; and they were treated like males by other males. In contrast, the ERKO males were not aggressive. These phenotypes resulted from changes in the brain, not just in abnormal levels of hormones circulating during behavioral tests. This was shown by surgically removing the gonads and gaining experimental control over hormone levels, after which one could still observe the reversals in sex roles.

## Mechanisms

Some of the differences in hormone/behavior relations between males and females may be due to differences in sex hormone effects on gene expression

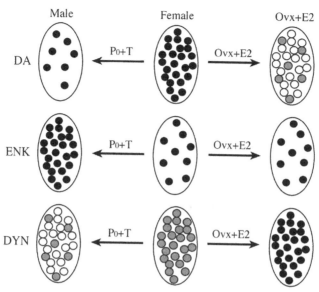

**FIGURE 9.4** Sex steroid hormones direct differentiation of hypothalamic neurons. Schematic diagram to illustrate the influence of sex steroid hormones on dopamine (DA), dynorphin (DYN), and enkephalin (ENK) peptide-containing neurons in the AVPV nucleus of the hypothalamus, a sexually dimorphic nucleus that plays a critical role in controlling gonadotropin secretion and ovulation. Treatment of newborn ($P_0$) female rats with testosterone (T) results in a male pattern of neurotransmitter expression (left column). Treatment of ovariectomized adult females (Ovx) with estrogen (E2) increases cellular levels of DYN, decreases DA, and does not affect ENK (DeVries and Simerly, 2002).

in specific brain regions. The molecular neurobiological data demonstrating sex hormone effects on opioid peptide genes only in the female, for example, could obviously contribute to the determination of female sex behavior not shared by male rodents; however, other molecular findings could bear on a wider range of social behaviors and even on sex differences in neuronal controls over hormone secretions from the anterior pituitary gland (Fig. 9.4). Sex steroid hormones also direct development of neural inputs to the AVPV from sexually dimorphic parts of the limbic region of the forebrain, and regulate gene expression within these same pathways in adults (Simerly, 1990; 2002). Such divergent patterns of neuronal development and gene expression illustrate the cell-type specificity that is characteristic of hormonal regulation of neuroendocrine circuits and provide a glimpse of the complexity of neurobiological events underlying physiological differences between males and females. The precise cellular and molecular mechanisms utilized to specify these hormone-dependent cellular phenotypes remain to be discovered (Simerly, 1990, 2002; DeVries and Simerly, 2002).

So far we have talked about sexual differentiation in the brain and behavior as though it was due just to testosterone. But Arnold and

Burgoyne (2004) had a new thought. Consider this: the Y chromosome houses the sex-determining region Y (SRY) gene that leads to testes formation, but it also has genes whose expression is quite independent of the SRY gene; likewise, the X chromosomes have lots of genes important for brain development, and are not simply to be thought of as "not SRY." Hence Arnold and Burgoyne's question: could the sex chromosomes, Y and X, contribute directly to sexual differentiation of the brain *independent* of their roles related to the testes and testosterone? To answer that question, Arnold and Burgoyne came up with the "four core genome" sets of animals: the regular XY male; the regular XX female; an XX animal with the SRY gene inserted to chromosome #3; and an XY animal with the SRY gene deleted. Using these animals, they came up with a clear answer: The presence of particular sex chromosomes X or Y, *independent of testes formation*, can influence brain sexual differentiation. For example, XY females (that is, mice with a Y chromosome lacking an SRY gene) attacked more than XX females, suggesting that the sex chromosome complement has an effect, at least when ovaries are present. Using this model, sex differences not mediated by early gonadal hormones have been discovered in behaviors including aggression, parenting, habit formation, nociception (pain), and social interactions. In addition, there are differences due to the number of X chromosomes in gene expression of vasopressin (AVP) and in susceptibility to disease, including neural tube closure and autoimmune disease. The four core genotypes can unmask effects of genes that are sometimes obfuscated by the effects of early hormones.

## CLINICAL EXAMPLES

Throughout the previous discussion, the term *gender differences* has not been used. This is because to appreciate the full range of biological and behavioral differences between men and women and their social manifestations, different terms must be employed to keep things straight (reviewed by Oakley, 2015). Consider that the neuroscientist is dealing with chromosomal differences (XY versus XX); epigenetic differences in the chromatin; hormonal differences; consequent differences in organ and tissue structure and function; differences in frank sexual behaviors; differences in mate choice; differences in psychological gender self-identification; and differences in social roles. The most elaborate expressions of all these variations between XY and XX individuals can be observed in the behaviors of human beings.

In this and other chapters we refer to the *organizational* versus *activational* effects of hormones. Organizational effects are those that are exerted early in life, during brain development, which are permanent, and which affect later responses to hormones and other stimuli in adulthood. Activational effects are exerted postnatally and facilitate or repress specific classes of behaviors in animals and humans. As a result, as noted above, genetic sex and social gender roles are usually consonant, but in modern human societies various combinations are possible.

As mentioned, early exposure to androgens (*in utero* in humans; *in utero* and shortly after birth in laboratory animals) can result in a "masculinized" brain in phenotypic females. In contrast, early in development biochemical events contravening normal androgenic hormonal actions can result in a "feminized" brain in a phenotypic male. The effects of an altered prenatal androgen milieu can be observed when there are mutations that alter steroid metabolism. The resulting gender identity and sexual orientation have been interpreted differently by biologists and psychologists. One example in humans is androgen insensitivity resulting from a mutation in the gene that encodes ARs. Children with an XY chromosome complement are born with feminized external genitalia, often indistinguishable from those of XX females. They are most often raised as girls and as adults report a female gender identity and heterosexual orientation, with an incidence no different from that of the general population. This can be seen as support for the organizational hypothesis, which states that in the absence of prenatal androgen action the female phenotype will result. Others have considered this as support for a psychological development hypothesis, which states that the female phenotype results from the reaction of the child and its family to having female genitalia. In another example, an XY individual is born with female-typical genitalia due to a deficiency of 5-alpha reductase, which converts testosterone to DHT. The lack of DHT results in failure to develop male genitalia, so genetic males may be raised as females, but they readily change to males when their phallus develops at the time of pubertal androgen secretion (Gooren and Byne, 2009) (see also Chapters 9 and 10). Again, psychologists note that the individual gender identity forms in response to the reaction to the external genitalia, and thus the child identifies at first as a girl and then later as a man. By contrast, the biologist notes that the eventual adult gender identity is likely a result of the prenatal hormone milieu, which has presumably prepared the brain for the increase in androgens that occur at puberty. Another social factor is that in families predisposed to 5-alpha reductase deficiency, ambiguous (i.e., not entirely female) genitalia are recognized early, and the child is not forced into a female gender identity, because pubertal changes resulting from increased androgen secretion are expected. These complications make it difficult to unravel prenatal from environmental effects in clinical case studies. Despite biochemical versus behavioral influences having been portrayed historically as opposing camps in medical science, they are not mutually exclusive and interact throughout the lifespan.

II. HISTORY: HORMONE EFFECTS CAN DEPEND ON FAMILY

Future research in this area will be aided by the discovery of sexually differentiated traits that can serve as a proxy for early androgen exposure. For example, the ratio of the length of the first to the fourth fingers (2D:4D digit ratio) differs between men and women, is masculinized by high levels of prenatal androgens, is masculinized in lesbians, and is more masculinized in "butch" compared to "fem" lesbians. The 2D:4D digit ratio is also masculinized in XX women who have congenital adrenal hyperplasia with early fetal androgen exposure. It is interesting that the 2D:4D ratio is completely masculinized in gay males (their 2D:4D digit ratio does not differ from heterosexual males), suggesting that biological mechanisms that influence male homosexuality might be independent of early androgen exposure. Thus the mechanisms that result in female homosexuality might differ from the mechanisms that lead to male homosexuality. The sex difference in the digit ratio is present soon after birth in humans, nonhuman primates, and laboratory rodents.

Similarly, there is a sex difference in click-evoked otoacoustic emissions, with the strength of these emissions significantly greater in female compared to male humans and nonhuman primates. Otoacoustic emissions are sounds that come from the sensory cells for hearing (the cochlear hair cells) and propagate outward to the inner ear, where they can be recorded by audiologists. It is extremely unlikely that sex differences in these emissions are learned or affected by the postnatal environment, because the sex difference is present at birth. The strength of otoacoustic emissions is decreased in males and in individuals with prenatal exposure to high levels of androgens. These sex differences in otoacoustic emissions and digit ratio do not cause sex differences in physiology and behavior, but they can be used to assess the extent of prenatal androgen exposure retrospectively. They are currently used to study the origins of sex differences in eating disorders and body fat distribution, and they emphasize that sex differences involve not just the brain but the entire body (reviewed by de Vries and Forger, 2015; Breedlove, 2010).

As mentioned above, the incidence of major depression is twice as great in adult women as in adult men. Prior to puberty the incidence is about the same in girls and boys, and it is not until puberty and adulthood that women have a twofold greater risk of this illness. This implies a hormonal factor that increases risk among sexually mature women. Understanding sex differences in central nervous system (CNS) neurotransmitter function is providing information relevant to major depression, because antidepressant treatments alter the dynamics of several neurotransmitters, including serotonin, norepinephrine, and acetylcholine. Serotonergic function is known to be affected by sex hormones, and cholinergic function also may show a sex difference. For example, Fig. 9.5 shows plasma ACTH responses to cholinergic stimulation by low-dose physostigmine

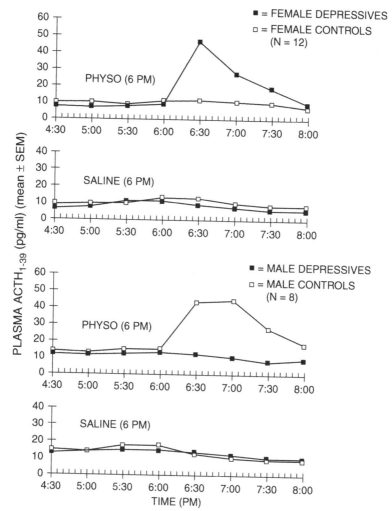

**FIGURE 9.5** Plasma adrenocorticotropic hormone (ACTH) profiles before and after saline and physostigmine (PHYSO 8 μg/kg) administration at 6 p.m. in 12 premenopausal women and 8 men with major depression and in 12 premenopausal female and 8 male normal control subjects. *Adapted from Rubin, R.T., O'Toole, S.M., Rhodes, M.E., Sekula, L.K., Czambel, R.K., 1999. Hypothalamo-pituitary-adrenal cortical responses to low-dose physostigmine and arginine vasopressin administration: sex differences between major depressives and matched control subjects. Psychiatry Res. 89, 1–20.*

(PHYSO), a cholinesterase inhibitor that blocks the hydrolysis of acetylcholine, in 12 women and 8 men of similar age with major depression, and in 12 female and 8 male control subjects individually matched to the patients (Rubin et al., 1999).

Normal men had a greater ACTH response to PHYSO than did normal women, whereas depressed women had a greater ACTH response than did depressed men—a significant sex by diagnosis interaction. These findings suggest that estrogen may have augmented already hypersensitive CNS cholinergic systems in the depressed women, which may bear some relationship to the increased incidence of this illness in women. Estrogen enhances CNS cholinergic neurotransmission, and sex differences have been reported for virtually all cholinergic markers. The net effect of estrogen on cholinergic stimulation of the HPA axis is likely a result of a complex set of factors, including estrogen's influence on specific steps in acetylcholine synthesis and metabolism, the role of sexually dimorphic areas of the brain that result from the organizational and activational effects of estrogen and testosterone (as noted above and discussed in other chapters), and, in the depressed women, the overlay of a psychiatric illness that itself may involve abnormalities of CNS cholinergic neurotransmission. Gaining an understanding of these factors is an important task for future research.

Earlier we briefly summarized differences between men and women in the power of $\mu$-opioid agonists to reduce pain. With modern scanning techniques applied to the human brain, Zubieta et al. (2001) compared men and women with respect to $\mu$-opioid receptor dynamics. Men showed significantly larger magnitudes of $\mu$-opioid receptor activation in response to sustained pain than did women in the thalamus, basal forebrain, and amygdala. In contrast, women had reductions of $\mu$-opioid activation during pain in the nucleus accumbens, a forebrain cell group closely connected to both negative and positive reward. These differences in magnitude and direction of $\mu$-opioid responses in specific brain regions could contribute to differences in pain experience between men and women.

From this type of scientific work, social and environmental factors that influence sex differences, as well as chromosomal determinants of differences, can be used to inform several areas of medicine, notably the growing field of women's health.

## OUTSTANDING CLINICAL AND BASIC SCIENTIFIC QUESTIONS

1. Given that in neuroscience male studies have outnumbered female studies by a factor of 5.1:1 (Beery and Zucker, 2011), should research funding expand to correct this bias?
2. Given that the immune system is intimately involved in sexual differentiation, what are the roles of inflammation and disease in sexual differentiation?

3. What are the most functionally significant sex differences outside the hypothalamus and limbic system? For example, are there sexual dimorphisms in the lower brainstem circuitry that mediates arousal?

4. Regarding the most sophisticated differences in gender role behaviors in humans, to what extent do they depend on the individual's appreciation of biological states and changes of state in his or her body? By analogy to the James–Lange theory of emotion, a person may reach an emotional state because of the brain's recognition of changes in the viscera. Is there an equivalent situation with sex differences in behavior among humans?

II. HISTORY: HORMONE EFFECTS CAN DEPEND ON FAMILY

# Hormone Actions Early in Development Can Influence Hormone Responsiveness in the CNS During Adulthood

Near the end of World War II, the German occupation of Holland (now the Netherlands) resulted in a devastating famine known as the Dutch Hongerwinter (hunger winter). A German blockade prevented the import of food and fuel from the farmlands to a densely populated area. The result was a severe food shortage during a very cold winter. Each of the estimated 4.5 million residents was forced to subsist on about 400–800 calories per day for over a year (compared to the normal average daily intake of 2000–3600 calories per day consumed by moderately active adults). An estimated 22,000 people died of starvation and exposure, but many survived, including pregnant women. What hormonal changes were these residents and their offspring exposed to?

The offspring of the mothers who were exposed to food shortages became a famous and unusual study population because, after they matured to adulthood, these offspring developed many strange characteristics. They showed a significantly higher incidence of atherosclerosis, a high incidence of altered blood clotting, and a threefold increase in cardiovascular disease. Daughters of mothers pregnant during the famine grew up to have significantly more body fat accumulated in the abdominal-visceral region (around the stomach and intestine). The sons of mothers pregnant during the famine had higher rates of obesity, schizophrenia, and an exaggerated response to stress. The effects of having a famine-exposed mother varied according to sex and the phase of gestation during which the mothers were exposed to food shortages (e.g., Ravelli et al., 1976).

Since that time, many other studies have examined the offspring of undernourished mothers. Depending on the phase of gestation and the level and

211

type of nutritional deprivation, the offspring of deprived mothers suffer from an increased incidence of obesity and its metabolic sequelae. Even more surprising, the daughters tend to develop into women who give birth to underweight infants, who in turn grow up to have an increased incidence of obesity and its metabolic consequences. These studies set the stage for a line of research into the effects of fetal experience on adult physiology and behavior. The field of research that stems from these early studies is variously known as "fetal programming," "maternal programming," or "gestational programming." It is related to a larger field of research known as "epigenetics," loosely defined as transgenerational changes in individuals caused by modification of gene expression (epigenetics is explained in a bit more detail in Chapter 11).

As Chapter 9 illustrates, the brains of females are both similar to and different from those of males. The factors that determine sex interact with the factors that program body weight during fetal development, as hinted by the Dutch Hongerwinter studies and demonstrated in the first laboratory animal studies to address this issue. In pregnant rodents that are food-restricted during the first 2 weeks of gestation and then allowed to eat ad libitum after birth, obesity occurs in adult sons, but not adult daughters, of food-restricted dams. As explained in earlier chapters, the presence of pre- or neonatal androgens masculinize/defeminize, and the absence of the same hormones feminize/demasculinize, during early critical periods of brain development. In this chapter, we note that the presence or absence of androgens may interact with other environmental factors that program adult traits. These other environmental factors extend beyond the energetic status of the mother (food availability) to include many other stressors. Was the pregnant mother subjected to predation or harsh ambient temperatures? Did the pregnant mother survive a war or natural disaster? Did the pregnant mother live in a dangerous neighborhood, an abusive marriage, or was she a caregiver to aging parents while working and raising many other offspring? Was she exposed to bacterial, viral, or fungal infections? Did the mother suffer from depression or diabetes, and did she take prescription or nonprescription drugs during pregnancy? Was she forced to survive on polluted water or exposed to pesticides and/or herbicides? This chapter examines the fetal hormonal milieu and how it might interact with the experiences of the mother, her general health and level of nutrition, her ingestion of food and drugs, and her exposure to chemical pollutants.

## BASIC EXPERIMENTAL EXAMPLES

### Prenatal and Neonatal Nutritional Stress

Based on the study of children born during the Dutch Hongerwinter and using laboratory rats as a model system, Jones et al. (1982, 1984) tested the hypothesis that food restriction of the pregnant mother programs an obese phenotype in the adult offspring. Pregnant rats underfed in the first 2 weeks of gestation had male offspring that showed abnormally rapid gains in body weight beginning at 5 weeks of age. They developed obesity in adulthood, and at the time of death, they had significantly larger adipose tissue pads and significantly larger fat cells. The obesity was exaggerated in males fed a calorically dense diet (such as laboratory chow supplemented with sucrose and dietary fat). The females did not show the same effects on body weight and adiposity, but they showed alterations in specific aspects of their adipocytes (body fat cells). Perhaps the most striking result was that diet-induced obesity and increased body fat content occurred in males without increased caloric intake. The ability to store more fat (by the biochemical process of adipogenesis) without eating more calories implies that the animals were able to conserve available metabolic fuels by lowering their energy expenditure and/or rate of metabolism. That fact that these obesity-prone individuals appear to do more with less (i.e., more adipogenesis with lower food intake) is implied by the word *thrifty* in the name used to describe this syndrome, "the thrifty phenotype." In obesity research, the thrifty phenotype is characterized by a high rate of body weight gain despite little or no increase in food intake. In other words, there is some truth to the claim that some of us gain weight more easily than others. For some obesity-prone individuals, the predisposition to gain weight can be traced to the experience of the individual's mother or grandmother.

After the work of Jones et al. (1982, 1984), many investigators joined the study of maternal diet and its effects on energy metabolism. In particular, a physician by the name of David Barker studied effects of early environment on adult diseases and popularized a general theory that become known as "the fetal origins hypothesis" or "The Barker Hypothesis". The Barker Hypothesis posits that many chronic adult diseases result from a fetal environment characterized by deprivation followed by a postnatal environment characterized by abundance. In other words, in a deprived environment, the offspring are programmed to be thrifty, i.e., to be able to make the most out of whatever energy becomes available. The strategy of the thrifty phenotype is to conserve energy by low metabolism and/or sedentary behavior while eating and storing as much body fat as is possible. The opposite of the thrifty phenotype is the wasteful phenotype, seen in individuals with a high metabolic rate and/or high level of energy expenditure who are predisposed to high levels of heat production. Heat production, exercise, and nonexercise activity thermogenesis (NEAT) use up the available fuels that might otherwise be turned into the storage molecules that make up body fat cells. NEAT occurs when organisms make repetitive movements (e.g., twitching, jerking, ankle shaking, fidgeting, or rocking to and fro) that are not part of a conscious effort toward locomotion or training, and these involuntary movements make up a substantial portion of total energy expenditure. An extreme example of a wasteful phenotype is found in the very lean person who typically has a relatively high body temperature and appears to be able remain lean no matter how much he or she eats. When a mother gestates her offspring in an energetically challenging environment (such as the Dutch Hongerwinter) and then gives birth to the same offspring in an environment of abundance, the mismatch of environments is thought to predispose the individual toward the thrifty phenotype, i.e., excess eating, low metabolic rate, low levels of energy expenditure, body fat accumulation, and many adverse metabolic consequences. The idea is based on observations of men and women whose mothers experience various adversities during pregnancy. The effects are often transgenerational, that is, they are apparent in the children, grandchildren, and great-grandchildren of mothers who experienced adversity during pregnancy. The effects can persist for many subsequent generations. It is clear that gestational programming depends upon a wide array of maternal experiences, including the mother's degree of control over her environment, her diet, exercise requirements, lactational demands, exposure to changes in ambient temperature, the light–dark cycle, injury, drug use, exposure to pathogens, and many other environmental stressors.

Some patterns have emerged from the study of laboratory and farm animals. In rats, mice, and sheep, prenatal calorie or protein restriction often, but not always, leads to low birth weight, usually followed by a period of rapid catch-up growth and body fat deposition, followed by adult obesity in one or both sexes. In some species and in some types of gestational restriction, offspring develop obesity even though they were not born with a low birth weight. Prenatal undernutrition during some, but not all, stages of pregnancy leads to newborns with a lower-than-average birth weight. If offspring born to undernourished mothers are allowed to eat as much as they want, they tend to gain weight rapidly during lactation. These effects are similar to what is seen in human babies. For example, catch-up growth of babies from underfed mothers is exaggerated in formula-fed compared to breast-fed babies, and both groups from undernourished mothers are prone to develop metabolic syndrome, a loosely defined syndrome that includes insulin resistance (decreased cellular response to insulin leading to elevated blood glucose concentrations, i.e., hyperglycemia), hyperinsulinemia, hypertension (abnormally high blood pressure), atherosclerosis (hardening and narrowing of the arteries), high levels of low-density lipoproteins, high levels of abdominal, visceral adipose tissue, and obesity as adults. Metabolic syndrome puts one at a higher risk for coronary heart disease.

An interesting question is whether the development of offspring obesity is more likely when an undernourished, impoverished mother fosters her offspring to a mother living in relative affluence. The mechanisms might include changes in energy storage and/or expenditure, and these effects might be mediated by changes in the hormones that control body fat storage and oxidation. For example, most offspring are able to use both fat and carbohydrate fuels for energy by oxidation of free fatty acids or glucose. The oxidation of free fatty acids requires triglycerides in lipids to be broken down to fatty acids and glycerol, in a process known as lipolysis. Lipolysis is stimulated by the secretion and increased activity of an enzyme known as hormone-sensitive lipase. Obese offspring of nutritionally deprived mothers (humans and rats) show deficits in the oxidation of free fatty acids and a predisposition toward utilization of glucose. In addition, obese offspring of undernourished mothers show deficits in circulating levels of insulin-like growth factor 1 (IGF-1), a hormone that promotes lipolysis by increasing the activity of hormone-sensitive lipase. Currently, this is an active, fast-moving field of research.

Conversely, some individuals are programmed by overnutrition of the mother. In some model systems, the offspring of overnourished, obese mothers are prone to develop metabolic syndrome and cardiovascular disease later in life. In nonhuman primates such as the Japanese macaque, mothers fed a high-fat diet during gestation become obese and give birth to offspring that develop metabolic syndrome as adults. The effects on the adult offspring include altered expression of genes that encode putative orexigenic and anorectic peptides, increased secretion of cortisol, and increased proinflammatory signals. Furthermore, as adults, the offspring develop preferences for high-fat diets, and the obesity of offspring of overnourished mothers can influence subsequent generations, leading to obesity in the grandchildren and great-grandchildren (reviewed by Schneider et al., 2014).

The mechanisms of maternal programming are unknown, but they include effects on the utilization of metabolic fuels, as well as effects on brain mechanisms that control food intake. Some experimental results have implicated altered hormone secretion during an early critical period that might change the neural circuitry of the arcuate nucleus of the hypothalamus. This can be observed in a line of obese mice that are homozygous for a mutated gene that encodes the adipocyte hormone leptin (this line of mice was formerly known as the *ob/ob* line and is now known as the *Lep^{ob}/Lep^{ob}* line). Mice from this mutant line are hyperphagic, obese, and diabetic. In normal, wild-type mice, a large increase, or spike, in plasma leptin concentrations occurs during the first week after birth, usually around postnatal day (PND) 5. In contrast, *Lep^{ob}/Lep^{ob}* mice lack the spike in leptin secretion and consequently develop permanent deficits in the hypothalamic neural projections from the arcuate nucleus to the paraventricular nucleus (PVH), dorsomedial nucleus (DMH), and lateral hypothalamus (LH). These deficits last until adulthood and are thought to account for altered adult ingestive behavior. When *Lep^{ob}/Lep^{ob}* mice are treated with leptin during a critical window of arcuate development (PND 4–12), the deficits in arcuate projections are fully reversed, leading to a partial reversal of the obesity. When leptin is given in adulthood to $Lep^{ob}/Lep^{ob}$ mice, there is no effect on arcuate nucleus projections to PVH, DMH, and LH, suggesting that leptin acts neonatally as a neurotrophic factor to determine the fate of these neural structures involved in adult energy balance. If the process whereby leptin organizes arcuate nucleus projections is analogous to the process whereby gonadal hormones organize the neural substrates for male- and female-typical sex behavior, then perinatal leptin surges might render the brain more sensitive to the adult effects of leptin (Bouret et al., 2004; reviewed by Schneider et al., 2014). There is a great deal to be learned about how the neonatal leptin surge alters the brain and leads to changes in behavior and metabolism, but it gives us a hint about how the early hormonal milieu might program later morphology and behavior via the growth of neuronal projections.

### Interim Summary 10.1

1. The effects of gestational food restriction differ according to the severity and timing of gestational restriction and include low birth weight, rapid catch-up growth, adult obesity, and a wide array of metabolic consequences in the offspring of food-restricted mothers.
2. Possible mechanisms of such gestational programming due to food restriction might include decreased IGF-1, which is a hormone that promotes lipolysis and the breakdown of triglycerides in adipose tissue to oxidizable free fatty acids.
3. Another possible mechanism is the elimination of the neonatal leptin surge that purportedly organizes the neural projections from the arcuate nucleus of the hypothalamus to other brain areas. These projections might determine, in part, adult ingestive behavior, energy storage, and energy expenditure.

## Stress Hormones

Stress hormones can have profound effects on fetal development. The effects can occur either indirectly, from the mother's adrenal glucocorticoid secretion, which reaches the baby across the placenta, or directly, by altered stress hormone secretion in the fetal adrenal glands or changes in sensitivity to stress hormones in the fetal brain and body. Both indirect and direct prenatal effects can alter the stress responses later in adult life. For example, pregnant rats, after exposure to restraint stress, show elevated levels of circulating corticosterone compared to mothers not restrained at all. The offspring of these chronically stressed mothers have a greater increase in the secretion of stress hormones in response to restraint stress compared to the offspring of unstressed mothers (Nathanielsz, 1999). This effect can be mimicked, at least partially, by removing newborn offspring from the nest. Those who are removed briefly and returned to the nest where they receive expert maternal licking and grooming grow up to show a modulated stress response as adults.

The effects of stress and maternal care on the offspring can be permanent (i.e., the neonate can be "programmed" for life). The long-term effects of briefly removing the pup from the nest on the adult offspring vary, depending on how long the pup is isolated from its mother and the level of maternal care received when the pup is returned to the mother. Nathanielsz (1999) suggested that organisms have something like a set point in the hypothalamic–pituitary–adrenal (HPA) system, and this set point determines the level of external stress that can be tolerated before the onset of a stress response. That is, the sensitivity curve is "shifted to the left," in that a lower than usual amount of stress can activate the HPA axis. Anxious individuals in this scenario are thought to have a stress response that is more easily triggered. Resilient individuals have less sensitivity, and thus, their stress response is not so readily initiated.

According to this hypothesis, the sensitivity of stress tolerance is altered by early experiences with stressors, accompanied by elevations in neonatal adrenal hormones. The result is that offspring who are exposed to different degrees of stress grow up to be adults with altered stress responses. To demonstrate this idea, rats were treated within the first 4 days of birth with high doses of a synthetic glucocorticoid. Following neonatal glucocorticoid treatment, adult brain, pituitary, and adrenal hormonal responses were blunted (Fig. 10.1). In the offspring of rhesus monkeys, prenatal stress also caused reduced attention span and neuromotor ability (Schneider et al., 1999), and the offspring were less exploratory and showed what were interpreted as disturbance behaviors.

Nathanielsz (1999) speculated that this type of early hormone exposure contributes not only to the formation of temperament, but also to the likelihood of depression and other mood disorders. It also might be possible that, in some cases, a brief stress during early development could be adaptive, by making the adults more stress-resistant; this is an important consideration for future research.

Stressful (i.e., adrenal cortex activating) social experiences early in life also can lead to a tendency toward depression (see Chapter 8). Among rhesus monkeys, even relatively brief separation from the mother is followed by a marked tendency toward social isolation and behaviors that are indicators of depression. The endocrine concomitants of this important behavioral change are under investigation.

One possibility for a causal route from earlier stress to later behavioral change is an alteration in the stress hormone–receiving systems in the brain. Mohammed et al. (1993) used an environmental enrichment program that might be conceived of as altering stress hormone levels.

II. HISTORY: HORMONE EFFECTS CAN DEPEND ON FAMILY

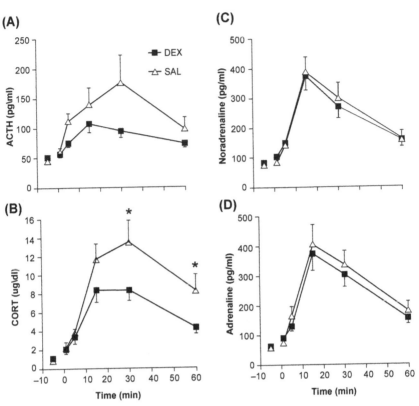

FIGURE 10.1 Neonatal treatment with a synthetic stress hormone, dexamethasone (DEX), compared to neonatal vehicle control treatment (SAL) affected responses to stress as measured by (A) adrenocorticotropic hormone and (B) corticosterone, but not as measured by (C) noradrenaline or (D) adrenaline. *From Kamphuis, P.J., Bakker, J.M., et al., 2002. Enhanced glucocorticoid feedback inhibition of hypothalamo–pituitary–adrenal responses to stress in adult rats neonatally treated with dexamethasone. Neuroendocrinology 76 (3), 158–169. With permission.*

Enrichment consists of adding complexity to an otherwise monotonous environment. Compared to control rats, rats raised in an enriched environment (Fig. 10.2) have higher glucocorticoid levels in their hippocampal neurons. Both the behavioral and the neurochemical consequences of early glucocorticoid administration have been demonstrated widely (see Fig. 10.3A and B). These phenomena of early stress hormones affecting later responses could provide mechanisms for the early experience "familial" effects mentioned in Chapter 8.

In Chapter 7, early deprivation of maternal care and neonatal handling of experimental animals was discussed. Both are presumed to have long-term effects resulting from the alterations in neonatal stress hormone release causing later changes in the same neuroendocrine system. In addition, the fetal HPA axis is very sensitive to programming

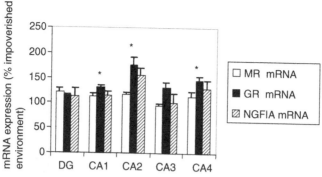

**FIGURE 10.2** Transcript levels for glucocorticoid receptors (GRs), mineralocorticoid receptors (MRs), and nerve growth factor receptors (NGF1As) in the hippocampus. One group of rats had been exposed to an enriched environment early in life, while the others had been restricted to an impoverished environment. The enriched group had significantly greater levels of mRNA for GRs in specific parts of the hippocampus (e.g., Ammon Horn pyramidal cells in area 2; CA2). *From Mohammed, A.H., Henriksson, B.G., Södeström, S., Ebendal, T., Olsson, T., Seckl, J.R., 1993. Environmental influences on the central nervous system and their implications for the aging rat. Behav. Brain Res. 57, 183–192. With permission.*

by high maternal glucocorticoid levels, whether caused naturally or experimentally. Elevated adrenal hormone levels during gestation result in increased basal corticosterone levels neonatally and increased stress responses in adulthood. Perhaps as a consequence, there is a greater stress sensitivity in the sympathetic nervous system and altered learning capacities. One possible mechanism connected with a specific example is that prenatal stress hormones may increase neuronal vulnerability to oxidative stress. For example, in some experiments, rats are exposed prenatally to high levels of a synthetic stress hormone, dexamethasone, whereas others are treated with a vehicle; then in adulthood, experimenters measure their auditory brainstem response (ABR) to sound. The prenatally treated and control animals are exposed to acoustic trauma as adults, and their recovery is monitored. The rats treated prenatally with dexamethasone show little recovery of the ABR, whereas normal controls recover completely (Fig. 10.4). Furthermore, the prenatal stress hormone treatment reduces the ability of the adult central nervous system (CNS) to deal with oxidative stress. This is just one example of effects of prenatal stress hormones on sensory function and behavior.

Some of the mechanisms by which early hormone exposure could influence adult hormone/behavior relationships are just beginning to be explored. One set of possible alterations resides in the lower brainstem. When the mild stress of neonatal handling is imposed upon newborn rats, fewer neurons in the locus ceruleus are seen at later developmental stages in both male and female offspring of stressed compared to control dams. This is potentially important because neurons in the locus ceruleus

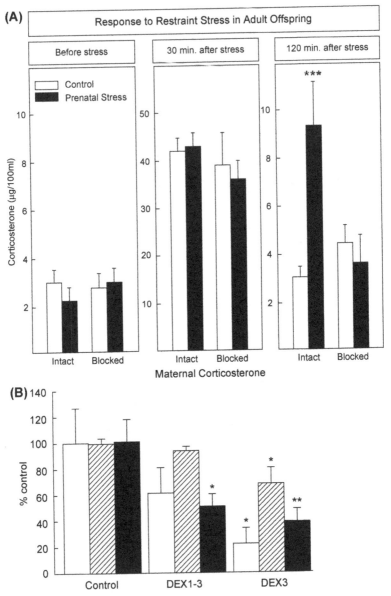

FIGURE 10.3   Prenatal stress and prenatal exposure to stress hormones influence neuro-endocrine and behavioral responses in later life. (A) Prolonged responses to restraint stress (right panel, *black bar*) were evident in the adult offspring of mothers who had been intact and secreted stress hormones during gestation, but not in adult offspring of mothers whose gestational stress secretions had been blocked (comparison on far right). The inability to turn off the stress response has been linked with various behavioral maladies. (B) Prenatal exposure to the synthetic stress hormone dexamethasone (DEX) throughout gestation (DEX 1–3) or only in the last week of gestation (DEX 3) influenced certain anxiety-related behavioral responses after the baby rats had grown up. Number of rears in a 12-min open field test (*black bars*) were very sensitive as behavioral indicators, as were open-arm entries in the elevated plus maze assay (*open bars*). Less sensitive was an assay that is supposed to model depression: time spent floating in a forced-swim test (*striped bars*). (A) From Barbazanges, A., Piazza, P.V., et al., 1996. Maternal glucocorticoid secretion mediates long-term effects of prenatal stress. J. Neurosci. 16 (12), 3943–3949. With permission. (B) From Seckl, 2001. J. Neuroendocrinol. With permission.

## Noise Exposure (110 dB SPL, 4 hours)
### Without PBN

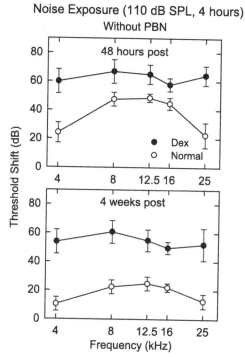

**FIGURE 10.4** Fetal exposure to high levels of glucocorticoids can alter nervous system development. In this experiment, rats were treated prenatally with a synthetic glucocorticoid (dexamethasone, DEX) or control (normal). The recovery of response of brainstem auditory neurons following damage of the auditory system due to high amplitude noise exposure was harmed in DEX-treated rats compared to controls tested as young adults; that is, the thresholds for electrophysiological response to a wide variety of frequencies of auditory stimuli were higher in the DEX rats either 48h or 4weeks after damage. SPL 1/4 Sound Pressure Level. *From Canlon, B., Erichsen, S., Nemlander, E., Chen, M., Celsi, G., Ceccatelli, S., 2004. Alterations in the intrauterine environment by glucocorticoids modifies the developmental programme of the auditory system. Eur. J. Neurosci. (in press). With permission.*

respond strongly to salient stimuli (including, of course, stressful stimuli) and then report to the rest of the brain through far-ranging adrenergic projections.

Neuronal mechanisms in the forebrain are also candidates for remembering the effects of early stress. When the synthetic glucocorticoid dexamethasone is injected into the mothers of fetal rats, the prenatal hormone injection not only reduces adrenocorticotropic hormone (ACTH) levels but also, in the PVH, reduces (1) expression of corticotropin-releasing hormone (the main neuropeptide stimulating ACTH release) and (2) reduces glucocorticoid receptors and vasopressin (reviewed by Maccari et al., 2003). Decreased biosynthetic activity of the hypothalamic neurons controlling pituitary and adrenal stress-related hormone release may account, at least partially, for the behavioral effects of early stress, mentioned earlier and in Chapter 7. Further forward in the brain, in the prefrontal cortex, lie other molecular mechanisms that could participate in altered hormone-behavior relations. Chronic maternal stress during pregnancy significantly reduces gene expression for brain-derived neurotrophic factor (BDNF) specifically, without affecting other telencephalic regions In contrast, postnatal stress, imposed in the form of maternal separation, causes a decrease in BDNF expression in the adult hippocampus. These changes in BDNF expression are potentially important because BDNF is a neurotrophic molecule that might promote permanent structural changes in neural circuitry that creates specific behavioral adaptations.

Most of the effects of early life stress are not independent of the topic introduced at the beginning of this chapter: maternal programming of body weight and other aspects of energy balance. Early life stressors and both endogenous and exogenous elevations of glucocorticoids in the fetus and neonate have permanent effects on energy balance, i.e., on food intake, energy expenditure, and body fat storage and distribution. In turn, these changes are linked to adult obesity and the metabolic syndrome. It also is important to note that these effects differ between male and female offspring, related to their different gonadal steroid milieus.

## Effects of Early Sex Hormones

As discussed in Chapter 9, the brain and body are subject to the effects of the Y chromosome. In mammalian fetuses, before hormonal differentiation, Wolffian and Mullerian ducts, the precursors of the internal genitalia, are present in both sexes. In the male, the trigone of the bladder, seminal vesicles, epididymis, and vas deferens are derived from the Wolffian ducts. The Y sex chromosome carries a gene or genes coding for the production of testis-determining factor (TDF), enabling an XY individual to become phenotypically (gonadally) male. This is because the testes, unlike the ovaries, secrete high levels of androgens pre- and neonatally. Perhaps the most startling effects of neonatal hormone exposure are in the realm of sex behavior (whether the adult shows male-typical sex behavior in response to adult exposure to testosterone or female-typical sex behavior in response to adult exposure to estradiol followed by progesterone).

In rats, the presence of high levels of circulating androgens from the testes during the first 2 or 3 days after birth largely determines whether the neuroendocrine and behavioral tendencies of the animal during adulthood will fit those expected of a normal female or those expected of a normal male. The masculinizing effects extend to the sexually differentiated hypothalamic–pituitary–gonadal system that creates either cyclic ovulation or continuous sperm production. If testosterone is injected into the newborn female (an animal with an XX chromosome complement), she loses the ability to ovulate during adulthood, as well as the ability to perform female-typical courtship and sex behaviors, such as lordosis, in response to adult sex hormone treatment and in the presence of an adult male rat. In other words, she is defeminized by early androgens. In addition, she shows male-typical sex behavior (mounting and intromission) in response to adult treatment with testosterone: she also is masculinized by early androgens. Conversely, if the male (an animal with an XY chromosome complement) is deprived of circulating testosterone by surgical castration on the day of birth, his feminized brain will create an ovulatory luteinizing hormone (LH) surge in response to adult treatment with estrogen plus progesterone. Correspondingly, he exhibits female-like lordosis behavior in response to adult estradiol plus progesterone treatment and the presence of an adult, sexually experienced male rat. In addition, he will fail to show normal, male-typical mounting and intromission in response to adult testosterone treatment and a female sex partner. In sum, he is demasculinized and feminized by the lack of neonatal androgens. Thus, the most visible features of adult sexual life are essentially determined by neonatal androgen exposure (Fig. 10.5) (reviewed are by Wallen and Baum, 2002).

| | Ability to generate ovulatory surge of LH release? | Sex Behaviors? | |
|---|---|---|---|
| | | Male-like? | Female-like? |
| Normal male | NO | YES | NO |
| Neonatally castrated male | YES | NO | YES |
| Normal female | YES | Minimal | YES |
| Female given testosterone neonatally | NO | Increased | NO |

FIGURE 10.5 Removal of testicular androgens from the neonatal male rat permits his neuroendocrine system to generate a pulse of GnRH sufficient for ovulation (if he has been transplanted with ovaries) and to engage in female sex behaviors such as lordosis. Conversely, injection of testosterone neonatally into the female rat abolishes her ability to ovulate and to respond normally to estrogens with lordosis behavior. Male-like sex behaviors change, some less dramatically, in exactly the opposite direction from female behaviors.

As noted earlier, in a normal male, XY animal, the testis secretes testosterone and Mullerian duct-inhibiting hormone (also known as Mullerian regression factor). In males, in response to TDF, the formation of the testes, and testicular androgen secretion, the Wolffian duct forms the epididymis, seminal vesicle, and vas deferens. In addition, in response to the testicular secretion of Mullerian-duct-inhibiting hormone, the feminine Mullerian ducts disappear; the testes descend into the scrotum, and the kidneys ascend into the upper abdomen. In a normal XX female, the absence of TDF results in development of the ovary, which secretes comparatively little in the way of steroid hormones and no Mullerian regression factor. The Mullerian ducts develop into the fallopian tubes and uterus. This is often referred to as "the default phenotype," in that the female phenotype will develop in the absence of the high levels of androgens necessary to produce the male phenotype. It should be noted, however, that even in the absence of high levels of androgens that would masculinize/defeminize sex behaviors, some small levels of ovarian steroids may be necessary for complete sociosexual development of the female repertoire of behavior. That is, the "default female phenotype" may be anatomically complete, but is not behaviorally complete, in the absence of active ovarian steroid secretion at some point in postnatal development.

In the absence of exogenous interference, most sex-specific behaviors are consonant with genotypic and phenotypic sex. Experimentally, these aspects can be dissected, to understand the specific components and timing of developmental milestones. For example, in female rats, testosterone treatment on postnatal day 4 (PND4), during the critical period of female sexual differentiation results in a low number of hypothalamic preoptic area spine synapses, acyclic LH secretion, and low female and high male sexual behaviors. In contrast, testosterone treatment of females on PND16, after the critical period for sexual differentiation, has no effect on these adult traits. Correspondingly, in males, castration on PND1, during the critical period of male sexual differentiation, produces a high number of preoptic area spine synapses, cyclic LH secretion, and low male and high female sexual behaviors. Castration on PND7, after the critical period, has no effect on these traits.

Rodents are altricial (i.e., helpless, naked, and blind at birth), and their CNS-critical periods are still occurring after birth. In contrast, primates are precocial (i.e., more developed at birth), and their CNS-critical periods already have occurred in utero. In humans, therefore, the critical milestones for normal, concordant genotypic and phenotypic sex development, including sexually dimorphic development of CNS regions that influence neuroendocrine function, as well as sexually diergic (functionally different) behavior patterns, occur before birth. Rodents are similar to primates, in that prenatal hormones masculinize the genitalia, but in rodents, the critical period for sexual differentiation extends into the first few days of life. In addition, as discussed in Chapters 4 and 8, male rodents are masculinized because testosterone is secreted from the testes, escapes the binding of alpha-fetoprotein in circulation, and reaches the brain, where it is aromatized to estradiol. Females are protected from masculinization and are feminized owing to the lack of steroid secretion from the ovaries and because alpha-fetoprotein binds any estradiol from the mother.

In many aspects of behavioral masculinization, it is estradiol binding to estrogen receptor α that is critical. Estradiol is thought to be important in the fetal masculinization of primates, such as the monkey, whereas in humans, the little data we have suggest that androgens are more important than estrogens in forming male gender identity and other sexually differentiated behaviors. As described in previous chapters, the effects of early estradiol on the brain in rodents is mediated in some ways by changes in prostaglandin secretion, and by a wide array of redundant neurotransmitter systems. In essence, different steroids have different effects at different critical periods. Many of these differences depend upon the part of the brain and the behavioral effects under study; thus, each individual can be considered a mosaic of masculinized and/or feminized, and demasculinized and/or defeminized, neurons. It will be important to include prenatal and neonatal sexual differentiation as a variable that can have important interactions with the effects of prenatal and neonatal stress and other early-life factors that program adult behavior. This perspective is of clear relevance to our better understanding of the biological bases of the spectrum of human gender identities.

### Interim Summary 10.2

1. Both environmental stressors and environmental enrichments have effects on early levels of stress hormones that alter the adult responsiveness to stress response.
2. These changes in responsiveness might be mediated by levels of glucocorticoids and their effects on the development of the HPA system.
3. Androgens and androgen metabolites act during a critical period of early development to masculinize the body, the brain, and behavior.

## Sexual Differentiation Meets Maternal Programming

Understanding the mechanisms of sexual differentiation is imperative to understanding maternal programming because these processes overlap. As discussed earlier in this chapter, energy-balancing characteristics can be programmed not only by perinatal steroid action but also by hormones originating from the fetus, including leptin. As explained in Chapter 9, in most vertebrate species, males differ from females in aspects of ingestive behavior and energy balance. These include daily food intake, preference for sweet and creamy flavors, binge eating, susceptibility to diet-induced obesity, energy restriction-induced hyperphagia (overeating after a period of deprivation), energy restriction-induced decreases in energy expenditure, resistance to leptin- and insulin-induced hypophagia (decreased food intake after treatment with anorexigenic peptides), resistance to ghrelin-induced hyperphagia (increased food intake after treatment with orexigenic peptides), gluco- or lipoprivation-induced hyperphagia (overeating in response to treatment with inhibitors of glucose and free fatty acid oxidation), proopiomelanocortin (POMC) gene expression in the arcuate nucleus of the hypothalamus, seasonal food intake responses to energetic challenges and leptin, and other traits (reviewed by Asarian and Geary, 2013; Shi and Clegg, 2009; Shi et al., 2009; Schneider et al., 2014). In addition, males tend to store fat in the abdominal, visceral area, whereas females tend to store fat subcutaneously (in a layer under the skin) and in the gluteofemoral area, i.e., the hips and thighs. There is therefore a significant sex difference in the waist-to-hip ratio. The male-typical, or "apple," shape (high waist-to-hip ratio) is more closely linked to metabolic syndrome, whereas the female-typical "pear" shape is cardioprotective.

Because males and females differ in their food intake and body fat distribution, an interesting question is whether there are there sex differences in the response to leptin neonatally, and if so, do gonadal hormones rewire the response to leptin? The answer is not known at this time. Some experiments have been designed to address the effects of perinatal gonadal steroids on the neural circuitry of ingestive behavior. These experiments show that perinatal androgens masculinize the circuitry of the arcuate nucleus of the hypothalamus. As noted in Chapter 8, male mice have a significantly higher daily food intake throughout life than do female mice. In addition, they show higher levels of postfast hyperphagia. Previously food-deprived mice treated with leptin, however, tend to decrease food intake relative to saline-treated control mice (leptin-induced hypophagia). Male mice show a higher sensitivity to leptin-induced hypophagia compared to female mice. Furthermore, males show significantly lower POMC gene expression and neuronal projections from POMC cells to the arcuate nucleus. This might be significant, because the POMC molecule is the precursor for the anorexigenic peptide alpha-melanocyte-stimulating hormone. Daily food intake, postfast hyperphagia, and POMC expression and projections are fully masculinized in females treated perinatally with either testosterone or dihydrotestosterone (DHT), both of which act via androgen receptors. This is consistent with the idea that testosterone masculinizes the energy-balancing system during early development via action on the arcuate nucleus. There are some surprising twists to these experimental outcomes, suggesting that testosterone can act through different receptors to act on different parts of the energy-balancing system. Males, remember, are more sensitive than females to leptin-induced hypophagia. Females that are neonatally treated with testosterone or estradiol, but not DHT, are not masculinized in their leptin-induced hypophagia, but they are instead ultrafeminized; that is, they show male-like sensitivity to leptin-induced hypophagia. Thus, different aspects of the energy-balancing system might be masculinized, feminized, or hyperfeminized by different steroid hormones acting on different steroid receptors (reviewed by Schneider et al., 2014). Together, these results indicate the importance of continuing research on the interface of sexual differentiation and maternal programming.

### Interim Summary 10.3

1. Males and females differ in many aspects of ingestive behavior (e.g., taste preferences and binge-eating tendencies) and in body and brain morphology, and these are programmed by pre- and/or neonatal androgens.
2. Ingestive behavior responses (sensitivity) to anorexigenic and orexigenic peptides differ between males and females, and these differences are programmed by pre- and/or neonatal androgens.
3. Adult ingestive behavior and body morphology are programmed by spikes in leptin secretion and IGF-1 during an early critical period; thus, one important topic of research concerns the interactions between pre- and neonatal androgens/estrogens and pre- and neonatal anorexigenic and orexigenic hormones in determining adult sex differences in behavior.

# CLINICAL EXAMPLES

## Consequences of Abnormal Fetal Hormone Exposure in Humans

Gender identity (i.e., one's sense of self as female or male) and sex-specific behaviors normally develop concordantly with genotypic and phenotypic sex. Several genetic abnormalities can alter this course of events. For example, crossing-over can occur in a homologous region of the X and Y chromosomes, so TDF becomes translocated on the X chromosome. If this occurs in a male gamete (sperm), the offspring could be either an XY female having no Y-chromosome TDF region, or an XX male having an abnormal X-chromosome TDF region.

Both genetically determined deficiencies of enzymes involved in sex-steroid production and genetically determined deficiencies in steroid hormone receptor function can produce disconnects between genotypic and phenotypic sex. An example in XX girls is virilizing congenital adrenal hyperplasia (CAH), in which genetically based deficiencies in enzymes involved in glucocorticoid synthesis lead to a shunting of steroid precursors into androgenic pathways. The most common form of the CAH results from a deficiency in 21-steroid hydroxylase, the enzyme that converts 17-hydroxyprogesterone to 11-deoxycortisol. 17-hydroxyprogesterone instead is shunted to androstenedione and then to testosterone. Excess testosterone in utero results in female pseudohermaphroditism and precocious puberty (see Chapter 9), with an enlarged phallus, a common urogenital sinus at the base of the phallus, and fused and pigmented labioscrotal folds. The deficiencies in glucocorticoids and mineralocorticoids are fatal if not treated, because these hormones are critical for normal cellular metabolism and salt and water balance. Fortunately, this condition is almost always recognized at or shortly after birth, and the glucocorticoid/mineralocorticoid deficiencies can be successfully treated with exogenous hormone administration, which suppresses inappropriate secretion of adrenal androgens. The physical effects of prenatal androgens in an XX individual can be treated with reconstructive surgery to produce female-typical genitalia.

Females with the XX chromosome complement with CAH have a masculinized body fat distribution, e.g., a masculinized waist-to-hip ratio closely correlated with the digital 2D:4D ratio (as explained in Chapter 8, the ratio of the second to the fourth digit is a reliable indicator of prenatal androgen exposure). As a result of their masculinized body fat distribution, XX individuals with CAH have an increased incidence of abdominal-visceral obesity and metabolic syndrome. In addition, they have an increased incidence of polycystic ovarian syndrome (PCOS), a syndrome linked to early androgen exposure.

Another example of the disconnect between genotypic and phenotypic sex is complete androgen insensitivity (AIS) in individuals with the XY chromosome complement, a genetically determined failure of androgen receptor function (see also Chapter 9). All patients are XY, their testes are capable of normal testosterone secretion, and they undergo normal fetal regression of Mullerian derivatives. However, because of the absence of androgen receptor function, their external genitalia remain female. These genetic males come to medical attention in infancy because of undescended, inguinal testes, or at puberty because, as phenotypic girls, they do not develop menstrual cycles. As adults, they have a paucity of pubic hair, normal female breast development secondary to unopposed estrogen action, and generally tall stature. They are always raised as girls and can have satisfactory vaginal intercourse without surgery, but, lacking ovaries, they are sterile. Fig. 8.3 in Chapter 8 illustrates an XY male with complete AIS. In most cases, XY individuals with AIS show a female-typical body fat distribution, including increased subcutaneous adipose tissue and a low waist-to-hip ratio. It will be important to understand the effects of early androgen exposure and its absence on the development of metabolic and energy-balancing traits.

## Sex, Gender Identity, and Behavior Are Usually Consonant but May Not Always Be

During the past few years, considerable variability and flexibility with respect to gender identity have been adopted in many societies (Erickson-Schroth, 2016). Facebook started long ago with three categories ("male, female, it's complicated") but now has more than 50. In this text, we neuroscientists cannot deal adequately with the full complexity of choices made in the world of the internet and social media. Therefore, we concentrate on those simpler aspects of gender identity that relate to basic neurobiology, but we do keep in mind that human behavior is determined by a complex mosaic of mechanisms (see Chapter 9).

Consistent differences in gender role behaviors across cultures appear to be related to whether or not the brain has been masculinized during development. As indicated earlier, in the absence of male hormones, estrogens promote female CNS development. In addition to in utero masculinizing factors and postnatal androgen-organizing influences on the CNS, the interactions of parents and others with the infant based on appearance of the infant's external genitalia, the infant's developing self-awareness of testes and male external genitalia, environmental influences such as sex of rearing, and societal norms for sex-appropriate behavior are all important in human gender identity and sexual behavior.

Verified normal sex differences in behavior include the following: Males are more aggressive than females across cultures and from childhood through adulthood, more so in children. Boys are more active and display more playful aggression than do girls. For toys, boys prefer vehicles, weapons, and building toys, whereas girls prefer dolls, kitchen accessories, and cosmetics. As adults in most cultures, women are more interested in parenting than are men. Males and females also differ in dietary preferences, body fat distribution, and the propensity to develop metabolic syndrome, obesity, and/or eating disorders, as we have discussed earlier.

Men are more likely to be left-handed than women, and women are less likely to have exclusive left cerebral language dominance. Whereas there is no sex difference in general intelligence, some differences in specific cognitive measures have been found. Females have slightly better overall verbal ability, but this has come into question in recent years. Males are slightly better on analogies, whereas females are slightly better in speech production and verbal fluency. From childhood through adulthood, males rotate mental images more accurately and rapidly than do females. Females solve mathematical problems slightly better in childhood, whereas males' problem-solving ability is slightly better in adulthood.

There is no sex difference in comprehension of mathematical concepts at any age. Finally, females show better perceptual speed and accuracy than do males, but the magnitude of the difference has been declining since the 1940s, likely related to evolution of the tests used to measure this performance characteristic. It is important to emphasize that within male and female populations, the distribution in these traits is wide, and although mean differences are sometimes statistically significant with high enough sample sizes, there is a great deal of overlap in the distributions. Thus, for traits in which the mean is higher in men, women with the highest scores rank higher than a large proportion of the men, and men with the lowest scores score lower than a large proportion of the women. The utility of such scores for practical purposes and further study thus is quite limited.

As illustrated by the previous examples of CAH in XX females and complete AIS in XY males and the example of 5-alpha-reductase deficiency in Chapter 4, there can be hormone-induced discordances among sex, gender identity, and sexually diergic cognitive and emotional behaviors. XY individuals with complete AIS have concordant female phenotypic sex, sex of rearing, and gender identity. In contrast, as indicated in Table 4.1, XY individuals with 5-alpha-reductase deficiency have female-appearing or ambiguous genitalia at birth, so some have been raised as girls. At puberty, however, due to an overriding increase in testosterone production, many have developed male secondary sex characteristics and have experienced a male gender identity awakening. More difficult to explain on an endocrine basis is male transsexualism, where there is a pervasive female gender identity, in spite of unambiguous male genital sex and sex of rearing. No abnormalities of CNS morphology or function and no endocrine abnormalities have been verified in either transsexual males or females.

XX individuals with CAH have masculinized external genitalia at birth but, as indicated previously, are now usually quickly recognized and treated. XX children with CAH are most often altered surgically and raised as girls because with treatment they will develop a female gender identity and a normal reproductive life. Fig. 10.6 illustrates the external genitalia of an XX female infant with virilizing CAH. Their psychosocial development tends toward some masculinizing behavioral effects of excess androgen exposure. Girls with CAH show more rough-and-tumble and a preference for "boys" toys, and during adolescence, their interests and activities are intermediate between the distributions typical of unaffected girls and boys (Berenbaum et al., 2000). Nevertheless, they may have higher than usual energy levels and behaviors that in the 1950s were referred to as "tomboyish." These include the propensity toward rough-and-tumble play and participation in sports that involve physical contact (e.g., boxing, football, rugby, or martial arts). As well, they may have a higher than usual

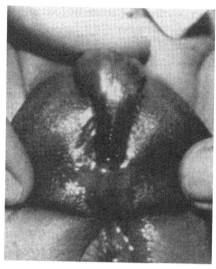

FIGURE 10.6 Some of the effects of early hormone exposures—or lack of same—on adult behavior may be related to the peripheral anatomical effects of such exposure. Shown here are the genitalia of an XX female infant with 21-hydroxylase deficiency. The phallus is enlarged, below which is a common urogenital sinus. The labioscrotal folds are pigmented and fused. *From Forest, M.G., 2001. Diagnosis and treatment of disorders of sexual development. In: DeGroot, L.J., Jameson, J.L. (Eds.), Endocrinology, fourth ed. W.B. Saunders, Philadelphia, PA, p. 1997. With permission.*

incidence of homosexual fantasies and arousability, but not necessarily a higher incidence of homosexual orientation than is found in the general population. Some older individuals who were raised as boys because of the extent of masculinization of their external genitalia have successfully maintained a male gender identity. These are profound examples of the effects of early hormone milieu on the development of the brain and behavior.

## Thyroid Hormones

If a child develops without circulating thyroid hormones for prolonged periods, his/her adult performance will be marked by disastrous cognitive failures (Chapter 5). Abnormalities in morphological brain development accompany and, in fact, can explain the cognitive problems. These comprise the definition of cretinism (congenital hypothyroidism). What about brief or subtle decreases in thyroxine during early development? Might one of the implications of early thyroxine loss be an abnormality in later responses to this and other hormones?

Even relatively subtle changes in thyroxine in pregnant women can have consequences for their children's neurological development. Later psychological performance can be affected, including the well-documented significantly lower IQ. The precise type of behavioral deficit depends on the time of thyroid hormone loss: Early in pregnancy, problems with visual attention and processing have been reported. Later in pregnancy, decreased visuo-spatial skills and slower motor responses may result. The latter can include deficits in eye–hand coordination, motor imitation, and object manipulation. With thyroid insufficiency extending until after birth, problems in linguistic and memory skills are evident. Zoeller and Rovet (2003) reflect that these clinical observations far exceed our neurological understanding. By exactly what mechanisms do these losses come about? Do some of the causal routes involve later responses to thyroxine (T4)?

Dowling and Zoeller (2000) used the differential display technique to examine the genomic consequences of acute T4 administration to the mother just before the normal onset of fetal thyroid function. Several of the thyroxine-sensitive genes discovered were expressed selectively in brain areas that contain thyroid hormone receptors (TRs). One of them, transcription factor Oct-1, shows how interesting these genes of potential importance for cognitive development can be. How are the effects of early hormonal changes stored in such a way that the individual's later responses, genomically and behaviorally, to both hormones and environment are changed permanently? In this era of discerning causal relations between gene expression and behavior, a new set of possible mechanisms has emerged. Epigenetic mechanisms are defined as those closely related to gene expression without changes in the primary nucleotide sequences themselves. We know now that DNA can be methylated, and such a chemical alteration changes the sensitivity of a gene to transcriptional enhancers throughout the life of the animal. Further, changes in the chromatin can be long lasting. Histone acetylation or methylation provide additional possible mechanisms by which prenatal or neonatal hormone exposures could alter the animal's hormone responsivity for the rest of its life. As might be expected from the discussion earlier in this chapter, early exposure to high or low levels of thyroid hormones has the potential to influence many aspects of energy balance, body weight, adiposity, and especially glucose and lipid metabolism. Furthermore, these effects are likely to differ according to pre- and neonatal levels of androgens.

## SOME OUTSTANDING QUESTIONS

1. In most experiments that examine the effects of undernutrition during gestation on the programming of energy balance and ingestive behavior, the experimenter must handle the newborn pups for the purposes of measuring body weight and hormone levels. In addition, the mothers used in these experiments are almost always purchased

and transported from the vendor to the laboratory during the week before parturition. The offspring under study are typically cross-fostered to mothers that are not their birth mothers. In a separate line of research, however, experimenter handling of infants is used as a stressor in experiments that have revealed profound effects of these effects of early-life stressors on the adult HPA system. How should we interpret the data on nutritional gestational programming collected using these methods?

2. The incidence of transsexualism, in which gender identity is discordant with phenotypic sex and sex of rearing, is about 1:25,000 in males and 1:75,000 in females. These ratios are reasonably consistent across cultures. In contrast, the incidence of bisexuality/homosexuality, in which there is concordance among gender identity, phenotypic sex, and sex of rearing but same-sex partner preference, is about 1:25 in males and 1:20 in females. Are there identifiable hormonal and/or CNS determinants or correlates of male or female transsexualism and male or female homosexuality?

3. Given that there are endocrine-disrupting compounds in the environment that bind to steroid receptors and alter sexual differentiation, how do congenital and environmental influences interact to affect gender identity and/or sexual orientation? Should the biological and environmental determinants be separated from or blended with psychological determinants to understand gender identity?

4. To what extent do the effects of early exposure to high and prolonged levels of stress hormones such as cortisol relate to the psychological effects of low socioeconomic status early in life? Do they represent some of the mechanisms by which early feelings of deprivation and loss have their eventual behavioral effects?

5. Considering the effects of environmental toxins on the mother, what are some circumstances where these act through hormonal mechanisms, be they thyroidal, sexual, or adrenal?

# 11

# Epigenetic Changes Mediate Effects of Hormones on Behavior

In 2014, it was estimated that the genome of *Homo sapiens* contains about 19,000–30,000 protein-encoding genes. This number is not significantly larger than that of many less complex organisms, such as the nematode roundworm, *Caenorhabditis elegans* (19,735–20,470 protein-encoding genes). How then is our behavior so much more complex than that of a worm? One answer is related to the topic of this chapter, epigenetics. *Epi* is from the Greek for "over, outside of, or around." The word *epigenetics* simply implies something outside of the genetic information provided in the nucleotide sequence. This includes factors that alter the transcription of DNA into messenger RNA. Modulation of gene transcription, known as gene expression, can occur in response to aspects of the environment and in response to actions of non-protein-encoding regions of DNA occurring during specific phases of development. The scientific study of epigenetics stems in part from the early science of embryology, a subset of developmental biology concerned with development prior to birth. Since the discovery of gene expression, embryology has been concerned with the developmental program that determines differential gene expression during different stages of life. One of the earliest uses of the word, *epigenesis*, was by Waddington, who studied the effects of the environment on the developing embryo, effects that lasted into adulthood and appeared in the next generation (e.g., Waddington, 1942). According to this earlier meaning, before the explosion of molecular biology, epigenetics encompassed almost any maternal, neonatal, or pubertal effect that lasted into adulthood and could be passed to future generations (reviewed by Crews, 2008). Epigenetic effects also were of interest to early ecologists, evolutionary biologists, and psychologists, who were intrigued by the fact that many phenotypes, including behaviors and mental disorders, were altered by the lasting effects of early experience. (Early effects of experience on behavior have been discussed in earlier chapters, especially Chapters 7 and 8.) Throughout history, there have been many different uses for the term *epigenetics*. The present chapter, however, is concerned mainly with molecular epigenetics.

In this chapter, we consider the instructions in the genetic code of DNA and how the DNA itself is instructed to become a protein. Currently, *molecular epigenetics* is used to describe various processes that regulate genome activity, especially gene expression. First and foremost, this process relies on the DNA sequence of the four nucleotide bases: adenine, cytosine, guanine, and thymine. The word *epigenetic* does not imply changes to the fundamental process of genetic inheritance or evolution. Epigenetic changes, by definition, do not change the nucleotide sequence or the allele frequencies in the population; rather, epigenetic changes alter the strength of the genes' influence within cells. In this way, epigenetic effects can alter the observable phenotype and thereby affect how organisms perceive and interact with each other and their environment.

In the normal process of genetic inheritance, after genes in the gametes are combined to form new zygotes, those genes are transcribed and translated. The initiation of gene transcription depends on additional chemical changes in the nuclei of all cells, including nerve cells. The DNA and RNA are associated with important proteins (e.g., histones) that participate in transcription; together, the DNA, RNA, and proteins are known as chromatin. Through these histones and other molecules, individual nucleotides or sequences of nucleotides within the nuclei can be designated for higher or lower levels of transcription. That is, via histones and other molecules, environmental and/or developmental events can cause chemical tags or labels to be added to or subtracted from certain nucleotides in the DNA sequence.

Some chemical tags mark the nucleotide sequence so that, when transcription of the gene is initiated, there will be more transcription, whereas other tags mark other nucleotides for less transcription. Epigenetic tag changes may last through the process of mitosis (cell division), they may last for the duration of the cell's life, and they may even last for multiple generations, even though they do not involve changes in the underlying nucleotide sequence. Furthermore, these epigenetic changes can affect both somatic cells and germline cells. Thus, even though gene frequencies are not changed, phenotypic frequencies are changed epigenetically. Molecular epigenetics therefore includes external or developmental modifications of DNA that turn gene transcription on or off without alterations in gene frequency. Following are three types of epigenetic processes and a few examples of their importance in hormone-behavior relations:

1. Histone protein modification. There is no such thing as "naked" DNA. DNA exists within chromatin, a complex of macromolecules along with RNA intermingled with proteins known as histones (Fig. 11.1). The histone proteins act as spools around which DNA strands are wound in structures called nucleosomes. The N-termini

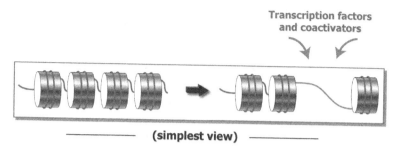

FIGURE 11.1   Inactive DNA is wound around histone proteins in nucleosomes.

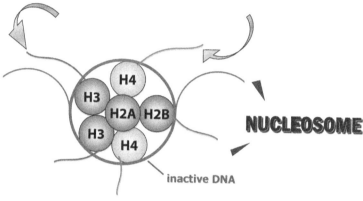

FIGURE 11.2   Looking at the nucleosome from the side: the N-termini of some of the histone proteins stick out and can be chemically modified.

of the histone proteins stick out of the nucleosomes, and they can be modified in ways that either enhance or suppress transcription (Fig. 11.2). There are five major families of histones. Acetylation (adding a chemical $CH_3CO$ group of two carbons, three hydrogens, and one oxygen) and deacetylation of histones H3 and H4 have been shown to induce a relaxed and competent or a condensed and inactive chromatin, respectively. Remember from earlier chapters that steroid hormone receptors, such as estrogen receptors, can act as transcription factors, binding to their response elements segments of DNA that can promote transcription. In the case of transcription enhancement, the result of histone protein modification is that behaviorally important transcription factors gain access to their corresponding DNA response elements. Thus, in response to sexual stimuli, estrogen receptors can be the object of histone protein modification that increases their transcription and increases receptor access to the estrogen receptor response elements on the genes for other proteins. Analogously, in response to early life stressors, glucocorticoid receptors can be the object of histone protein modification, which increases transcription of the receptor and allows access to the glucocorticoid receptor response elements and enhances transcription (Fig. 11.3).

Gagnidze et al. (2013) used assays of histone protein N-terminus modification to chart epigenetic modifications of possible importance for female reproductive behaviors. For example, estrogen treatment, important for female mating behavior, as we have discussed earlier, leads to transcription-enhancing changes in histone H4 in the ventromedial hypothalamus, a cell group crucial for mating behavior (Fig. 11.6). Further specificity can be given to the histone protein assay by using a technique called chromatin immunoprecipitation ("Chip") as illustrated in Fig. 11.7. This technique has dual specificity: First, it uses antibodies to select for specific histone protein modifications, and second, it uses DNA probes to query where on the DNA (e.g., of a neuropeptide gene promoter) that histone is attached. For example, a Chip assay has shown how the oxytocin receptor gene can be activated by hormonal treatment (Fig. 11.8). The oxytocin receptor, in turn, is one of several transcriptional systems that lead from estrogen treatment to female reproductive behavior (Fig. 11.9).

Evidence that such histone changes can cause behavioral change comes from experimentally inhibiting the removal of histone acetyl groups in the ventromedial hypothalamus (Fig. 11.10). Such inhibition led to greater acetylation of gene promoters in the ventromedial hypothalamus and enhanced levels of female-typical

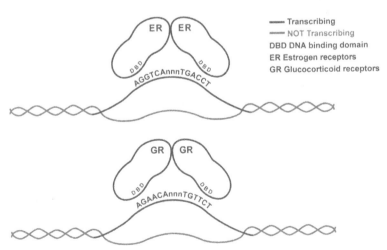

FIGURE 11.3    Estrogen receptors (ER) bind to different DNA response elements than do glucocorticoid receptors (GR).

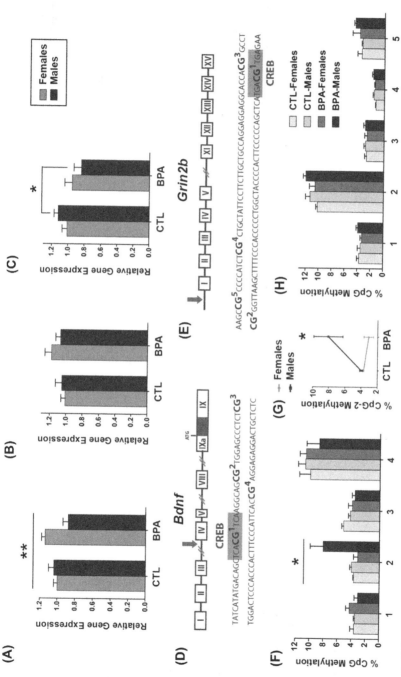

**FIGURE 11.4** Sex-specific changes in hippocampal *Bdnf* gene expression and DNA methylation at postnatal day 28 after prenatal bisphenol A exposure. Gene expression analysis of (A) *Bdnf*, (B) *Grin2b*, and (C) *Gadd45b* in the P28 hippocampus of males and females prenatally exposed to BPA are plotted, and coordinate with changes in DNA methylation at four different sites (F). Adapted from Kundakovic, M., Gudsnuk, K., Herbstman, J.B., Tang, D., Perera, F.P., Champagne, F.A., 2015. DNA methylation as a biomarker of early-life adversity. PNAS 112, 6807–6813.

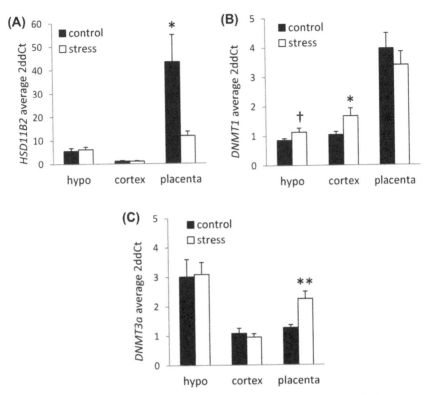

FIGURE 11.5 Changes in the expression of a steroid hormone metabolic enzyme (HSD11B2) and DNA methyltransferase (DNMT3a) in tissues (placenta, hypothalamus, and cortex) of the offspring following prenatal stress. *Adapted from Jensen Peña, C., Monk, C., Champagne, F.A., 2012. Epigenetic effects of prenatal stress on 11β-hydroxysteroid dehydrogenase-2 in the placenta and fetal brain. PLoS One 7, e39791.*

## E treatment acetylates H4

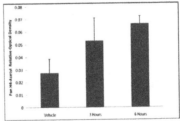

## ERα cells contain H4 Acetylated Histones

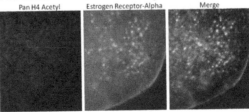

FIGURE 11.6 In ventromedial hypothalamic neurons, estrogen treatment is followed by increased acetylation of the N-terminus of histone H4.

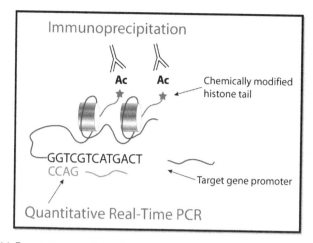

FIGURE 11.7   Illustration of the dual specificity of chromatin immunoprecipitation. A specifically modified histone protein can be shown to bind to a particular part of a gene's promoter.

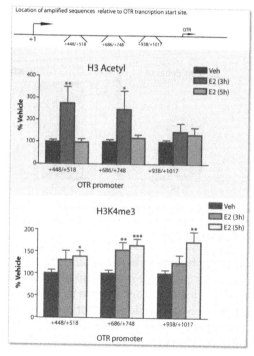

FIGURE 11.8   Estrogen treatment is followed by particular increases in histone 3 acetylation and histone3/lysine 4 trimethylation in ventromedial hypothalamic neurons. Both of these histone modifications are pro-transcriptional.

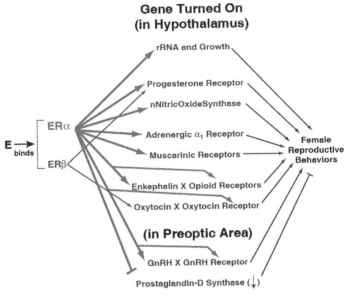

FIGURE 11.9   Estrogens, having bound to estrogen receptor alpha or estrogen receptor beta, increase messenger RNA levels for several genes whose products foster female reproductive behavior.

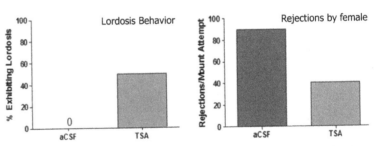

FIGURE 11.10   Increasing histone acetylation in the ventromedial hypothalamic nucleus by microinjecting a deacetylase inhibitor there increases female reproductive behavior (lordosis) and decreases rejections of the male.

sex behavior. As described in earlier chapters, the consummatory aspects of sex behavior in female rats are quantified as the frequency of the arched-back lordosis posture relative to the number of times the female is mounted by the male. In addition, a lack of sexual motivation is indicated by other postures associated with rejection of the male. Inhibition of deacetylation of histones in the ventromedial hypothalamus significantly increased the incidence of lordosis behavior and decreased the incidence of rejections of the male. These experiments directly link molecular epigenetic mechanisms involving histone protein modification to hormonal effects on behavior.

Another application of histone protein analysis tied to hormone effects in the brain concerns stress. Hunter et al. (2015) found that acute stress had opposite effects on two different locations of histone N-terminus methylation in hippocampal neurons. A still more exotic effect of acute stress was the finding that a specific form of histone methylation, amino acid lysine #9 on protein histone #3, was associated with the repression of transposons ("jumping genes") that would be deleterious to the health of hippocampal neurons. Both the chemical details and the behavioral consequences of these discoveries remain to be worked out, but they provide more examples of histone protein modification effects on hormone-modulated behavior.

2. DNA methylation. Here, a methyl group (one carbon atom and three hydrogen atoms) is added to a specific cytosine base, particularly in the promoter of the gene. Initially, in every instance, such a chemical change was assumed to repress transcription of the methylated gene. Recently, this interpretation has been called into question. In general, methylation is the most frequent epigenetic modification.

Many natural and synthetic compounds mimic the action of hormone ligands acting on their receptors. These environmental "endocrine disruptors" can act on hormone receptors during gestation and influence developmental processes. Endocrine-disrupting compounds interfere with hormone action, including the actions of estrogens, androgens, thyroid hormones, and glucocorticoids. They affect steroid biosynthesis, steroid-sensitive neurotransmitter systems, hormone degradation and/or elimination, reproductive development, physiology, and behavior (reviewed by Gore and Patisaul, 2010). The earliest phases of fetal and neonatal development are the periods of greatest cellular plasticity; thus, organisms are most susceptible to endocrine-disrupting compounds during early development. Endocrine-disrupting compounds can affect sexual differentiation, i.e., masculinization and demasculinization, and feminization and defeminization, of morphology, physiology, and behavior. The mechanisms are of great interest in biomedical research, ecology, and environmental biology. Many endocrine-disrupting compounds have adverse effects on fertility, and some endocrine-disrupting compounds are known to have feminized most of the members of entire populations (e.g., the American alligators in Lake Apopka, Florida), which has caused alarming crashes in population size.

Some endocrine-disrupting compounds have brought species to the brink of extinction (e.g., the bald eagle). Effects of endocrine disruptors on the mammalian reproductive system include, but are not limited to, inhibited spermatogenesis and ovulation, cryptorchidism, hypospadias, decreased semen quality, premature ovarian failure, polycystic ovary syndrome, endometriosis, accelerated puberty, masculinized genitalia in genetic females, feminized yolk production (vitellogenesis) in males, initiation of mitosis in estrogen-sensitive breast cancer cells, disrupted hypothalamic–pituitary–gonadal function, and altered sex and social behavior (Gore and Patisaul, 2010). Endocrine-disrupting compounds have widespread effects on energy metabolism, energy expenditure, energy intake, and energy storage, especially the processes involved in adipocyte differentiation and body fat distribution (reviewed by Schneider et al., 2014). Endocrine-disrupting compounds can alter the sex of human babies; for example, they can masculinize the genitalia of an otherwise healthy XX child. With this wide array of effects on early aspects of sexual differentiation, there is no question that endocrine disruptors influence behavior. One heavily investigated endocrine disruptor is bisphenol A, and there is evidence that this and other endocrine-disrupting compounds have epigenetic effects on brain and behavior via DNA methylation. Kundakovic et al. (2013) have taken the lead in exploring how DNA methylation can translate the results of early life experiences into behavioral change, showing that prenatal exposure of laboratory animals to bisphenol A results in sex-specific changes in hippocampal neuronal gene expression for the neuropeptide, brain-derived neurotrophic factor. These sex-specific changes in the nervous system correlate with DNA methylation (Fig. 11.4).

Jensen Peña et al. (2012) investigated the consequences of prenatal stress for DNA methylation in the placenta, an organ of obvious importance for fetal development (Fig. 11.5). In the placenta, prenatal stress is associated with (1) increased DNA methylation at specific CpG sites within the promotor region of an important steroid hormone metabolizing enzyme, 11β-hydroxysteroid dehydrogenase type 2 (HSD11B2); (2) a significant decrease in HSD11B2 mRNA levels; and (3) increased mRNA levels of the DNA methyltransferase DNMT3a. In contrast, within the fetal hypothalamus, prenatal stress resulted in decreased CpG methylation within the HSD11B2 promoter and increased methylation at sites within exon 1, and it had no effect on HSD11B2 mRNA levels. Additionally, within the fetal cortex, prenatal stress did not alter HSD11B2 mRNA or DNA methylation. DNA methylation chemistry thus can be used to investigate early environmental and hormonal events that could influence a variety of behaviors later in life.

3. Noncoding RNAs. Both long, noncoding RNAs and short RNA sequences, only about 20–25 bases in length, can have global structural consequences in the nerve cell nucleus. Rinn et al. (2014)

observed how a long, noncoding RNA can enforce the colocalization of genetic loci, even across different chromosomes. That is, gene expression is determined not only by close range interactions with the gene promoter, but also by long range interactions with adjacent chromosomes.

The multiplicity and complexity of noncoding RNAs and the recency of their discoveries tell us that these are "early days" in this aspect of hormone/brain/behavior relations. Suffice it to say that noncoding RNAs are differentially expressed in the brain, both in time and in neuroanatomical locations. And, their deletions have proven neurochemical consequences. How these phenomena play into hormone effects on behavior remains to be discovered.

Many of the developmental processes discussed in previous and subsequent chapters are known to involve epigenetic mechanisms. Certainly, many of the effects of prenatal and childhood stress on adult hypothalamic–pituitary–adrenal function and behavioral responses to stress can be categorized as epigenetic. As discussed in Chapter 9, the quality and quantity of parental care a rodent pup receives during the first few days of life modulates its susceptibility or resilience to stress during adulthood. This occurs largely through changes in the glucocorticoid receptor in the hippocampus, thought to be epigenetic, at least in part, because the effects of postnatal experience persist into adulthood and are passed to the next generation (or generations) of pups born to the offspring of the original mothers.

Strong evidence for the role of epigenetic mechanisms in resilience to stress is mounting. For example, offspring that receive high-quality care during an early, critical period of development show increased expression of a nerve growth factor-inducible protein, NGFI-A. NGFI-A binds to the first exon of the glucocorticoid receptor gene, thereby increasing the expression of the glucocorticoid receptor. The mechanism whereby NGFI-A increases expression of the glucocorticoid receptor involves demethylation of NGFI-A and the acetylation of histones. Additionally, cross-fostering newborn offspring to mothers that give low-quality maternal care can reverse these molecular and behavioral changes. And, the effects of cross-fostering to low-quality care-giving mothers can be mimicked by hippocampal infusion of methionine, a histone deacetylase inhibitor. It is important to note that these are context-dependent changes, in that having a particular maternal environment alters the later maternal behavior of the offspring; these offspring grow up to have a similar level of maternal behavior, and so on in each generation. The effects seldom persist for more than five generations and can be at least partially reversed by early cross-fostering to a mother that offers a different level of care. In other words, the manifestation of the epigenetic change depends upon the context of rearing. Similarly, in human beings, early adoption into a home that provides a healthy and supportive rearing environment often can counter the effects of violent and abusive parental patterns.

Gestational programming of adult characteristics, including obesity or leanness, body fat distribution, and dietary preferences also can be traced to epigenetic mechanisms. The normal onset of puberty also is modulated by epigenetic changes, some of which are upstream and some of which are downstream from the initiation of the GnRH pulse generator. In Chapter 12, you will find that puberty is initiated upon DNA methylation of the genes for two key members of the Polycomb complex (*Eed* and *Cbx7*). The result of DNA methylation is a decrease in the expression of these genes. Through a series of steps, this decreased expression leads to increased expression of the gene that encodes kisspeptin. Increased secretion of kisspeptin stimulates GnRH secretion, and thus the entire HPG system, at the time of puberty (reviewed by Ojeda and Lomniczi, 2014). In Chapter 7, you learned of two other gene products, MKRN3, which delays puberty by silencing the gene for kisspeptin, and FGF21, which represses the expression of vasopressin from the suprachiasmatic nucleus, thereby decreasing kisspeptin secretion. The structural features of MKRN3 and its pattern of expression in the developing hypothalamus suggest that MKRN3 might function much like Polycomb group proteins. Thus, recent lines of evidence indicate that initiation of puberty is another developmental process controlled largely by transcriptional repression.

In conclusion, there is no doubt that all the epigenetic chemistries illustrated earlier have roles in the development of hormone/behavior relations. Some are context-dependent, such as the effect of maternal care on subsequent maternal behavior, and others, such as the effects of some endocrine-disrupting compounds and diet, are passed through the germline (cells that are designated in the first mitotic division of the zygote to become gametic cells). Germline epigenetic changes can last more than five generations and do not depend on context. Some are as simple as histone modification or DNA methylation. Others likely are involved with the genetic complexity of "genomic imprinting," in which the allele of a specific gene coming from the father's chromosome or the allele coming from the mother's chromosome is preferentially expressed (Keverne et al., 2015).

# SOME OUTSTANDING QUESTIONS

Despite the emerging use of epigenetic techniques for investigating maternal behaviors, reproductive behaviors, and responses to stress, as illustrated before, we are just beginning to face questions such as the following (Fig. 11.11):

1. Among at least four levels of regulation (transcription, translation, protein location in the nucleus, and availability of binding surfaces on each protein), how can we investigate more than one level in the same experiment?
2. Among the different kinds of epigenetic alterations, since they would rarely happen at the same time, which happens first in any given, behaviorally important event?
3. How can we specifically inhibit a single epigenetic alteration underlying a behaviorally important event, to show that such an epigenetic alteration is required for the behavior?
4. Epigenetic effects on stress responses as programmed by early life experience can be partially reversed by cross-fostering to a family that provides a different early life experience. How might we develop pharmacological therapeutic treatments that counteract harmful epigenetic effects?
5. How might epigenetic phenomena alter the effects of genes that influence reproductive success in particular environments?

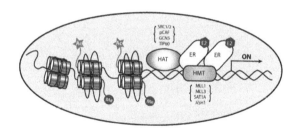

- At least 4 levels of regulation (transcription, translation, position of protein in cell nucleus, availability of binding surfaces. Which coactivators involved in ER/VMH?
- Order of events (acetylation, methylation, phosphorylation and their interactions).
- Specific inhibitors of coactivator/receptor interactions? Different coactivators/different receptor protein surfaces?

FIGURE 11.11 Conceptual questions about hormonal regulation of gene transcription that are currently faced by neuroscientists and by "hormone action" labs.

# Puberty Alters Hormone Secretion and Hormone Responsivity and Heralds Sex Differences

To appreciate the profound transformation that occurs at puberty, we need only compare ourselves as children to ourselves as young adults. Perhaps most memorably, puberty brings a heightened awareness of our own developing bodies and a newfound fascination with the bodies of others. Some of us develop a distracting attraction to members of the opposite sex. Others become aware of a sexual and romantic attraction to those of the same sex or to both sexes. While we fulfill the stereotype of the distracted pubescent teen, we also make great leaps in our intellectual abilities and expand our repertoire of social skills. In fact, most of us progress from being mostly dependent upon our parents to being fully independent individuals. Puberty is fraught with contradiction and complexity. We take pride in our newborn autonomy, but we are plagued by anxieties and self-doubt. We are surging haphazardly into adulthood, but all the while, we are developing important decision-making and coping strategies. Not coincidentally, these changes occur in tandem with our metamorphosis from a sexually immature, infertile child to a sexually mature, fertile adult.

These changes give undeniable evidence that our behavior and our perspectives on the world are shaped by two biological phenomena. First, the development of our physiology and behavior is tied to the development of the reproductive system and its gonadal steroid hormones. Changes in hormone secretion alone, however, do not control our full transformation from childhood to adulthood. For example, treatment with high, peripubertal levels of gonadal steroids do not elicit full-blown adult sexual behaviors in children or in juvenile nonhuman animals. Thus, the prepubertal brain is not yet ready for steroid

*Principles of Hormone/Behavior Relations*
http://dx.doi.org/10.1016/B978-0-12-802629-8.00012-7

activation of reproductive motivation and behavior. Appropriate maturation of steroid-sensitive behavioral circuits must occur during early adolescence to allow the brain to respond to the change in gonadal steroid secretion that occurs at the time of puberty.

The second biological phenomenon involved in puberty is that we are subject to an inborn, internal clock that determines the timing of development over our lifespan. Our responses to the pubertal rise in gonadal hormones include neural events that are activational (transient) and others that are organizational, i.e., they involve permanent changes in neural architecture and function, but they only occur at the critical phase of development between childhood and adulthood. In Chapter 9, we introduced the notion of a critical period, defined as a prenatal time-window, during which steroids can permanently masculinize the brain and behavior. In this chapter, we provide evidence that another important developmental phase occurs at the time of puberty. Thus, puberty is not only a time of sociosexual awakening but also a time to solidify and refine the organized neural substrates that influence masculinity and femininity.

Instead of dividing this chapter into discussions of basic and clinical examples, we have treated the material in a way more convenient to its nature, first illustrating some phenomena and then exploring their mechanisms.

## THE PHENOMENA OF PUBERTY

Puberty, the time when individuals become reproductively competent, represents one of the more obvious transitions in the endocrine environment of the brain and pituitary gland. In humans, over a period of several years, changes in the gonads and other organs associated with reproduction and secondary sexual characteristics offer the clinician clear developmental stages to track (Table 12.1). Puberty often coincides with adolescent development, the time between childhood and adulthood when there are changes in brain development that underlie many cognitive, social, and emotional maturational processes. With regard to behavior, hormonal changes during puberty impose dramatic changes. The word *dramatic* is appropriate, both figuratively and literally, to describe these changes. Puberty is a time when emotions run high. Part of this is related to the emergence of sexual awareness, thoughts, motivation, reactions, and behaviors. The emergence of sexuality in the adolescent provokes new reactions from adults and peers. The dramas of puberty depend in part upon the societal attitudes surrounding sexuality, but even in nonhuman primates, the first sexual advances of peripubertal females is a time of widespread social disruption.

**TABLE 12.1**   Human Development Stages

*FEMALE DEVELOPMENT*

| Breast Stages | Description | Pubic Hair Stage | Description |
|---|---|---|---|
| B1 | Prepubertal: elevation of the papilla only | PH1 | Prepubertal; no pubic hair |
| B2 | Breast buds are noted or palpable with enlargement of the areola; this is quite subtle and often missed on examination | PH2 | Sparse growth of long, straight, or slightly curly minimally pigmented hair, mainly on the labia; this stage is very subtle and sometimes missed on cursory examination |
| B3 | Further enlargement of the breast and areola with no separation of their contours | PH3 | Considerably darker and coarser hair spreading over the mons pubis |
| B4 | Projection of areola and papilla to form a secondary mound over the rest of the breast | PH4 | Thick adult-type hair that does not yet spread to the medial surface of the thighs |
| B5 | Mature breast with projection of papilla only | PH5 | Hair is adult in type and is distributed in the classic inverse triangle |

*MALE DEVELOPMENT*

| Genital Stages | Description | Pubic Hair Stages | Description |
|---|---|---|---|
| G1 | Preadolescent | PH1 | Preadolescent; no pubic hair |
| G2 | The testes are more than 2.5 cm in the longest diameter excluding the epididymis; and the scrotum is thinning and reddening | PH2 | Sparse growth of slightly pigmented, slightly curved pubic hair mainly at the base of the penis; this stage is very subtle and may be missed on cursory examination |
| G3 | Growth of the penis occurs in width and length and further growth of the testes is noted | PH3 | Thicker, curlier hair spread laterally |
| G4 | Penis is further enlarged and testes are larger with a darker scrotal skin color | PH4 | Adult-type hair that does not yet spread to the medial thighs |
| G5 | Genitalia are adult in size and shape | PH5 | Adult-type hair spread to the medial thighs |

*Modified from Marshall, W.A., Tanner, J.M., 1969. Variations in pattern of pubertal changes in girls. Arch. Dis. Child. 44 (235), 291–303; Marshall, W.A., Tanner, J.M., 1970. Variations in pattern of pubertal changes in boys. Arch. Dis. Child. 45 (239), 13–23.*

## II.  HISTORY: HORMONE EFFECTS CAN DEPEND ON FAMILY

As well, adolescence is the time for organization of neural circuits that underlie enormous cognitive and intellectual development. Adolescents seek and ultimately achieve autonomy, and this, too, provokes both positive and negative reactions from surrounding adults. Puberty thus is a span of years in which adolescent boys and girls are vulnerable to some of the most delicate and sometimes fraught developmental events. How young adults adapt both emotionally and physically to these changes can have life-long consequences.

Whereas previous chapters emphasized gestational critical periods in which gonadal steroid hormones masculinize/defeminize or feminize/demasculinize neural circuits and behavior, this chapter recognizes the peripubertal period as another phase of masculinization/defeminization or feminization/demasculinization. Sexual differentiation of brain and behavior thus can be seen as a multistage process of behavioral development in which an important "wave" of adolescent organization builds upon the traditional prenatal period of sexual differentiation. During the adolescent phase, the structures that were organized during fetal development are solidified and refined by the pubertal increases in sex-specific steroid hormones, i.e., elevated androgens in males and elevated and cyclic increases in estradiol and progesterone in females (reviewed by Sisk and Zehr, 2005).

Of the many pubertal changes, the most obvious are the changes in the hypothalamic–pituitary–gonadal (HPG) system. Prior to puberty, gonadotropin-releasing hormone (GnRH) mRNA and protein are expressed at very low levels that are insufficient to initiate secretion of the gonadotropins, luteinizing hormone (LH), and follicle-stimulating hormone (FSH). Without appropriate stimulation from the pulsatile secretion of FSH and LH, the gonads produce very low levels of gonadal steroids. The onset of puberty is characterized by an increase in the pulse amplitude and frequency of GnRH, LH, and FSH secretion, and a resulting increase in gonadal steroid synthesis and secretion. In female puberty, the hypothalamus and pituitary develop the capacity to show estradiol-positive feedback, i.e., the GnRH cells respond to high, prolonged levels of estradiol and progesterone by releasing a surge of GnRH. The GnRH surge initiates the ovulatory surge of LH and ovulation. Postpubertal women, unlike prepubertal adolescent girls, show a 28-day fluctuation of estradiol and progesterone secretion. Peripubertal boys undergo a rise in GnRH, LH, FSH, and testosterone secretion. Under the influence of elevated levels of gonadal steroid hormones, libidinous interests in both sexes undergo a marked rise over a period of several years.

There has been a trend in Western, industrialized nations toward an earlier onset of puberty. Sex differences occur as the consequences of the pubertal rise in gonadal hormones. Early puberty may have more deleterious effects on the psychosocial development of girls, whereas late puberty may be more deleterious for boys. Puberty also is marked by elevations in aggressive behavior that are significantly more common in boys than in girls under a wide variety of circumstances. The easiest form of aggression to quantify is the rate of homicides. Wilson et al. (2002) charted murders of unrelated persons of the same sex as a function of the age of the killer (Fig. 12.1). Two facts are immediately clear. First, males kill more frequently than females. Second, the homicide rate rises tremendously at puberty and falls gradually after the early 20s. These data transcend national boundaries and are not limited to Western societies.

These correlations have been further explored in experiments with laboratory rodents, and there is some support for the idea that peripubertal increases in aggression are at least partially accounted for by increases in circulating androgens and their metabolites, including estrogen and dihydrotestosterone. These steroids interact with serotonergic and other neurotransmitter systems in the forebrain to facilitate aggressive behaviors. In addition, genetic predisposition and experience can increase or decrease the probability of pubertally increased aggressive behavior. Experiments that examine aggression in animals that lack steroid receptors illustrate both the effects of steroid action on aggression and the importance of the

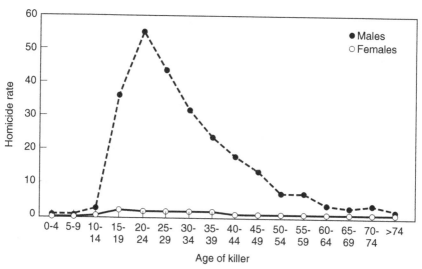

**FIGURE 12.1**   Age-specific homicide rates (homicides per million persons per annum) for men and women who killed an unrelated person of the same sex in Canada, 1974–92 (upper panel), and in Chicago, 1965–89 (lower panel). Data include all homicides known to police in which a killer was identified. *From Wilson, M., Daly, M., Pound, N., 2002. An evolutionary psychological perspective on the modulation of competitive confrontation and risk-taking. In: Pfaff, D.W. (Ed.), Hormones, Brain, and Behavior, vol. V. Elsevier, San Diego, p. 389. With permission.*

developmental period. For example, male mice that have the gene for the estrogen receptor knocked out show high levels of vicious biting attacks when tested at puberty, but not later in life. Whereas genetic predisposition and experience are important, the causal effects of androgens and estrogens on aggression is worthy of further study, especially given the increased use of anabolic androgenic steroids in adolescent boys and even in girls.

We hasten to point out that there are documented effects of socioeconomic status, size of schools attended by the boys, gang membership, initiation rites, and gun availability on aggressive behavior. In addition, these factors might interact with rising levels of androgens and/or low levels of serotonin. Just as crucial is the treatment of children by parents, siblings, peers, and teachers. Food availability, and the consequent level of calories and nutrition, are also important during the prenatal, neonatal, and pubertal periods (see Chapter 11). In particular, maternal or nutritional deprivation during gestation, which is sometimes associated with low birth weight, affects later aggressive tendencies. Certainly, early life stressors have a key influence on juvenile and adult behavior. More recently, it has become clear that levels of testosterone and other hormones have a reciprocal interaction with experience (discussed in Chapter 7). For example, experience with winning, experience with success, and experience with power can increase levels of testosterone in both men and women. Thus, cultural and social environments that promote these experiences in men and women might also increase levels of circulating androgens and their metabolites (which in turn might affect future assertive and/or aggressive behaviors). All these social and psychological factors (acting in a manner consistent with the focus of Chapter 21) must be considered together with androgenic hormone action for a comprehensive understanding of violence exhibited by teenage boys and young adult men.

Pubertal years are the time during which sex/gender differences in several major psychiatric disorders emerge. For example, in several nations, mid-to-late pubertal and adult women are more likely than mid-to-late pubertal and adult men to be diagnosed with depression. As mentioned in Chapter 9, in the United States the ratio of depression in adult women to men is about 2:1, and in Denmark, it is about 6:1. Mechanisms underlying the peripubertal appearance of this sex difference in rates of depression are unknown, and theories span the range from hormonal (increasing estrogen vs. increasing testosterone) to psychosocial (greater peer pressure for girls to conform socially, with consequently greater opportunity for lower self-esteem).

## Interim Summary 12.1

1. Puberty is characterized by profoundly increased activity of the HPG system, including increased gonadal steroid secretion and consequent

maturation of the body, secondary sex characteristics, and sexual motivation, awareness, and behavior.

2. The onset of puberty heralds the appearance of many differences between males and females that include differences in levels of aggression and the incidence of various physical and psychiatric diseases.

3. Hormones and environment interact to either accentuate or diminish the sex differences in behavior that arise at puberty.

## MECHANISMS OF PUBERTY

As noted earlier, puberty is heralded by large increases in the synthesis and secretion of hormones associated with growth and reproduction (Fig. 12.2). The pituitary hormones, growth hormone (GH), LH, and FSH, must be secreted in pulsatile form to stimulate somatic growth and gonadal steroid secretion (see related material later). In general, their biological actions during puberty depend upon greater pulse amplitudes, not pulse frequency.

What are the neuroendocrine mechanisms that allow these drastic transitions? It is likely that puberty begins in the brain because the gonads are not required for pubertal maturation. Animals that lack gonads from birth show pubertal maturation of the GnRH system regardless of the lack of steroid secretion. The initiation of puberty requires increases in the secretion of GnRH from neurons in the caudal preoptic area and mediobasal hypothalamus. During the prepubertal juvenile period, neurons secreting GnRH are subjected to persistent transsynaptic inhibition. When this inhibition is lifted, GnRH secretion increases, which leads to puberty. Some of this inhibition is from various neurotransmitters. For example, at the onset of puberty in rhesus monkeys, there is decreased release of the inhibitory transmitter $\gamma$-aminobutyric acid (GABA) (Mitsushima et al., 1994; Fig. 12.4). In support of this idea, blockade of synthesis for a synthetic enzyme for GABA induces a pubertal-like GnRH release in monkeys. Furthermore, blocking GABA receptors causes precocious puberty (see Chapter 8).

In addition to neurotransmitter suppression and release, there are other neurotransmitters and neuropeptides with stimulatory effects on GnRH secretion, and these may play a role in the onset of puberty. Some of the prepubertal inhibition of GnRH secretion stems from modulators of the expression of genes that encode these stimulatory neurotransmitters and neuropeptides. GnRH secretion is stimulated by excitatory transmitters and stimulatory neuropeptides impacting GnRH neurons. For example, puberty in rhesus monkeys can be advanced by treatment with N-methyl-D-aspartate, which stimulates a subset of receptors for the excitatory transmitter glutamate (Plant et al., 1989; Fig. 12.3). The

II. HISTORY: HORMONE EFFECTS CAN DEPEND ON FAMILY

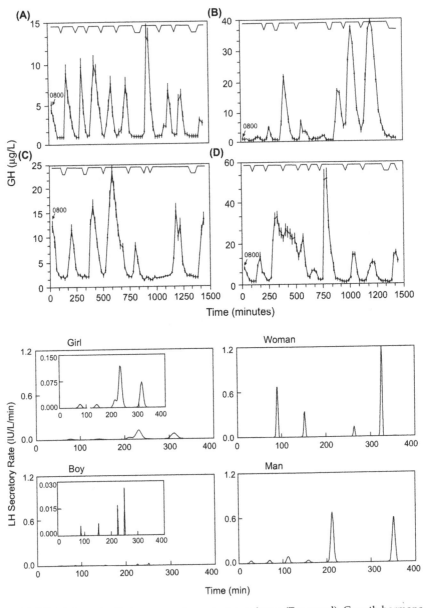

**FIGURE 12.2** Pulsatile hormone release during puberty. (Top panel). Growth hormone (GH) secretory profiles of three boys. (A) A 14-year-old boy, prepubertal. Note the pulses. (B) Another 14-year-old boy going through puberty. Note the larger numbers on the Y-axis. (C) A hypogonadal 14-year-old boy before testosterone therapy. If a boy is allowed to grow up hypogonadal, there are serious implications for his level of libido. (D) The same boy as in C, after testosterone administration. Note the larger numbers on the Y-axis. (Bottom panel). Luteinizing hormone (LH) secretory profiles. Note the larger Y-axis numbers for the postpubertal woman and man, compared to the girl and boy, respectively. *(Top panel) From Mauras, N., Blizzard, R.M., et al., 1987. Augmentation of growth hormone secretion during puberty: evidence for a pulse amplitude-modulated phenomenon. J. Clin. Endocrinol. Metab. 64 (3), 596–601. (Bottom panel) From Clark, P.A., Iranmanesh, A., et al., 1997. Comparison of pulsatile luteinizing hormone secretion between prepubertal children and young adults: evidence for a mass/amplitude-dependent difference without gender or day/night contrasts. J. Clin. Endocrinol. Metab. 82 (9), 2950–2955.*

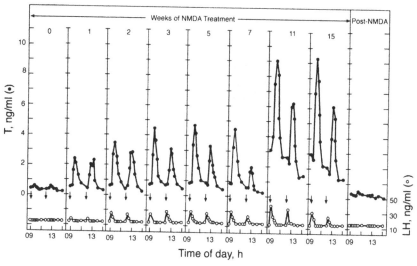

**FIGURE 12.3** Premature activation of the hypothalamic–pituitary–testicular axis in immature male rhesus monkeys induced by repetitive stimulation of the hypothalamus with N-methyl-D-aspartate (NMDA), an excitatory amino acid receptor agonist, administered once every 3h for 15 weeks. The treatment was initiated 1.52 years before the normal age of puberty in this species. The *arrows* indicate the time of injection. *LH*, luteinizing hormone; *T*, testosterone. *From Plant, et al., 1989. Puberty in monkeys is triggered by chemical stimulation of the hypothalamus. Proc. Natl. Acad. Sci. U.S.A. 86 (7), 2506–2510. With permission.*

neuropeptide kisspeptin and its receptor, GPR5, are important permissive factors for pulsatile GnRH secretion, and there is support for the idea that kisspeptin, encoded by the *Kiss1* gene, is a permissive factor in the onset of puberty. One hypothesis is that, prior to puberty, *Kiss1* gene expression in the arcuate nucleus is under the suppressive influence of a group of proteins known as the Polycomb complex. According to this idea, puberty is initiated upon DNA methylation of the genes for two key members of the Polycomb complex (*Eed* and *Cbx7*). The result of DNA methylation is a decrease in the expression of these genes and a consequent decrease in the association of the protein products with the *Kiss1* promoter, resulting in an increase in the expression of *Kiss1*. Increases in synthesis and secretion of kisspeptin stimulate GnRH secretion, which marks the onset of puberty (reviewed by Ojeda and Lomniczi, 2014). Of interest is that these influences on GnRH are not limited to neurons (e.g., Ojeda et al., 2000). Glial cells (astrocytes) also control GnRH secretion through growth factor pathways, including transforming growth factor-alpha (TGF-alpha) and the neuregulins, both of which act through tyrosine kinase receptors such as erbB-1 and erbB-4. The full complexity of the mechanistic steps involving astrocytes can be appreciated from Fig. 12.5.

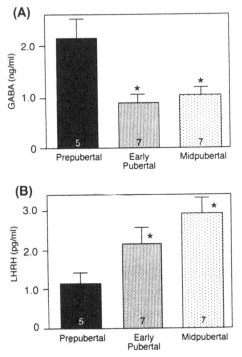

**FIGURE 12.4**   Decrease in γ-aminobutyric acid (GABA) release (A) from the rhesus monkey hypothalamus at the time of puberty. The decrease coincides with the pubertal increase in luteinizing hormone releasing hormone (LHRH) release (B). Both GABA and LHRH were measured in conscious animals, via push-pull perfusion of the stalk-median eminence of the hypothalamus. Numbers inside each bar represent number of animals per group. *, Significantly different from prepubertal value. *From Mitsushima, D., Hei, D.L., Terasawa, E., 1994. Gamma-aminobutyric acid is an inhibitory neurotransmitter restricting the release of luteinizing hormone-releasing hormone before the onset of puberty. Proc. Natl. Acad. Sci. U.S.A. 91 (1), 395–399. With permission.*

Puberty is a continuation of the process of sexual differentiation that began prenatally and involves the maturation of both reproductive and nonreproductive behaviors. Testicular hormones secreted at the time of puberty can both masculinize and defeminize behavioral responses in adulthood. To examine the importance of pubertal testosterone, comparisons have been made among groups of males castrated just before puberty and treated either with testosterone or with vehicle. A third group of gonadally intact males serves as controls. As would be expected, male Syrian hamsters castrated prior to puberty and treated with vehicle receive little or no testosterone stimulation at puberty. They are then tested later in adulthood for their behavioral response to testosterone. Castrates treated with vehicle show lower levels of testosterone-induced, male-typical sex behavior compared to males that retained their testes during puberty and compared to castrated males treated with testosterone during puberty.

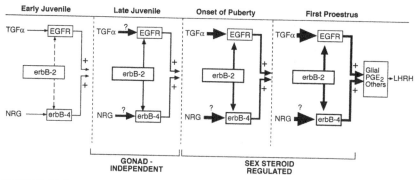

**FIGURE 12.5** Changes in hypothalamic erbB receptor expression during juvenile and peripubertal development of the female rat. The first increase in synthesis occurs at the end of the juvenile period in the absence of changes in gonadal steroid secretion. This first increase is postulated to be a component of the centrally activated, gonad-independent process that sets in motion the initiation of puberty. Once puberty is initiated, erbB receptor synthesis is further increased, first by rising estrogen levels, and then by the concerted action of estrogen plus progesterone on the day of the first preovulatory surge of gonadotropins. Whether there is a sex-steroid-independent increase in transforming growth factor (TGF) and/or NRG gene expression during juvenile development is not yet known. *From Ojeda, S.R., Ma, Y.J., Lee, B.J., Prevot, V., 2000. Glia-to-neuron signaling and the neuroendocrine control of female puberty. Recent Prog. Horm. Res. 55, 197–223. With permission.*

Both gonadally intact males and pubertally castrated males treated with testosterone during puberty improve their sexual proficiency with repeated sexual experience, whereas castrated males treated with vehicle (no testosterone at puberty) fail to show this improvement. Similarly, adult testosterone treatment does not activate male-typical aggressive behavior (flank marking) in either prepubertal hamsters or in hamsters castrated prior to puberty and treated with vehicle throughout puberty. These results suggest that peripubertal androgens play a strong role in masculinizing adult aggressive behaviors. In other species, including rats and humans, similar effects of peripubertal hormones occur with regard to the development of a number of sociosexual behaviors, including sexual preference for a male versus a female mating partner (reviewed by Sisk and Zehr, 2005).

There are sexually dimorphic brain regions that are involved in male-typical sexual behavior and the HPG system in rats. The sexual dimorphism is explained in part by differences in cell number and morphology that arise at puberty. The sexually dimorphic region of the hypothalamus (SDN) is larger in males than in females, whereas the anteroventral periventricular nucleus (AVPV) is larger in females than in males. The size difference is apparent in neonates, but it is greater after puberty. Furthermore, manipulation of pubertal gonadal hormones can produce the predicted changes in cell number in theses brain regions. Prepubertal gonadectomy decreases the number of new cells in the male but not the female SDN, which suggests that the masculinization of this area is enhanced by the pubertal rise in testosterone in males. Prepubertal gonadectomy significantly decreases the number of new cells in the female but not the male AVPV, which suggests that the morphology of this area is further feminized by ovarian estradiol and/or progesterone secreted at puberty. A wide array of studies show that the hormones of puberty stimulate the addition and/or loss of new neurons and glia in hypothalamic, limbic, and cortical regions of the brain. This architectural remodeling might underlie sex differences in male-typical and female-typical sex behavior, but it also might underlie, in part, sex differences in vulnerability to depression and other disorders (reviewed by Juraska et al., 2013). The fact that adult behaviors and brain development are deficient in laboratory animals that miss the appropriate pubertal rise in gonadal steroids should caution parents, physicians, and patients who are faced with issues regarding adolescent dieting, eating disorders, and pharmaceutical interventions that tend to delay the onset of puberty.

## Interim Summary 12.2

1. Puberty begins in the brain and occurs even in gonadectomized animals.
2. The GnRH pulse generator in the hypothalamus is under inhibitory control prior to the onset of puberty by inhibitory neurotransmitters and by proteins that suppress expression of genes that encode stimulatory neuropeptides, such as kisspeptin.
3. The pubertal increase in gonadal hormones, testosterone in males and estradiol and progesterone in females, not only stimulates neural structures involved in adult behavior, but it also solidifies and refines the process of sexual differentiation of the brain. When the rise in hormones does not coincide with the pubertal window of development, males and females show long-term deficiencies in adult, sexually dimorphic behaviors and in cell numbers in specific brain regions. Thus, missing out on the regularly scheduled steroid stimulation at the time of puberty can permanently alter aspects of sexually differentiated behaviors.

### Growth and Feeding in Puberty

Puberty is a period of dynamic changes in body shape, size, and composition. The changes are sexually dimorphic, so in our species, males tend to be taller, heavier, leaner, more muscular, and have a tendency for adipose tissue deposition in the abdominal visceral area.

Females tend to have a higher body fat content, and fat deposition is concentrated subcutaneously, especially on the hips and thighs, the gluteofemoral region. These differences arise during pubertal development. The adolescent growth spurt and alterations in body composition depend on the release of gonadotropins, sex steroids, GH, and leptin. Bone mass is maintained by a constant yin/yang relationship between osteoblasts, which form bone, and osteoclasts, which reabsorb bone. In addition, there are cartilaginous plates at the growth ends of long bones; closure of these plates as they become ossified by the actions of osteoblasts defines the length of the bone. Testosterone plays a role in the growth spurt by increasing bone size and closing the growth plates. Nutrition, which provides both energy and specific nutrient intake, is a major determinant of pubertal growth because poor nutrition inhibits linear growth.

The relations between reproductive hormone secretion and energy balance are reciprocal. It has long been known that food deprivation delays pubertal development and inhibits GnRH and LH pulsatile secretion. The effects of food deprivation on the HPG system can be handily reversed by treatment with GnRH pulses at the species-specific frequency and amplitude. Thus, the effects of food deprivation or nutritional deficit occur primarily at the level of the hypothalamic GnRH pulse generator, although, depending on the degree of starvation, there may be effects on the pituitary and gonads as well. Conversely, a high level of energy availability is stimulatory for the HPG system and increases fertility and reproductive success. Increases in body weight, body fat, and obesity are thought to be among the factors that contribute to an acceleration of puberty (reviewed by Bronson, 1989; Sanchez-Garrido and Tena-Sempere, 2013).

The effects of food deprivation on reproduction, including GnRH secretion, can be reversed by treatment with various factors that are elevated in well-fed animals. For example, food deprivation-induced anestrus, hypogonadotropism, and the delay of puberty are reversed by treatment with the brain peptide kisspeptin. Kisspeptin is synthesized in two major brain areas in rodents, the AVPV and the arcuate nucleus of the hypothalamus. In sheep, kisspeptin is synthesized in the preoptic area of the hypothalamus and arcuate nucleus of the hypothalamus. Levels of *Kiss1* mRNA are reduced, and levels of Kiss1 receptor (Kiss1r) are increased, in animals that are food restricted to the point of delayed puberty or inhibited LH pulses. Kisspeptin is thought to act on GnRH neurons, most of which contain Kiss1r. In some model systems, however, pubertal development and reproduction continue without functional Kiss1r. Together, these results suggest that kisspeptin might be only one of a number of factors that influences puberty onset and might mediate nutritional effects on puberty (reviewed by De Bond and Smith, 2014; Bellefontaine and Elias, 2014). Other results address mechanisms by which a high level of body fat and food intake can be stimulatory for kisspeptin-induced pulsatile GnRH secretion and fertility.

In females, puberty involves the commencement of pulsatile GnRH secretion, and the mechanisms are similar in males and females. In females, however, puberty is not completed until the first LH surge, which is triggered by a massive release of GnRH. Thus, an important part of female puberty is the ability to respond with high levels of estradiol with the GnRH and LH surges that induce ovulation. Kisspeptin secreted from the AVPV is critical for the GnRH surge. AVPV kisspeptin secretion is stimulated by arginine vasopressin input from the suprachiasmatic nucleus (SCN) of the hypothalamus. Puberty is delayed by deficits in the availability of metabolic fuels occurring during food restriction or deprivation. Among the many changes in the food-restricted female are increases in the secretion of a growth factor produced in the liver, fibroblast growth factor 21, FGF21. In laboratory rodents, increases in FGF21 inhibit expression of the gene that encodes arginine vasopressin in the SCN, thus removing the stimulatory input to AVPV kisspeptin neurons (reviewed by Ojeda and Lomniczi, 2014). These results suggest that a nutrition-related signal from the liver tells the female SCN that there is insufficient energy to commence pubertal maturation, thereby inhibiting the kisspeptin and GnRH secretion that initiates puberty. Inhibition of the reproductive system by a peripheral nutritional signal may be adaptive if it allows the female to conserve the energy necessary for survival during periods of harsh energetic conditions. If conditions improve, levels of FGF21 decrease, and the reproductive system is allowed to mature on schedule.

Prolonged food deprivation of young animals inhibits the increase in HPG function that is the hallmark of puberty, and treatment with the adipocyte hormone leptin can prevent the effects of food deprivation. Leptin treatment, however, does not drive puberty *per se*, although the evidence suggests that leptin might be another permissive factor that allows the increase in HPG function when the time of puberty arrives. Leptin is a hormone secreted from adipocytes, and leptin levels are positively correlated with levels of adiposity. Ghrelin is an orexigenic hormone secreted by cells in the gut in response to fasting, and circulating levels of ghrelin fall when the stomach stretches during eating. During fasting and body weight loss that inhibits the HPG system, plasma leptin concentrations fall, ghrelin levels increase, and the resulting lack of leptin and high level of ghrelin are thought to be a powerful stimulus for hunger and overeating in food-deprived animals. Ad libitum feeding is characterized by the opposite leptin:ghrelin ratio. In support of a role for these peripheral molecules in control of reproduction, leptin is stimulatory and ghrelin is inhibitory for *Kiss1* gene expression and for GnRH and LH secretion. Treatment with leptin prevents the inhibitory effects of fasting on pubertal development, the HPG system, and over-eating in rats and other species. There are leptin receptors and ghrelin receptors on at least a portion of the arcuate nucleus kisspeptin cells, providing circumstantial evidence for a mechanism whereby high levels of leptin and low levels of ghrelin can link a high level of nutrition with kisspeptin-induced HPG function. If high levels of leptin and/or low levels of ghrelin are important for kisspeptin-induced HPG function, the level of peripheral hormones might be detected by intermediate cells that project to kisspeptin cells (e.g., those that secrete neuropeptide Y or alpha-melanocyte-stimulating hormone).

Whereas the previous scenario supports a critical role for arcuate nucleus kisspeptin in control of puberty, at least one experimental result, from mice homozygous for a mutation in the leptin receptor, calls this idea into question. As would be expected if leptin is critical for pubertal development, mice homozygous for a mutation in the leptin receptor fail to develop a mature reproductive system. The unexpected result is that, in mice in which leptin receptors have been reactivated in the kis-speptin cells of the arcuate nucleus, puberty is not activated. Thus, in certain genetically engineered mice, leptin action on kisspeptin cells of the arcuate nucleus is not sufficient for normal reproduction. Still other results suggest that the critical location of leptin effects on reproduc-tion occur in a different brain area, the ventral premammillary nucleus, because lesions of the ventral premammillary nucleus prevent the rescue of puberty by leptin treatment. The ventral premammillary nucleus has leptin-responsive cells with direct synaptic connections to GnRH cells. This supports the idea that there is a more direct brain mechanism by which leptin can tell the HPG system that the animal is prepared for the energetically expensive processes related to reproduction (reviewed by Bellefontaine and Elias, 2014). By contrast to leptin's permissive effects on reproductive development, leptin's effects on body weight and bone mass may be mediated by different brain areas. Understanding the effects of the environment (including nutrition, calories, social cues, and seasonal changes) requires that we learn much more about the molecular events underlying normal pubertal maturation.

In summary, the mechanisms that initiate and modulate the onset of puberty are not fully understood, and only a few hypotheses have been provided in this chapter. Many different pubertal changes, some of which are only partially understood, affect reproductive behaviors, aggressive behaviors, territorial behaviors, cognition, and emotion, as well as the changes in body weight, height, adiposity, and fat distribution. These energy-balancing and reproductive traits are interrelated because success-ful reproduction is facilitated if the offspring are born to mothers that have reached an adequate size, level of strength, and plateau of maturity. In addition, reproduction is more likely to be successful if mating is timed so that offspring are born at times of energy abundance, at optimal ambient temperatures, and during mild weather conditions. Thus, the onset of puberty is influenced by age, speed of maturation, nutrition, the presence or absence of mating partners, social cues, and other environmental factors such as day length and season. Conversely, pubertal development affects growth, energy expenditure, sensitivity to sexual stimuli, and the disposition of metabolic fuels and body fat content.

## SOME OUTSTANDING QUESTIONS

1. What are the interactions among social and hormonal influences on libidinal changes during puberty?
2. How do sex differences in eating disorders arise? For example, *anorexia nervosa* is more common in teenage girls than in teenage boys. Is this disorder influenced by prenatal and/or pubertal estrogens and/or androgens?
3. What have been the causes of the trend toward early onset of puberty, and how do we determine which were the most important?
4. What accounts for the rise in depression diagnosed in girls during and after puberty, but not in boys?
5. Boys with short stature may be treated with a GnRH antagonist to delay the onset of puberty and prolong the growth of the long bones, so that they might grow a few more inches before fusion of the osteoplates. If the pubertal rise in testosterone does not occur within its normal developmental time frame, what are the effects on sociosexual development?

# Changes in Hormone Levels and Responsiveness During Aging Affect Behavior

*The world has gone daft with the latest new fad*
*For the fountain of youth has been found...*
*The latest sensation's the Sequard Elixir*
*That's making young kids of the withered and gray*
*There'll be no more pills or big doctor bills*
*Or planting of people in churchyard clay*

*19th century popular song.*

In these lyrics, the fountain of youth is Sequard's Elixir, a mixture of ground testicles from dogs and guinea pigs. The formula was published in 1889 by the physiologist/neurologist, Charles Edouard Brown-Sequard. The elderly Brown-Sequard injected himself with the aqueous solution (i.e., ground testicles and a little water), and he reported that within days he experienced an increase in his appetite, sleep, strength, and stamina that lasted for weeks. In addition, he reported improved mental energy and concentration, increased force of his urinary stream, and greater regularity of his bowel movements. Subsequent to Brown-Sequard's publication and public presentation of his findings, unscrupulous opportunists marketed the tonic without the consent or knowledge of Brown-Sequard. Sales were brisk, but most customers were disappointed. There were widely publicized protests, and the fiasco diminished the reputation of an otherwise esteemed scientist/physician and inspired many unflattering articles, cartoons, and jokes, including the preceding song lyrics (Aminoff, 2011). The putative therapeutic effect of the elixir was never supported by experimental evidence (i.e., it was a strong placebo effect), but the story of Brown-Sequard's Elixir is testimony to the power of an idea (or wish) that would not die: Circulating factors are secreted from various organs that travel through the blood stream in high concentration during youth, and high levels of these factors support the youthful function of the nervous system, metabolism, and other physiological processes.

*Principles of Hormone/Behavior Relations*
http://dx.doi.org/10.1016/B978-0-12-802629-8.00013-9

Brown-Sequard's focus was on hormones from the testes, and in fact, there are high circulating concentrations of gonadal steroid hormones in young men and women that decline with age. Research in this area is not restricted to gonadal steroid hormones or even to hormones. The circulating and cerebral concentrations of many different chemical messengers increase or decrease with age. In 2011, Villeda et al. provided initial evidence that a circulating factor might control aging of the brain. Aging in the nervous system includes a plateau or reversal of neurogenesis, i.e., growth and development of neurons and their dendrites. Most neural dendrites are covered with dendritic spines, and these structures are critical for synapse formation. The density of dendritic spines increases with the level of innervation, i.e., the number and strength of synapses. The young brain displays rapid learning and lasting memory associated with neural plasticity, the ability to respond to experience by strengthening and increasing the number of synapses on dendritic spines, as well as pruning, the ability to decrease spines and clean out inefficient connections.

Decreases in neurogenesis during aging can lead to lack of neural plasticity. During the aging process, cessation of neurogenesis and low dendritic spine density in particular brain areas are associated with declining learning and memory. To study the importance of blood-borne factors in the aging of the nervous system, researchers connected the circulatory systems of young and old mice. Young mice that received transfusions of the blood plasma from elderly mice showed a decrease in synaptic plasticity and an impaired ability to associate certain spatial stimuli with an aversive, fear-inducing stimulus. The effect was mimicked by infusion of eotaxin, a chemotaxin known as CCL-11 (Villeda et al., 2011). Eotaxin is a type of cytokine, or signaling molecule, that is increased in the plasma and cerebrospinal fluid of healthy aging men and women. This is a fascinating example of the reciprocal relations between chemical messengers and aging, but more research is necessary to determine cause and effect, and whether removing such circulating factors can slow the aging process without compromising immunity or other important processes necessary for survival.

Hormone-behavior relations during aging, and behaviorally relevant effects of hormones on diseases of the aged, are taking on new dimensions because of our increasing life span and growing elderly population (Fig. 13.1). In contrast, in a species that is likely to die at a young age due to predation, inanition, accident, and disease, these issues would be moot. This chapter documents significant changes over the lifespan in several hormones, their receptors, physiology, and metabolism, and reviews some of the initial investigations. Because of the plethora of possible clinical examples, we focus on those of importance for aging humans.

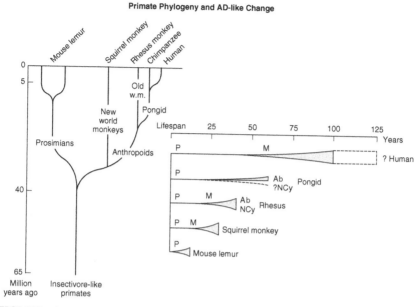

**Primate Phylogeny and AD-like Change**

FIGURE 13.1 Behaviorally relevant effects of hormones on the diseases of the aged. In this case, effects of hormones include those on menopause (M) and the production of the protein fragment A-beta (beta-amyloid), crucial in Alzheimer disease (AD).

## CLINICAL EXAMPLES

Aging in humans has been considered by some to begin at birth. While this might be a somewhat overstated perspective, it is true that some tissues begin the process toward senescence early in the life cycle. Consider the difference between an 18-year-old dog and an 18-year-old man. The dog's hair has turned gray; it has become mostly sedentary and arthritic, and it has lost much of its sight and hearing. The 18-year old dog is at the end of its life. The 18-year old man has only just begun his adulthood and is likely at the height of his physical strength and sensory acuity. Life span (the time from birth until death) and the process of aging (changes in organs or tissues over time that increase the probability of death) are programmed, at least partly, by the genes of each species. Hundreds of genes have been identified that can modify life span in a wide variety of organisms, including yeast, roundworms, flies, mice, and humans (reviewed by Tissenbaum, 2012). Across all these species, longevity is linked to the genes for the peripheral protein hormones insulin and insulin-like growth factor 1 (IGF-1). Mutations in insulin/insulin-like growth factor signaling pathway lead to impressive increases in lifespan as well as the "healthspan." In mammals, lifespan-extending mutations often reduce growth hormone and/or IGF-1 signaling (reviewed by Bartke, 2011). Related effects on lifespan and aging are environmentally produced. Lifespan is extended by restricting caloric intake, which produces many of the same downstream effects as mutations in the insulin/IGF-1 signaling pathway. Effects include increased stress resistance, improved insulin sensitivity, a shift from a pro- to an antiinflammatory profile of circulating adipokines, reduced mammalian target-of-rapamycin-mediated translation, and altered mitochondrial function, including greater utilization of lipids relative to carbohydrate fuels. It is not surprising that these effects reduce food intake, body weight, and body fat content. For example, exposure of skin to excess ultraviolet radiation (i.e., sunburn) in childhood can induce changes that result in prematurely aging skin and skin cancer in adulthood. With reference to endocrine influences on behavior, it is important to recognize that both neurotransmitter and neuroendocrine systems continually change during the life cycle, at different rates and in very different ways.

The human life span is increasing, so we now need to consider elderly people as the "old" and the "oldest old," the latter including those in their 80s and 90s and even over 100 years of age. These individuals comprise an ever-increasing percentage of the population; it has been estimated that half of the female children born in industrialized countries in the 21st century will live to be 100 years of age or more. Thus, studying the physiology of the oldest old, in particular their endocrinology, becomes increasingly important as this sector of the population grows.

After middle age, there are many hormonal changes in men and women, but not all hormone levels fall with advancing age. With regard to the hormones of the hypothalamic–pituitary–gonadal (HPG) system, there are well-documented declines in the peripheral levels of estrogens and androgens, but increases in levels of luteinizing hormone (LH), follicle-stimulating hormone (FSH), and sex hormone-binding globulin. The absolute levels of specific estrogens and androgens in aging individuals are highly variable, as is the decline with advancing age. The clinical significance of these changes in reproductive hormones is also quite variable. The effects include, but are not limited to, reduced protein synthesis, decrease in lean body mass and bone mass, increased fat mass, increased insulin resistance, higher cardiovascular disease risk, increases in vasomotor symptoms, fatigue, and depression, anemia, diminished libido, erectile dysfunction, and a decline in immune function. None of these effects, however, is guaranteed. In particular, sexual libido and erectile function are not uncommon, even in octogenarians. Additionally, there are declines in serum concentrations of growth hormone, insulin-like growth factor-I, and dehydroepiandrosterone (DHEA) and its sulphate-bound form dehydroepiandrosterone sulphate (DHEA-S). With regard to the hormones of the hypothalmic–pituitary–adrenal (HPA) and hypothalamic–pituitary–thyroid (HPT) systems, there are lesser changes with aging. These are frequently studied changes, but there continues to be controversy over their clinical importance and whether treatment is needed (e.g., restoration of hormone levels to those of an earlier period in life: replacement therapy).

Several neurotransmitter systems in the central nervous system (CNS) become less active with advancing age, including norepinephrine, serotonin, dopamine, and acetylcholine. Decreases in neurotransmitter synthesis, uptake, receptor binding, and responsiveness to stimulation all occur. As noted, neuroendocrine activity in general also declines with age, and an underlying factor may be a decline in CNS hypothalamic catecholamine neurotransmitter activity. One example is likely to be familiar, due to the ubiquitous advertising and popular press that has surrounded this hormone. The concentrations of the steroid DHEA-S decline rapidly with age. It diminishes by 95% by the age of 85, and therefore some have speculated that DHEA-S deficiency has a causal effect on the process of aging. This endogenous steroid hormone is readily available as a nutritional supplement and is widely purported to slow the aging process. To explore this putative link, researchers measured plasma DHEA-S concentrations and 2.5 million genetic variants in 14,846 people from Europe and the United States. They found at least eight genes that vary with DHEA-S levels, some of which are linked to the aging process and/or age-related diseases, including type 2 diabetes (Zhai et al., 2011). However, there is little evidence that a sustained increase in DHEA-S forestalls the aging process. As well, the fact that DHEA-S is a precursor to other steroid hormones leads many behavioral endocrinologists to question the relative safety of this supplement.

Loss of hormone receptor sensitivity, as well as a reduction in the output of target endocrine glands, also is a likely contributing factor to aging, and if so, supplemental hormone treatments would be of limited value. One familiar pattern, as first seen in the instance of Sequard's Elixir, is the initial exploration into the restorative effect of the hormone, followed by a muddle of information, most of which is conveyed most effectively by those who would profit from sales of the hormone. The ability to detect conflicts of interest is therefore essential to progress in understanding the effects of hormones on behavior (as it is now in all of science), especially with regard to the process of aging, which is dysphoric for many people. Following are examples of endocrine changes in later life that can have major influences on behavior.

## Hypothalamic–Pituitary–Gonadal Function: Menopause

Perhaps the most profound endocrine change to occur with aging is the female menopause. The pool of ovarian follicles becomes exhausted, estrogen and progesterone secretions by the ovary decline, and gonadotropin

(LH and FSH) secretion from the pituitary increases, owing to the lack of gonadal steroid feedback to the hypothalamus and pituitary. The events occur at several levels of the HPG axis, including early changes in the secretion of LH and FSH, coincident, in some women, with the onset of hot flashes.

Behavioral disturbances vary in severity and frequency, and include impaired sleep, reduced energy, fatigability, difficulty concentrating, increased sensitivity to pain, and mood changes, most often depressive in nature. The overall incidence of major depression, however, does not seem to be increased after menopause. Measurable cognitive changes are often minor, although estrogen replacement can result in improved memory function. Estrogen plus progestin replacement in some normal postmenopausal women has salutary effects on hot flashes, night sweats, disturbed sleep, bone mineral density, and a general sense of well-being, but at the possibly increased risk of hormone-sensitive malignancies (e.g., breast), blood clots, and stroke. The recommendations, therefore, are not straightforward, and many doctors and patients do not choose hormone replacement therapy. Estrogen replacement by itself, without a progestin, may have a different risk/benefit ratio and is still being studied. Another important consideration is the timing of hormone replacement therapy. Initial studies that showed an increased risk of cardiovascular disease with hormone replacement therapy examined women who began hormone replacement postmenopausally. In many women, however, ovarian steroid hormone levels decline slowly over many months. It might prove healthier to prevent the decline in ovarian hormones rather than replace the hormones after they have already reached their nadir. Furthermore, some have questioned whether ovarian steroid replacement should be continuous or whether it should mimic the 28-day hormone fluctuations that were present prior to the onset of menopause, and this question is under study.

In addition, selective estrogen receptor modulators (SERMs) are being designed to maximize beneficial effects, such as increasing bone density, while not activating receptors associated with malignancies. For example, both tamoxifen and raloxifene prevent bone loss, but tamoxifen increases the risk of uterine cancer, and raloxifene is an estrogen antagonist in the uterine lining. Raloxifene, however, does not alleviate hot flashes. SERMs with a broad spectrum of efficacy and with few or no unwanted side effects are an area of active drug development.

Estrogen is not an effective antidepressant by itself, but because it enhances serotonergic neurotransmission, some have suggested that it may be a useful adjunctive therapy in postmenopausal women who are only partially responsive to standard antidepressant drugs. Dementias in the elderly have become an increasing public-health problem as the old and oldest old become an ever-increasing percentage of the population, and estrogen therapy also may have a role in addressing these illnesses. Alzheimer disease accounts for about 50% of dementias in the elderly, its incidence rises dramatically after age 70, and its clinical course is unremittingly downhill. A prominent neurochemical change in Alzheimer disease is the loss of cholinergic neurons in the brain. Currently available treatments are mainly directed toward enhancing the neurotransmitter effect of whatever acetylcholine is still being produced by these neurons. Estrogen has significant neuroprotective effects (reviewed by Luine, 2014), so an important question is whether estrogen replacement in postmenopausal women can prevent the onset and slow the progression of Alzheimer disease. Although research data are still being gathered, evidence to date suggests that estrogen replacement may be protective against disease onset as well as being effective in slowing progress of the disease when used as an adjunct to treatment with drugs that enhance cholinergic neurotransmission (Henderson and Reynolds, 2002; Luine, 2014).

Menopause also is accompanied by changes in body weight and energy balance. As with all changes owing to menopause, it is difficult to separate those related to aging *per se* and those related to hormonal deficits. In fact, some have argued that in women over 50 years old, hormonal changes cannot account for increases in body weight and body mass index (a measure of weight relative to height). Rather, the decline in ovarian steroid concentrations is linked to a masculinization of body fat distribution and metabolic parameters. The postmenopausal period is associated with increased abdominal visceral body fat and decreased subcutaneous body fat, and this in turn is associated with increased insulin resistance and the development of Type II diabetes, although there is a great deal of variability in this effect. It would not be surprising if the strength of the effects of postmenopausal hormone deficits on body fat distribution are attributable to levels of prenatal androgen exposure (i.e., prenatal androgen exposure might predispose some women to a greater level of masculinization of their body fat distribution at the time of menopause). In postmenopausal women, increases in abdominal visceral body fat might be at least partially ameliorated by hormone replacement therapy, depending on the type of hormone replacement. For example, estradiol replacement tends to decrease abdominal visceral fat accumulation, whereas the addition of progesterone can prevent the positive effects of estradiol on energy balance. There still is much to be learned about healthy menopause.

## Androgen Decline in Aging Men

The relatively abrupt and profound neuroendocrine changes associated with menopause do not occur in men; rather, throughout their adult lives, with advancing age, men experience a gradual decline in circulating testosterone concentrations and blunting of its circadian rhythm.

Because protein binding of testosterone increases somewhat with age, the net result is a faster decline in circulating free (unbound) testosterone, the bioactive fraction, than in total testosterone. Properly termed androgen decline in aging men, this slow decrease in testosterone with age once was referred to as "andropause," implying a male counterpart to menopause (i.e., occurring universally in men and requiring some degree of androgen replacement), which was a misnomer. In fact, the decrease in androgen with age varies from person to person, and it is now medically accepted that only men with documented subnormal circulating testosterone and who have symptoms attributable to low testosterone should be considered for replacement therapy.

Symptoms associated with low testosterone include reduced sex drive, reduced ability to maintain erections, low energy level, fatigability, sleep disturbance, difficulty concentrating, and depressive mood. These symptoms can respond dramatically to hormone replacement. Many of these nonspecific symptoms, however, also occur in both young and elderly patients with major depression who have normal androgen levels. Therefore, these symptoms alone cannot be used as indications for hormone replacement therapy; rather, abnormally low circulating testosterone must be documented to justify androgen replacement. Androgen treatment of men with borderline-low testosterone may have some salutary effects on well-being, but the effects are modest at best and often not worth the risk of unwanted side effects. As with estrogen replacement in postmenopausal women, there are risks associated with testosterone replacement in elderly men, including possibly accelerating the development of prostatic cancer and increasing circulating red blood cells, blood viscosity, and therefore the risk of stroke.

## Aging Women

It is clear that, at least in a subset of older women, estrogen administration might protect against or delay some cognitive and/or emotional decline during aging, and at least some of these considerations are independent of menopause. Cognitive decline is in part related to age, regardless of reproductive hormone levels. Cognition is roughly defined as the ability to learn, retain, and recall information and includes processes related to attention, memory, reasoning, comprehension, and language production. Chapter 8 describes a central role for aromatization of testosterone to estradiol in the brain in the acquisition of the language of bird song. In fish, birds, and mammals, estradiol is neuroprotective, i.e., high levels of aromatase and estradiol prevent long-term neural damage after brain injury (Duncan, 2015). Some of these health-preserving actions in women may be on mechanisms of associative memory; others may be on more primitive cognitive functions such as arousal, alertness, and attention.

TABLE 13.1 Protective Actions of Estrogen

| Short-Term Protective Effects | Preventive Effects | Miscellaneous Effects |
|---|---|---|
| Oxidative stress | β-Amyloid precursor | Neurotrophism |
| Excitatory neurotoxicity | Protein metabolism | Synaptic plasticity |
| β-Amyloid toxicity | Apolipoprotein E | Neurotransmitter systems: acetylcholine, noradrenaline, serotonin, dopamine, etc. |
| Hypoglycemia | Vascular system | |
| Ischemia | | |
| Apoptosis | | |
| Hypothalamic–pituitary–adrenal axis reactivity | | |
| Inflammation | | |
| Cerebral blood flow | | |
| Glucose transport | | |

*From Henderson, V.W., Reynolds, D.W., 2002. Protective effects of estrogen on aging and damaged neural systems. In: Pfaff, D.W. (Ed.), Hormones, Brain, and Behavior, vol. 4. Elsevier, San Diego, p. 823 (Chapter 80). With permission.*

Still others may be emotional effects, which in turn could affect cognitive performance (reviewed by Luine, 2014). Henderson and Reynolds (2002) reviewed how, under some circumstances, estrogen treatment can delay some of the symptoms of Alzheimer disease. From a clinician's point of view, the number of possibilities is large (none of them mutually exclusive and some of them synergistic) as to how estrogens could protect mental functions in the aging brain (Table 13.1).

The beneficial effects of estrogen may involve protection from oxidative stress, trophic support, and maintaining $Ca^{2+}$ homeostasis. Detailed mechanistic investigations have been carried out in experimental animals and point to rapid estrogenic actions on neural tissue through several distinct mechanisms, including activation of estrogen nuclear receptors and membrane interactions that influence second-messenger signaling cascades (Fig. 13.2). Classical genomic effects depend on estrogen binding to either the alpha or beta estrogen receptor. The hippocampus exhibits relatively low expression of estrogen receptors, which is limited to a small number of cells. Nevertheless, most hippocampal cells exhibit rapid changes in several physiological processes upon estrogen exposure, indicating powerful estrogenic influences through nongenomic mechanisms (reviewed by Luine, 2014). Memory-enhancing effects of hormone replacement can be observed under specific experimental conditions, particularly

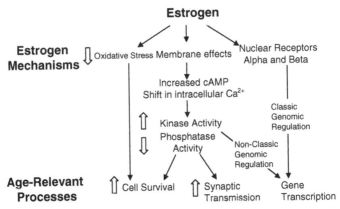

FIGURE 13.2   A growing body of research indicates that estrogen can influence hippocampal anatomy and physiology, lending support to the idea that estrogen may influence
memory function. Interest in estrogen effects on memory and the hippocampus has increased,
owing to evidence that hormone replacement can delay the progression of Alzheimer disease
and reduce memory decline during normal aging.

in animals whose learning ability has been pharmacologically impaired.
Conversely, treatment with tamoxifen or other drugs that block estradiol
binding to its receptor has adverse effects on cognition.

Nongenomic estrogen effects are rapid, independent of RNA or protein synthesis, and usually require the continued presence of estrogen.
Interestingly, many of the rapid effects are diametrically opposite to
changes observed in aged, memory-impaired animals. For example, aging
is associated with modifications that reduce transmission through the hippocampus (Foster, 1999). In contrast, estrogen rapidly increases cell excitability and the strength of synaptic transmission (Kumar and Foster, 2002;
Sharrow et al., 2002). In addition, estrogen adjusts synaptic plasticity processes, facilitating the induction of long-term potentiation and impairing
the induction of long-term synaptic depression, forms of $Ca^{2+}$-dependent
plasticity that are modified by aging and thought to play a role in the
establishment and maintenance of memory (Sharrow et al., 2002).

The exact mechanism for nongenomic estrogen effects is not established. Research indicates that estrogen can influence processes that are
regulated by G-protein activity and intracellular $Ca^{2+}$. A shift in these signaling cascades changes the phosphorylation state of proteins, increasing
or decreasing protein kinase and phosphatase activity. Estrogen-induced
changes in the phosphorylation state of transcription factors, such as
CREB, regulate gene transcription and provide a nonclassical mechanism
for genomic regulation. As well, estrogen is a highly lipophilic steroid
and changes the conformation of the cell membrane, a lipid bilayer with
a hydrophilic center, perhaps thereby changing cell membrane function.

Of clinical importance, the efficacy of estrogen replacement therapy may be reduced if therapy is not initiated until several years after the onset of menopause or after the symptoms of Alzheimer disease are manifest. It thus will be important for future research to understand age-related changes in estrogen signaling pathways.

In these experiments, virtually every estrogenic action is in a direction opposite to that evident during aging. Further, a reasonable possibility for one component of the antiaging effects of estrogen is that estrogen reduces the consequences of vascular accidents in the brain, large and small, whose cumulative effects over time reduce cognitive performance. An impressive body of data demonstrates that chronic estrogen administration can reduce cerebral damage following strokes (see later). Thus, through genomic and metabolic actions on positive anabolic functions (e.g., neuronal growth; reviewed by Woolley and Cohen, 2002; Luine, 2014) and through the reduction of negative effects, estrogens can, in easily understandable ways, delay degeneration in the aging brain.

## Estrogens and Neuroprotection

We now appreciate that estrogens are pleiotropic gonadal steroids that exert profound effects on plasticity and cell survival of the adult brain. Over the past century, the lifespan of women has increased dramatically from 50 to over 80 years, but the age of onset of menopause has remained constant at approximately 50 years. This means that women may now live over one-third of their lives in a hypoestrogenic, postmenopausal state. The impact of prolonged hypoestrogenicity has many repercussions, as these women may suffer an increased risk of cognitive dysfunction and neurodegenerative diseases, including Alzheimer disease and stroke. Accumulating evidence from both clinical and basic science studies indicates that estrogens exert critical protective actions against these conditions.

Studies using animal models of cerebral ischemia provide strong evidence that estradiol is a neuroprotective factor that profoundly attenuates the degree of ischemic brain injury. Experiments conducted by Wise (Fig. 13.3) showed that low physiological concentrations of estradiol replacement exert dramatic protective actions in the brains of young and middle-aged female rats, as well as in cultured explants of the cerebral cortex (reviewed by Down and Wise, 2009). The results emphasize that estradiol attenuates delayed, apoptotic cell death by acting at multiple levels to decrease cell death and promote cell survival. It enhances the expression of genes that attenuate apoptosis, diminishes the expression of genes that promote apoptosis, decreases activation of caspases, increases activation of signaling pathways that promote cell survival, promotes the proliferation of neurons in the region of the brain that may migrate to

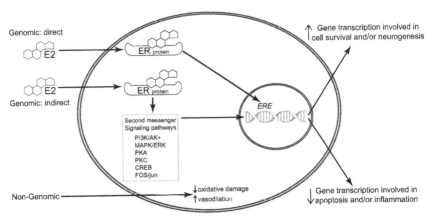

FIGURE 13.3    Both genomic and nongenomic mechanisms activated by treatment with estrogens (1) limit damage due to stroke, and (2) tend to counteract the effects of aging (see text). *Illustration courtesy of Prof. Phyllis Wise.*

the cortex, and attenuates the immune response. Together, these actions can dramatically decrease the extent of brain damage after stroke injury.

The protective actions of physiological levels of estradiol replacement require the presence of estrogen receptor alpha because estradiol was unable to exert a protective effect in estrogen receptor alpha knockout mice. Intriguingly, this receptor is not normally expressed in the adult cortex, but only in the developing cortex. After injury in the adult, it reappears and plays a pivotal functional role in neuroprotection. Pharmacological concentrations of estradiol, on the other hand, appear to protect against neurodegenerative agents by very different mechanisms that do not require estrogen receptors. At these concentrations, estrogens act on the vasculature to influence blood flow and endothelial cell function, are effective antioxidants and free radical scavengers, and can alter ion channel biophysics (Golden et al., 1999).

As pointed out earlier, it is important to emphasize that estrogens do not always protect. Under some circumstances, hormone replacement therapy does not afford protection and may increase risk. Together, these clinical studies indicate that estrogen replacement therapy can slow or prevent the onset of many neurodegenerative disease processes, but it does not protect against or reverse a disease process that already has been initiated. Furthermore, these studies show that the effects of hormone replacement depend greatly on the formulation of the therapy, i.e., whether estrogen is combined with progestin replacement, how the hormone is administered, and the doses that are used. Clearly, much more work is necessary before we gain a more complete understanding of the spectrum of estrogen actions and the multiple interacting and diverse mechanisms underlying those actions.

While the implications of estrogenic treatment for the reduction of stroke damage and the delay of Alzheimer's symptoms would seem to apply to men as well as to women, more controversial are the conditions under which testosterone replacement therapy is appropriate in older men. As noted before, testosterone levels, on the average, decline steadily, by about 1% per year, in aging men. The autonomic nervous system equivalent of the hot flash in menopausal women occurs rarely. Nevertheless, the decreasing physical and mental energy and reduced libido in older men call attention to the possibility of hormonal therapy to deal with behavioral changes as well as with other physiological declines.

Normal levels of free and total testosterone exist for groups of men at different ages, including the elderly, but these are averages, and cutoff points for low testosterone that may require androgen replacement therapy are based on these averages. The problem remains of the elderly man who has symptoms suggestive of low testosterone, who is not suffering from major depression, but who has circulating testosterone above the established cutoff point for hypogonadism in men his age. What cannot be established is what that individual's starting testosterone concentration was as a young man and how much it has fallen in the intervening decades. The percentage decline may have been severe, even though the current concentration is above the average cutoff point. Strategies for addressing this dilemma need to be developed.

In aging men with confirmed low testosterone, transdermal treatments with androgens in gels, and injections of testosterone or its reduced metabolite, dihydrotestosterone, can improve the ability to initiate and maintain penile erections; however, reports of improved feelings of general well-being have been modest and have been clouded by significant placebo effects. While positive effects of systemic androgen treatment on sexual function and physical energy do occur, the chance of exacerbating prostate cancer growth and the increased risk of stroke in aging men, as noted earlier, require that these behavioral benefits be pronounced, quite certain, and worth the risk.

## MECHANISMS

The field of neurochemical and cell biological examination of how hormones can be used to combat aging of the brain has been growing rapidly. One extensive set of cellular and molecular investigations was referred to in Fig. 13.2; in addition, nonreproductive hormones are important and should be considered.

II. HISTORY: HORMONE EFFECTS CAN DEPEND ON FAMILY

## Hypothalamic–Pituitary–Adrenal Function

In contrast to the marked changes in the pituitary–gonadal axis, plasma glucocorticoids change little with age. Some studies indicate no change, whereas others suggest a slight increase. As well, there may be a blunting of the glucocorticoid circadian rhythm and a shift in its timing. The increase in circulating glucocorticoids with age appears to be based on the reduced sensitivity of hippocampal negative feedback on corticotropin-releasing hormone (CRH) and adrenocorticotropic hormone. Some investigators have proposed a glucocorticoid-induced loss of hippocampal glucocorticoid receptors (GRs), such as from early stress-induced increases in HPA axis activity and continuing stress activation of this axis throughout adult life. In older individuals, stress and other provocations of the HPA axis result in a prolonged cortisol response, in contrast to younger individuals. This, in turn, could contribute to the major problem of depression among the elderly.

From a human behavior standpoint, one current mechanistic hypothesis of major depression is that increased CRH, through its anxiogenic properties, produces some of the symptoms, and increased glucocorticoids, which impair hippocampal functioning, contribute to other symptoms. Drugs that interrupt these neuroendocrine processes, such as CRH receptor blockers, GR blockers, and glucocorticoid synthesis inhibitors, have been proposed as treatments for major depression, but none has yet shown sufficient therapeutic activity to be clinically useful. As well, the hypothesis upon which these treatments is being developed (i.e., increased HPA axis activity in major depression) is weak, in that the majority of such patients, including elderly depressives, have normal HPA axis activity.

## Hypothalamic–Pituitary–Thyroid Function

In the elderly, the half-life of thyroxine (T4) is decreased, but circulating T4 usually is within the normal range, because its production is correspondingly reduced. The production of triiodothyronine (T3) may be reduced as well, but its half-life is not increased, leading to somewhat reduced circulating T3. Basal thyroid-stimulating hormone concentrations are more variable, some being below and some above the normal range for younger individuals, perhaps representing the early stages of hypo- and hyperthyroidism, respectively. What is important is that the symptoms of both hypo- and hyperthyroidism are frequently more muted in the elderly, so clinicians need to carefully screen their geriatric patients for thyroid abnormalities. This is particularly important in elderly patients who present with symptoms of major depression, since there are symptoms in common between depression and hypothyroidism.

## Steroid Receptors and Memory Loss

Aging is commonly accompanied by mild hypercorticism, with implications for both stress responses and fluid balance. Owing to the catabolic effects of elevated glucocorticoids, GRs may be lost in hippocampal neurons. This parallels the concept of Sapolsky (2000) that stress, which causes increased exposure to glucocorticoids, produces irreversible loss of GRs in the hippocampus. In addition to GRs, there also are mineralocorticoid receptors (MRs) in the hippocampus, and a strong structural homology exists between GRs and MRs (see Chapter 4). This allows for MRs and GRs to recognize identical hormone-responsive elements within the promoter region of genes activated by corticosteroid receptor complexes. Thus, there is a binary hormone response system for one single hormone, namely corticosterone/cortisol. The loss of excitability and even the loss of neurons in the hippocampus could, if sufficiently critical in number, lead to losses in memory, because hippocampal neurons are involved in memory functioning. And, the hippocampus contains the highest concentration of MRs in the brain. In aging, the number of hippocampal MR binding sites is reduced, and the resulting imbalance between GR and MR impairs their capacity to control neuroendocrine feedback, so corticosterone/cortisol levels tend to become higher with aging.

## SOME OUTSTANDING QUESTIONS

1. What is the optimal clinical paradigm for postmenopausal female hormone replacement? Does it vary among individuals?
2. The usual combination therapy of estrogen plus a progestin has shown increased risk of breast cancer and stroke. (1) Will estrogen alone be sufficient to reduce hot flashes and night sweats, to increase bone density, and to have other salutary effects without these risks? (Studies are underway to address this question.) (2) Will SERMs prove to have the desired spectrum of beneficial effects without detrimental side effects? (Drug development in this area is ongoing.) (3) What is the usefulness of dietary phytoestrogens in this regard (e.g., from soybean products)? They have weak estrogen-like activity, and some may antagonize the actions of estrogen itself.
3. What is the optimal clinical paradigm for androgen replacement in aging men? Does it vary among individuals?
4. How can we best address the issue of hormone therapy for the prevention and treatment of dementias such as Alzheimer disease?
5. Does a possible hormonal preventive therapy for Alzheimer disease increase the risk of other complications, such as cancer, heart disease, or stroke, similar to hormone replacement therapy in menopausal women?

II. HISTORY: HORMONE EFFECTS CAN DEPEND ON FAMILY

SECTION III

# TIME: HORMONAL EFFECTS ON BEHAVIOR DEPEND ON TEMPORAL PARAMETERS

# 14

# Duration of Hormone Exposure Can Make a Big Difference: In Some Cases, Longer Is Better; in Other Cases, Brief Pulses Are Optimal for Behavioral Effects

## BASIC EXPERIMENTAL EXAMPLES

### Sex Steroids

In many cases, the behavioral effects of a given hormone are greater when the hormone is present at optimal levels for a longer duration. A case in point is the effect of testosterone on male-typical sex behaviors in rodents. Male sex behavior is comprised of many measurable components, including but not limited to the number of times the male mounts the female, thrusts his pelvis, inserts his penis, and ejaculates. All of these behaviors wane after castration and gradually return, beginning several days after the start of testosterone treatment. In general, the longer the duration of treatment, the higher the level of these male-typical sex behaviors, until they reach a plateau. In contrast to sex behaviors in males, sex behaviors in female laboratory rodents are more complicated. In response to an adult male, female rodents display the lordosis posture, a stiff, arched-back posture that allows the male to gain entry. Lordosis is infrequent or absent in ovariectomized females because they have no circulating ovarian steroids, estradiol and progesterone. To restore lordosis, a long (almost 2-day) "priming" period with estradiol must be given, followed by a shorter (4- to 6-h) exposure to another ovarian hormone, progesterone. After ovariectomy, the longer the time the female goes without ovarian steroid treatment, the longer the estradiol-priming period required to induce lordosis. The effects of progesterone are even more complicated than the effects of estradiol: a 6-h exposure to high levels of progesterone initially facilitates lordosis, but a longer duration of exposure has the opposite effect.

A molecular interpretation of the mechanism is the following: prolonged absence of estrogens allows a decline of nuclear coactivator protein levels (see Chapters 11 and 18), which are needed to transduce the nuclear binding of estrogen receptors into behaviorally relevant transcriptional facilitators. An important event resulting from estradiol priming is the appearance of progestin receptors. Without estradiol-induced progestin receptors, treatment with progesterone does not facilitate lordosis. A surprising development with respect to long priming actions came from biochemical work with the uterus by Jack Gorski, who found that two brief exposures to estrogen, suitably timed, could be substituted for a priming period of 24 h. Parsons et al. (1982) showed the same sequence is effective in inducing central nervous system and behavioral effects. Again, an estrogen priming of 24 h could be replaced by two 1-h exposures, the first from hour 0 to hour 1 and the second beginning between hour 4 and hour 13. This pulse schedule was effective for inducing both progestin receptors and lordosis behavior (Fig. 14.1).

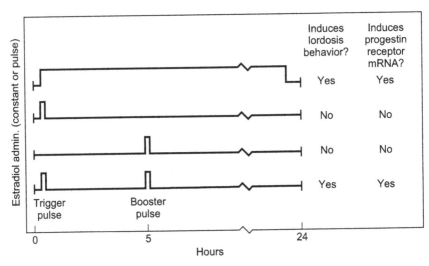

**FIGURE 14.1**   (Top line) Prolonged high levels of estrogens are sufficient for both the female reproductive behavior, lordosis, and for the induction of progestin receptor mRNA (whose protein product, in turn, permits progesterone to amplify estrogen action). (Middle lines) A single brief pulse of estradiol is insufficient to induce either response. (Bottom line) In contrast, a trigger pulse followed by a booster pulse is sufficient for both responses. For the lordosis behavior test, estrogen action has been amplified by progesterone treatment on the second day of the experiment. *Adapted from Parsons, B., Rainbow, T.C., Pfaff, D.W., McEwen, B.S., 1981. Oestradiol, sexual receptivity and cytosol progestin receptors in rat hypothalamus. Nature 292, 58–59; Parsons, B., McEwen, B.S., Pfaff, D.W., 1982. A discontinuous schedule of estradiol treatment is sufficient to activate progesterone-facilitated feminine sexual behavior and to increase cytosol receptors for progestins in the hypothalamus of the rat. Endocrinology 110, 613–619.*

In female laboratory rodents, estrogen priming must be followed by progesterone for optimal behavioral facilitation (refer to Chapter 2). The progesterone, if timed correctly, amplifies the estrogenic effect. The required temporal parameters for progesterone administration are different from those for estradiol, because there is a biphasic action of progesterone both on pituitary release of luteinizing hormone (LH) and on behavior. In female rats, for example, 2–5 h after progesterone injection, LH release and lordosis behavior are strongly facilitated. On the other hand, the continued presence of progesterone, several hours later, inhibits both the endocrine output (LH) and the behavioral response (lordosis).

## Protein Hormones From Pituitary

In marked contrast to the situation with testosterone and estradiol, peptide and protein hormones secreted by the hypothalamus and pituitary gland do not require long priming periods for their effects. Instead, they require rapid, pulsatile exposure of their target tissues for maximum effectiveness. The classic example is the sudden pulse of gonadotropin-releasing hormone (GnRH, also known as LHRH) from the hypothalamus, which is necessary to trigger the LH surge from the anterior pituitary, which itself is an obligatory pulse for ovulation (Fig. 14.2). In addition to the LH surge, a surge of brain angiotensin II lasting about 30 min occurs 1 h before the LH surge, and there also is a surge of prolactin. As well, a steady buildup of neuropeptide Y (NPY) begins some hours before the LH surge. Thus, several events, arranged in an ordered sequence, are necessary to prime the sudden release of a large amount of LH required to initiate the ovulatory surge.

### Interim Summary

1. Testosterone has a greater effect on male-typical sex behavior with a longer duration of exposure, and the duration of the effect is often several days.
2. To facilitate female sex behavior in laboratory rodents, treatment with estradiol must be given for 36–48 h prior to exposure to progesterone for 4–6 h. Longer durations of progesterone are inhibitory for female sex behavior.
3. Many protein and peptide hormones must be secreted in a pulsatile manner to induce changes in physiology and behavior. These same hormones have the opposite or no effect when given continuously.

## CLINICAL EXAMPLES

As mentioned earlier, GnRH and other hypothalamic peptide hormones are secreted in pulses. On the one hand, it is fascinating that populations of identical GnRH neurons can manage pulsatile output.

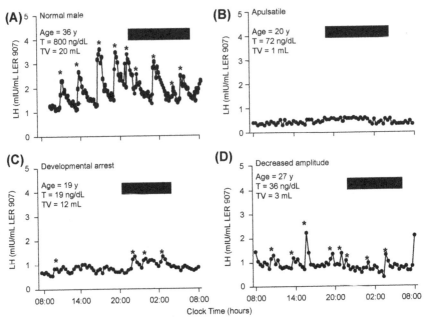

**FIGURE 14.2** Gonadotropin-releasing hormone (GnRH) is necessary both for normal LH release from the pituitary gland and for normal sexual behaviors. The effective mode of GnRH administration is not constant, but pulsatile; that is, infertile men can be brought into a state of fertility by pulsatile GnRH administration. The normal mode of LH release also is pulsatile. (A) Spontaneous pulsatile LH release in a normal man; (B) a male patient who lacks normal GnRH and LH secretion; (C) a male patient with a developmentally arrested, low amplitude LH pattern; and (D) a male patient who has the same type of pulses as the normal man in (A) but whose amplitudes are significantly decreased. The *black bar* indicates the hours of sleep; *asterisks* (\*) signify statistically significant LH pulses. *From Seminara, S.B., et al., 1998. Endocr. Rev. With permission.*

Pulsatile output of GnRH is necessary for the pituitary to respond with substantial gonadotropin release into the blood. A steady, high level of GnRH administration turns off the gonadotropin system and thus can be used either as a birth control device or to damp down the system in the treatment of gonadal hormone-sensitive cancers. For example, long-acting LHRH agonists are used to suppress LH and testosterone secretion in the treatment of prostate cancer in men and in the treatment of endometriosis in women.

Some of these phenomena are beginning to be understood at the molecular level. The GnRH promoter is activated in an episodic fashion in GnRH neuronal cultures. A specific 410-base region of the GnRH promoter is required for pulsatile GnRH promoter activity. Within that region, a small, 5-base site that represents a binding site for the transcription factor Oct-1

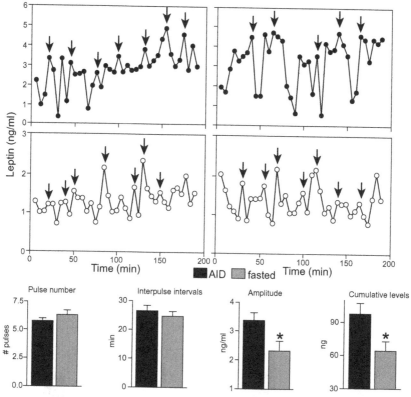

**FIGURE 14.3** Pulses of leptin release had higher average amplitudes, causing higher circulating levels, in fed rats than in fasted rats. Top of the figure shows raw data from individual animals. *Filled circles*, fed; *open circles*, fasted. Note differences in numbers on the y-axis. Bottom of the figure shows average values. *From Kalra, S.P., Bagnasco, M., Otukonyong, E.E., Dube, M.G., Kalra, P.S., 2003. Rhythmic, reciprocal ghrelin and leptin signaling: new insight in the development of obesity. Regul. Pept. 111 (1–3), 1–11. With permission.*

is likewise, required. In sum, to achieve fertility, LH must be given in pulsatile fashion.

Referring to the ability of leptin to inhibit energy intake (feeding), what is the optimal temporal schedule for its application? Leptin secretion in adult rats also is pulsatile (Fig. 14.3), as is the secretion of another hormone related to food intake: ghrelin (Fig. 14.4). Ghrelin effects could help account for the pulsatility of growth hormone (GH) release (Fig. 14.5), an effect that operates through GH-releasing hormone (GHRH) release, as shown by the blocking effect of a GHRH antibody (Fig. 14.6). In males, leptin pulses are low amplitude and high frequency. In females, high-amplitude pulses dominate, leading to a higher average plasma concentration of leptin. Moreover, removal of the ovaries reduces leptin pulse

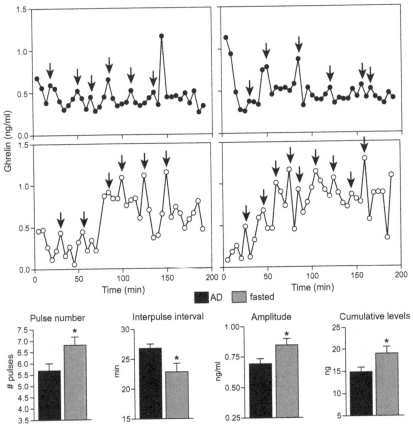

**FIGURE 14.4** Pulses of ghrelin release were different between fed and fasted rats. Top of the figure shows raw data from individual animals. *Filled circles,* fed; *open circles,* fasted. Bottom of the figure shows average values. Ghrelin pulse numbers were higher in the fasted animals, and their interpulse intervals were reduced. Amplitudes were, on the average, higher in the fasted animals, leading to higher circulating levels. See Chapter 13 for additional material. *From Kalra, S.P., Bagnasco, M., Otukonyong, E.E., Dube, M.G., Kalra, P.S., 2003. Rhythmic, reciprocal ghrelin and leptin signaling: new insight in the development of obesity. Regul. Pept. 111 (1–3), 1–11. With permission.*

frequency and amplitude, perhaps explaining the higher body weights of ovariectomized female rats (also see Chapter 15).

Finally, to again emphasize the importance of the duration of hormone exposure in a clinical context, not only is hormone replacement therapy used in endocrine practice, but blocking of hormone secretion and/or its effects also is used when the body is producing an excess of hormone. For example, as mentioned earlier in this chapter, long-acting LHRH agonists are given to suppress LH and testosterone secretion in the treatment of

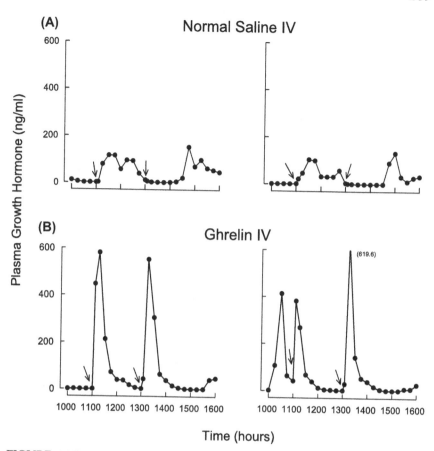

FIGURE 14.5 Pulsatile release of growth hormone (GH) from the pituitary gland. (A) Normal saline given intravenously cannot stimulate large pulses of GH from somatotroph cells of the anterior pituitary. (B) The stomach hormone ghrelin given intravenously is able to stimulate significant GH pulses. *From Tannenbaum, G.S., et al., 2002. In: Kordon, et al. (Eds.), Brain Somatic Cross-Talk and the Central Control of Metabolism, Springer-Verlag, Heidelberg. With permission.*

prostate cancer in men (so-called "chemical castration"), and to suppress LH and ovarian hormone secretion in the treatment of endometriosis in women. An area of continuing controversy is the use of long-acting LHRH agonists for testosterone suppression in male sex offenders, to reduce both their psychological preoccupations (sexual fantasies) and their physical sex drive. There is little evidence, however, that suppression of androgen secretion has significant influence on their violent tendencies or their general lack of empathy for their victims.

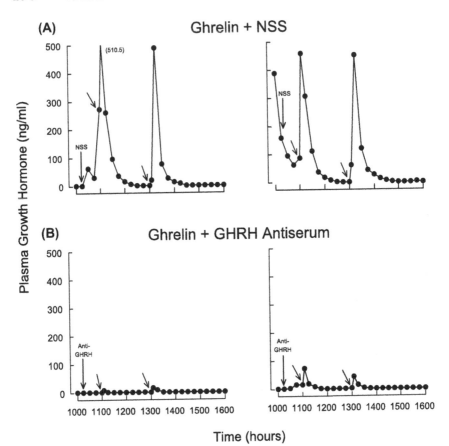

FIGURE 14.6    One mechanism for pulsatile release of growth hormone (GH). (A) The stomach hormone ghrelin stimulates pulsatile release of GH from the pituitary, and the effect is not blocked by normal saline (NSS). (B) Ghrelin apparently works through GH-releasing hormone (GHRH) because a GHRH antiserum blocks its effect. *From Tannenbaum, G.S., et al., 2002. In: Kordon, et al. (Eds.), Brain Somatic Cross-Talk and the Central Control of Metabolism, Springer-Verlag, Heidelberg. With permission.*

## OUTSTANDING NEW BASIC OR CLINICAL QUESTIONS

1. What are the optimal methods of application of hormone replacement therapy for older women and androgen supplementation for older men? Reports from Fan et al. (2015) are relevant to this question.
2. When two pulses of hormone treatment can substitute for a longer, constant exposure, can the time period be broken up even more finely into three or four well-timed pulses as a route to finer dissection of the underlying mechanisms? For example, it is known that a first pulse

of estrogen, acting at the membrane, can augment the action of a later pulse, acting in the nucleus.

3. When long-term hormone priming is required, what molecular mechanisms are changing during that long time period? Conversely, when, as in older adults, there is a prolonged absence of a certain hormone, what cellular components are declining, to account for reduced medical effectiveness of hormone treatment? Is DNA methylation involved? Histone modifications?

4. How do rapid steroid hormone actions (Kow et al., 2016a,b) potentiate later actions in the cell nucleus ?

5. How do short-term, protective actions of stress hormones (McEwen, 1998) turn into damaging, long-term actions?

6. What are the detailed molecular and biophysical mechanisms that allow a neuroendocrine population of cells to generate a pulse of secretion?

# Hormonal Secretions and Responses Are Affected by Biological Clocks

In the summer of 1938 a University of Chicago professor, Nathaniel Kleitman, and his student assistant, Bruce Richardson, spent more than a month deep inside Mammoth Cave, a cold, dark, solid-rock chamber 130 ft underground. Living for a month in underground isolation was their way of answering questions about sleep patterns. Why do we sleep at night and awake each morning? Do we learn it from our parents? Is the sleep–wake cycle a response to the day–night differences in light and noise in our environment? Are sleep–wake cycles controlled by the sunrise or the smell of coffee in the morning? Alternatively, are daily sleep–wake cycles driven by the workings of an internal biological clock that keeps time on its own? Are changes in hormone levels involved? In 1938 no one knew the answer. While living above ground, Kleitman and his student took initial round-the-clock measurements of their own body temperatures and sleep times. Their body temperatures fell each night during the time they were sleeping and rose again predictably after they awoke each morning. Furthermore, their body temperature exhibited a repeating cycle with a *period of* 24 h, i.e., the time interval from one peak in body temperature to the next peak was consistently about 24 h. If circadian rhythms in body temperature were driven *only* by external cues (the onset of morning light and/or by the sound of birds and traffic), the rhythms would be expected to disappear in constant conditions (e.g., continuous darkness and quiet). Kleitman and Richardson wanted to know whether it was possible to alter the period of their rhythm purposefully, and so they decided to isolate themselves in a place devoid of cues that differentiate night from day. They set up camp in Kentucky's Mammoth Cave, where there is no day–night difference in darkness, temperature, or sound. They furnished their underground lair with everything they

*Principles of Hormone/Behavior Relations*
http://dx.doi.org/10.1016/B978-0-12-802629-8.00015-2
293

needed: a table, chairs, a washstand, and special beds that protected them from pesky cave rats. Food was periodically brought in to replenish their supplies. They discovered that even under these constant conditions, their daily sleep–wake cycles and body temperature rhythms persisted with a period of about 24 h. After they set up camp, they tried to alter the period of their daily rhythms from the normal 24–28 h. The young student, Richardson, succeeded in shifting his rhythm to a 28-h period by intentionally waking, eating, and remaining active for 19 h, and then sleeping 9 h. By contrast, the older Kleitman, after trying for over a month, failed to shift his own body temperature rhythms or his sleep–wake cycle. The period of his rhythm stubbornly remained at 24 h. Kleitman and Richardson failed to find a definitive answer to their burning questions about the human ability to reset daily rhythms purposefully. Because one of the two subjects succeeded and the other failed to shift to the 28-h period, they considered their work to be inconclusive (Kleitman, 1939).

Kleitman and Richardson's work was important for another reason. The fact that Keitman's 24-h rhythm persisted for more than 30 days in constant conditions was a profound discovery. It provided some of the first evidence in humans for the existence of an innate, self-sustaining timekeeper that can drive rhythmic changes in the absence of external cues. The experiment also suggested that these rhythms could be set, i.e., *entrained*, at least in some individuals, by an external cue (Kleitman, 1939).

Since then, many endogenous biological rhythms have been discovered in behaviors, physiological processes, susceptibility to disease, and of course hormone secretion. Many of these rhythms are innate and self-sustaining. Levels of many hormones, blood glucose, other metabolic substrates, and gene transcription fluctuate markedly over the day and night. Physicians must therefore consider the timing of blood, urine, and tissue sampling to make correct diagnoses. There is also a circadian rhythm in the incidence of many diseases, e.g., the incidence of heart attack is highest in the early morning and early evening. There are circadian rhythms in the efficacy of anesthesia and cancer treatments. Indeed, aberrations in biological clocks play a role in sleep disorders, diabetes, bipolar disorder, and many types of cancer. Animals used in biomedical research are nocturnal and engage in their important activities during darkness. Their daily rhythms are entrained to the light–dark cycle and can be shifted by exposure to light during the dark period of the 24-h light cycle. As a result of accidental light exposure in experiments, laboratory animals can exhibit shifts in the timing of their hormone secretion, sensitivity to hormones, and behavior, sometimes confounding the experimental results. These are just a few examples of the importance of circadian clocks in health and disease (reviewed by

Silver and Kriegsfeld, 2014), emphasizing the principle that biomedical researchers must always consider biological rhythms in both theory and practice.

Every biological rhythm has quantifiable characteristics, including its period (the time it takes to complete one cycle), its frequency (the number of cycles per unit of time), its peak (highest measure), its nadir (lowest measure), its amplitude (difference between peak and nadir), and its phase (relationship of the rhythm to clock time and/or to other rhythms) (Wirz-Justice, 2007). Circadian rhythms, for example, occur with a frequency of once per day and with a period of 24 h. *Circa* comes from the Latin for "about," and *dia* from the Latin for "day," and so *circadian* means "about a day." Circadian rhythms in body temperature occur with an amplitude of only about 1–1.5°Fahrenheit (F), and total variation in body temperature over the day is between 2°F and 3°F. Rhythms with a period of less or more than a day are termed ultradian and infradian respectively. For example, one specific type of infradian rhythm is the seasonal rhythm of breeding and migration. Seasonal (annual) rhythms allow temperate zone, Arctic, and Antarctic animals to take advantage of optimal weather patterns, especially for reproduction. In some species these yearly rhythms are driven by an endogenous circannual clock. Other species show circatidal rhythms (with a frequency of twice per day), which allow them to survive and reproduce in the changing conditions of salinity and nutrient availability created by high and low ocean tides. Still other species show circalunar rhythms (with a frequency of once per lunar cycle).

Biological rhythms illustrate the link between astrophysical phenomena and biological life. With regard to underlying molecular and neural mechanisms that drive the rhythms, we know most about circadian rhythms. Circadian rhythms occur in all organisms, from bacteria to humans, and nearly half of all mammalian genes are expressed rhythmically in one or more tissues. These ubiquitous circadian rhythms developed as a response to the regular cyclic changes in environmental conditions created by the Earth rotating on its axis, an astrophysical phenomenon that places organisms living at most points on Earth in sunlight for a portion of each day. For life forms that evolved on Earth, solar cues from the Earth's rotation on its axis, sunrise and sunset, are the most reliable of external rhythmic cues available. Endogenous circadian clocks are thought to have increased survival and reproductive success by allowing organisms to synchronize their activity to changing environmental conditions that are a consequence of the Earth's rotation and the 24-h day–night cycle (reviewed by Hastings et al., 2014). For example, bats and owls feast on mosquitoes and mice at night and sleep during the day, as do many species that hunt for prey at night. Seabirds and humans tend to fish or hunt in the day, when they can see their prey. Endogenous circannual clocks (endogenous rhythms with a period of 12 months) allow organisms to anticipate and make predictions about the availability of food, mates, predators, and other aspects of their ever-changing environment that result from the Earth's orbiting the Sun while remaining tilted on its axis (the astrophysical phenomenon that creates the changing seasons of the year).

This chapter concerns internal biological clocks, endogenous oscillators that drive hormone secretion, the responsiveness or sensitivity of the brain and body to hormones, and the effects of hormones on the clock mechanisms.

## BASIC SCIENTIFIC EXAMPLES

In a wide variety of experimental animals and in humans, very regular 24-h fluctuations are observed in circulating hormone levels (Fig. 15.1). In many diurnal species (those that are awake and active during daylight) secretion of glucocorticoids, e.g., cortisol, begins to increase during the

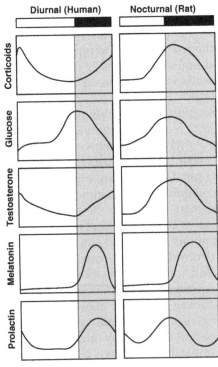

FIGURE 15.1   Normally, there are very regular 24-h fluctuations in circulating hormone levels. *White bar* = daytime; *black bar* = nighttime. Amplitudes of hormone rhythms are not to scale.

nighttime sleep phase and peaks every morning about the time of awakening and the start of daily activities (Fig. 15.1). In contrast, in nocturnal species the secretion of glucocorticoids, e.g., corticosterone, begins to increase during the day and peaks at dusk, at the start of the nocturnal activity phase. Whether the animal is nocturnal or diurnal, the rhythm in plasma glucocorticoids has a period of 24 h. Testosterone levels also fluctuate in a circadian fashion. In diurnal males, such as humans, plasma testosterone concentrations increase in the middle of the night and peak around dawn, in a pattern similar to that of cortisol. Nocturnal animals show the opposite pattern.

Steroid hormones are not the only hormones secreted in a circadian fashion. Thyrotropin, a glycoprotein, and growth hormone, a peptide, also show 24-h rhythms; their plasma concentrations peak during the middle of the sleep period. The anterior pituitary hormone, prolactin, is also at its highest level during the nighttime sleep phase and lowest during the daytime active phase. Again, this pattern is reversed in nocturnal species; prolactin levels are highest in the daytime sleep phase.

Melatonin is different. Concentrations of melatonin are high during the dark period in both nocturnal and diurnal animals. This indoleamine, which is secreted by the pineal gland, rises at night coincident with the onset of the sleep phase, peaks soon after darkness falls, remains high during the dark of night, decreases rapidly at dawn, and remains low throughout the day. Melatonin is a special case, in that its rhythm is determined by an endogenous master oscillator—the master clock in the suprachiasmatic nucleus (SCN) of the brain—and its rhythmic secretion drives circadian and circannual rhythms in many other tissues (reviewed by Hastings et al., 2014; Erren and Reiter, 2015). Some scientists (and purveyors of neutraceuticals) suggest that the high levels of melatonin during the dark portion of the daily light–dark cycle are important for healthy sleep patterns. It is true that melatonin is involved in mediating the effect of light and dark on seasonal rhythms in some species, but the idea that high melatonin is necessary for sleep is not consistent with the fact that all nocturnal animals which have been studied show high circulating concentrations of melatonin during their nighttime activity period and very low concentrations of melatonin during their daytime sleep period. Nonetheless, melatonin is often prescribed to humans with insomnia as a safe and gentle sleep aid, and it is effective in some individuals.

Melatonin plays a primary role in many seasonal rhythms that are entrained by changes in day length, especially reproduction in seasonal breeders, because the duration of the melatonin pulse is longer in animals exposed to short day lengths (long nights) of winter and, conversely, the duration of the melatonin pulse is shorter in animals exposed to long day lengths (short nights) of summer. Winter-like behaviors and physiological responses can be initiated by providing long-duration melatonin infusions to animals housed in summer-like (long) days or in constant conditions. Some seasonal traits are melatonin dependent, whereas other traits are melatonin independent.

Rhythms in hormones are best studied in the laboratory, where the experimenter can control the light–dark cycle. A mechanical timer switches the lights on at the same time each day, for example at 8 a.m. If the lights are then set to go off at 10 p.m., the animals are exposed to a 14L:10D light–dark cycle (known as a long-day photoperiod). On a strict light–dark cycle, the animals' rhythms in circulating hormone concentrations and behavior show a period of 24 h, and the rhythms remain synchronized among all individuals in the population. The endogenous nature of the rhythm can be tested by placing the animals in constant dark or constant light. If the hormonal rhythm is truly driven by an endogenous clock, the rhythms will persist under constant conditions with a consistent period. Like the daily rhythms in Professor Kleitman's sleep–wake cycle, daily rhythms in plasma concentrations of glucocorticoids, testosterone, thyrotropin, and growth hormone persist in constant conditions, with some important caveats. The period of the rhythms of each individual animal housed in constant conditions will be close to, but not exactly, 24 h. Some individuals exhibit a repeatable period of 23.9 h, others exhibit a repeatable period of 24.2 h, and so on. In constant conditions, such as constant light and temperature, each individual's own rhythms are consistent from cycle to cycle, but the rhythms are no longer synchronized among individuals. In constant conditions, such as 12D:12D (constant darkness), any animal with an endogenous period shorter than 24 h will wake slightly earlier each subsequent day, and hormone levels also will peak slightly earlier each day. Under the same constant conditions, an animal with an endogenous period longer than 24 h will wake slightly later each subsequent day, and hormone levels will also peak later. Such rhythms are said to "free run," and "free running under constant conditions" is one of the criteria by which we determine whether a rhythm is driven by an endogenous oscillator. When animals are returned to a long-day photoperiod (a 14L:10D light–dark cycle), the rhythms of all the individuals will again become synchronized (entrained) to the light–dark cycle. Whenever a new rhythm in behavior or hormone secretion is discovered, the true test of whether the rhythm is driven by an endogenous oscillator is to place the animals in constant conditions to determine whether the rhythm free runs. If a free-running rhythm can be resynchronized to a particular light–dark cycle, this is evidence that the free-running rhythm can be entrained by the photoperiod.

Even in the controlled constant darkness and temperature of the modern laboratory, some researchers have speculated that persistent rhythms are not endogenous but rather driven by other geophysical forces (e.g., gravity, magnetic fields) that are as yet undetected. To test this idea, a variety of different species, including bread molds, fruit flies, and crabs, have been shot into space and maintained in constant conditions aboard space shuttles. In most cases the free-running period of the circadian or circatidal rhythms persists in constant conditions, even in the low gravity of outer space. While the period of the rhythms remain intact under constant conditions, the amplitudes are sometimes altered by context. For example, in the low gravity of outer space and during periods of sleep deprivation, the amplitudes of body temperature rhythms and hormonal rhythms are altered, and these conditions, plus the odd light–dark cycles in the space shuttle, produce serious sleep disturbances in astronauts.

Another criterion for an endogenous rhythm is sensitivity to a *zeitgeber*. *Zeit* is German for "time" and *geber* is German for "giver." For many circadian rhythms, the zeitgeber is the onset of light or dark. Circadian rhythms in many hormones are driven by endogenous circadian clocks and entrained by daily patterns of light intensity. Thus the daily rhythm in cortisol free runs under constant darkness, but when the animal is exposed to 16 h of light on a regular daily schedule, the rhythm settles in to the new schedule (e.g., there are peaks of plasma cortisol concentration at the time of awakening, the time of "lights on"). There are many other zeitgebers, such as the fluctuation in ambient temperature and the availability of food. When these environmental cues occur in a regular daily pattern, their rhythm can act as a zeitgeber for an endogenous rhythm. Zeitgebers synchronize the multiple clocks within organisms and the clocks among individuals in a population.

A final criterion for an endogenous rhythm is temperature compensation. Because you know that the chemical reactions of hormone synthesis occur faster at high temperatures, you might reasonably suspect that living in warm temperatures would speed up and increase the frequency of hormone rhythms. This is not the case. Whereas many biological processes are accelerated at high temperatures and decelerated at cold temperatures, endogenous oscillators are not affected in this way. Thus we find that a rhythm in behavior or a physiological process, if driven by an endogenous oscillator, will persist with the same period in warm or cold temperatures as long as those temperatures are compatible with life. Daily fluctuations in ambient temperature, like the light–dark cycle, can act as a zeitgeber, shifting the phase of the rhythm, and constant extreme temperatures can affect the amplitude of the rhythms. That being said, the ambient temperature itself does not affect the period of the rhythm by acceleration or deceleration of chemical reactions.

III. HORMONAL EFFECTS ON BEHAVIOR DEPEND ON TEMPORAL PARAMETERS

## Interim Summary 15.1

1. There are circadian, ultradian, and infradian rhythms in many different activities, behaviors, physiological traits, and circulating hormone concentrations.
2. Rhythms are characterized by the following parameters.
   a. The period is the time interval between a point on the cycle and the same point on the next cycle, and frequency is its inverse, the number of cycles per unit of time. Endogenous rhythms show a free-running period in constant conditions.
   b. The amplitude is the difference between the peak (highest point) and the nadir (lowest point). The amplitude of an endogenous rhythm can be affected by environmental perturbations.
   c. The phase is a point on a cycle, often expressed in relation to the time frame of the particular rhythm, e.g., the peak or nadir in relation to sunrise outside or "lights on" in the laboratory. Phases can be altered experimentally or naturally.
3. A *bona fide* endogenous oscillator must meet the following criteria.
   a. Over multiple sequential cycles it drives a constant period that otherwise free runs in constant conditions.
   b. Its period is set by a zeitgeber.
   c. It is temperature compensated.

## MECHANISMS

All living organisms that have been studied have endogenous clocks, including invertebrates, plants, single-celled organisms, and bacteria. Thus clocks do not require a brain or nervous system and are a manifestation of the metabolic biochemistry of the organism. Many organisms have more than one endogenous oscillator controlling different traits. In general, the more complex the organism, the more oscillators occur in different tissues. In some animals peripheral oscillators are coordinated by a master circadian oscillator in the brain. The SCN serves this purpose and is well suited for the job; it is privy to information from the visual system about light and dark and sends projections and chemical messengers to many parts of the brain and body. It receives information about day–night cycles via the retino-hypothalamic tract, which receives information from the sensory receptors (rods and cones) of the retina and melanopsin-producing photosensitive retinal ganglion cells. Light entrainment of free-running circadian rhythms requires either visual receptors (the eye) or photosensitive retinal ganglion cells.

Retinal ganglion cells receive information from the retina of the eye, are themselves sensitive to light, and project to the SCN. These cells are located on the inner surface of the retina and produce a pigment known as melanopsin, which renders them photosensitive. One of these two sensory modalities (either the retinal ganglion cells or the rods and cones) must be intact for entrainment by light. Entrainment of rhythms is not abolished in mice that lack rods and cones (e.g., in mice with congenital retinal degeneration) or in mutant mice with intact retinal ganglion cells that fail to produce functional photosensitive melanopsin; entrainment is abolished only when both the retinal rods and cones and melanopsin expression are abolished. Intact retinal ganglion cells and their projections, the retino-hypothalamic tract, are necessary for light entrainment, because the retinal ganglion cells are the conduits through which signals from light received in the eye reach the SCN. In summary, information about environmental light is detected in the rods and cones and retinal ganglion cells, and reaches the SCN via the retino-hypothalamic tract.

By a circuitous route, SCN projections synapse on other neurons in other hypothalamic areas that leave the brain to reach the superior cervical ganglion and then return to the pineal body, an endocrine gland located deep within the brain. Temporal signals from the environment are thus sent from the SCN to stimulate nightly melatonin secretion. As noted, nighttime melatonin secretion occurs in animals that are awake at night (nocturnal) as well as in those that are typically asleep at night (diurnal). Similarly, melatonin may be a critical communicator of temporal information: Rhythms in melatonin secretion from the master circadian oscillator in the SCN are thought to synchronize many oscillators in other tissues (reviewed by Erren and Reiter, 2015; Silver and Kriegsfeld, 2014; Smale et al., 2003).

Mammals have clocks in peripheral organs such as the liver and stomach, and in fibroblasts, but these are all synchronized by the master circadian oscillator, the SCN. The SCN is a small, bilateral brain area near the front base of the hypothalamus next to the bottom of the third ventricle, just above the optic chiasm. Removal of the SCN (by a number of lesioning techniques) results in arhythmicity (a lack of coordinated rhythms) in many behaviors, physiological processes, and hormone secretion. Many behavioral and physiological traits will continue to be expressed without an intact SCN, but the rhythmicity of these behaviors is disrupted with removal of the SCN. Some neuroendocrine rhythms, such as those of gonadotropin-releasing hormone (GnRH) and gonadotropin secretion, are controlled by neural projections from the SCN. Others, such as locomotor rhythms, are controlled by chemical messengers released by the SCN (reviewed by Silver and Kriegsfeld, 2014). There still is much to be learned about the mechanisms by which the SCN communicates temporal information to the rest of the brain and body.

Other evidence that the SCN is the master circadian oscillator comes from a mutant hamster, the *tau* mutant (*tau* is the Greek letter used to designate period). This mutant is so named because it shows a shorter period of locomotor activity. Wild-type hamsters given running wheels begin running at the onset of darkness and end running at the time of "lights on." Their locomotor rhythm has a period of 24 h. In contrast, the *tau* mutant hamster runs with a period of only 20 h. When the SCN of the *tau* mutant is transplanted into the hypothalamus of a wild-type recipient, and the SCN of the wild-type hamster is transplanted into the hypothalamus of a *tau* mutant recipient, the transplanted donor SCN establishes control of the recipient animal's rhythms, so the postsurgical animals always exhibit the circadian period of the donor animal. A wild-type recipient with a *tau* mutant SCN will show rhythms of locomotor activity with a period of 20 h. This constitutes convincing evidence for an endogenous circadian oscillator that resides in the SCN (reviewed by Silver and Kriegsfeld, 2014).

Furthermore, cells of the SCN can be removed from an animal, kept alive, dissociated from each other, provided with growth factors and nutrients, and grown in a dish. Even when the cells in culture are dispersed in such a way that they cannot form synapses with each other, individual SCN neurons generate independent oscillations of gene expression and neuronal firing.

Whereas isolated SCN neurons show rhythmic patterns of gene expression, the rhythms are not the same as those that occur when the intact SCN is functioning in the brain. *In vivo*, some 20,000 SCN cells produce precise, endogenous, free-running rhythms in electrical activity and gene expression that require reciprocal synaptic connections within the nucleus. Most cells in the SCN are glutamatergic. The core SCN neurons synthesize and secrete vasoactive intestinal peptide, while the shell SCN neurons synthesize and secrete arginine vasopressin. In addition, the SCN contains receptors for melatonin and thus can receive feedback from the pineal gland (reviewed by Erren and Reiter, 2015). Owing to the presence of these hormonal factors, rhythms in gene expression and electrical activity produced by SCN cells *in vivo* are far more organized and precise than those expressed *in vitro*.

The cellular components of the SCN contain a molecular clock composed of genes whose transcription is alternately initiated and repressed. These have been discovered in a series of brilliant experiments with fruit flies (Fig. 15.2) and mice (Fig. 15.3). The translational products of some clock genes influence the transcription of other genes, forming one or more feedback loops that create daily rhythms of gene expression. The core proteins in fruit flies contain morning and nighttime feedback loops: CLK and VR1 in the morning, and CLK, PERIOD, and TIMELESS in the afternoon. Initiation of the circadian rhythm in the morning begins with CLK, which stimulates synthesis of VR1, which provides inhibitory feedback to CLK.

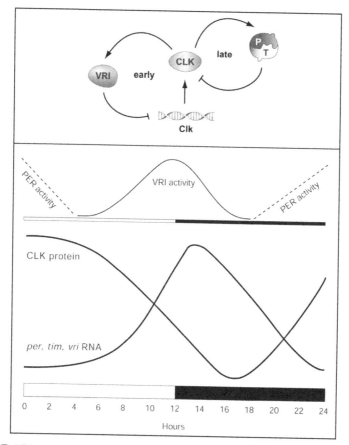

**FIGURE 15.2** The fly clock: a simplified diagram of some of the transcriptional steps underlying the circadian clock in *Drosophila*. Notice the conceptual similarity and conservation to a mammalian clock (see Fig. 15.3). (Top) Negative feedback cycles with a time delay make a rhythm. *P*, period gene; *T*, timeless gene. (Middle) High levels of VRI activity block the expression of the genes for period (per) and timeless (tim), and entrainment by the light cycle occurs. (Bottom) During the day, clock (CLK) mRNA and protein levels decline, while VRI mRNA and protein accumulate. Because T is quickly degraded by light, the absence of T after the lights go on allows CLK to switch on its repressor, VRI. This simplified summary does not include every autoregulatory link conceivable from current data; in particular, it puts emphasis on transcriptional feedbacks but does not fully represent posttranscriptional steps. *Adapted from Young, M.W., Kay, S.A,. 2001. Time zones: a comparative genetics of circadian clocks. Nat. Rev. Genet. 2 (9), 702–715.*

A similar feedback loop occurs later in the day and involves another gene, *Tim*, which encodes the TIMELESS protein (Fig. 15.3).

In mammalian species there are similarities in feedback mechanisms with different gene players (Fig. 15.4): the transcriptional activators CLOCK and BMAL1, and the suppressors, PERIOD and CRYPTO-CHROME. These proteins are encoded by the *Per* and *Cry* genes respectively. At the

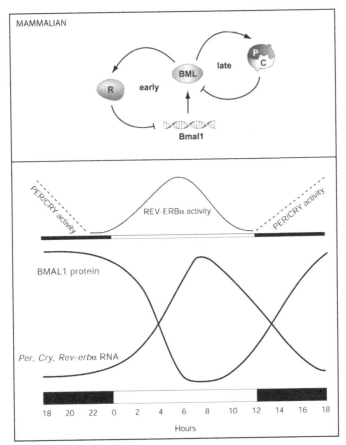

**FIGURE 15.3**　The mouse clock, simplified. Note the similarity to the fly clock (Fig. 15.2); some of the abbreviations are repeated. (Top) Negative feedbacks at the transcriptional level, coupled with time delays, make a rhythm. R, orphan nuclear receptor REV-ERB-α. (Middle) During the day, REV-ERB-α protein activity restricts the BMAL–protein product function, thus keeping *Per/Cry* activity low. (Bottom) In the morning, high BMAL protein levels induce a wave of *Per* (P), *Cry* (C), and REV-ERB transcription. Accumulation of *PER* and *CRY* proteins will be delayed until after dark. As with Fig. 15.2, this simplified summary does not include every autoregulatory link conceivable from current data; rather, it emphasizes transcriptional feedbacks. *Adapted from Young, M.W., 2002. Big ben rings in a lesson on biological clocks. Neuron 36 (6), 1001–1005.*

time of "lights on," CLOCK and BMAL1 activate the transcription of *Per* genes 1–3 and the *Cry* genes 1 and 2. As the day wears on, *Per* and *Cry* genes are translated to their protein products, PERIOD and CRYPTO-CHROME. The proteins in turn inhibit transcription mediated by CLOCK and BMAL1. The period of the clock is further stabilized by other transcriptional feedback loops. A similar mechanism is present in nonmammalian species, although in some species different genes and gene products play the role of CLOCK, BMAL1, *Per*, and *Cry* (reviewed by Silver and

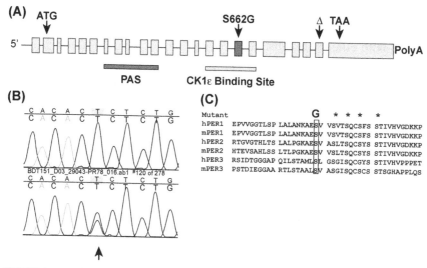

**FIGURE 15.4**   Genetic structure of the human period two gene, the mutation of which produces a loss of function in patients' daily activity rhythms. (A) The gene has 23 exons (*gray rectangles*), shown with the translation start site (ATG) and the signal for the polyadenosine tail (TAA). The *triangle* above exon 22 shows the location of the mutation. (B) The upper DNA sequencing record is from a normal control; the lower record is from an individual patient. The place where the records do not match shows the site of the base substitution that comprises the mutation. (C) The amino acid substitution, S (serine) to G (glycine), causes a protein not to function properly. *From Toh, K.L., Jones, C.R., et al., 2001. An hPer2 phosphorylation site mutation in familial advanced sleep phase syndrome. Science 291 (5506), 1040–1043. With permission.*

Kriegsfeld, 2014; Erren and Reiter, 2015). Mutations that prevent transcription of clock genes or synthesis of their protein products render the animal arrhythmic. Additionally, transcription of the clock genes is influenced by circadian rhythms in cytoplasmic metabolism, which in turn can be influenced by food availability. One of the frontiers of circadian rhythm research is the study of nuclear and cytoplasmic clock mechanisms and how they interact to produce organismal rhythms.

## Interim Summary 15.2

1. Virtually all organisms that have been studied show circadian rhythms in various traits, including hormone secretion.
2. In mammals there are a number of different endogenous circadian oscillators, and these are synchronized by a master circadian oscillator in the SCN. Cells in the SCN show endogenous rhythms in gene expression and electrical activity; ablation of the SCN results in arhythmicity; and a transplanted SCN confers its own endogenous period on to the recipient of the transplant.

3. SCN rhythmicity is in part related to an autonomous nuclear, molecular clock that interacts with cytoplasmic rhythm generators and with other SCN cells.

## HORMONE–CLOCK INTERACTIONS

Changes in glucocorticoid secretion can trigger circadian gene expression in cells, and dexamethasone, a glucocorticoid receptor agonist, can cause phase shifting for peripheral oscillators. This provides some hint of the ways in which acute and chronic stress can influence circadian patterns of behavior, especially sleep–wake cycles.

In animals that are restricted to less than their usual daily food intake, timed meals can act as a zeitgeber for the onset of activity that occurs in anticipation of meal times, and for changes in gene expression in liver, pancreas, and heart. Even SCN-lesioned animals that are arrhythmic for many behavioral and physiological traits show circadian patterns of clock gene expression in the liver and gut. Food and food metabolites, or hormones that are stimulatory for food intake such as ghrelin, can act as cues for peripheral (slave) oscillators in these organs. Further evidence is provided by the fact that clock genes are rhythmically expressed in stomach cells that secrete ghrelin, and the pattern of gene expression matches the pattern of meals. This might be part of the mechanism whereby peripheral metabolic clocks synchronize meals to the time of optimal metabolism. The pattern of meal times has long been known to modulate the secretion of glucocorticoid hormones from the adrenal gland (see Chapter 5).

The SCN controls circadian corticosteroid levels via its influence on the release of adrenocorticotropic hormone. There are also afferent and efferent neural connections from the SCN to the adrenal gland. In addition to the effects of peripheral clocks on hunger for food, the SCN has afferent and efferent communication with adipose tissue, providing a mechanism whereby disrupted sleep, shift work, and jet lag might increase obesity and body fat storage. Other links between glucocorticoids, rhythms, and food intake are well known. Nocturnal animals feeding in the day or diurnal animals eating at night show altered rhythms in glucocorticoid secretion that might promote increased adipose tissue accumulation and insulin resistance, ultimately leading to the development of type II diabetes (reviewed by Silver and Kriegsfeld, 2014).

Rhythm desynchronization, unhealthy preference for sweet and fatty foods, lack of concentration, disrupted sleep, and indigestion are among a suite of traits characteristic of jet lag. This happens when we fly from one time zone to another, e.g., from the United States to Europe.

Similarly, when laboratory animals living in one consistent light–dark cycle are exposed to a new light–dark cycle (e.g., the lights come on 6 h later each day), they experience the same symptoms (an increased preference for sweet and fatty foods, lack of concentration, disrupted sleep). When we fly to a new time zone it takes several days for our sleep–wake cycle and rhythms of cortisol secretion and body temperature to adapt to the change in phase: about a day for every hour of time zone shift for most traits. Some aspects of our physiology take more or less time to adjust. Separate rhythms, and perhaps separate endogenous oscillators, are desynchronized after a phase shift, and this is further exaggerated by a flight in the opposite direction within days or weeks after the first flight.

Altered glucocorticoid rhythms are one reason why this adaptation takes so long: Secretion of glucocorticoids stimulated by a phase shift uncouples the SCN's influence over individual cells, so they become desynchronized (Schibler et al., 2001). Similar increases in glucocorticoids occur in response to major life stressors, such as a car accident or becoming the victim of violence. Readjustment entails more phase shifting and resynchronization. To facilitate the resynchronization of various hormonal rhythms, travelers are advised to eat and drink minimally when in transit, and then begin eating normal-sized meals in phase with the new light–dark cycle. Resynchronization is hastened by phase shifting both meals and sleep–wake cycles together during entrainment and reentrainment. This idea stems from the knowledge that both light and food act as zeitgebers via the hypothalamo–pituitary–adrenal system, and that glucocorticoid receptors are expressed in a wide array of cells in the brain and peripheral nerves, giving them broad influence over peripheral and central oscillators.

Transcriptional feedback loops affecting multiple hormone systems are of central importance for clinical medicine and human hormone/behavior relations. This can be illustrated by a *Per* gene mutation. An individual with this mutation begins the sleep cycle in the early afternoon and awakens very, very early each morning (Fig. 15.4). Sleep physiology research continues to elucidate molecular mechanisms related to multiple hormone interactions, e.g., a microarray study with DNA chips has shown that gene expression for the synthetic enzyme that produces a prominent sleep-inducing substance, prostaglandin D, is significantly reduced by estrogen administration in a basal forebrain region in the preoptic area that governs sleep (reviewed by Hayaishi, 2000). It can be speculated that estrogens acting via inhibition of prostaglandins reduce the drive for sleep, thus increasing opportunities for courtship behaviors and sexual arousal. All this data is related to changes in behavior throughout the 24-h day.

## Seasonal Rhythms

As shown in Fig. 15.5 and discussed in Chapter 21, seasonal rhythms have a period of 12 months. They allow animals to synchronize their reproduction with optimal levels of energy supply and demand, nest sites, melted snow and ice, rainfall, and predation. Small mammals, because of their high surface-to-volume ratios that exaggerate heat loss, must dedicate the bulk of their energy stores to thermogenesis, even in mild ambient temperatures. Some small mammalian species have evolved mechanisms to enable them to survive in geographic areas with harsh winter conditions. Reproduction is energetically costly, especially to females, and in small mammals living in habitats where energy supply and demand fluctuate, the reproductive system is inhibited to conserve energy necessary for winter survival. Some seasonal breeders conserve energy by hibernation, a regulated drop in body temperature, and may gain weight in the shortening days of autumn so they can use their stored body fat to fuel arousal from hibernation in the spring. Nonhibernating seasonal breeders use a different strategy: some lose body weight in winter, presumably because a small size requires less energy intake, and a decrease in hunger and foraging activity promotes survival when the food supply is under cover of snow and ice. Seasonal rhythms of reproduction have allowed small mammals to occupy habitats with cold winters and enable behavioral endocrinologists to understand hormonal control of seasonal rhythms.

Seasonal rhythms can be classified into three distinct types (Prendergast et al., 2002): type I rhythms are driven by an interval timer initiated by an environmental cue, such as day length, type II rhythms are driven by an endogenous oscillator, and type III rhythms are driven strictly by exogenous cues.

Type I rhythms involve an interval timer that is initiated by decreasing day length, typical of autumn in the northern hemisphere. These rhythms disappear when the animals exhibiting them are housed under constant long-day/short-night conditions; the seasonal rhythm must be initiated by a decrease in day length. For example, in response to the decreasing day lengths of autumn in Syria and Turkey, Syrian hamsters accumulate adipose tissue and become reproductively inhibited. About 5–6 months later they spontaneously lose body weight and become reproductively active. The inhibition of the hypothalamic–pituitary–gonadal (HPG) system is triggered by a decrease in day length, but stimulation of the HPG system in long days is not triggered by increasing day length; rather, in spring the HPG system automatically restarts by a process known as *spontaneous recrudescence*, whereby the system is initiated by an interval timer that renders hamsters refractory to the short day length. In the

## Type I Rhythm
Mixed: Driven by endogenous and exogenous components
(e.g., Syrian hamster reproduction)

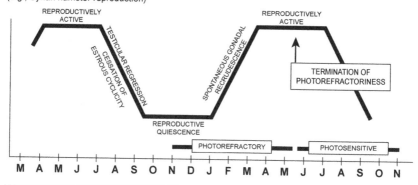

## Type II Rhythm
Endogenous: Product of circannual clock(s)
(e.g., ground squirrel body weight)

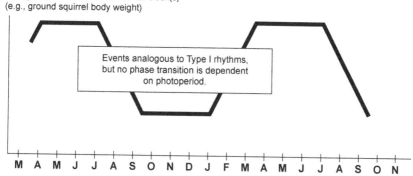

## Type III Rhythm
Exogenous: Environmentally triggered
(e.g., hay fever)

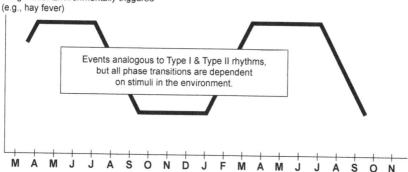

**FIGURE 15.5** Three distinct types of seasonal rhythms. *From Prendergast, B.J., Nelson, R.J., Zucker, I., 2002. Mammalian seasonal rhythms: behavior and neuroendocrine substrates. In: Pfaff, D.W. (Ed.), Hormones, Brain and Behavior, vol. 2. Elsevier Science, pp. 93–156.*

laboratory, even if the spring day lengths never arrive (the day length remains set on a short-day photoperiod), Syrian hamsters will become reproductively active after about 20 weeks of exposure to day lengths shorter than 10 h. Exposure to long day lengths breaks refractoriness and resensitizes the neuroendocrine axis to the short-day signal that initiates gonadal regression and the interval timer each year (reviewed by Prendergast et al., 2002).

As discussed in Chapter 21 and earlier in this chapter, melatonin secretion occurs at night during the dark phase of the light–dark cycle and is suppressed by light. Thus in the short days of winter, nighttime secretion of melatonin lasts longer. The duration of the melatonin signal mediates the effect of day length on annual reproductive cycles in many mammalian and avian species. Long-duration melatonin pulses (typical of short days) are inhibitory for the HPG system, suppressing hypothalamic GnRH, pituitary secretion of gonadotropins, and gonadal steroids. In contrast, high levels of gonadal steroids drive sexual motivation and performance, increasing the number of matings, pregnancies, and births in spring and summer. Experimental infusions of melatonin have been used to test this idea. Long-duration melatonin pulses that mimic the short days of winter increase the hypothalamic sensitivity to negative feedback from the gonadal steroids, thereby inhibiting the HPG system. In addition, exposure to short day lengths decreases behavioral responsiveness to sex hormones. Males housed in conditions of short day lengths and treated with androgens show fewer mounts, intromissions, and ejaculations. Females housed in conditions of short day lengths and treated with ovarian steroids show a lower incidence of lordosis (Fig. 15.6). Furthermore, long-duration melatonin pulses alter neuropeptide secretion and receptor density, sympathetic tone, and the innervation of adipose tissue. When the animals are pinealectomized, the rhythms in reproduction disappear and the animals return to the summer condition. Some of these winter responses, however, are independent of melatonin and gonadal steroids.

Natural cycles of body-weight loss and gain in seasonal species might help elucidate the mechanisms that control obesity and leanness. For example, marked body-weight loss occurs in Siberian hamsters exposed to short days (long nights), related to increased lipolysis, i.e., the mobilization and breakdown of triglycerides stored in lipids in adipose tissue. Triglycerides are broken down to glycerol and free fatty acids. The free fatty acids are metabolized during body-weight loss to provide energy for thermogenesis. While some humans struggle and sacrifice to lose weight, Siberian hamsters lose a large proportion of their body weight naturally when exposed to the short days of winter. These changes in metabolism are not direct effects of changes in melatonin, insulin, leptin, or epinephrine; rather, the effects of short day length are mediated by increased secretion

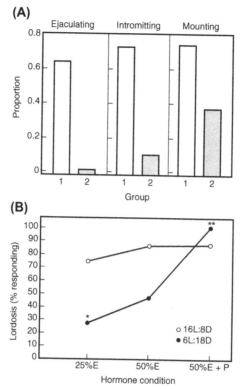

**FIGURE 15.6** Short day lengths can decrease responsiveness to sex hormones. (A) Shaded bar indicates short days. (B) E, estradiol; P, progesterone. *From Prendergast, B.J., Nelson, R.J., Zucker, I., 2002. Mammalian seasonal rhythms: behavior and neuroendocrine substrates. In: Pfaff, D.W. (Ed.), Hormones, Brain and Behavior, vol. 2. Elsevier Science, pp. 93–156.*

of norepinephrine by the sympathetic nervous system, and the effects of norepinephrine are mediated by direct two-way connections between the SCN and adipose tissue (reviewed by Bartness et al., 2014). In the wild, seasonal changes in reproduction are synchronized with seasonal changes in behavior and metabolism to increase reproductive success. In the laboratory, the mechanisms that control seasonal changes in reproduction can be dissociated experimentally from those mechanisms that control body weight to reveal important differences in the underlying neuroendocrine responses to changes in day length.

In addition to day length, food and water availability, the presence or absence of mating partners, and ambient temperature can provide additional cues to fine-tune the response to seasons. This occurs in a variety of mammalian and avian species. In Siberian hamsters, while gonadal regression occurs in short day lengths (less than 10h of light), at intermediate day lengths (neither short nor long, as would occur in spring or autumn) gonadal regression is delayed by the availability of an abundant food supply. Conversely, in animals housed in conditions of intermediate day lengths, a scarcity of food can accelerate gonadal regression. These effects of food availability in Siberian hamsters are associated with changes in the expression of the gene for kisspeptin, whereas the effects of day length are associated with changes in the expression of the gene for RFamide-related peptide, the mammalian ortholog of avian GnIH (Paul et al., 2009). Thus the day length can be regarded as an initial predictive cue that allows animals to anticipate a likely change in energy supply and demand, while food availability provides a proximate cue to allow animals to adjust according to energetic reality. Intermediate day lengths can be used to parse mechanisms involved in effects of different zeitgebers (reviewed by Wingfield and Farner, 1980).

In contrast to type I rhythms, type II rhythms are driven by an endogenous oscillator with a period of 12 months. The evidence for this mechanism is gathered by placing animals (some rodent, ungulate, and avian species) on a constant photoperiod (e.g., 14h of light: 10h of darkness) and temperature. Endogenous circannual rhythms persist for more than 3 years, with a free-running period of a little over or under 12 months. Depending on the species, type I endogenous oscillators are entrained to the changing day length and the underlying rhythm of melatonin secretion. Golden-mantled ground squirrels and woodchucks, for example, show yearly changes in body weight and reproductive status, and these persist for 3 or more years under the constant conditions of the laboratory (the same ambient temperature, food availability, and light–dark cycle). In fact, type II annual rhythms continue even in pinealectomized animals that have no melatonin secretion; the rhythms persist with a free-running period of about 12 months. The site of the circannual oscillator is unknown; animals with bilateral SCN ablation continue to show free-running circannual rhythms. The pineal gland and the SCN are required for entrainment to a light–dark cycle, but they are not required for free-running circannual rhythms, which include annual cycles in body weight, HPG function, and hibernation.

# IN HUMAN BEINGS

In humans, as in experimental animals, daily changes in hormone secretions and behavior are governed by at least two types of forces: light-entrained 24-h rhythms in which the underlying circadian mechanisms, referred to earlier, are synchronized with the environment by visual stimuli; and the drive for sleep, which can be gated by light-entrained rhythms but works by different mechanisms. With respect to daily changes in hormone secretions by human volunteers, for example, many investigators have documented very clear rhythms in cortisol, growth hormone, and thyroid-stimulating hormone (Fig. 15.7) (reviewed by Czeisler and Klerman, 1999).

In humans, seasonal changes in moods can be quite striking. Depression, alcohol use, violence against women, breast feeding, food consumption, and other behaviors associated with hormone-influenced neurochemical systems in the midbrain and forebrain during winter months, particularly in extreme latitudes with long, dark winters, tell us that day-length-dependent and temperature-dependent mechanisms in the human brain are important factors in hormone-controlled behaviors (Zordan and Sandrelli, 2015). Circadian factors are also being increasingly recognized as factors in metabolic disorders such as diabetes, in which regulation of the hormone insulin is so important (Challet, 2015).

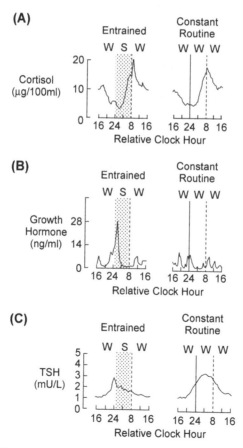

FIGURE 15.7  Human circadian rhythms in cortisol (A), growth hormone (B), and thyroid-stimulating hormone (C). W, awake; S, sleep (Czeisler and Klerman, 1999).

III. HORMONAL EFFECTS ON BEHAVIOR DEPEND ON TEMPORAL PARAMETERS

Disruptions in circadian rhythms, sleep deprivation, sleep interruptions, and altered sleep patterns are common in adults and children suffering from mood disorders. Sleep disorders often precede clinical diagnoses of mental illness, and in many cases restoration of regular nocturnal sleep patterns can lead to improved mental health. For example, manipulation of the timing of sleep can temporarily ameliorate symptoms of major depressive disorder and bipolar disorder. Some mood disorders are influenced by clock genes that drive endogenous oscillators. People with mutations in the genes for casein kinase 1 epsilon and PER2 exhibit familial advanced sleep phase disorder (a tendency to fall asleep several hours earlier than normal). Polymorphisms in the *Cry2* genes are linked to bipolar disorder and depression. Polymorphisms in the *Per3*, *Cry1*, and *Tim* genes are linked to schizophrenia and schizoaffective disorder (reviewed by Silver and Kriegsfeld, 2014). There is no doubt that the rhythmic patterns of neuroendocrine processes are central to mental and physical health.

## SOME OUTSTANDING QUESTIONS

1. Now that we know that endogenous oscillators can occur in bacteria and can be synthesized by placing the key metabolic components of the clock in a test tube, how do biochemical clocks interact with endogenous oscillations in gene transcription, and how do the biochemical and molecular clocks interact with and/or determine the workings of the neural clock?
2. What is the relationship of entrainment of hormonal biological clocks by light/dark cycles to the development of behavioral pathology in humans (e.g., seasonal affective disorder syndrome)?
3. What is the effect of time of day upon responsiveness to exogenously administered hormones, medications, anesthetics, and analgesics?

# SPACE: SPATIAL ASPECTS OF HORMONE ADMINISTRATION AND IMPACT ARE IMPORTANT

# Effects of a Given Hormone Can Be Widespread Across the Body; Central Effects Consonant With Peripheral Effects Form Coordinated, Unified Mechanisms

Near the beginning of this millennium, obesity researchers noted a striking difference in body weight between two populations of laboratory mice. The lean and obese populations did not differ in genotype or diet. Both populations were raised in rooms with the same ambient temperature, humidity, and photoperiod. They were raised on diets with the same macronutrient content and had the same free access to water. The only differences between populations were in their body weight and body fat content (level of adiposity). The fatter mice were raised in the conventional way whereas the leaner mice were raised to be germ-free. On average, the conventionally reared mice were 42% fatter than the same-aged germ-free mice, even though the conventionally reared, fatter mice ate 29% *less* food per day than their lean, germ-free counterparts (Backhed et al., 2004). These lean mice were categorized as "germ-free" because they were not colonized by the microorganisms that inhabit our bodies and the bodies of most mammals. Germ-free mice are born in sterile conditions, removed from their mothers' wombs by sterile surgical techniques, and raised by germ-free foster mothers in sterile isolators in sterile laboratories. To enter the laboratory to weigh the germ-free mice, researchers and caretakers must don a sterile gown, gloves, and mask, just as if they were entering a surgical suite.

What are these fattening "germs" in mice that are raised conventionally? Many microorganisms naturally inhabit the bodies of most animals, including humans. These microorganisms live on the skin and in the hair, nose, ears, and mouth, but the vast majority are anaerobic bacteria that live in the intestines. They make up the gut flora (also known as the gut microbiome or gastrointestinal microbiota), a complex community of microorganisms that influence the amount of energy and nutrients that can be derived from digested food. The gut microbiome of adult humans is remarkably large; we are composed of approximately 10 times more microbial than human cells.

When researchers discovered that germ-free mice were leaner than those with a conventional gut microbiome, they hypothesized that obesity was caused by the bacterial microbiome in the conventionally raised mice. They tested this idea by harvesting bacteria from the intestines of conventionally raised mice and then applying the harvested bacteria to the fur of the germ-free mice (where the germ-free mice would be sure to lick them off and thereby ingest the bacteria). Within 2 weeks of treatment, the mice gained 60% more body fat compared with germ-free mice that were not treated with the intestinal bacteria, and the treated mice grew fatter while eating significantly less food. The increase in body fat content in germ-free mice was partially replicated by infusing a single strain of bacteria directly into their intestines (Backhed et al., 2004). The researchers hypothesized that gut flora can influence metabolism, nutrient absorption, and obesity.

Next, investigators searched for and discovered a set of identical human twins who would allow them to test their hypothesis further. One twin was obese and the other was lean. Could gut flora from these individuals reverse or cause obesity? The researchers harvested fecal microbiota from the obese human twin and transplanted it into the intestines of germ-free mice. The mice rapidly gained body fat while decreasing their food intake. The researchers then harvested and transplanted microbiota from the lean twin and transplanted it into germ-free mice. Those mice remained lean. All mice were raised on the same food, but those with microbiome transplants from the obese human twin rapidly gained body fat without increasing food intake, whereas those with the transplants from the lean human twin remained lean. Once the researchers had produced two mouse populations with different gut flora and levels of adiposity, they housed some of the lean and obese mice together to determine whether the lean gut microbiome and its effects were transmissible. The obese mice lost body weight and body fat content. These results are consistent with the idea that the lean microbiome and its effects can be transferred from one individual to another (Ridaura et al., 2013).

The results are astounding for many reasons. First, they suggest that mammalian bodily processes are linked to those of symbiotic and infectious microorganisms that inhabit the intestine. Second, they suggest that one possible cause for the increase in obesity in modern western societies is related to the gut microbiome and its influence on the rest of the body. It is possible that some obese individuals harbor an intestinal microbiome that increases the efficiency of absorption of fuels and nutrients from digested food. Third, it shows that events in the body, in this case the intestines, have a profound influence on metabolism, physiology, and behavior (e.g., food intake). Fourth, the fact that obesity develops in animals that decrease, rather than increase, their food intake, implies that overeating is not the primary cause of obesity. The results lead to many questions. How are increases or decreases in energy from the gut sensed, and how does the signal travel to the brain? How does information about food absorbed by the intestine arrive at the brain areas that determine food intake, energy storage, and energy expenditure? Are hormones released in the gut in response to changes in levels of nutrient absorption? Are hormones released from the liver, pancreas, or fat cells in response to levels of metabolic fuels that are synthesized there? If so, where and how are increases and decreases in hormone levels detected? This chapter concerns the many ways in which hormone-behavior relations are determined by two-way communication between brain and periphery.

Every chapter of this book has illustrated examples of the brain–body connection. Our brains receive information from the external environment via peripheral sensory receptors in the retina, cochlea, taste buds, olfactory epithelium, and skin. Different brain circuits send back neural and chemical messages from the brain, for example to the pituitary, which in turn directs hormone secretion in other peripheral organs, including the adrenal and gonads. Hormones from peripheral organs modulate the secretion of the hypothalamus and pituitary by both negative and positive feedback. Other chemical messengers from the immune system and from peripheral biological clocks influence the brain. This chapter emphasizes peripheral mechanical, hormonal, and metabolic signals that have a profound influence on brain and behavior.

# BASIC EXPERIMENTAL EXAMPLES

## Control of Food Intake

In Chapter 2, chemical messengers that influence food intake were used to illustrate how hormones work in combination to affect behavior. For example, the peripheral hormone leptin, which is secreted from adipocytes, travels through the bloodstream, binds to leptin receptors in the arcuate nucleus (ARC) of the hypothalamus and thereby increases the secretion of α-melanocyte-stimulating hormone (α-MSH). α-MSH is secreted from axonal projections to other brain areas, including the lateral hypothalamus, to decrease food intake. In Chapter 3, chemical messengers that influence both ingestive and sex behavior were used to illustrate the concept that most hormones influence more than one behavior and often affect two different behaviors in opposite directions. For example, by stimulating α-MSH secretion, leptin both suppresses food intake and increases reproductive processes including sexual motivation. Ghrelin, on the other hand, stimulates the secretion of neuropeptide Y (NPY) and agouti-related protein (AgRP), which inhibit reproductive processes. These are only a few of the chemical messengers that have opposite effects on food intake, sexual motivation, and the hypothalamic–pituitary–gonadal (HPG) system (Table 3.2).

The current chapter adds a new layer of complexity to the control of reproductive and ingestive behavior: Signals that influence sex and ingestive behavior can act in the periphery and be transmitted to the brain via the vagal, splanchnic, and spinal afferent neurons, and neural communication by these routes can be elicited by changes in peripheral (1) hormone secretion, (2) mechanical stimuli, and (3) metabolic stimuli.

## Humoral Signals

Many humoral signals can affect the size of meals. In laboratory rodents, food is ingested in discrete meals over several minutes. The end of the meal is heralded by a process known as *satiation*, mediated by the secretion of hormones and other chemical messengers carried by the circulatory system. For example, the end of each meal is mediated by decreased secretion of ghrelin and increased secretion of cholecystokinin (CCK) and glucagon-like peptide I (GLP-I) from the stomach and intestines. Treatment of hungry mice and rats with CCK or GLP-I decreases meal size. Treatment of well-fed mice and rats with ghrelin significantly increases meal size. Ghrelin, CCK, GLP-I, and other signals can act in the periphery and affect the brain via the primary gut–brain neural route and the vagus nerve, as well as by another route, the celiac-superior mesenteric ganglion, a division of the splanchnic nerves (reviewed by Shechter and Schwartz, 2016). Afferent fibers of the vagus nerve carry information from the visceral organs (the gut) to the caudal brain stem, entering via the nucleus of the solitary tract (NTS). Evidence comes from experiments that examine the effects of meals and gut peptides on vagal electrophysiology and experiments in which neural communication between peripheral organs and the brain are eliminated. For example, CCK and leptin are two peripheral hormones that decrease meal size and activate vagal afferents, whereas ghrelin is a peripheral hormone that increases meal size and inhibits vagal afferents. When these vagal afferent connections are experimentally abolished, meal size increases (Schwartz et al., 1999). These results are consistent with the idea that the vagus nerve is critical for negative feedback on the control of eating by the stomach and intestines.

The vagus is not the only important neural mediator of gut signals. Gut infusions of carbohydrate solutions before presentation of a test meal decreases meal size in sham-operated rats, but not in rats with celiac-superior mesenteric ganglion (part of the splanchnic nerve) transection. Thus, signals from both the vagus and the celiac arm of the splanchnic nerve participate in the termination of meals. These signals limit the size of singular meals but do not affect the total amount of food eaten per day, because in response to the decrease in the size of the meal, animals compensate by increasing the number of meals. Thus, it is doubtful that these peripheral afferents alone prevent the development of obesity. Furthermore, vagal, splanchnic, and central nervous system signals are integrated. Future research will focus on the mechanisms by which these signals are integrated to coordinate food intake, energy expenditure, and energy storage.

## Mechanical Signals

A second type of peripheral signal comes from the feeling of fullness that occurs at the end of a meal. This sensation is mediated by mechanoreceptors that detect the stretching of smooth muscle when the stomach and intestines fill with food. To examine mechanoreceptors and their effects on food intake, artificial distension of the stomach or intestines is accomplished by inflating a small balloon surgically implanted in targeted parts of the gastrointestinal tract or by infusing water. When a small balloon is inflated in the upper intestine just before a meal is presented to human subjects, the size of the meal is significantly decreased. When the small balloon is inflated before a meal in laboratory rodents, there are increases in the expression of genes for α-MSH and decreases in the expression of NPY and AgRP in the ARC of the hypothalamus (similar to those observed in animals after a meal). These results support the idea that peripheral signals for a decrease in the size of a meal are the mechanical sensation when the gut is filled with food at the end of a meal and the hormonal changes that ensue (reviewed by Shechter and Schwartz, 2016).

## Metabolic Signals

A third type of peripheral signal is generated by the availability of metabolic fuels. Glucose and free fatty acids are considered to be metabolic fuels because they can be oxidized, and the resulting products of oxidation can be used to make adenosine triphosphate (ATP), the energy currency of the cells. One likely site of detection of metabolic substrates is the liver, the peripheral organ that processes metabolic substrates during fasting and feeding. After meals, the liver stores nutrients in the form of glycogen (glycogenesis) and triglycerides; during food deprivation, the liver releases glucose by breaking down glycogen (glycogenolysis) and releases free fatty acids and glycerol by breaking down triglycerides (lipolysis). During prolonged food deprivation, the liver is the primary site of ketogenesis, a by-product of fatty acid oxidation, and gluconeogenesis (the production of new glucose from amino acid backbones and glycerol). Like the pancreas, the liver has the ability to sense increases and decreases in circulating glucose concentrations.

There is considerable evidence that food intake can be altered according to the availability of fuels sensed by the liver. First, food intake is decreased in rats provided with glucose by systemic injection, gastric infusion, hepatic portal vein infusion (which supplies blood and blood contents directly to the liver), or jugular infusion (which supplies blood and blood contents to the brain). Hepatic portal vein infusion is more effective at decreasing food intake than infusion of the same dose of glucose into the jugular vein. Furthermore, rats that receive hepatic portal infusions of glucose will learn to associate the effects of the infusion with a novel flavor and develop a strong preference for that flavor; thus, the decrease in food intake that accompanies hepatic portal infusion cannot be explained by sickness caused by the infusion. The hepatic glucose-infused rats show no signs of sickness and decrease food intake to compensate for the glucose that is apparently detected by the liver (reviewed by Friedman, 1995). These results suggest that a potent stimulus for eating is a deficit in peripheral fuel availability detected in the liver, and that the detection of metabolic fuel availability by the liver provides a means to balance energy needs with both energy intake and energy storage. It is plausible that metabolic signals for energy deficit override hormonal, pharmaceutical, or prefrontal impulse controls for the sake of survival.

More evidence for control of food intake by detection of metabolic fuels in the liver comes from experiments that used the fructose analog 2,5-anhydromannitol (2,5-AM). 2,5-AM acts in the periphery to block the use of glucose, so that although glucose is present, it cannot be metabolized to increase the synthesis of ATP. Systemic treatment with 2,5-AM creates a peripheral deficit in the availability of oxidizable fuels and causes significant increases in food intake, even though this agent does not reach the brain in appreciable quantities. Furthermore, hepatic portal vein infusion of 2,5-AM increases food intake more effectively than does jugular vein infusion, and the effects of 2,5-AM on food intake are significantly attenuated by diaphragmatic vagotomy (reviewed by Friedman, 2008).

### Neural Signals From Adipose Tissue

Another important peripheral link to the brain comes directly from body fat. Body fat comes in two main types, brown and white. White adipose tissue (WAT) is mainly a storage depot for triglycerides, which can be broken down into glycerol and oxidizable free fatty acids by the process of lipolysis. For example, food-deprived animals survive by increasing lipolysis and the oxidation of free fatty acids. Brown adipose tissue (BAT) has triglyceride stores; however, BAT also has the capability of generating significant amounts of heat by virtue of its high mitochondrial content and unique biochemical properties. Both BAT and WAT send and receive information to and from the brain via spinal neurons of the sympathetic nervous system (but little if any input or output comes from the parasympathetic nervous system). The main neurotransmitter of the sympathetic nervous system is norepinephrine (also called noradrenaline). Sympathetic innervation of WAT by β-noradrenergic cells is necessary and sufficient for WAT lipolysis. Norepinephrine is thought to be far more important than epinephrine (also called adrenaline) in the control of lipolysis. For example, complete sympathectomy (elimination of β-noradrenergic input to WAT) blocks lipolysis, whereas complete removal of the adrenal medulla, the source of epinephrine, fails to block lipolysis. The noradrenergic control of WAT and BAT is responsible for naturally occurring seasonal changes in body weight in Siberian hamsters (*Phodopus sungorus*); thus, adipose tissue sends and receives information from the brain (reviewed by Bartness et al., 2014). Brain adipose tissue signals are relevant for guiding therapeutic obesity treatment such as liposuction and other procedures that remove adipose tissue. Communication between adipose tissue and brain allows the regrowth of adipose tissue to compensate for loss of adipose tissue by liposuction. Thus, removal of too much subcutaneous adipose tissue in one location leads to growth of adipose tissue at other sites in the body. This is significant because adipose tissue in the visceral, abdominal region is associated with insulin resistance and type 2 diabetes, whereas adipose tissue in subcutaneous areas such as the gluteofemoral region is cardioprotective. People who undergo lipectomy of gluteofemoral adipose tissue might experience an unwanted growth of adipose tissue in the abdominal region, which also might increase their susceptibility to disease (Van Pelt et al., 2005).

### Interim Summary 16.1

1. Ingestive and reproductive processes are sensitive to peripheral signals that come in three types, all of which can be transmitted via the vagus and/or splanchnic nerves to the caudal hindbrain via the NTS.
   a. Humoral signals, including the secretion of leptin, GLP-I, and CCK, are involved in meal termination and limitation of meal size.
   b. Mechanical signals from gut distension participate in negative feedback on food intake at the end of meals.
   c. Metabolic signals result from deficits in the availability of oxidizable fuels and ATP, and these signals increase ingestive behavior.
2. Both afferent and efferent communication to and from the brain from adipose tissue are critical for seasonal changes in adiposity in hamsters.

### Food Intake Fluctuates to Maintain Homeostasis in the Availability of Metabolic Fuels

It has been hypothesized that the signals that terminate meals interact with other signals to affect long-term food intake: for example, the amount of food eaten per day or per season. Some of these other signals might be related to environmental cues such as seasonal changes in the length of day, the ambient temperature, and food quality and availability. Other signals are detected directly by the brain. One pervasive idea is that these signals are integrated to "regulate" body weight and food intake, but this idea is not consistent with all of the evidence. Mice and rats housed in typical laboratory cages and allowed to eat as much food as they want accumulate adipose tissue over their life span and develop insulin resistance and hypertension (Martin et al., 2010). In other words, homeostasis in food intake and body weight is not maintained, but rather, food intake, body weight, and metabolic abnormalities increase with age. This process can be accelerated by high-calorie diets but it occurs even in animals fed standard laboratory chow. The development of obesity and insulin resistance is not always apparent in the short time frame of most laboratory experiments (2–6 weeks), but over the long term, homeostasis in food intake and body weight is not maintained.

It is therefore important to take an evolutionary perspective, which is based on the notion that most phenotypes that we observe in extant species influenced survival and reproductive success in the habitats in which those phenotypes evolved. For most species, the primary biological mandate is to survive and reproduce in environments where energy availability fluctuates. The neuroendocrine systems that control food intake did not evolve in an environment where food was available everywhere continuously. In addition, most organisms did not evolve in small cages isolated from conspecifics; most had to expend energy to obtain energy. To survive in environments in which energy is not available continuously, it is necessary to anticipate future energy shortages by increasing food intake and body fat storage when food is available. This suggests that many species have mechanisms that allow them to anticipate energy shortages.

Furthermore, empirical evidence shows that food intake and body weight are not "regulated" at a particular set point. Within individuals, body weights and levels of food intake fluctuate greatly to maintain homeostasis in another variable, the availability of metabolic fuels. Homeostasis in the availability of metabolic fuels is an absolute mandate for cellular life, because all cells require a constant supply of ATP for cellular respiration, ion flux, growth, etc., and the formation of ATP requires the oxidation of metabolic fuels such as glucose and free fatty acids. Metabolic fuels come from the macronutrient content of food. Glucose comes from carbohydrate foods, and free fatty acids come from fatty foods. Alternatively, or in addition, we get metabolic fuels from molecules stored in body tissues. For example, free fatty acids can be mobilized from body fat by lipolysis, and glucose can be derived from glycogen in muscle and liver by glycogenolysis. To provide a constant supply of oxidizable metabolic fuels, we either eat food or break down our own tissues, usually body fat, which results in loss of body weight.

In other words, eating food and performing lipolysis are the means by which we regulate the availability of metabolic fuels.

The cellular mandate for metabolic fuels is that food intake *must* fluctuate markedly to maintain life in an environment in which energy supply and demand are not continuous. Food intake is low at times when the demand for energy expenditure is low. Food intake increases sharply when the demand for energy expenditure increases: for example, after prolonged exercise or when the ambient temperature drops, triggering increases thermogenesis (heat production). Most animals eat more food per day at colder temperatures. If the ambient temperature were to drop and animals were to try and maintain "homeostasis" in food intake, they would become hypothermic and die. Another example is during pregnancy (in some species) and lactation (in most mammalian species). Many female mammals increase food intake during lactation. If wild female mammals were to maintain prepregnant food intake during lactation, their offspring would most likely starve. Thus, females of many species will double their food intake during lactation to meet the energetic demands of milk production to feed their offspring. The ability to change rather than maintain the level of food intake is a biological imperative in many species. The mandate is illustrated when two energetic challenges are combined. For example, lactating female mice, *Peromyscus leucopus*, increase their food intake 230% when housed at cold ambient temperature compared with their prepregnant food intake at higher ambient temperatures. Furthermore, when the total caloric density of laboratory chow is diluted with nonnutritive bulk, laboratory rodents do not maintain their food intake at a particular set point. Rather, they show a controlled increase in food intake in proportion to the caloric dilution. Food intake is not regulated (i.e., homeostatically maintained within a certain range); rather, it is controlled to meet metabolic needs (Friedman, 2008).

Similarly, body weight is not regulated; rather, body fat levels rise and fall to meet the need for homeostasis in the level of metabolic fuels. When food is not available, body weight is lost when animals release triglycerides from lipids in adipose tissue and break down triglycerides and oxidize the resulting free fatty acids. This perspective will facilitate biomedical researchers in forming testable hypotheses about controlling food intake and help consumers avoid oversimplified explanations of antiobesity drugs, diet, and exercise regimens.

Given these considerations, it is obvious that the purpose of so-called "feeding hormones" is not to maintain homeostasis or a set point in food intake and body weight, but rather to orchestrate the appetites for food and motivational states in environments in which energy availability fluctuates (reviewed by Schneider et al., 2013). These concepts are illustrated in Fig. 16.1A–C. Reproductive processes are permitted

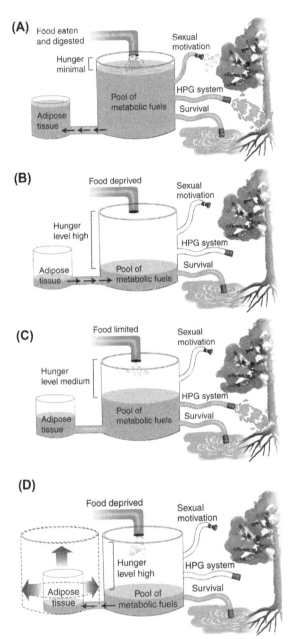

FIGURE 16.1   Behavioral priorities are set by the general availability of oxidizable meta-bolic fuels (represented by the *large tank* in the middle of each figure), which is the balance between the food eaten, digested, and absorbed and the energy expended. Excess can be stored in adipose tissue (represented by the *small tank* on the left), and deficits can be amelio-rated by breaking down triglycerides in adipose tissue to release oxidizable fuels. When food

and hunger and ingestion are minimized when the pool of oxidizable metabolic fuels (glucose and free fatty acids) is stable (Fig. 16.1A). In this scenario, the individual experiences minimum food insecurity, plentiful body fat stores, and low to moderate demands for energy expenditure (for example, ambient temperature is moderate and the individual has low demands for energy expended on foraging and hoarding food). The pool of oxidizable fuels is symbolized by the full tank in the middle. The low level of hunger is represented by the small distance between the top of the tank and the level of metabolic fuels in the big tank. Adipose tissue stores are symbolized by the small, full tank on the far left. The different types of energy expenditure are symbolized by the tree on the right. The roots represent processes critical for immediate survival, the trunk represents the HPG system, and the leaves represent reproductive behaviors (courtship, sex, and parental behavior). According to this model, when the middle tank is filled with oxidizable metabolic fuels, levels of the hunger-producing hormone ghrelin are relatively low, and levels of the anorexigenic hormone leptin are relatively high. This hormone profile supports the HPG system and sexual motivation and tends to increase overall energy expenditure. The center of this illustration is not the brain, but rather the level of oxidizable metabolic fuels. This high level of fuels provides a sensory signal that increases the secretion of hormones of the HPG system and leptin, whereas hormones of the hypothalamic–pituitary–adrenocortical (HPA) system and ghrelin are inhibited. This figure emphasizes the multiple communications between brain and periphery.

During food deprivation and/or when energy demands are high, the individual begins to break down the contents of adipose tissue and muscle to obtain glycogen. Glycogen is broken down into glucose, an oxidizable metabolic fuel. If food deprivation continues, the animal begins to break down lipids in adipose tissues; i.e., triglycerides are broken down into glycerol and free fatty acids, an oxidizable metabolic fuel. The tank

---

is abundant, hunger levels are low and there is ample energy for all biological processes, including reproduction (A). When food is scarce and adipose tissue stores are depleted, hunger increases and individuals can conserve energy by decreasing reproductive process (B). Most animals do not have unlimited quantities of food, but must expend energy to obtain food; thus, they remain vigilant about foraging and food hoarding and limit reproductive activity to the most fertile periods (C). In some types of obesity, a predisposition toward excess fuel storage depletes the supply of oxidizable metabolic fuels, thereby increasing food intake and decreasing the energy available for reproductive processes. This can occur as a result of chronically elevated cortisol and/or insulin (D). *HPG*, hypothalamic–pituitary–adrenocortical. *(A–D) Drawing by Jay Alexander adapted from Schneider, J.E., Wise, J.D., Benton, N.A., Brozek, J.M., Keen-Rhinehart, E., 2013. When do we eat? Ingestive behavior, survival, and reproductive success. Horm. Behav. 64, 702–728.*

of oxidizable fuels is not refilled and adipose tissue stores are depleted (Fig. 16.1B). When fat stores are depleted, the overall levels of metabolic fuels, mechanical signals from the gut, and secretion of leptin, CCK, and GLP-I begin to fall, and levels of ghrelin, NPY, AgRP, and other orexigenic neuropeptides begin to rise. This profile of hormones and peptides stimulates ingestive behavior, including increased hunger for food, increased foraging, and food hoarding. The high level of hunger is represented by the long distance between the top of the tank and the level of metabolic fuels in the big tank (Fig. 16.1B). Under these conditions, the animal will try to conserve energy by decreasing energy expenditure. One way to conserve energy for survival is to inhibit the HPG system. The resulting low levels of gonadal steroids (estradiol and progesterone in females) will preclude the occurrence of reproductive behavior. Again, the center of the illustration is the pool of oxidizable metabolic fuels, which initiates a host of hormonal responses when depleted. A primary function of the system is to restore homeostasis in the availability of oxidizable metabolic fuels.

In these two scenarios, the organism responds to an abundance of food (Fig. 16.1A) or a lack of food (Fig. 16.1B). Fig. 16.1A illustrates an unusual situation for wild animals and is more representative of animals living in the laboratory, which lead a relatively sedentary lifestyle with food available in unlimited quantities. A more realistic scenario is depicted in Fig. 16.1C, in which food is available, but in limited quantities, and/or the individual must forage for food or work to obtain food. In this scenario, the HPG system is not compromised but the animals are highly motivated to engage in vigilant foraging and hoarding behavior. The individual will engage in reproduction only if it has just eaten, and in the case of females, they are in the most fertile phase of the estrous cycle (reviewed by Schneider et al., 2013).

Fig. 16.1D illustrates an individual with a strong disposition toward energy storage, which can cause body fat accumulation with or without increased food intake. Any food that is eaten is rapidly broken down into macronutrients, which are broken down into fuels, which are converted into storage molecules, especially triglycerides stored in adipose tissue. In other words, metabolic fuels are shunted away from tissues where they can be oxidized and into tissues where they are stored. When this predisposition to store fuels is strong, it creates a deficit in the pool of oxidizable fuels. Hunger and food intake can be stimulated when the pool of oxidizable fuels is depleted by the transformation of these fuels into storage molecules (Fig. 16.1D). Thus, someone who is obese with a high body fat content can be voraciously hungry. When energy expenditure is inhibited, the accumulation of adipose tissue can increase without increased food intake. In some forms of obesity, excessive weight gain is accompanied by inhibition of the HPG system and sexual libido (reviewed by Ludwig and Friedman, 2014).

In many cases, excessive body fat accumulation precedes overeating. In practically every animal model of obesity in which overeating is present, animals show a predisposition toward energy storage that is independent of the increase in food intake. This is true in genetic models of obesity (mutant animals) and in wild-type animals after lesions of the ventromedial hypothalamus (VMH). VMH-lesioned rats rapidly gain weight even when they are restricted to their original food intake. As noted by Friedman and Stricker, "an animal with VMH lesions may not increase its food intake in order to gain weight, but because it is gaining weight" (Friedman and Stricker, 1976).

This can be demonstrated by treatment with hormones that promote body fat storage. The main function of the pancreatic protein hormone, insulin, is to remove glucose and free fatty acids from circulation and shunt those fuels into tissue where they can be oxidized or stored. At high doses, insulin "pushes" circulating fuels into storage in adipose tissue. Wade et al. (1991) treated adult female hamsters with supraphysiological doses of insulin or the saline vehicle (Fig. 16.2). The insulin-treated hamsters significantly increased their food intake while gaining a significant amount of body fat. When these hamsters were given an unlimited amount of food, they increased their food intake and continued to show normal estrous cycles. A second insulin-treated group was not permitted to overeat but was limited to its preinsulin food intake. This insulin-treated, food-limited group also gained a significant amount of body fat, although it did not overeat. When an animal can gain body fat without increasing its food intake, it is reasonable to wonder where the extra energy comes from. In this case, the hamsters conserved energy by inhibiting the HPG system. The hamsters that were treated with insulin and limited to their preinsulin level of food intake became anestrous and failed to show normal HPG function (Wade et al., 1991). This reflects the fact that insulin-induced body fat accumulation creates a deficit in the availability of oxidizable fuels that must be balanced by a decrease in the energy expended on reproduction. Furthermore, these results demonstrate that the inhibitory effects of insulin on reproduction are not direct, but act by modulating the availability of oxidizable metabolic fuels. Both insulin-treated groups received the same dose of insulin, but one had sufficient oxidizable metabolic fuels to maintain full reproductive function whereas the food-limited group became anestrous (Fig. 16.2). This refutes the notion that high doses of insulin are sufficient to inhibit reproduction. Insulin only inhibits reproduction in food-limited females, because these females cannot compensate for the fuels directed toward body fat storage. High doses of insulin

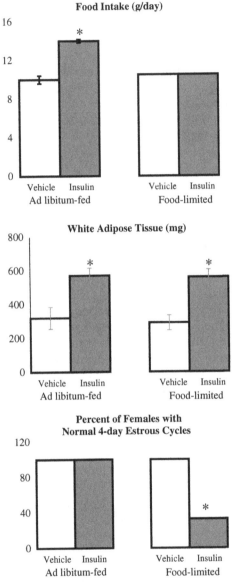

FIGURE 16.2   Effects of high daily systemic doses of insulin on food intake (top), white adipose tissue weight (middle), and the percentage of female hamsters showing regular 4-day estrous cycles (bottom). Levels of food intake were either ad libitum or limited to the level of saline-treated hamsters. Insulin treatment increased both food intake and body weight relative to the saline-treated controls. Body fat content was increased even in hamsters that were not allowed to overeat, and a greater percentage of these food-limited, insulin-treated hamsters became anestrous compared with those treated with insulin and allowed to increase their food intake. *P < .05. *Adapted from Wade, G.N., Schneider, J.E., Friedman, M.I., 1991. Insulin-induced anestrus in Syrian hamsters. Am. J. Physiol. 260, R148–R152.*

illustrate how hormones can modulate the peripheral metabolic stimulus for ingestive behavior. This perspective should facilitate biomedical researchers' forming testable hypotheses about control of food intake and should help consumers avoid oversimplified explanations of anti-obesity drugs, diet, and exercise regimens.

## Anticipatory Food Intake and Appetite

As noted earlier in this chapter, the gut signals that limit meal size do not account for obesity. For example, elimination of vagal signals produced by these gut peptides does not result in perpetual overeating and total bilateral vagotomy does not result in the acceleration of obesity. Rather, vagotomized animals compensate for their larger meals by decreasing the frequency of those meals (Schwartz et al., 1999 and reviewed by Shechter and Schwartz, 2016). The primary function of the mechanisms that limit meal size might be to match meal size to the ability of the gut to digest food and absorb nutrients. Although these mechanisms do not regulate body weight, understanding the mechanisms by which the peripheral visceral organs communicate with the brain might conceivably lead to therapies that allow us to intensify signals that prevent overeating.

Again, an evolutionary perspective is helpful. Animals living in environments in which energy availability fluctuates require mechanisms that promote overeating and energy storage. Energy can be stored in the home or on the body as glycogen in muscle or triglycerides in the lipids in adipose tissue. Mechanisms abound that promote overeating. One such mechanism is termed *appetition*, which involves the development of preferences for high-calorie foods. Peripheral signals are involved in appetition. This process has been studied by allowing laboratory animals to develop learned preferences for specific flavors. For example, when a mouse or rat is provided with a glucose solution and then provided with a choice between plain water and the glucose solution, it will choose to drink the glucose solution, as shown by a significantly greater ratio of glucose to water ingestion. The key to forming the preference is the availability of fuels from the glucose. For example, rats can be conditioned to drink a nonnutritive sweet cherry flavor by pairing that flavor with infusion of glucose directly into the stomach or upper intestine (the duodenum). After pairing the conditioned flavor (cherry) with the unconditioned stimulus (duodenal glucose infusion), the rats are allowed to choose between flavored and unflavored solutions. After presenting the cherry flavor paired with the glucose infusion, laboratory rodents invariably prefer the flavor that had been paired with the glucose infusion. The conditioning process is accompanied by increased dopaminergic activity in the mesolimbic system of the brain. Conditioned flavor preference does not require either the vagal or splanchnic nerves to be intact, but conditioned flavor preference requires an intact upper intestine near the exit of contents from the stomach (the duodenum–jejunal juncture). For example, gastric bypass surgeries that prevent nutrients from reaching the duodenum–jejunal area decrease the appetite for food and prevent conditioned preferences for sweet flavors. Infusion of glucose, but not nonnutritive control solutions, into the duodenum–jejunal area increases dopamine release in the ventral striatum of the brain; this effect on dopamine release is absent in animals with duodenal–jejunal bypass surgery (reviewed by Shechter and Schwartz, 2016). These results are consistent with the idea that flavor preferences akin to a "sweet tooth" can be learned when flavors are associated with glucose uptake in the gut. Furthermore, these experiments provide a clue to the mechanisms that underlie a tendency toward overeating highly palatable foods. It suggests that the signals that normally terminate meals might be overridden by the positive feedback effects of learned flavor preferences, especially when those flavors have been associated with a peripheral stimulus that provides metabolic fuels such as oxidizable glucose.

The mechanisms of appetition and their integration with mechanisms of satiation and satiety are unknown; this is an active area of research. It is clear that signals from mechanical gut distension can be transmitted neurally and/or humorally (e.g., via increases in CCK) and it has been hypothesized that these signals are integrated with other signals. Gut hormones such as CCK, GLP-I, and ghrelin are thought to provide short-term signals that influence meal size, whereas other peripheral signals are thought to provide long-term signals such as those that come from the individual's prior food intake and body fat content. For example, levels of oxidizable metabolic fuels and plasma leptin concentrations decrease after food deprivation and with decreasing levels of adiposity. Integration of peripheral and metabolic signals might occur in the NTS, where there are receptors for leptin, ghrelin, CCK, and GLP-I. A fall in adiposity and low levels of metabolic fuels might alter the sensitivity to gut signals such as those that arise from the peripheral secretion of CCK. In support of this idea, previously food-deprived animals are leaner and less sensitive to the inhibitory effects of CCK than are animals fed ad libitum. NTS neurons of food-deprived animals are less responsive to CCK, and the effects of food deprivation on neural and behavioral sensitivity to CCK are prevented by treatment with the adipocyte hormone leptin (reviewed by Schwartz, 2006). These and other studies provide initial evidence that peripheral short-term signals can be overridden by long-term signals in the NTS to affect behavior. These interactions might be involved in the overeating illustrated in Fig. 16.1D, and suggest possible mechanisms by which animals anticipate future energetic requirements.

## Gaining Body Weight Can Increase Food Intake and Inhibit Reproduction

In addition to signals that arise from food deprivation (as illustrated in Fig. 16.1B) a strong predisposition toward storage of metabolic fuels in adipose tissue might create a metabolic stimulus that overrides signals for satiation (Fig. 16.1D). This might involve the peripheral hormone, insulin. One possibility is that excessive exposure to sweet, high-calorie foods increases the secretion of insulin, which has four consequences: High levels of peripheral insulin (1) increase the flow of metabolic fuels from circulation into adipose tissue, (2) increase body fat accumulation, (3) create a deficit in the availability of fuels for oxidation, and (4) increase hunger and food intake.

This process is illustrated by experiments that compared food intake in female hamsters treated systemically with high doses of insulin or saline (Fig. 16.2). As noted earlier, Wade et al. (1991) treated adult female hamsters with supraphysiological doses of insulin or the saline vehicle, and the animals were allowed to eat ad libitum (as much food as they wanted). Compared with saline-treated, fed hamsters fed ad libitum, those treated with high doses of insulin ate significantly more food and increased in body weight and body fat content. A third group of hamsters was treated with insulin but was not allowed to increase food intake; this group was fed the same low amount of food eaten by the saline-treated hamsters. The food-limited, insulin-treated hamsters gained body weight and body fat content even though they were not allowed to overeat (Fig. 16.2), which illustrates that insulin promotes body fat accumulation even in the absence of overeating (Wade et al., 1991).

The process of appetition might involve high levels of insulin stimulated by the sweet flavor of food previously associated with a high-calorie infusion. For individuals who have experienced a flavor previously paired with a glucose infusion, the flavor elicits a spike in insulin secretion. Insulin promotes the storage of excess glucose in adipose tissue, which is often followed by a drop in blood sugar, i.e., a deficit in the availability of this oxidizable fuel. Thus, it is possible that the preference for sweet, high-calorie, insulin-stimulating foods exaggerates excess fuel storage (body fat accumulation), increases subsequent hunger, and sets into motion a vicious cycle of overeating and body weight gain (Ludwig and Friedman, 2014). In addition, a number of other environmental factors might increase energy storage, thereby decreasing the availability of metabolic fuels, increasing appetite and food intake, and promoting obesity. These include exposure to endocrine-disrupting compounds (reviewed by Schneider et al., 2014), maternal programming (epigenetic effects as discussed in Chapter 11), jet lag, shift work, light pollution (discussed in Chapter 15 and reviewed by Fonken and Nelson, 2014), and alterations in the microbiome. Thus, obesity is likely related to many factors that act via a common energy storage mechanism.

## Interim Summary 16.2

1. Changes in food intake and body fat content are the means by which animals maintain homeostasis in the availability of oxidizable metabolic fuels. Hormones that control food intake and body weight serve to maintain a level of oxidizable metabolic fuels compatible with cellular life.
2. When energy is abundant, individuals have low levels of ghrelin and high levels of leptin, CCK, and GLP-I, which decrease food intake and increase energy expenditure, including the energy expended on reproduction. When energy is scarce, the opposite hormonal profile increases hunger, food intake, foraging, and food hoarding, and decreases energy expenditure, particularly energy expended on reproduction.
3. Excessive body fat accumulation causes increases in hunger and food intake; many individuals are predisposed toward gaining excess body fat, with a resulting increase in hunger, food intake, and preference for calorically dense foods. This fat phenotype can be accompanied by low levels of energy expenditure and reproductive deficits.

### Inhibition of Energy Expenditure, Sexual Motivation, and the Hypothalamic–Pituitary–Gonadal System

Fig. 16.2 illustrates that an excessive increase in peripheral insulin secretion promotes food intake and body fat storage, and the increase in body fat storage can occur without increases in food intake. Where does the increased energy come from? One way that animals can gain body fat content without overeating is to decrease energy expenditure. In Fig. 16.2, hamsters that received injections of insulin but were not allowed to increase their food intake also became anestrous; that is, they conserved energy by inhibiting the energetically costly process of reproduction (Wade et al., 1991). This illustrates the idea that increased hunger and inhibited reproduction can result from deficits in the availability of metabolic fuels.

Fig. 16.2 also illustrates that the inhibition of reproduction is not a result of excess levels of insulin. Both the ad libitum–fed, insulin-treated hamsters and the food-limited, insulin-treated hamsters had the same level of insulin. The difference between the fertile and infertile groups is that insulin-treated hamsters fed ad libitum were able to garner enough energy for reproduction by increasing their food intake whereas the insulin-treated hamsters that were food-limited were not able to garner enough energy for reproduction.

The HPG system is inhibited by a variety of factors that decrease the availability of metabolic fuels, including food deprivation, treatment with inhibitors of glucose or fatty acid oxidation, and housing under cold ambient temperatures, or by an increase in energy expended on exercise without increases allowed in food intake (reviewed by Schneider, 2004;

Schneider et al., 2013; Wade et al., 1996). In many cases, the primary locus of inhibition occurs at the level of the gonadotropin-releasing hormone (GnRH) pulse generator. Without GnRH, there is no secretion of luteinizing hormone and follicle-stimulating hormone, and no synthesis of gonadal steroids and little or no sex behavior. GnRH secretion is inhibited by gonadotropin-inhibiting hormone in birds and the mammalian analogue, RFamide-related peptide 3, in mammals (reviewed by Tsutsui et al., 2010). In addition, inhibition of GnRH secretion might be mediated by decreases in kisspeptin action (reviewed by Clarke and Arbabi, 2016; De Bond and Smith, 2014). The mechanisms by which deficits in energy availability inhibit reproduction might be adaptive if they allow animals to survive metabolic challenges and thus increase the likelihood of successful reproduction if and when energetic conditions improve.

Peripheral and central inhibition of reproduction shares a great deal in common with peripheral and central stimulation of food intake. This might be adaptive when individuals can conserve energy by inhibiting the reproductive system and increasing their chances of survival by foraging for and eating food. For example, reproduction can be inhibited by peripherally secreted hormones known to stimulate food intake. Ghrelin is stimulatory for food intake and inhibitory for reproductive processes. Reproduction can be stimulated by peripherally secreted hormones known to inhibit food intake. For example, leptin is inhibitory for food intake and stimulatory for reproductive processes. Furthermore, mechanoreceptors that are stimulated by gut distension create neuropeptide profiles known to be inhibitory for the HPG system. For example, increased NPY/AgRP and decreased $\alpha$-MSH are stimulatory for food intake and inhibitory for reproductive processes. Metabolic inhibitors that block glucose and fatty acid oxidation stimulate food intake and inhibit estrous cyclicity and the HPG system. The role of the vagus is similar in food intake and reproduction: The effects of deficits in free fatty acid availability on reproduction are blocked by total, bilateral subdiaphragmatic vagotomy, whereas the deficits in glucose availability are not blocked by the same surgical procedure (reviewed by Wade and Jones, 2004; Schneider, 2004).

If food intake, body weight, and reproduction are controlled by the availability of metabolic fuels, what are the roles of hormones and neuropeptides? Hormones and neuropeptides can either mediate the effects of changes in fuel availability or modulate the metabolic stimulus. An example of hormonal mediation would be the sharp drop in plasma leptin concentrations that occur 12–24h after the start of food deprivation, long before any detectable changes in body fat content occur. The fall in leptin is stimulated by the drop in availability of metabolic fuels, and might participate in stimulating food intake. An example of hormonal modulation of the metabolic stimulus would be the increase in glucose and free fatty acid oxidation caused by leptin in many different tissues (Bryson et al., 1999; Ceddia et al., 1999; Minokoshi et al., 2002; Shimabukuro et al., 1997). Another example of hormonal modulation of the metabolic stimulus would be the increase in peripheral lipolysis that is stimulated by increases in estradiol levels (Nunez et al., 1980; Wade et al., 1985). Gonadal steroids have many effects on peripheral metabolism that are too numerous to recount in a book focused on behavior (Rettberg et al., 2014). In addition, gonadal steroids influence the gut microbiome and its ability to provide fuels and nutrients and to increase energy expenditure (Cox-York et al., 2015). These changes in peripheral fuel oxidation, in turn, might be detected peripherally and transmitted neurally, via the vagus or sympathetic nervous system.

One way in which hormones and fuels interact is that hormones modulate the metabolic stimulus for behavior. The stimulatory effect of leptin on the reproductive system in female hamsters depends on leptin's ability to increase metabolic fuel oxidation. When female hamsters are deprived of food on the first 2 days of their estrous cycle, they fail to ovulate and show estrous behavior, events that normally occur on the fourth day of the estrous cycle. This is termed "food deprivation-induced anestrus." Treatment with leptin during food deprivation prevents food deprivation–induced anestrus, and the restorative effects of leptin can be blocked by treatment with inhibitors of either free fatty acid or glucose oxidation (Schneider and Zhou, 1999). In addition, treatment with inhibitors of free fatty acid oxidation inhibits the HPG system (Sajapitak et al., 2008a,b).

A growing body of research has confirmed that leptin and other putative anorexigenic peptides influence behavior and physiology by their ability to influence the pathways for the use of metabolic fuel; leptin increases the oxidation of glucose and free fatty acids in various tissues. The ability of leptin and melanocortin agonists (such as α-MSH) to inhibit food intake depends on the action on enzymes involved in the synthesis of ATP or the phosphorylation potential: that is, the potential for the activation of kinase cascades. Enzymes that are sensitive to ATP/adenosine monophosphate, such as adenosine monophosphate kinase (AMPK), a serine/threonine kinase, are found in all species from yeast to humans, in both the brain and the periphery. These enzymes are activated by metabolic challenges that deplete ATP, such as exercise, fasting, and thermogenesis. Fasting increases AMPK activity in brain and has the opposite effect in the periphery. Experimental manipulation of hypothalamic AMPK activity can change food intake and body weight: Treatments that increase AMPK activation in the brain increase food intake and body weight, whereas inhibition of AMPK activation decreases food intake and body weight. For example, treatment with leptin or melanocortin agonists decreases food intake and inhibits AMPK activity in the ARC and paraventricular nucleus of the hypothalamus, and this inhibition is necessary for the decreases in food intake (reviewed by Minokoshi et al., 2004, 2008). It remains to be determined how these changes in brain metabolism are integrated with changes in peripheral metabolism (reviewed by Schneider et al., 2012); future work must take into consideration both peripheral and brain effects of chemical messengers that affect behavior.

## Clinical Example: Obesity

Biomedical researchers often refer to homeostasis, i.e., a "set point" at which levels of food intake and body weight might be maintained by hypothetical neuroendocrine mechanisms. As described in this chapter, however, many mechanisms that inhibit food intake are overridden by mechanisms that promote food intake. An evolutionary perspective can be enlightening. One of the most important forces of evolution is natural selection, which favors survival and reproductive success. In many species including our own, there is no strict reproductive mandate to maintain homeostasis (a set point) in food intake or body weight, and therefore a wide range of body weights and levels of daily food intake are compatible with survival and reproductive success. The human population in most westernized, industrialized nations is composed of obese individuals and their many offspring, along with excessively lean supermodels, ballet dancers, and long-distance runners, and every level of body weight in between. Humans across the globe express a large variation in adult body weight and adiposity, from the Maasai of East Africa at one end of the spectrum to the American Samoans at the other.

Like obesity in laboratory rodents, household pets, and domestic animals, human obesity is common. By some estimates, about 65% of the global population is classified as either overweight or obese (reviewed by Ogden et al., 2002, 2013). The incidence of obesity has risen over the past 150 years, with a slow and steady increase before 1970 and a steep rise between 1970 and 2000. This situation raises two related questions: "What caused the rise in obesity?" and "How can obesity be halted or reversed?" These questions have not been answered, so it might be helpful to view this issue through a biological perspective.

Obesity has been defined by the National Institutes of Health as a body mass index (BMI) of 30 or above. BMI is calculated as body weight in kilograms divided by height in meters squared. The higher the BMI, the greater the risk of some diseases, including high blood pressure, coronary artery disease, stroke, osteoarthritis, some cancers, and type 2 diabetes. In actuality, these maladies are more closely associated with an accumulation of abdominal, visceral adipose tissue and a high waist-to-hip fat ratio than they are with a high BMI. In contrast, the accumulation of adipose tissue subcutaneously in the gluteofemoral region (hips and thighs) is not associated with these disease states, and in fact might be cardioprotective. A low-waist-to-hip ratio is significantly positively correlated with levels of ovarian steroid hormones and fertility, and a high waist-to-hip ratio can be achieved by low levels of body fat in the abdominal, visceral region or high levels of body fat in the gluteofemoral region (Singh, 1993).

The adaptive value of body fat should not be underestimated. Most organisms, certainly all mammalian species, have the capacity to survive during starvation by lipolysis, the mobilization and oxidation of fuels derived from lipids in adipose tissue (Wang et al., 2006). Furthermore, lipids are a critical structural component of cell and nuclear membranes and organelles, and the lipid molecule, cholesterol, is the precursor of all steroid hormones. It has been posited that the unique complexity and energetic requirements of the human brain were made possible during evolutionary history by the exceptional ability of human primates to store adipose tissue (Cunnane, 2005; Cunnane and Crawford, 2003). It is important to remember and understand the adaptive value of overeating and gaining body fat.

In the social and physical sciences, it is important to pose questions precisely. For example, the answer to the question "What causes obesity?" requires us to specify the population under study; depending on the particular population, the answer might be "It's in their genes." For example, patients with Prader–Willi syndrome develop a ravenous appetite and obesity because of the epigenetic deletion of genes on the paternal copy of chromosome 15. Other individuals develop obesity because they have a mutation in a specific gene such as the gene for leptin, the leptin receptor, or the melanocortin receptor. Point mutations account for only a small fraction of the obese human population, however (Garver et al., 2013; Hinney and Hebebrand, 2008); many genes contribute to heritable variation in body weight and adiposity. In some cases, obesity results from the effects of multiple genes that confer a propensity to prefer a high-calorie diet and to respond to this type of diet by overeating and accumulating adipose tissue. Some genes that influence obesity affect food intake, others affect energy expenditure, still others affect energy storage, and most affect all three aspects of the energy equation. Although genes clearly have a major role in the development of obesity, this cannot entirely explain the rise in obesity that has occurred in human populations over only two or three generations. Some researchers have suggested that epigenetic effects might be involved (reviewed in Chapter 11).

Many environmental factors also are under study. It has been noted that obesity in some individuals results from chronic stress or a congenital abnormality in the HPA and/or hypothalamic–pituitary–thyroid systems. In a subset of this group, obesity might stem from acute stressors that occur during fetal development that predispose the individual toward body weight gain in adulthood (reviewed by Grattan, 2008). Childhood trauma, other environmental stressors, gestational programming, and genes also might interact with the gut microbiome to exaggerate obesity.

The plausible explanations for obesity are numerous and include genetic predispositions that interact with many environmental factors.

Some evidence suggests that increased peripheral energy storage is the most important determinant of appetite, and contrary to expectation, the process of gaining fat can increase, rather than decrease, the appetite for food. For example, within a few years after the end of a calorie-restricting diet, exercise regimen, or drug treatment, body weight and levels of adiposity often rebound and overshoot their former levels, which can occur without overeating or increasing daily caloric intake. The repeated cycles of dieting and gaining weight might decrease energy expenditure and/or promote energy storage, and thus exaggerate the propensity for body weight gain, although this is difficult to document. The effects of high-calorie, insulin-stimulating food, chronic stress, childhood trauma, other epigenetic effects, dim light at night, and endocrine-disrupting compounds do not exclude increased peripheral energy storage as the main cause of obesity. It is possible that many causes of obesity and/or overeating involve mechanisms that first increase energy storage and body fat accumulation before increases in food intake. For example, a diet high in simple sugars increases insulin secretion, which predisposes an individual toward body fat accumulation accompanied by a ravenous appetite (Fig. 16.2). This mechanism (predisposition toward energy storage) explains the paradoxical occurrence of an intense appetite for food in an individual with a high body fat content.

There is no doubt that hormones have a powerful role in controlling the appetite for food, the propensity to expend energy, and the propensity to accumulate adipose tissue. However, hormones are not triggers or determinants that act in isolation; they increase the probability that certain behaviors occur in a certain context. As illustrated in Fig. 16.1, their main function might be to increase survival and reproductive success in environments where energy availability is unreliable or fluctuates. If so, it is not surprising that there is no mechanism that easily keeps rising body fat accumulation in check in the presence of readily available, calorically dense food. In Chapter 1, it was noted that many drugs have been developed to treat obesity, but the drugs are effective only as long as they are taken, and patients often develop resistance or insensitivity to them. The drugs are often withdrawn from the market after patients develop dangerous side effects, which is most certainly related to the principle described in Chapter 3 (one hormone can have multiple effects) and the food–sex connection. The current high level of global obesity is a complicated puzzle with some but not all of the pieces in place. At the heart of this puzzle is the fact that hormones and therapeutic treatments can act in the brain to affect fuel metabolism and storage in the periphery. Furthermore, calorie restriction can decrease the metabolic rate and increase fuel storage. Similarly, effects of some hormones and therapeutic

IV. SPACE: SPATIAL ASPECTS OF HORMONE ADMINISTRATION

treatments alter peripheral fuel availability (for example, by altering the gut flora to increase calorie and nutrient absorption), which in some cases can increase the availability of oxidizable fuels from a given mass of digested food and in other cases can increase peripheral energy storage in adipose tissue and thus create a strong stimulus for increased appetite and overeating. The key to this puzzle will be to understand the coordination among peripheral and central systems.

## Other Examples

Whereas metabolic signaling and food intake constitute a wonderful example of hormone actions widespread across the body, there are other examples as well. Oxytocin works on breast tissue to facilitate lactation, and on the uterus to facilitate labor, as well as in the central nervous system. Corticotropic releasing hormone (also called corticotropin-releasing factor) works in the pituitary gland as well as the brain. GnRH works in the anterior pituitary as well as centrally; likewise, thyrotropin releasing hormone. Gonadotropin-inhibiting hormone acts in the hypothalamus and pituitary, but also has important effects on steroid synthesis by acting in the gonad (Bentley et al., 2017). In all cases, these widespread hormonal actions guarantee that behavioral responses are consonant with the state of the body, and vice versa.

## SUMMARY

1. Many hormones that decrease body weight and adiposity have documented effects on the HPG system and sexual libido. Their effects are mediated by peripheral signals detected in the gut, liver, stomach, and adipose tissue and are sent to the brain by the vagus, splanchnic, and spinal nerves. The brain, in turn, influences peripheral processes including metabolism, energy expenditure, and energy storage.
2. The myriad causes of obesity include a genetic predisposition, many different environmental effects, and interactions among genes and environment. Knowledge of the peripheral control of appetite and food intake have influenced current treatments for obesity. Conversely, obesity treatments such as bariatric surgery have enlightened obesity researchers about the role of peripheral signals that influence food intake.
3. Specific environmental factors that affect obesity and reproduction include increases in environmental stressors, dim light at night, disrupted circadian rhythms, endocrine-disrupting compounds, diet, sedentary behavior, decreased metabolic rate, learned associations between flavors and the postingestive metabolic consequences of

eating, and the gut microbiome. It is possible that the primary result of many environmental and genetic factors is an increase in the storage of metabolic fuels, which causes an internal deficit in the availability of fuels for oxidation. One final consequence of this fuel deficit might be a ravenous appetite and overeating. This explains why an intense appetite for food can accompany a high body fat content. Thus, body fat storage might increase food intake, and not the other way around. A future challenge for obesity researchers will be to craft experimental designs that differentiate between this hypothesis and the one stating that obesity is caused primarily by lack of discipline and overeating.

IV.  SPACE: SPATIAL ASPECTS OF HORMONE ADMINISTRATION

# Hormones Can Act at All Levels of the Neuraxis to Exert Behavioral Effects; The Nature of the Behavioral Effect Depends on the Site of Action

Are there neuroanatomical "centers" that control behaviors such as eating and drinking? Decerebrate rats, i.e., those that have had the neural connections between the midbrain (mesencephalon) and forebrain (diencephalon) completely severed, have been used to answer this question. Decerebrate rats do not show voluntary locomotion or foraging behaviors, but they accept or reject liquid meals provided via a feeding tube that delivers liquids to the palate. Normal rats eat good-tasting food (sweetened with sugar) at a faster rate than they eat laboratory chow. Furthermore, they refuse to eat bad-tasting food (food adulterated with an aversive flavor) until they become desperate for nourishment. Similarly, decerebrate rats ingest palatable food more readily than unpalatable food, and they reject unpalatable food until they become desperate for nourishment. Both normal rats and decerebrate rats eat less food after their stomach has been prefilled and distended than when they have an empty stomach. The intake of a sucrose solution is increased in normal rats that have been pretreated with a certain type of benzodiazepine. Despite having no neural connection between the forebrain and midbrain, like normal rats, decerebrate rats increase their intake of a sucrose solution after they have been pretreated with the same type of benzodiazepine. These findings suggest that at least some of the mechanisms controlling eating in response to taste, smell, and gastric distension occur outside the forebrain. These extrahypothalamic, evolutionarily older brain areas are integral to our experience of hunger for food.

*Principles of Hormone/Behavior Relations*
http://dx.doi.org/10.1016/B978-0-12-802629-8.00017-6

343

As mentioned in Chapters 1 and 2, feeding and hunger are affected by actions of neuropeptide Y, leptin, α-melanocyte–stimulating hormone, cocaine- and amphetamine-regulated transcript, agouti-related peptide, melanin-concentrating hormone, and other hormones. For decades, most scientists interested in peptide control of food intake have focused on these peptides and their action in the hypothalamus, particularly in the arcuate nucleus. The distribution of identified neurons and their peptide receptors is now intricately mapped with regard to the arcuate nucleus of laboratory animals. The traditional focus on the well-studied arcuate nucleus as the center for control of food intake is like looking for your lost car keys at night. If you limit your search to a small area lighted by a street lamp, you are aided by the light but remain blind to the much larger area likely to contain your keys. Analogously, the peptides and receptors for most if not all of these peptides are in not only the hypothalamus but also the midbrain and hindbrain, especially in the parabrachial nucleus and nucleus of the solitary tract. Understanding neuroendocrine control of food intake therefore requires us to study not only the arcuate nucleus but also the distributed neural network that includes the mid and hindbrain (reviewed by Grill, 2006). This chapter will explain some of the distributed networks that control several different aspects of physiology and behavior.

## BASIC EXPERIMENTAL EXAMPLES

### Sex Hormones

Many hormones, in particular sex hormones, can act at multiple levels of the neuraxis, with the exact behavioral effect depending on their site(s) of action. Estradiol is synthesized from testosterone in a chemical reaction catalyzed by the enzyme aromatase. Aromatization occurs in many tissues including the gonads, brain, and adipose tissue. Estrogen receptors alpha and beta are located throughout the neuraxis. Working through neurons of the preoptic area of the hypothalamus, estrogens facilitate locomotor, courtship, and maternal behaviors and decrease food intake. Cells in the preoptic area are inhibitory for the lordosis posture in female rodents, but estradiol action in the preoptic area counteracts this inhibition. Through ventromedial hypothalamic neurons, estradiol and progesterone facilitate lordosis behavior in the presence of an adult male. In the midbrain, with estrogen receptors expressed in neurons of the dorsal raphe nucleus, estrogenic effects on serotonergic neurons influence mood (depression, anxiety, vitality, or resilience). Further posteriorly in the neuraxis, estrogens act on medullary reticular neurons, which result in ascending arousal pathways (Fig. 17.1). In the spinal cord, estrogen binding at specific sites in the dorsal horn (in Rexed layer II) are involved in the perception of pain. In the hippocampus, estradiol has rapid effects on the growth of dendritic spines and increases synaptic plasticity, with positive

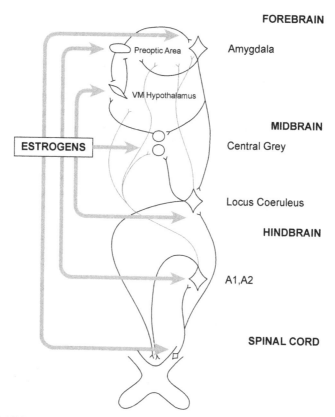

**FOREBRAIN**

Preoptic Area
Amygdala

VM Hypothalamus

**MIDBRAIN**

ESTROGENS
Central Grey

Locus Coeruleus

**HINDBRAIN**

A1,A2

**SPINAL CORD**

**FIGURE 17.1** Estrogens act at several levels of the neuraxis. In the spinal cord and hind-brain, fundamental arousal mechanisms and pain are affected. In the hypothalamus, estrogen-dependent lordosis behavior circuitry in rodents is completed in the ventromedial (*VM*) nucleus. In the forebrain, both neuroendocrine cells and emotional states are hormone-dependent. Many other hormones (testosterone, progesterone, stress hormones, and thyroxine) also have wide-spread targets of action in the central nervous system. Estrogens are the example here, because cellular and molecular mechanisms are understood in considerable detail.

effects on learning and memory. Androgen, progesterone, and glucocorticoid receptors have their own distribution spanning the neuraxis, and receptor distributions overlap with estrogen receptors. Like estrogen receptors, the effects of ligand binding to these receptors depend on their location.

## Arginine Vasopressin

The main action of the peptide, arginine vasopressin (AVP) (vasopressin; antidiuretic hormone), is to retain body fluid by reducing the production of urine by the kidney. In the kidney, vasopressin binds to receptors in the collecting duct and the distal convoluted tubule, increasing their permeability to water by causing translocation of proteins called aqua-porins into the plasma membranes (see also Chapter 5). The aquaporins permit water to flow through the lipid membranes down the channel core in single file and prevent the passage of ions or larger molecules. Whereas there are a number of homologous aquaporins, vasopressin in vertebrates activates aquaporin 2 to produce its water-saving functions. Vasopressin is synthesized in magnocellular neurons in the paraventricular (PVN) and supraoptic (SON) nuclei of the hypothalamus. These neurons have axonal projections to the posterior pituitary, where vasopressin is released into the bloodstream in response to increased plasma osmolarity. The recep-tors for plasma osmolarity (osmoreceptors) are located in the organum vasculosum of the lateral terminalis and subfornical organ in the brain. The osmolarity signal is relayed by projections to the vasopressin-contain-ing magnocellular neurons in the PVN and SON nuclei of the hypothala-mus. Thus, during a stress response that requires water conservation and/or preservation of blood volume, vasopressin acts in both the brain and periphery.

In addition to osmolarity, other stimuli influence vasopressin release from the pituitary. Neural connections to the vasopressin-secreting neu-rons include baroreceptor inputs that synapse first in the nucleus of the solitary tract. When blood pressure is elevated, they signal the inhibition of vasopressin secretion via a γ-aminobutyric acid (GABA) interneuron. A decrease in blood pressure (hypotension) reduces cardiopulmonary and arterial baroreceptor excitatory inputs to the nucleus of the solitary tract, thus activating noradrenergic A1 neurons in the caudal ventrolateral medulla. These neurons have ascending projections to the vasopressin-secreting neurons in the hypothalamus. Hypotension, for example, occurs during severe hemorrhage, and the vasopressin response can prevent a fatal drop in blood pressure.

Hypothalamic magnocellular neurons release vasopressin not only into the bloodstream but also throughout the central nervous system, where it modulates behavior and other functions appropriate for situations of low body fluid levels. Magnocellular neurons project throughout the central nervous system, including the spinal cord, and vasopressin V1 receptors are widely distributed in the hippocampus, hypothalamus, and cerebral cortex (Mavani et al., 2015). In this way, vasopressin is able to influence diverse functions including social behavior, anxiety, aggression, pain threshold, and cognition (Mavani et al., 2015). These actions may have evolved to facilitate survival, particularly in adverse situations such as transition to drier climates for terrestrial animals. They also may have assisted recovery from blood loss from aggressive behaviors, reduction in pain from wounds, and facilitation of bone growth.

Vasopressin also is involved in a number of affiliative behaviors such as pair bonding, maternal behavior, and paternal defense of the young. These effects parallel effects of the closely related peptide oxytocin in fostering maternal social behavior related to nursing and nurturing. Both oxytocin and vasopressin, which differ by only two amino acids, evolved from a common ancestral molecule, vasotocin. In a major review, Carter (2014) suggested that "Vasopressin is critical to social adaptation in a demanding world, with a behavioral profile that is associated with attachment to and defense of self, family, and other members of our social networks." An example is pair-bond formation, for which both vasopressin and oxytocin have a role. The neuroanatomical organization of the vasopressin-producing neurons is well conserved across mammalian species, and vasopressin is typically increased during social interactions. In contrast to the distribution of vasopressin-producing cells and vasopressin secretion, the distribution of the V1a receptors differs widely among species, and differences in V1a receptor distribution are associated with different affiliative behavior patterns and mating systems.

A monogamous mating system is one in which the members of a male–female pair prefer to mate with each other, engage in mutual affiliative behaviors, and live in proximity to each other. In species that form monogamous pair bonds, there is a tendency for the males to participate in parental care. Monogamy in male prairie voles is associated with high levels of the V1a receptor in the ventral pallidum and low levels in the lateral septum. In contrast, promiscuous montane voles have low levels of V1a receptor in the ventral pallidum and high levels in the lateral septum. Receptor distribution is associated with a genetic polymorphism in the gene that encodes the V1a receptor. Mice and montane voles (which normally show a distribution of the V1a receptor different from that of the prairie voles) can be genetically engineered to express the prairie vole gene for the V1a receptor. Such genetically engineered mice and montane voles have the prairie vole distribution of the receptor and display high levels of affiliative behavior.

Monogamy requires the abilities to discriminate among different conspecific individuals, remember which individuals are familiar, and recognize strangers. It requires cognitive processes involved in learning and memory. Animals that are unable to make active vasopressin, such as the Brattleboro rat, which has a mutation in the vasopressin gene, have disturbed cognition. Furthermore, these animals spend a considerable part of the night drinking because they must consume close to their entire body weight each day in fluid to maintain fluid balance, because without functional vasopressin they cannot concentrate urine. Among other things, Brattleboro rats have an exaggerated response to pain. The analgesic effects of vasopressin appear to be mediated through V1a receptors not only in the forebrain but also in the spinal cord. These examples illustrate the extent of the diverse effects of vasopressin and its V1a receptor.

IV. SPACE: SPATIAL ASPECTS OF HORMONE ADMINISTRATION

In a broader sense, vasopressin is an important modulator of the stress response (Mavani et al., 2015). It acts synergistically with corticotropin-releasing hormone (CRH) to increase adrenocorticotropic hormone (ACTH) secretion and modulate the release of corticosteroids from the adrenal gland. Mice lacking V1b receptors have lower resting levels of plasma corticosterone (Mavani et al., 2015). Vasopressin is anxiogenic both endogenously and when administered exogenously. Overexpression of V1a receptors has been linked to greater anxiety in mice; conversely, mice lacking V1a receptors have reduced anxiety and reduced memory. Mice lacking both V1a and V1b receptors have impaired social interaction (Mavani et al., 2015). Eliminating the V1b receptor also reduces aggression and social memory (Carrasco and Van de Kar, 2003). Vasopressin thus is an excellent example of a hormone that has multiple actions at most levels of the neuraxis and that modifies multiple personal and social behaviors associated with aversive conditions, providing the species with a better chance of survival.

## Angiotensin

Angiotensin II is a good example of a hormone that acts at multiple levels throughout the body, including the central nervous system. Angiotensin II maintains blood volume and fluid and electrolyte balance through a range of actions in the kidney, vasculature, and brain. Dipsogenic (thirst-provoking) effects of low doses of angiotensin II are observed when given into the cerebral ventricles or into brain tissue (Fitzsimons, 1998). The effects are accompanied by an increase in blood pressure through increased sympathetic vasomotor tone and the release of vasopressin, which is a potent stimulus of thirst. The dipsogenic effects of angiotensin are among a wide range of actions that maintain body fluid status and, importantly, protect the central nervous system from low blood pressure (McKinley and Johnson, 2004). Treatment with angiotensin-converting enzyme inhibitors at low doses increases fluid intake in volume-replete animals by accumulating angiotensin I in the blood, which then enters the brain and is converted to angiotensin II in regions that are not affected by the inhibitor. At higher doses, angiotensin-converting enzyme inhibitors also penetrate the brain and prevent this conversion, leading to normalization of thirst (Evered and Robinson, 1984). Many species, including dogs, rats, cats, sheep, mice, pigeons, ducks, and goats, drink in response to central angiotensin administration; however, rabbits are an exception. In rabbits, even when its cardiovascular effects are blocked, angiotensin strongly activates the brain but does not induce drinking behavior (Badoer and McKinlay, 1997). The precise mechanism for the awareness of thirst is still to be elucidated, but via neural imaging it has been shown that specific regions of the cerebral cortex become activated during increases in thirst.

*Interim Summary*

1. With regard to ingestive behavior, many different peptides and receptors for those peptides are located throughout the brain, from the caudal brain stem and in the nucleus of the solitary tract to the midbrain and the forebrain.
2. Gonadal steroid hormones and their receptors have a wide array of effects that depend on where the receptors are located in hindbrain, midbrain, and forebrain.
3. Vasopressin and its receptors are synthesized and secreted from many different cells of the neuraxis and affect osmolarity, blood volume, and thirst, as well as a wide array of social behaviors, including monogamy, affiliation, territoriality, and aggression. Angiotensin is another peptide that acts throughout the body and brain to coordinate behaviors and physiological processes related to fluid balance and thirst.

# CLINICAL EFFECTS

## Arginine Vasopressin

The basic aspects of AVP physiology have been discussed previously. From a clinical standpoint, polydipsia is the condition in which patients drink copious amounts of liquid although their volume status is normal. In some cases, polydipsia is caused by diabetes insipidus (DI). This form of diabetes results when there is an absence of AVP release from the brain (central DI) because of a mutation in the AVP gene, or when AVP is released but is not effective in the kidney (nephrogenic DI). We elaborate on this distinction subsequently. AVP provides an excellent example of a hormone's actions at multiple levels of the neuraxis. Not only do axonal projections from the SON, PVN, and other magnocellular anterior hypothalamic cell groups reach the posterior pituitary for release of AVP into the bloodstream, AVP-bearing axons also project to forebrain cell groups, the lower brain stem, and the spinal cord.

### Diabetes Insipidus

DI is a clinical syndrome involving the lack of AVP effects either centrally or in the kidney. As mentioned, the syndrome is characterized by polyuria and polydipsia, because AVP is either not being produced in the brain (central DI) or AVP is being released from the brain but the kidney is not responding to it (nephrogenic DI). The most frequent type is central DI, resulting from an inadequate secretion of AVP. There can be various causes for this, such as tumors in the pituitary stalk, loss of expression of the AVP gene, and a low amount of AVP synthesis in the SON and

PVN cells of the hypothalamus. This is the case in an animal model, the Brattleboro rat. The Brattleboro rat was discovered by an observant technician who noted that the rats had constant polydipsia and polyuria, and reported this observation to a professor, Heinz Valtin, who understood what it meant. When he treated the rats with AVP, they stopped overdrinking and urinating, and diluted urine became normally concentrated. In the Brattleboro rat, the cause of DI is a lack of expression of AVP in the magnocellular cells of the PVN and SON of the hypothalamus. The gene mutation that causes it is a frameshift mutation that prevents synthesis of the complete prohormone.

### Central Diabetes Insipidus

In humans, central DI must be differentiated from psychogenic polydipsia, a condition in which some mentally ill patients seize every opportunity to drink, sometimes to the point of fatal water intoxication. In psychogenic polydipsia, AVP levels are normal. Restricting water in these patients eliminates the diluted urine, which shows that AVP is working normally. Such patients do not respond to treatment with an AVP-like hormone. Polydipsia in dialysis patients, however, is a different problem; it results from increased angiotensin secretion and can be treated with angiotensin-converting enzyme inhibitors (Kuriyama, 1996).

In the clinic, central DI must be differentiated from nephrogenic DI because the treatments are different. To determine whether the DI is caused by a lack of AVP, the patient is tested with desmopressin, a synthetic form of AVP. In the case of central DI, replacing AVP by treatment with desmopressin restores the antidiuretic effects. Among the causes of brain lesions that lead to central DI are germinoma, craniopharyngioma, and sarcoidosis of the central nervous system. Autoimmune and vascular diseases, trauma after surgery, and genetic autosomal-dominant or X-linked recessive traits also may lead to a defect in AVP biosynthesis and central DI.

### Nephrogenic Diabetes Insipidus

If desmopressin does not work, attention turns to the kidney. The kidney may not respond because it lacks either vasopressin $V_2$ receptors or $V_2$ effects on cells. In nephrogenic DI, mutations occur in the AVP receptor type 2 gene. Genetic defects may also occur in the signal transduction pathway after $V_2$ receptors have been stimulated. Normally, $V_2$ receptors are linked to cyclic adenosine monophosphate (cAMP), and their stimulation by AVP induces cAMP to mobilize aquaporins. As we noted earlier, aquaporins are proteins that insert water channels into the membrane. $V_2$ stimulation causes one type of aquaporin ($AQP_2$) to insert water channels into the apical membrane of kidney cells in the collecting duct. This enables water to be reabsorbed directly back into the cells. Aquaporins are proteins that have a structure that forms a narrow channel only big enough for one molecule of $H_2O$ to pass through at a time. With millions of aquaporins in the membrane, the passage of water in or out of the cell is efficient. Different aquaporins at the basal membrane escort water out of the cell and back into the plasma. Studies in mice that have severe DI ($DI^{+/+}$) have low levels of cAMP, which is the second messenger for $V_2$ receptors, and low expression of $AQP_2$, which is the water channel protein in the apical cell membrane of the collecting duct cells. cAMP stimulates $AQP_2$ expression by activating cAMP-responsive elements on the $AQP_2$ gene.

## Progestins

The class of steroid hormones known as progestins, the action of which is to maintain pregnancy, also have effects upon mood, acting through midbrain reticular system neurons. Progestins depress the activity of neurons associated with arousal of the entire forebrain, perhaps because of their ability to potentiate the effectiveness of the inhibitory neurotransmitter, GABA. These cellular actions may be related to the severe and sometimes incapacitating mood changes that are experienced by some adult women taking sequential hormone replacement therapy. In experimental animals, progestins acting through ventromedial hypothalamic neurons enhance sex behavior, and when they act through basal forebrain neurons, they facilitate courtship behaviors. In addition, progestins have an important role in the cerebral cortex in preventing tissue swelling after injury. Thus, for progestins, at least four different sites of action have four types of functional consequences.

When peptide hormones produced in the nervous system have both endocrine effects (of behavioral importance by indirect routes) and straight behavioral actions, a question arises: Are the extrahypothalamic, behavioral, and autonomic effects of the pituitary hormone-releasing and inhibiting factors coordinated with their effects on pituitary hormone release? As documented in previous chapters, many of the hypothalamic-releasing and inhibiting factors for pituitary hormones also act as neurotransmitters and neuromodulators in other areas of the central nervous system. Some examples are provided subsequently.

## Corticotropin-Releasing Hormone

CRH provides another clear example of a hormone acting in different places, with different clinical consequences. Through CRH receptors in the hypothalamus and pituitary gland, CRH controls release of ACTH, which itself has behavioral consequences, but which primarily causes the secretion of stress hormones (adrenal corticosteroids) into the blood. ACTH also has major effects on conditioned fear and avoidance. We were reminded of this in Chapter 2. In addition, prolonged high levels of CRH expression in the amygdala are thought to be associated with anxiety and depression. In parallel, activation of CRH receptors in the forebrain increases anxiety. Also, CRH-activating noradrenergic systems that act in the hindbrain through locus ceruleus neurons have effects on alertness and fear. CRH thus has widespread sites of actions with different behavioral consequences.

## Thyrotropin-Releasing Hormone

Thyrotropin-releasing hormone (TRH) occurs in high concentrations in motor nuclei of the brain stem and spinal cord and in lower concentrations in the amygdala, mesencephalon, and cerebral cortex. It is colocalized with serotonin and substance P in brain stem raphe nuclei and the medulla. Behaviorally, TRH is activating and mildly euphorigenic, and it has a brief antidepressant effect when administered to humans, primarily in women.

## Somatostatin

Somatostatin is widely distributed outside the hypothalamus, with relatively high concentrations in amygdala and certain brain stem regions and lower concentrations in limbic structures and cerebral cortex. It is colocalized with GABA in the hippocampus, thalamus, and cortex and with norepinephrine in the medulla and sympathetic ganglia. When administered into the cerebrospinal fluid in humans, somatostatin ameliorates intractable pain (e.g., cluster headache).

Little is known about possible coordination between the extrahypothalamic, direct behavioral effects of these releasing and inhibiting factors and the behavioral effects of the hormones they regulate via their secretion into the pituitary portal circulation. If, for example, hypothalamic–pituitary–adrenocortical (HPA) axis hyperactivity is a contributing factor in major depression, as discussed in Chapter 2, the extrahypothalamic and neuroendocrine effects of CRH likely are consonant in promoting some of the clinical components of major depression, such as heightened anxiety. On the other hand, most patients with major depression do not have increased HPA axis activity, although they still could have increased activity of extrahypothalamic CRH systems. CRH receptor antagonists have suggestive antianxiety effects in depressed patients who have, on average, normal HPA axis activity. In addition, clinical improvement in anxious, depressed patients after antidepressant treatment can occur whether their plasma ACTH and cortisol concentrations are normal or elevated. These findings support the concept of the independence of extrahypothalamic and neuroendocrine CRH systems, a hypothesis that needs to be verified experimentally.

# SOME OUTSTANDING QUESTIONS

1. To what extent can it be shown that actions of a given hormone at different levels of the neuraxis are orchestrated to produce synergistic effects?
2. Are there positive feedback mechanisms in which different actions multiply each other? Or, in contrast, do different actions at different levels of the neuraxis counterbalance and check each other? This could include well-engineered negative feedback effects.
3. What other molecules are the size of water, and why do they not pass through the aquaporins?

# MECHANISMS: MOLECULAR AND BIOPHYSICAL MECHANISMS OF HORMONE ACTIONS GIVE CLUES TO FUTURE THERAPEUTIC STRATEGIES

# 18

# Hormone Receptors Act by Multiple Interacting Mechanisms

Steroid hormone receptors affect brain and behavior by different mechanisms at different rates and at different locations on cells. The investigation of steroid action began by studying steroid effects on the cell membrane, perhaps because cell membranes are composed of a lipid bilayer and steroids are soluble in lipids. In the late 1950s, however, Jensen and Jacobson (1962) labeled steroids with radioisotopes to allow their visualization after they were applied in physiological concentrations to various tissues. They found that cells in steroid target tissues retained and accumulated the radiolabeled steroids unaltered within the cell nucleus. Later, it was discovered that steroids bind to intracellular receptors, and these receptors are proteins (Toft and Gorski, 1966). Tremendous excitement was generated by the discovery that some intracellular hormone receptors act as transcription factors, proteins that initiate the transcription of DNA into messenger RNA (mRNA) (reviewed by Tsai and O'Malley, 1994). With this discovery, endocrine investigations were ushered into the era of molecular biology.

Many steps are involved in intracellular steroid action. First, the steroid must diffuse passively through the cell membrane to bind to intracellular receptors. Next, conformational changes take place in the receptor, inducing the ligand–receptor pair to form a dimer with another ligand–receptor pair. These dimers interact with other molecules known as coactivators and corepressors (discussed in detail in Chapter 20). Finally, in the nucleus, the complex binds with high affinity to the steroid receptor response element, a regulatory region on the DNA, which activates or represses gene transcription (Tsai and O'Malley, 1994).

Compared with other types of hormone action, classical genomic steroid action is slow. Days of steroid treatment are required for measurable effects on some behaviors. After many years of studying classical genomic effects of steroids on behavior, some behavioral biologists were puzzled by the rapid effects of steroids on particular behaviors. Some effects occur far too rapidly to be explained by classical nuclear transcriptional mechanisms. They hypothesized that steroids can act on membrane receptors that initiate rapid signal transduction pathways similar to those initiated by protein hormones. Alternatively, they hypothesized direct effects of steroids on ion channels that alter neural action potentials. Within the field of molecular endocrinology, questions about the importance of slow versus rapid steroid action led to conflicts and controversy. The idea of membrane-initiated hormone action was often seen as antithetical to nuclear transcriptional facilitation. The known steroid receptor mechanisms have since expanded and can be summarized as (1) classical genomic; (2) ligand-independent receptor signaling (Weigel and Zhang, 1998); (3) rapid, nongenotropic effects through a membrane-associated receptor coupled to stimulation of cytoplasmic signaling pathways, including the phosphatidylinositol-4,5-bisphosphate 3-kinase, mitogen-activated protein kinase K (MAP-K), and cyclic adenosine monophosphate–response element binding protein signaling cascades (reviewed by Kelly and Levin, 2001); and (4) genotropic, steroid receptor response element–independent signaling, in which steroid–receptor complexes act on other transcription factors such as activator protein-1, specificity protein-1, and nuclear factor κ–light chain enhancer of activated B cells (Gaub et al., 1990). This chapter illustrates rapid actions of hormones in the central nervous system (CNS) and shows how rapid, membrane-based effects can facilitate later genomic actions. Furthermore, it has been hypothesized that the same receptor could produce both rapid and genomic responses. Despite many years of controversy and shifts in the dogma, when viewed as a whole, a uniform view of steroid action has emerged.

## ESTROGENS

A landmark in the field of rapid steroid action came with the observation that multiple brief pulses of steroid exposure are sufficient for effects that had previously been induced only by long-duration, continuous exposure to steroids. For example, large-scale cell division in the uterus, previously thought to require constant exposure to estrogens for at least 24 h, is triggered by two 1-h pulses of estrogens properly spaced in time (e.g., Anderson et al., 1974). The same is true of behavior. Female-typical sex behavior, as well as induction of the progestin receptor (PR) in the hypothalamus, is stimulated by two 1-h estrogen pulses almost as well as by continuous estrogen exposure (Parsons et al., 1981) (see also Chapter 14, Fig. 14.1).

| PULSATILE ADMINISTRATION | | RESULTS | |
|---|---|---|---|
| Membrane-Limited Estrogen Admin. | Cell Nuclear Estrogen Access | Estrogen-Facilitated Transcription? | Estrogen-Facilitated Lordosis Behavior? |
| NO | NO | NO | NO |
| YES | YES | YES | YES |
| AS PULSE 1 | NO | NO | NO |
| NO | AS PULSE 2 | NO | NO |
| **AS PULSE 1** | **AS PULSE 2** | **YES** | **YES** |
| AS PULSE 2 | AS PULSE 1 | NO | YES |

**FIGURE 18.1** What schedules of neuronal cell membrane application and cell nuclear access allow estrogens to achieve the effects equal to unlimited estradiol administration (Admin.) (second row in figure)? For estrogen-facilitated transcription, membrane mechanisms in the first pulse and nuclear access in the second pulse are necessary and sufficient. For female sex behaviors, surprisingly, either order of pulsatile access (membrane and then nuclear, or nuclear and then membrane) is sufficient (Vasudevan et al., 2001). For other hormones as well, choreographing membrane mechanisms and nuclear mechanisms of cellular actions is essential to obtaining the full panoply of central nervous system hormone effects. Timing and details depend on the hormone administered and the cell involved.

What is the relation of the first, early pulse to the later pulse? This question was answered by using the two-pulse approach in experiments on estrogen-facilitated gene transcription in neuroblastoma cells. An early pulse, limited by chemical linkage of estradiol to the membrane, amplified the transcription-facilitating action of the second pulse (Fig. 18.1). The behavioral mechanisms in the hypothalamus were more permissive: The membrane-limited actions of estradiol could be brought into play as either the first or second pulse. Signal transduction pathways such as those involving both protein kinase A (PKA) and protein kinase C (PKC) are necessary and sufficient for membrane-limited estrogen amplification of the genomic effect of estrogens.

How and where in the membrane does that early action take place? One possibility is that the estrogen effect is transduced by *caveoli*, specialized domains of membrane that are attached to a wide variety of signal transduction initiators. In endothelial cells, estradiol working through estrogen receptor-$\beta$ (ER$\beta$) in membrane caveoli rapidly activates nitric oxide synthase (Fig. 18.2). The response to endogenous estrogen receptor activation is inhibited by the ER$\beta$-selective antagonist RR-tetrahydrochrysene (Chambliss et al., 2002). A related possibility is that rapid hormone effects

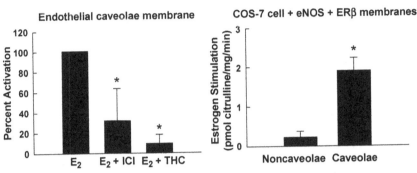

FIGURE 18.2  Rapid actions of steroid hormones. (*Left*) Enzymological activation of endothelial nitric oxide synthase (eNOS) by estradiol ($E_2$) can be obtained by using endothelial cell membranes in the regions that contain specialized pits called caveoli. Estrogen stimulation and overexpression of estrogen receptor (ER)-β enhanced rapid eNOS stimulation by $E_2$, and the response to endogenous ER activation was inhibited by the ERβ-selective antagonist RR-tetrahydrochrysene receptor blockers (THC). (*Right*) Estrogenic stimulation worked much better on membranous regions with caveolae than without. In this effect, ERβ is suspected to be the relevant molecule transducing the estrogenic effect. Such rapid actions are being explored in the central nervous system with experimental designs relevant to behavioral studies. *From Chambliss, K.L., Yuhanna, I.S., Anderson, R.G.W., Mendelsohn, M.E., Shaul, P.W., 2002. ERbeta has nongenomic action in caveolae. Mol. Endocrinol. 16 (5), 938–946. With permission.*

are organized by so-called *lipid rafts* on membranes. In addition, steroid hormones with different lipid solubilities concentrate in different parts of the cell membrane bilayer, changing the conformation of the membrane. For example, highly lipophilic estrogen concentrates in the lipid bilayer, whereas cortisol, which is more hydrophilic, localizes at the center of the cell membrane (Golden et al., 1998). The main lesson is that hormone action at the cell membrane does not occur at just any location. Rather, relevant receptors are organized by membranous structures, which in turn are linked physically to submembrane signal transduction molecules.

What then? Following cascades of protein phosphorylation by well-studied signal transduction pathways, such as those that involve PKC, PKA, MAP-K, and other messengers, the hormone receptors that will eventually act in the cell nucleus are phosphorylated and thus activated. An early experiment demonstrated that classical ERα is activated by prior delivery of a growth factor that acts at the cell membrane (Hewitt et al., 2010; Fig. 18.3). Moreover, ERs are activated by membrane events that result from synaptic input, and this synaptic activation of ER can lead to more efficient transcription. Similarly, dopaminergic agents can initiate ligand-free activation of steroid receptors. Second messenger systems and membrane-crossing ion fluxes are sensitive to estrogens (reviewed by Segars and Driggers, 2002). All of these studies show how events in the

# Estrogen Insensitivity in the αERKO Uterus

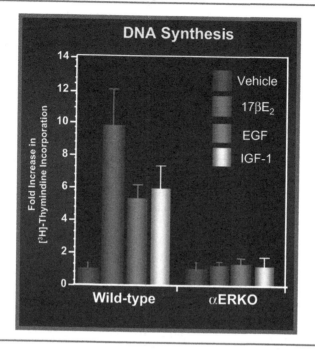

FIGURE 18.3  Estradiol treatment leads to a massive increase in the rate of cell division within the uterus of the wild-type female mouse. Growth factors also can have this effect. When the estrogen receptor-α (ERα) gene is disrupted (αERKO), all of these actions are abolished; therefore, even the growth factor effect relies on the gene product for ERα. *17βE₂, 17β-estradiol; EGF, epidermal growth factor; IGF-1, insulin-like growth factor-1. Courtesy of Korach, K., et al., NIEHS.*

environment of the cell interact with hormone administration, thus affecting the magnitude of the response of the cell to the hormone.

What do these molecular studies mean for behavior? Estrogens have rapid effects on many aspects of ingestive, social, sexual, and locomotor behavior. In addition, estrogens enhance the behavioral response to stimulants such as dopaminergic drugs (e.g., amphetamine and cocaine).

At least two nonclassical mechanisms are implicated in aspects of energy balance: nonclassical nongenomic and nonclassical genomic. Within hours of estradiol treatment, laboratory rodents decrease food intake. For example, ovariectomized rats decrease meal size when treated with either unconjugated estradiol (which can easily pass through the cell membrane) or estradiol–bovine serum albumin, a form of estradiol conjugated to a large molecule that prevents the estradiol from entering the cell (reviewed by Santollo and Daniels, 2015). In mice, estradiol also has genomic effects on body weight that are independent of the estrogen receptor response element (ERE). Genetic knockout mice that lack ERα (αERKO) are obese, and the obesity can be ameliorated in αERKO mice that have been engineered to express a form of ERα that is unable to bind to the ERE on DNA (KIKO mice). In the latter KIKO mice, classical genomic ERα signaling via the ERE is eliminated, whereas ERα ERE-independent signaling is intact. These KIKO mice do not show the typical obesity observed in αERKO mice, which suggests ERE-independent effects of estradiol on mechanisms that control energy storage and/or expenditure (Park et al., 2011).

In addition, steroids have rapid membrane-mediated effects on sensorimotor function and aspects of addiction. For example, rats traversing a narrow balance beam make many missteps known as *footfaults*. The number of footfaults is decreased in situations in which estradiol levels are higher; e.g., footfaults are low in (1) females compared with males, (2) females at the time of estrus compared with other times of the estrous cycle, and (3) ovariectomized females treated with estradiol compared with ovariectomized females treated with vehicle. Rapid effects of estradiol also facilitate the increase in locomotor behavior that occurs after treatment with amphetamine; i.e., greater dopamine-enhanced locomotion occurs when levels of estradiol are high. High levels of estradiol enhance dopamine release in the ventral striatum of the brain, and this effect has been implicated in sex differences in drug addiction (reviewed by Becker, 1999). Coinciding with these effects of dopamine and estradiol on behavior, electrophysiological recordings demonstrate PKA-dependent activation of hippocampal neurons immediately after estradiol administration (Gu and Moss, 1996). Estradiol also has rapid stimulatory effects on calcium mobilization in striatal neurons (Meitzen and Mermelstein, 2011). Steroids act rapidly at the cell membrane to modulate excitatory synaptic transmission in the brain, and these rapid actions are important at several levels.

## Interim Summary 18.1

1. Steroid action occurs via a number of different interacting mechanisms:
   a. classical genomic action
      The steroid ligand binds to intracellular receptors, and this complex binds to the steroid receptor–responsive element on DNA to regulate transcription.
   b. ligand-independent receptor signaling
      The steroid receptor affects cell properties in the absence of the steroid ligand.
   c. nonclassical genomic action
      The steroid ligand binds to intracellular receptors, and this complex binds to parts of the DNA independent of the steroid-receptor responsive element.

**d.** rapid nonclassical nongenomic action
The steroid ligand binds to a membrane-associated receptor, initiating intracellular signaling cascades.

2. These different mechanisms occur at many levels of the CNS and can interact to affect behaviors including food intake, social behavior, locomotor behavior, energy expenditure, energy storage, and sensitivity to drugs.

**a.** Effects of steroids on membrane receptors include activation of PKA, PKC, and MAP-K, which can lead to protein phosphorylation, changes in membrane ion flux, and altered neural transmission.

**b.** Previous or subsequent membrane receptor action can accentuate the genomic effects of these or other steroids.

## THYROID HORMONES

Thyroid hormones (TRs), which are not steroids but tyrosine-based molecules, have receptors with multiple mechanisms; i.e., they have both membrane (rapid) and nuclear (slow) receptors. TRs have rapid, physiologically important actions on a variety of cell types. For example, thyroxine (T4) can induce phosphorylation and nuclear translocation (activation) of the enzyme MAP-K within 10 min. As a result, MAP-K promotes the functional alteration of an important transcription factor protein, p53, within 10 min. Another consequence of this nongenomic activation of MAP-K after T4 administration is the phosphorylation of nuclear receptors for TRs, presumably because of a physical association of MAP-K with TRs. Of potential direct relevance to nerve cells, TRs can promote the opening of sodium channels, causing action potential bursting. Apparently because of changes in potassium channel currents, TRs can shorten action potential duration. Rapid, nongenomic actions on signal transduction pathways not only have important roles in TR effects, under some circumstances they may also potentiate later transcriptional effects (Fig. 18.4). There is every reason to expect that these or similar phenomena also occur in neurons governing T4-influenced behaviors, including emotional responses.

## ANDROGENS

At a molecular level, androgens, like estrogens, can exert rapid, nongenomic effects. First, they can activate kinase-signaling cascades with shorter latencies than transcriptional activation would require. Second, androgen effects can occur in cell types that lack functional androgen receptors. Early work implicated classical genotropic action of androgens on male sex behavior, but other androgenic effects on male sex behavior occur via rapid, nongenotropic effects.

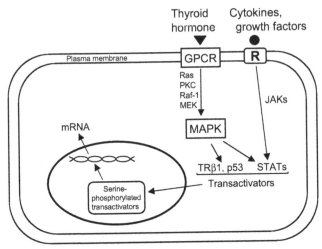

FIGURE 18.4    Rapid actions of thyroid hormones. Thyroxine, acting at the cell membrane and working through several signal transduction pathways, activates mitogen-activated protein kinase (MAPK). This kinase, in turn, phosphorylates transcription factors such as thyroid hormone β1 (TRβ1), p53, and certain Signal Transducer and Activator of Transcriptions (STATs), thus effecting changes in gene expression in the cell nucleus. In these rapid actions, thyroid hormones interact with cytokines and growth factors. *GPCR*, G protein–coupled receptor; *MEK*, mitogen-activated protein kinase/extracellular signal–regulated kinase; *mRNA*, messenger RNA. *From Lin, H.-Y., et al., 1999. Biochem. J. 388, 427–432; Lin, H.-Y., et al., 1999. Am. J. Physiol. 276, C1014–C1024; Davis, P.J., et al., 2000. J. Biol. Chem. 275, 38032–38039; Shih, A., et al., 2001. Biochemistry 40, 2870–2878. With permission.*

Many of the male-typical mating behaviors do not rise in frequency and intensity until long after initiation of testosterone treatment. However, testosterone injections can reduce anxiety-related behaviors within 30 min, which is too fast for most transcriptional effects (Aikey et al., 2002) (Fig. 18.5). Such effects also influence mating: Within 60 min of a subcutaneous injection of testosterone, latencies to mount females are significantly reduced (James and Nyby, 2002). Early electrophysiological recordings of single-unit activity from arcuate and preoptic area neurons suggest a mechanism for these behavioral effects. Testosterone application elevates neuronal activity markedly within minutes (Janson et al., 1993).

## PROGESTINS

The effects of progesterone on behavior depend on transcriptional activation by its nuclear receptor. Antisense DNA against PR mRNA microinjected into the hypothalamus significantly reduces progesterone-dependent female reproductive behaviors, including the lordosis quotient (the ratio of the incidence of lordosis to the incidence of mounting).

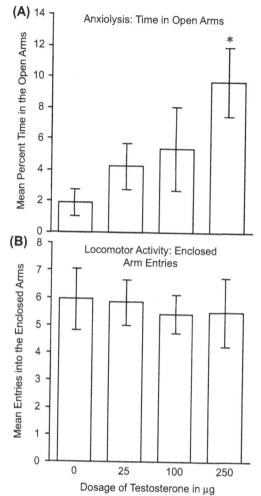

FIGURE 18.5    Only 30 min after receiving a subcutaneous injection of testosterone propionate (see x-axis for doses), male mice spent longer times in the open arms of an elevated plus maze (A), even though under these conditions general locomotor activity was not increased (B). *From Aikey, J.L., Nyby, J.G., Anmuth, D.N., James, P.J., 2002. Testosterone rapidly reduces anxiety in male house mice* (Mus musculus). *Horm. Behav. 42 (4), 448–460. With permission.*

Furthermore, progesterone amplification of estrogenic effects on lordosis disappears in female mice in which the gene for the PR has been knocked out.

In addition, there are rapid behavioral effects of progesterone and its metabolites. First, membrane-initiated signals, as could result from synaptic excitation, activate the PR with short latency in a manner important for behavior. Second, the effects of injected progesterone sometimes are exerted by its metabolites (see Chapter 4), and these metabolites are synthesized in the brain within neurons or glia. These brain-derived metabolites are called *neurosteroids*. In the midbrain, for example, reproductive behavior is facilitated by treatment with reduced progesterone metabolites, and this behavioral effect occurs too quickly to be accounted for by classical genomic mechanisms (Frye and Vongher, 1999b; Frye et al., 2000) (Fig. 18.6).

It is remarkable that in some cases, the steroid ligand is not necessary for profound effects of steroid receptors on behavior. When ligand-independent activation of steroids occurs during development, the effects can last into later childhood and/or adulthood. For example, neonatal treatment with a dopamine receptor agonist interacts with increased expression of the gene for PR and increases levels of juvenile social play. These ligand-independent effects on PR genes can be prevented by prior treatment with an estrogen receptor antagonist. It is well known that one important effect of estradiol is the induction of progestin receptors. Dopaminergic, ligand-independent action on progestin receptors alters estradiol-mediated PR gene expression, with lasting effects on social behavior (Olesen et al., 2005).

Knowledge of receptor mechanisms has expanded from one slow-acting genotropic mechanism to many interacting fast and slow mechanisms. The classical genotropic mechanism is characterized by the binding of the ligand-receptor complex to the steroid receptor response element. Subsequently discovered mechanisms include the rapid action of steroids on membrane receptors and the ligand-free action of chemical messengers on steroid receptors. In addition, some steroid effects are genotropic but do not involve the classical steroid receptor response element. Rather, these genotropic steroid effects occur via other protein–protein interactions and the initiation of transcription factors other than the steroid receptor response elements. Furthermore, different fast and slow mechanisms interact to influence brain and behavior. With the expansion of the known receptor mechanisms comes an expansion of possible drug targets and the hope that some specific drugs can ameliorate disease states without increases in oncogenic potential.

## OUTSTANDING NEW BASIC OR CLINICAL QUESTIONS

1. Are there cell membrane steroid and TH receptors still to be identified?
2. Are there behaviorally important physicochemical modulations of the cell membrane lipid bilayers by highly lipophilic steroids?
3. Where such modulations occur, do they make use of specialized membrane domains such as caveoli or lipid rafts?

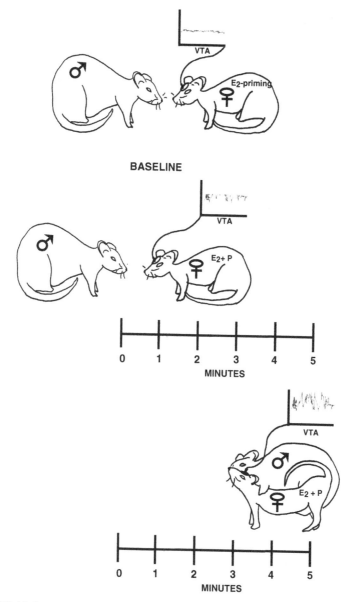

**FIGURE 18.6** Progesterone can exert rapid actions on electrophysiology and behavior. (*Top*) In the estrogen (E$_2$)-primed female laboratory rodent, progestins (P) given intravenously or into the ventral tegmental area of the midbrain (VTA) increase neuronal firing in the VTA within 2 min (*middle*). This is followed in another few minutes by increased lordosis behavior upon mounting by the male (*bottom*). From Frye, C.A., Bayon, L.E., Vongher, J.M., 2000. Intravenous progesterone elicits a more rapid induction of lordosis in rats than does SKF38393. *Psychobiology* 28 (1), 99–109; Frye, C.A., Vongher, J.M. 1999a. Progestins' rapid facilitation of lordosis when applied to the ventral tegmentum corresponds to efficacy at enhancing GABA(A) receptor activity. *J. Neuroendocrinol.* 11 (11), 829–837. With permission.

4. Under what circumstances do the rapid signaling pathways lead to nuclear events, and when do they lead more directly to alteration of ion channels in nerve cells or glia?

5. What are the stepwise chemical mechanisms of the fastest effects? In particular, because several signal transduction pathways have been identified (e.g., MAP-K, PKC, PKA, calcium mobilization), how do these signaling pathways relate to each other?

6. Are the rapid hormone actions, which are manifest in the first pulse of a two-pulse paradigm, an efficient site for cross-talk among biochemical systems of both (1) physiological and (2) medical importance?

7. In this chapter, several molecular phenomena have been cited, and rapid behavioral effects of hormones have been illustrated. How can we discover the exact routes from the former to the latter?

CHAPTER

# 19

# Gene Duplication and Splicing Products for Hormone Receptors in the Central Nervous System Often Have Different Behavioral Effects

In the 1960s, behavioral endocrinologists including W. C. Young noted that the same hormone at a given concentration will have different effects, depending on the genetic makeup of the individual. Specific changes in sequences for genes that encode receptors have been linked to simple and complex behaviors. Often this can be traced to a gene that encodes a hormone receptor. For example, territorial and parental behavior differ markedly in two types (morphs) of the white-throated sparrow. When circulating levels of testosterone are equalized in these two morphs by gonadectomy and testosterone treatment, the behavioral difference persists. The two morphs differ in their behavioral response to the same dose of testosterone. The difference in territorial or parental behavior cannot result from different steroid levels; rather, it is caused by a difference in a gene that encodes a steroid receptor, which alters the efficiency of transcription of the receptor gene. This chapter explains how various genetic rearrangements can alter steroid receptor function, thereby altering physiological and behavioral responses to hormones.

Hormone receptors are proteins encoded by genes on chromosomes, and chromosomes are strings of DNA that replicate and then divide twice, leading to haploid gametes (egg and sperm) during the process of meiosis. At the beginning of meiosis, the diploid chromosomes are replicated and attached by adhesion to form pairs of sister chromatids. The pairs of homologous chromosomes line up together before they undergo two rounds of cell division. In the first cell division, the homologues are separated. In the second cell division, the sister chromatids are separated. Before the first cell division, the closely aligned and paired chromosomes exchange some segments of DNA in a process called recombination. During this process, genetic information is shuffled among the homologous chromosomes. Other types of genetic rearrangements occur during this and other stages of meiosis. Rearrangement events include gene inversions, deletions, and duplications. Inversions occur when strands of DNA are broken, flipped, and inserted into the chromosome with the nucleotides in reverse order. Gene duplications include any doubling of a strand of DNA that includes a gene. Gene duplications can arise as products of several types of errors in DNA replication and repair. Duplicated sections of DNA may end up shifted in tandem positions on the same chromosome, or by transposition, they may be moved to a different chromosome. These events happen with low probability but they do happen, and over the extended periods involved in evolution, these events have great significance.

When one copy of the duplicated gene is shifted or otherwise altered, either of two consequences will occur. If the coding region is altered, the physical properties of the encoded protein are changed. If the promoter of the gene is altered, the temporal and spatial patterns of expression and/or the physiological regulation of expression are affected. The opportunities for small molecular change do not stop here. In the company of a cohort of RNA-binding proteins, nascent RNAs are spliced to make messenger RNAs, their half-lives are determined, their passage to the ribosome is controlled, the efficiency of their translation into protein is regulated, and elongation of the eventual chain of amino acids of the protein is completed. Then, the newly manufactured protein must be directed to the proper parts of the cell. The rates of degradation of proteins also are a factor in their levels of activity. These changes might be adaptive: that is, they might increase the chances of survival and reproductive success and be passed on from generation to generation, leading to evolutionary divergence. They also might be maladaptive, leading to reproductive decline and early mortality. In general, genetic rearrangements and duplication are important sources of genetic variability that allow populations to adapt to changes in the environment.

Reproductive behaviors are at the leading edge of evolutionary change because natural selection, which is thought to be a major force in the process of evolution, occurs when different genotypes confer differential levels of reproductive success. There is strong evidence for this idea. Traits associated with mating, fertility, and offspring care are often under strong selection pressure and evolve rapidly. During evolution, the appearance of sex-related traits, sexually differentiated traits, and sex chromosomes are attributable to gene rearrangements and duplications. For example, in some bird species, gene inversion might be an event that has led to the evolution of different mating strategies. For novel, sexually selected mating strategies to arise, there must be underlying genetic variability that influences their expression. These sexually selected traits often arise from gene duplication. For example, gene duplication provides the genetic variation in the genes for sensory receptors. These include receptors for olfactory cues, such as pheromones, receptors for visual cues, such as bright or drab coloration, and receptors for auditory cues, such as courtship song in birds. Many female birds and fish, for example, prefer to mate with males with bright plumage or scales, and gene duplications underlie changes in the photoreceptors that allow females to detect these stunning colors (Horth, 2007). In addition, gene duplications have led to different types of steroid receptors with both separate and overlapping functions (Thornton et al., 2003). All of the molecular steps (and accidents) mentioned here have implications for hormone–behavior relations. The principle is that closely related structures do not necessarily imply closely related functions, and even point mutations, inversions of small pieces of DNA, and duplicated genes can alter function.

## BASIC EXPERIMENTAL EXAMPLES

## Gene Duplication

### Estrogen Receptor Genes

The evolution of vertebrate steroid receptor genes is a story of gene duplication followed by divergence. Scientists interested in molecular evolution focus on six related nuclear steroid receptors: estrogen receptors α and β, progestin receptor (PR), androgen receptor (AR), glucocorticoid receptor (GR), and mineralocorticoid receptor. These are all present in early vertebrates such as lampreys, sharks, and skates. For many years, researchers failed to find evidence for steroid receptor genes in invertebrates, although they found orthologs for an estrogen-like receptor in fruit flies, *Drosophila melanogaster*. However, this ortholog did not act as a ligand-bound transcription factor. More recently, an ortholog of the estrogen receptor gene was discovered in an evolutionary group that appeared on the earth before insects: the annelids and mollusks. The *Aplysia californica* estrogen receptor is transcriptionally active, although unlike the vertebrate estrogen receptors, the effects of this mollusk receptor do not depend on binding to estrogens. The annelid estrogen receptor, however, is similar to the estrogen receptors found in vertebrates, including those in humans. Annelids such as *Platynereis dumerilii* and *Capitella capitata* are segmented worms thought to have first appeared on the earth over 500 million years ago. Functional assays show that annelid estrogen receptor orthologs bind with high affinity to estradiol and specifically activate gene transcription in response to low estrogen concentrations. Annelids synthesize estrogens, and these hormones regulate the provisioning of oocytes with vitellogenin during female reproduction. Together, these different types of evidence suggest that the earliest known steroid receptors have the ability to bind with high affinity to estrogens and to activate transcription.

Even before the appearance of the annelids, the first ancestral steroid receptor may have been present over 600–1200 million years ago and was produced by gene duplication and divergence. Evidence for this was obtained employing the technique of *inferred molecular phylogeny*, a combination of molecular biology and probability statistics used to understand the history of the evolutionary relationships among particular genes in different evolutionary groups. Evolutionary geneticists matched estrogen receptor genes from representatives of different animal taxa. Statistically inferred molecular phylogenies were used to make educated guesses about the ancestral genes that predate extant species. According to the statistically inferred phylogeny for the estrogen receptors, the appearance of the earliest steroid receptor likely preceded the evolutionary split of the two major clades (groups) of bilaterian (bilaterally symmetrical) organisms: the divergence of the protostomes from the deuterostomes. The protostomes are the superphylum that includes many invertebrates such as the arthropods, nematodes, annelids, and mollusks. The deuterostomes are the superphylum that includes the echinoderms and our own species among the chordates (vertebrates, animals with backbones). The inferred first steroid receptor, AncSR1, likely existed even before the common ancestor of the protostomes and deuterostomes (Thornton et al., 2003). This makes the first steroid receptor ancient indeed.

After the nucleotide sequence of the putative ancestral steroid receptor was statistically inferred, it was artificially synthesized (resurrected) by a process known as *ancestral sequence reconstruction*. Once the gene was reconstructed and expressed, the behavior of the resurrected gene product, AncSR1, was observed in cell culture, in which it binds estradiol and is transcriptionally active in response to estrogens. This provides additional support for the hypothesis that the first steroid receptor was an estrogen receptor (Thornton et al., 2003).

Studies of gene duplications in primitive organisms led to a profound hypothesis to explain how ligand-receptor systems evolve, the *ligand exploitation model*. This hypothesis suggests that biosynthetic intermediates in chemical reactions become true hormones only after their receptors evolved by gene duplication and divergence from the ancestral steroid receptor. You might remember from earlier chapters that all steroids are synthesized from cholesterol, which is metabolized in multiple steps to progesterone, which can be metabolized in multiple steps to either glucocorticoids or androgens. The last hormone in this biosynthetic pathway is synthesized by the aromatization of testosterone to yield estradiol.

The first ancestral receptor bound estradiol; thus, a final end product of steroid synthesis was the first to become a "hormone."

At that time, the other steroids in the chain of chemical reactions were probably mere biochemical intermediates. The ligand exploitation model purports that biochemical intermediates were coopted to perform hormonal functions only after genetic rearrangements and duplications resulted in their specific receptors. According to this idea, before the evolution of the first steroid receptor, natural selection must have favored individuals with all enzymes in the biosynthetic pathway for steroid synthesis. Then, in estradiol-producing organisms, there arose a relatively promiscuous molecule that formed the first steroid receptor. This receptor bound estrogens and/or estrogen-like compounds. The inferred phylogenies indicate that the estrogen receptor arose via duplication and divergence from the ancestral steroid receptor, AncSR1, which, as noted earlier, has been resurrected and found to bind estrogens. Next, progestin receptors evolved by gene duplication and divergence. According to this scenario, progesterone, formerly an intermediate to estradiol synthesis, became a bona fide hormone when its receptor evolved from the estrogen receptor. Evolutionary geneticists are looking for other steroid receptor genes among species of early protostomes. Orthologs of the progestin receptor have been found in the nematode worm, *Caenorhabditis elegans*. Androgen, glucocorticoid, and mineralocorticoid receptors are thought to have arisen later, but eventually they might be found among the protostomes.

Evolution of the other steroid receptors might have been gradual, with new receptors appearing many years apart by random events, or their evolution might have been abrupt and explosive, with many new receptors appearing after multiple genetic rearrangements. These rearrangements might have been brought about, and their divergence accelerated, by environmental perturbations. It is plausible that novel hormones arise by duplicating genes that encode receptors with affinity for biochemical intermediates and that are still evolving.

The evolution of an ancient steroid receptor with the ability to bind estrogens might provide hints to the ubiquitous effects of endocrine-disrupting compounds. These are natural or synthetic molecules that have the ability to either stimulate or antagonize hormone receptors. As discussed in other chapters, these are powerful chemicals that can alter any ligand-receptor process, but they have the most devastating consequences during the early critical period in which steroids normally masculinize brain and behavior. By interfering with estrogen receptors, endocrine-disrupting compounds can alter the development of internal and external genitalia, brain, metabolism, and behavior. Endocrine-disrupting compounds can also activate or antagonize the annelid ortholog of the estrogen receptor. This is an important warning that the effects of endocrine-disrupting compounds are not limited to vertebrates but could influence many different species across most animal taxa. Most of us do not often consider the health and well-being of the various worms, slugs, and snails in our soil, but these organisms have important consequences for natural ecosystems and agricultural success. Endocrine disruption of their basic physiological processes is an important concern.

Compared with the original gene, gene duplication products can have vastly different effects on physiology and behavior. For example, there are two or more types of estrogen receptors with different functional roles in different vertebrate taxa, but the roles can be different in individuals within one species, such as the laboratory mouse. After years of molecular endocrine work with the classical estrogen receptor (now called estrogen receptor-$\alpha$ [ER$\alpha$]), a new estrogen receptor was cloned. The second estrogen receptor, likely a gene duplication product, is called estrogen receptor-$\beta$ (ER$\beta$), and still other estrogen receptors are hypothesized to exist (referred to as ERX). ER$\alpha$ and $\beta$ bind ovarian estrogens with about the same affinity, yet their molecular properties in neurons and their behavioral effects are clearly different (Fig. 19.1). For example, on the promoter for the cyclin D1 gene (a major regulator of entry into the proliferative stage of the cell cycle), estrogen working through ER$\alpha$ induces transcription and ER$\beta$ inhibits the ER$\alpha$ effect. Furthermore, in neuroblastoma cell culture, the relative abilities of ER$\alpha$ and ER$\beta$ to manage transcriptional facilitations after estradiol treatment are dissimilar (Vasudevan et al., 2002). In addition, in kidney fibroblast cells, activation of the thyroid hormone receptor (TR), TR$\alpha$1, has inhibitory effects on ER$\alpha$-mediated transcription, whereas ER$\beta$ has the opposite effect in the presence of TR$\alpha$1 (ER$\beta$ stimulates transcription in the presence of thyroid hormone).

What about behavior? In laboratory mice, primary reproductive behaviors depend on ER$\alpha$; normal expression and regulation of the ER$\alpha$ gene is required for normal sex behavior in both males and females. On the other hand, social recognition and the suppression of aggression depend on ER$\beta$ working in concert with ER$\alpha$ (refer back to Fig. 3.5). Conceptual scenarios for the relations between ER$\alpha$ and ER$\beta$ have been tested extensively by comparing normal, wild-type mice with those in which either or both of the ER genes have been knocked out (Fig. 19.1).

In the history of biochemical genetics, the classic principle of "one gene, one enzyme" was derived from the pioneering, Nobel Prize–winning work of George Beadle and Edward Tatum. Now, however, in light of the complexities of the mammalian central nervous system and human behavior, we must say, "Combinations and patterns of gene expression influence combinations and patterns of behavior." For example, it is evident that different patterns of male sociosexual behavior depend on different relations between ER$\alpha$ and ER$\beta$ (Fig. 19.1). Simple male-typical mounting flourishes from a synergy between the two. In aggressive behavior tests, the suppressive effect of ER$\beta$ counteracts the influence of ER$\alpha$. As well,

| COMPARISONS of ER-α & ER-β FUNCTIONS in CNS | | | | |
|---|---|---|---|---|
| Necessary? | Sufficient? | | ASSAYS IN THE ♀ | ASSAYS IN THE ♂ |
| ER α & β | Neither α nor β | "ER α & β must synergize" | Social recognition | none |
| Neither α nor β | Either α or β | "ER α & β can subst for each other" | E induction of PR (ICC) E reduction of ER (ICC) | E induction of PR (ICC) E reduction of ER (ICC) Simple mounting |
| α, β each for its own | α, β each for its own | "ER α vs β contribs. not related" | Maternal behavior (α) Suppression of aggression (α) Reduction of food intake (α) Reduction of anxiety (α) | Ejaculation (α) Anxiety response (α) |
| ER α absent β | ER α absent β | "ER β can reduce α effect" | Lordosis behavior | Aggression |
| Optimal α/β balance | Optimal α/β balance | "ER α vs β always opposed" | none | none |

**FIGURE 19.1** Two likely gene duplication products, estrogen receptor (ER)α and ERβ, have significantly different behavioral effects in mouse brain. Theoretical scenarios for combinations and patterns of action between the two ERs are charted. Conclusions are derived from comparisons and double–estrogen receptor knockouts. It is also clear that different combinations of gene functions are required for different patterns of sociosexual behaviors. *CNS*, central nervous system; *ICC*, immunocytochemistry; *PR*, progestin receptor.

for aggression and female lordosis behavior, the "yin-yang" principle of Jan-Ake Gustafsson is important regarding the tendency of ERβ to oppose ERα. Finally, in the male, ejaculation simply depends on ERα. Across all behavioral functions tested, it appears that all possible combinations of ERα and ERβ effects can be found: dependence on ERα, on ERβ, on both, on neither, and finally, on ERβ actually opposing the influence of ERα. These complex behavioral effects can be traced to small genetic differences between ERα and ERβ.

Effects of rearranged genes for ERα have been demonstrated in wild animals living in their native habitats. In free-living white-throated sparrows, *Zonotrichia albicollis*, there are two different morphs within the same species. The different morphs appear together in the same geographic locations but they show clear differences in coloration, aggressive behavior (territorial song), and parental behavior, and these differences segregate with different arrangements of the ZAL2 chromosome. Sparrows with two copies of the ZAL2 chromosome, the tan-striped morph, have tan-and-brown striped heads. They sing very little and stay close to the nest, where they actively provision their baby chicks. In contrast, sparrows with at least one copy of the inverted ZAL2$^m$ chromosome, the white-striped morph, have starkly contrasting white-and-black stripes on their heads. They engage in frequent territorial song and aggressive territorial invasions, and they show less frequent parental behaviors.

In both morphs, the onset of aggressive territorial song is correlated with increases in testosterone secreted from the testes during the spring breeding season. You might suspect that the white-striped morph is more aggressive owing to higher levels of testosterone, but you would be wrong. When testosterone levels are equalized by gonadectomizing the birds and treating them with the same dose of testosterone, behavioral differences persist. Something else explains the differences in aggression and parental care. In the brains of sparrows and many other songbird species, testosterone is converted into estradiol. Territorial song is blocked by treatments that prevent conversion of testosterone to estradiol and by treatments that block estradiol binding to ERα. ERα-containing cells are located in brain areas involved in territorial aggression, including the medial amygdala. Parental behavior is related to ERα in other brain areas, including the medial preoptic area. Thus, it was hypothesized that differences in aggressive and parental behavior are related to differences in expression of the gene for ERα in these brain areas. The gene for ERα, *ESR1*, is contained within the ZAL2 chromosome, and thus in the inverted ZAL2$^m$ chromosome, *ESR1* is inverted. In other words, the white-striped, aggressive sparrows have an inverted *ESR1* gene. When levels of messenger RNA were measured in the two morphs, the white-striped sparrows' aggression was associated with a more efficient transcription of *ESR1* in a region of the medial amygdala associated with aggression. In addition, levels of parental behavior, e.g., nest provisioning, were closely associated with the level of *ESR1* expression in the medial preoptic area. Thus, a genetic rearrangement in the gene for the estrogen receptor was linked to a change in efficiency of gene transcription in particular brain areas, which in turn was linked to a change in naturally occurring behaviors in wild animals (Horton et al., 2014). This is a prominent example of differences in behavior produced by similar levels of hormones acting on different genetic backgrounds caused by gene inversion.

### Thyroid Hormone Receptor Genes

What about nuclear receptors for nonsteroid hormones? Two genes code for TRα and TRβ. They have different patterns of expression and ranges of physiological effects throughout the body. Of importance, animals with TRβ gene knockouts have greatly elevated circulating thyroid hormone levels because the TRβ receptor gene is essential for negative feedback of thyroxine upon the pituitary. Surprisingly, these animals are deaf. Mice with TRα1 knockouts are mildly hypothyroid and have heart defects and an abnormally low body temperature. If mice have both TRα1 and TRα2 knocked out, they are severely hypothyroid, have marked cognitive deficits, and do not survive until puberty. With regard to the hormone binding capacity of the splice variants of each gene, both TRβ1 and TRβ2 bind triiodothyronine (T3), the most effective thyroid hormone ligand. TRα1 also binds T3 but TRα2 does not.

Across an array of behavioral assays, there are significant differences among the effects of various types of thyroid hormone gene knockouts. Most striking are the results with lordosis behavior, in which (against predictions) the effects of a TRα1 knockout were opposite those of a TRβ gene knockout (Table 19.1). These effects remain to be explained. Also against expectations, the overall impact of these knockouts on tests of higher mental function, including learning and memory, turned out to be mild or nonexistent, representing another problem to be solved.

## Oxytocin Versus Vasopressin Genes

From an ancestral neuropeptide concerned with fluid distributions in various organs, two genes evolved (look ahead to Fig. 23.3). Both code for neuropeptide hormones that have nine amino acids that differ only in two of them, and both are shaped by an identically located disulfide bond. One gene is the oxytocin gene and the other is the vasopressin gene. Both genes are expressed by neurons in magnocellular groups in the

TABLE 19.1 Summary of Behavioral Phenotypes of Thyroid Hormone Receptor (TR)α Knockout (KO) and TRβ KO Male Mice

| | TRα KO | TRβ KO |
|---|---|---|
| Sexual behavior | Increased | No change |
| Activity | | |
| Open field test | No change | No change |
| Anxiety | | |
| Elevated plus maze | Increased | Decreased |
| Dark/light transition | No change | Decreased |
| Arousal (startle reflexes) | | |
| Baseline | Elevated | No change |
| Acoustic startle | Reduced | No change |
| Tactile startle | Reduced | No change |
| Passive avoidance | No change | No change |

TRα and TRβ gene deletions have differential behavioral effects.
*From Vasudevan, N., Morgan, M., Pfaff, D., Ogawa, S., 2013. Distinct behavioral phenotypes in male mice lacking the thyroid hormone receptor α1 or β isoforms. Horm. Behav. 63, 742–751.*

anterior hypothalamus (e.g., the paraventricular nucleus and the supraoptic nucleus), although only an extremely small percentage of cells express both genes. Both are still concerned with fluid distributions, albeit not in the same organs of the body; however, their roles in a variety of social behaviors are markedly different.

Oxytocin fosters a wide range of behaviors related to attachment. Many of these behaviors have direct effects on reproductive success: mating and parental behaviors. Oxytocin secreted from neurons of the posterior pituitary is directly involved in the physiological process of birth and lactation in all mammals. Oxytocin also is linked to a general class of behaviors known as affiliative behaviors. These behaviors may be species-typical, such as side-by-side sitting, and mutual licking and grooming by prairie voles. Oxytocin is central to the formation of partner preference and the monogamous mating systems in monogamous species of voles, mice, and monkeys. In humans, oxytocin is associated with increases in other species-specific affiliative behaviors such as trust, social attachment, and the tendency to "tend and befriend" (Carter, 2014).

In contrast, vasopressin has behavioral effects that are almost opposite those of oxytocin in some species. Vasopressin is integral to the formation of affiliative behaviors and partner preference in prairie voles, but high levels also are associated with intermale aggression. In other species, such as white-footed mice, house mice, hamsters, and humans, vasopressin is linked to high levels of agonistic behavior and/or aggression. Not only can vasopressin administration increase the probability of aggressive behavior, it can also intensify communicative behaviors that are preludes of aggression. This is of particular significance in solitary, territorial species such as the Syrian hamster, which prefers to live alone (with the exception of new mothers). Individual adult hamsters mark their territories by rubbing their flank glands on the substrate, which deposits an olfactory signal, a pheromone, on the territorial boundary. If these pheromonal signals are ignored by a neighboring hamster, a dangerous battle ensues. Flank marking is enhanced by high levels of vasopressin. Thus, depending on the species, gene duplication can lead to two hormone products (oxytocin and vasopressin), with strikingly different effects on behavior.

Even in a taxonomic group of species with the same level of hormone secretion, the behavioral response can differ according to the distribution of the hormone receptor. New receptor distributions have been linked to gene duplication. Prairie voles and meadow voles are monogamous, show strong preference for one mating partner, and have high levels of affiliative behavior. The males of these two species show high levels of parental behavior toward their own offspring. In contrast, montane voles and pine voles are promiscuous, show no preference for any particular mating partner, and have low levels of affiliative behavior, and the males do not show parental behavior even toward their own offspring. Both prairie and meadow voles have a distribution of oxytocin and vasopressin receptors that is similar between the two monogamous species and different from the two promiscuous species. The distribution of the vasopressin receptor V1AR is strongly linked to genetic polymorphisms in the gene that encodes this receptor (*avpr1a*). There are only minor differences in the coding sequence of *avpr1a* gene between prairie and montane voles, but the monogamous prairie vole displays a 428–base pair microsatellite sequence in the 5' flanking region that is not found in montane voles. When the *avpr1a* of the prairie vole is transgenically inserted into the brains of promiscuous house mice, the transgenic house mice show a V1AR distribution similar to that of the monogamous voles; when it is tested with opposite-sex conspecifics, it displays elevated levels of affiliative behaviors. Furthermore, variation in the 5' flanking region of prairie vole *avpr1a* affects brain expression of the gene and alters intraspecific variation in partner preference. Early studies emphasized that affiliative behavior was linked to variation in the *length* of this microsatellite. However, more recent evidence supports the idea that microsatellite length is less important than sequence variation within the microsatellite. These results exemplify the important role of hormone receptor gene arrangements in hormonally influenced behavior.

## Immature RNA Splicing Variants

Messenger RNAs coding for PRs can be spliced so that in addition to the larger sequence (PR-B), there is a smaller variant (PR-A). Both are present in the hypothalamus and pituitary and are induced by estrogens, but not to the same extent. Their transcriptional actions in cell lines are markedly different, as demonstrated by the results of Kate Horwitz and Donald McDonnell (Fig. 19.2). Most prominent among the results is the finding that there are molecular chemical conditions under which the

| | Spliced form of Progestin Receptor | |
| --- | --- | --- |
| | PR-B | PR-A |
| Supports ligand-dependent transcription? | 10 X | 1 X |
| Activation by Anti-Progestin? | Minimal | Absent |
| Trans-repression of ER-dependent transcription? | No | Yes |

FIGURE 19.2 Two splicing variants from the progestin receptor (PR) gene have significantly different molecular effects. PR-B is the transcriptional facilitator par excellence. In contrast, under some circumstances, PR-A can repress transcription. *Summarized from papers by Kate Horwitz and Donald McDonnell.*

transcriptional facilitation operating through PR-B was unable to be mimicked by PR-A, and PR-A would have a transcriptional repressing effect.
How this striking difference plays out in neuroendocrine systems is not yet clear. Leung and colleagues give an early hint. Under conditions in which expression of PR-B facilitates transcription through the promoter of the gonadotropin-releasing hormone (GnRH) receptor gene, PR-A expression actually decreases GnRH receptor transcription. If these data hold for neurons, the behavioral implications would be obvious. GnRH tends to facilitate certain aspects of courtship or mating responses in both males and females.
Vigorous transcription of the gene for its receptors in neurons would be a sine qua non of its behavioral influences. Different splicing variants are likely to exhibit different reactions to the variety of selective PR modulators of importance for medicine and behavior.

## Glucocorticoid Receptors

From the primary GR, now called GRα, a second form is spliced, GRβ. GRβ does not bind glucocorticoids such as corticosterone or cortisol, it does not activate transcription in a ligand-dependent manner, and in fact it can block the ability of GRα to activate transcription (Fig. 19.3). From the work of John Cidlowski et al. (National Institutes of Health), we see how a single gene coding for GRs can yield several proteins of differing functional capacities. This is just as important for the neuroscientist to consider as it is for the molecular endocrinologist.

## Androgen Receptors

Although only one AR has been cloned, genetic polymorphisms have been strongly associated with the risk for prostate cancer. For example, there is a tract of glutamines the length of which is inversely correlated with tumor size in men. It seems highly likely that genetic variations in AR, especially in a transactivation domain, would affect the influences of androgens on behavior, especially aggressive behaviors; however, these do not appear to have been tested as yet.
Several AR splice variants have been detected, some of which lack the ligand binding domain. As noted by Dehm and Tindall (2011), "Many of these truncated AR isoforms function as constitutively active, ligand-independent transcription factors that can support androgen-independent expression of AR target genes."

## Angiotensin Receptors

Receptors for angiotensin have evolved into four distinct receptors labeled AT1 through AT4. Many of the physiological effects of angiotensin

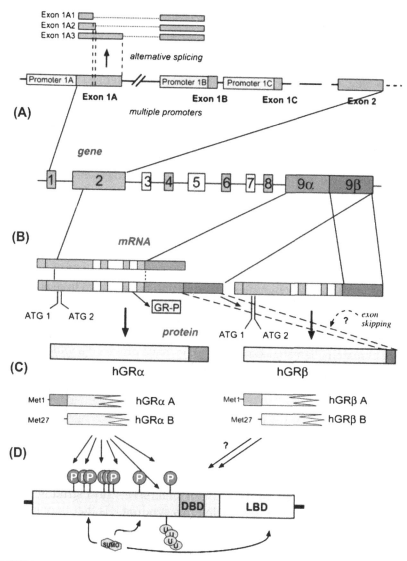

**FIGURE 19.3** Several isoforms of the glucocorticoid receptor (GR) (binds cortisol, corticosterone, etc.) come from the products of a single gene. The primary GR gene is composed of nine exons; there are three different promoter start sites (A). Alternative splicing of exon 9 at its 5′ end result in the GRα and GRβ families of proteins (B). Putting together the variations in parts (A) and (B), we see at least four proteins in (C). Then, posttranslational modifications by phosphorylation (P), ubiquitination (U), etc., of each of these proteins lead to different functional capacities (D). *DBD*, DNA-binding domain; *hGR*, human glucocorticoid receptor; *LBD*, ligand binding domain; *mRNA*, messenger RNA. *From Yudt, M.R., Cidlowski, J.A., 2002. The glucocorticoid receptor: coding a diversity of proteins and responses through a single gene. Mol. Endocrinol. 16 (8), 1719–1726. With permission.*

are mediated by AT1 receptors, including cardiovascular actions, thirst, and other behavioral effects. AT2 receptors are most prevalent in the fetus and developing animal and diminish considerably in favor of the AT1 receptor in the adult. The AT1 receptor is a seven-transmembrane G protein–coupled receptor that activates classical second-messenger systems, including phospholipase C. Cloning revealed that whereas the human has only one type of AT1 receptor, the rodent has two isoforms classed as AT1a and AT1b. The blood pressure effects of angiotensin are mediated by activation of AT1A receptors, whereas AT1B receptors are necessary for most of the drinking response to central angiotensin. In many cases, the AT2 receptor opposes the actions of AT1 receptors, and there is surprising little homology between the two (on the order of 32%). Furthermore, the AT1 receptor acts through Gq/11 proteins whereas the AT2 receptor works through Gi$\alpha$2 and Gi$\alpha$3. In laboratory rodents, disruption of the AT2 receptor gene reduces exploratory behavior and lowers body temperature. In rats, specific antagonists to the AT2 receptor, such as PD1233919, abolish angiotensin II–induced increases in the acquisition of conditioned avoidance responses. The AT3 binding site was found in cultured neuroblastoma cells and has received little attention. The AT4 receptor binds angiotensin IV and has been identified as an insulin-regulated aminopeptidase (IRAP). Deletion of the IRAP gene results in an accelerated decline in spatial memory with age. The duplication and divergence of mammalian angiotensin receptors have provided a wide range of behavioral and physiological adaptations.

# CLINICAL EXAMPLES

## Kallmann Syndrome

Kallmann syndrome is hypogonadotropic hypogonadism often coupled with anosmia (loss of the sense of smell). The most striking behavioral problem with men experiencing this syndrome is that they have no libido. This syndrome illustrates not how two similar genes have opposite effects, but rather how disruption of two different genes can have the same behavioral effect. The story starts with the discovery in 1989 by Schwanzel-Fukuda and colleagues that the neurons that control mammalian reproduction, those coding for GnRH (also referred to as luteinizing hormone-releasing hormone), are not born in the brain where other neurons are. Instead they are born in the developing olfactory pit, and during development migrate up the nose, along the bottom of the brain and into the basal forebrain. An individual with the X-linked Kallmann mutation at Xp22.3 displays a failure of neuronal migration: The GnRH gene is expressed normally; rather, the GnRH neurons never migrate out of the olfactory apparatus.

Because the Kallmann gene (called *anosmin*) codes for an extracellular matrix protein evidently connected with GnRH-neuronal migration, the molecular and endocrine facts permit the following explanation for the behavioral disorder: Men with X-linked Kallmann syndrome have no libido because they have low testosterone. They have testosterone deficiency because they have no gonadotropic hormones secreted from the pituitary to stimulate testicular androgen synthesis, which can be directly attributed to the lack of GnRH secretion from the hypothalamus to the pituitary. As noted previously, the lack of hypothalamic GnRH secretion occurs because the GnRH-producing neurons never make the migration from the olfactory bulb into the hypothalamus. The lack of migration is because of a deletion in the Kallmann gene. These facts have yielded the first proof that damage to an individual gene can drastically reduce an important human social behavior.

A second, different genetic rearrangement can lead to the symptoms of Kallmann syndrome (hypogonadotropic hypogonadism and loss of libido) but without anosmia. In these patients, familial hypogonadotropic hypogonadism resulted from multiple mutations widely distributed across the gene for the receptor for GnRH. Clearly, loss of function of the GnRH receptor could lead to the same downstream endocrine consequences as listed previously, including eventual loss of libido. The customary interpretation of GnRH receptor mutations would be that they reduce either the interaction of the receptor with its ligand or the ability of the receptor to trigger signal transduction pathways in the pituitary cell (or neuron). C. Michael Conn et al. (Texas Tech) presented a novel interpretation: Mutations leading to misfolding of the receptor protein and subsequent misrouting in the cell lead to the failure of GnRH signaling. Each of these interpretations, not mutually exclusive, provides an explanation of different forms of Kallmann syndrome.

## Androgen Receptors

An obvious behavioral effect of a steroid receptor gene is androgen insensitivity syndrome, in which female-typical external genitalia and female gender identity occur in chromosomally and gonadally male individuals who have mutations in the gene for the AR that render the receptor inactive. More subtle variations lead to less marked but detectible changes in morphology. Normal polymorphism in the *AR* gene correlates with digit ratios in men, sperm count, and the presence of female-like breasts (gynomastia). Males have a smaller ratio of the first and third digits (fingers) than do women, and women with androgen insensitivity syndrome (AIS) also have feminine (high) ratios of the first and third digits. This ratio is known as the 2D:4D ratio, and several lines of evidence show that the 2D:4D ratio is a proxy for prenatal androgen exposure in a wide variety of mammalian species including humans. The 2D:4D ratio is masculinized in prenatally and neonatally androgenized rodents and nonhuman primates, as well as in XX women exposed to prenatal androgens (e.g., in women with congenital adrenal hyperplasia). The AIS example is consistent with the hypothesis that the 2D:4D ratio is related to androgen exposure and sensitivity to those androgens as conferred by the AR. Consistent with this hypothesis, the greater the number of cysteine, adenine, guanine repeats in exon 1 of *AR* (coding for a longer polyglutamine stretch), the less effective the *AR* gene is in masculinizing the 2D:4D ratio and other male-typical characteristics such as sperm count, male-typical body fat distribution, and breast development. These results are consistent with the idea that prenatal androgen exposure and sensitivity to androgens are important for masculinization of brain, body, and behavior, and are influenced by polymorphisms in the human AR (reviewed by Breedlove, 2010).

## Resistance to Thyroid Hormones

Another reminder of how small changes in receptor proteins can make a large difference in endocrine and behavioral function comes from the work of Refetoff et al. (2000, 2014), cited earlier (see, for example, Figs. 8.4 and 20.5). Mutations that lead to a resistance of thyroid hormone effects should have clear behavioral consequences (e.g., see Chapter 1).

# OUTSTANDING NEW BASIC OR CLINICAL QUESTIONS

1. What can be done to distinguish the helpful effects of ER ligands from the harmful effects associated with estrogen-dependent cancers? For the central nervous system, for example, can ERβ be activated in a way that does not activate ERα, so that potential emotional and cognitive benefits of estrogenic ligands working through ER can be obtained without risk for breast cancer or endometrial cancer?
2. With respect to both PR-B (versus PR-A) and GRα (versus GRβ), can we see in neurons how the transcription-facilitating properties of the main isoform work, compared with the inhibiting properties of the splice variant? Are there consequent behavioral effects?
3. It is puzzling that the behavioral phenotypes of TR knockouts are as minimal and mild as they are. If there is compensation during development, what is it and how does it work?
4. Failures of parental care are rife in segments of modern industrialized societies. Are there any ways in which an understanding of oxytocin (vis-à-vis vasopressin) actions can either (1) contribute to an understanding of how such lapses come about, or (2) make inroads into the problem?

CHAPTER

# 20

# Hormone Receptors Interact With Other Nuclear Proteins to Influence Hormone Responsiveness

It is not surprising that steroid action is central to neuroscience and behavioral research. Steroid receptors are widely distributed in the brain, and steroid hormones are small lipophilic molecules that readily cross the blood–brain barrier and enter neurons and glia. Moreover, steroid receptors are transcription factors, proteins that influence the conversion of DNA into messenger RNA (mRNA) in the cell nucleus and are amenable to study with state-of-the-art molecular biology techniques. In fact, nuclear hormone receptors comprise some of the best-studied transcriptional systems in eukaryotic molecular biology. The steroid–receptor complex is necessary but not sufficient for the expression of many specific behavioral and physiological processes. Other molecules are needed for these processes to occur. For example, when a steroid or thyroid hormone binds to its receptor, a conformational change enables recruitment of molecules known as *coactivators*. Coactivators increase the efficiency of interactions between the ligand–receptor complex and the promoter regions of the gene, facilitating transcription. Other molecules, known as *corepressors*, are recruited by the ligand–receptor complex to bind to other molecules that act as repressors. Repressors bind directly to the promoter region of the gene or to coactivators, blocking or modulating gene expression.

## NUCLEAR RECEPTORS AS THEY INFLUENCE BEHAVIOR

The best-understood behavioral mechanisms are those that control the female-typical and male-typical behavior of laboratory animals such as rats

*Principles of Hormone/Behavior Relations*
http://dx.doi.org/10.1016/B978-0-12-802629-8.00020-6

385

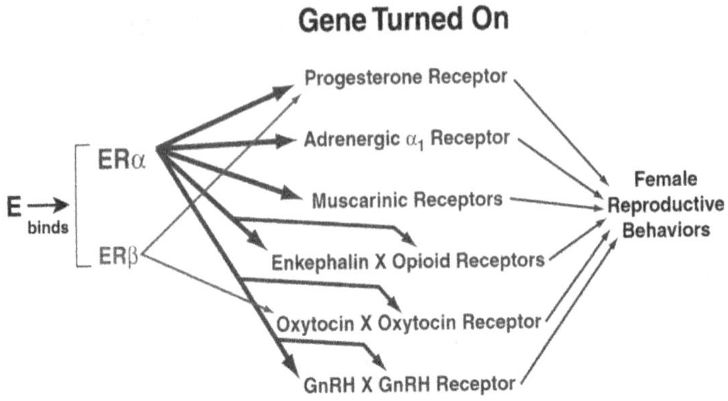

## Gene Turned On

FIGURE 20.1    Estrogens (E) binding to estrogen receptor (ER)α or ERβ turn on several genes that have been implicated in female sex behaviors. Considered as a logical syllogism, if E turns on a gene and the gene's product drives the behavior, then that genomic effect is one causal route by which the hormone influences the behavior. (1) A biologically sensible formulation of this list of genes is presented in Fig. 20.2. (2) We continue to add to the list of relevant genes by the use of microarrays. GnRH, gonadotropin-releasing hormone. *From Pfaff, D.W., 1999. Drive: Neurobiological and Molecular Mechanisms of Sexual Motivation, MIT Press, Cambridge, MA. With permission.*

and mice. With regard to female-typical reproductive behavior, a significant number of genes have the following properties: (1) they are turned on by estrogen administration, and (2) their products foster female-typical sex behaviors (Figs. 20.1 and 20.2). Years of research revealed a long list of genes, but two new questions were generated: would the list continue to grow, and would the number of genes eventually become as large as the number of reagents available? Indeed, the list will be long, but the magnitude of the problem created by this long list has been reduced by the use of microarray technologies and next-generation sequencing (see subsequent discussion). These techniques allow scientists to rank the significance of the genes in terms of function. Can the genes be categorized in a biologically sensible fashion? Yes, they can: Although the gene products in Fig. 20.1 have different biochemical roles, their routes of action in the organization of estrogen-dependent sex behaviors can be organized into modules that make biological sense (Fig. 20.2). The steps can be summarized as: growth of ventromedial hypothalamic neurons, amplification of the estrogen effect by progesterone, preparation for reproductive behaviors, permissive actions on hypothalamic neurons at the top of the neural circuit for lordosis behavior, and synchronization with ovulation (these steps thus forming the acronym GAPPS).

Within this set of molecular mechanisms, how do estrogen receptor (ER) α and ERβ interact? If we consider the roles of these two receptors in the development of sex-specific behavior over the life span, their action can be described as sequential, complementary, antagonistic, or any combination

## MODULAR SYSTEMS DOWNSTREAM FROM HORMONE-FACILITATED TRANSCRIPTION RESPONSIBLE FOR A MAMMALIAN SOCIAL BEHAVIOR: "GAPPS."

- *Growth* (rRNA, cell body, synapses).
- *Amplify* (pgst/PR →→ downstream genes).
- *Prepare* (indirect behavioral means; analgesia (ENK gene) and anxiolysis (OT gene)).
- *Permit* (NE alpha-1b; muscarinic receptors).
- *Synchronize* (GnRH gene, GnRH Rcptr gene —synchronizes with ovulation).

FIGURE 20.2   The GAPPS system (growth of ventromedial hypothalamic neurons, amplification of the estrogen effect by progesterone, preparation for reproductive behaviors, permissive actions on hypothalamic neurons at the top of the neural circuit for lordosis behavior, and synchronization with ovulation) for estrogen actions on sex behaviors. Estrogenic effects on neuronal growth (G) increase signaling capacity in the hypothalamic neurons that control lordosis. By facilitating transcription of progestin receptors (PR), estrogens arrange for their own amplification (A). Estrogens prepare (P) the female for strong stimulation from the male by enhancing transcription of an opioid peptide gene (enkephalin, ENK) and prepare for the anxiety of courtship behaviors in an open environment by increasing transcription of oxytocin (OT) and its receptor. At the top of the lordosis behavior circuit, estrogenic effects on neurotransmitter receptors increase electrical activity in the relevant hypothalamic neurons, thus permitting (P) the rest of the circuit to operate. Finally, it is biologically adaptive in a small animal that is prey to synchronize (S) sex behavior with reproduction; estrogenic effects on gonadotrophin-releasing hormone (GnRH) and its receptor have this effect. *NE*, norepinephrine; *rRNA*, ribosomal RNA.

of these actions. From studies of knockout mice, i.e., mice in which the genes for ERα, ERβ, or both types of receptors have been knocked out (see Chapter 19), it seems clear that early in development, ERα is essential for behavioral masculinization of the male (i.e., high levels of male-typical sex behavior in adulthood in response to high levels of androgens and the presence of a female stimulus), whereas ERβ is important for defeminization of the male (i.e., low levels of female-typical sex behavior in response to estradiol and progesterone and the presence of a stimulus male). This arrangement ensures that there are adult males and females with complementary mating behaviors. Females will prefer males and show far more female-typical than male-typical sex behavior. Males will prefer females and show far more male-typical than female-typical sex behavior. During later development, estradiol is essential for healthy bone growth, learning, and memory, which are traits that promote survival and successful competition. Estradiol acts via ERα on these traits, but ERβ may oppose the action of ERα, and the two receptors have been hypothesized to have balancing (or "yin-yang") effects.

After puberty, estrogens foster social recognition and affiliative behaviors that bring reproductively competent females and males to the same place at the same time (refer back to Chapter 3). Estrogens do this by working through ERα on the modules described earlier. In addition, ERα is necessary for both negative and positive feedback of estradiol on gonadotropin-releasing hormone (GnRH) and luteinizing hormone (LH) secretion. ERβ is found in high density in the ovary and is necessary for normal follicle development, gene expression (e.g., expression of the gene for gonadotropin receptors), and stimulation of the late follicular growth that must occur before ovulation can occur. During the estrous cycle of adult females, ERα mediates increased locomotor activity resulting from estrogenic action in the preoptic area that occurs during the periovulatory period. Such locomotion is an essential part of sexual motivation, mate searching, and courtship behaviors (i.e., direct approaches by the female to the male) that increase the likelihood of successful mating behaviors. ERβ comes into play during courtship, when it reduces aggression in males toward females, so that the animals will be more likely to "make love, not war."

For the consummation of sexual behavior, ERα is essential for the primary act of copulation (the lordosis posture) of the female. After the offspring are born, ERα subserves the positive effects of estrogens on maternal behavior. Thus, ERα is tied to an entire chain of behavioral steps essential to the nitty-gritty of reproductive physiology. This might be adaptive because estradiol and ERα are both critical for positive feedback of estradiol on the ovulatory GnRH and LH surges, and ERβ is critical for ovarian function. Thus, the same genes (those that encode the ERs) coordinate both the endocrine and behavioral aspects of reproduction, thereby (1) synchronizing the highest level of reproductive motivation with the highest level of fertility and (2) ensuring that the females do not risk predation or invest energy in courtship and mating until the gametes (i.e., eggs) are mature and available for fertilization. The late, great British neurophysiologist Sir Charles Sherrington opined that the neurobiologist should account for the flow of behavioral responses through time, and this set of roles for the ERα and β genes is a good example of how genes can be involved in a flow of behavioral responses.

Over the past 30 years, we have witnessed major advances in understanding how behavior and physiology are influenced by hormone receptors, including ERs, androgen receptors (ARs), glucocorticoid receptors, mineralocorticoid receptors, and thyroid hormone receptors (TRs). We have more to learn, however. One mystery that remains concerns the role of protein folding. The efficacy of nuclear hormone receptors depends on proper folding of all parts of the protein, not just in the ligand-binding domain but also in the transactivation domains (proven important for hormone-dependent facilitation of transcription) at the N-terminal and C-terminal of the proteins. It will be important to understand how proper folding allows for high-affinity binding of the behaviorally relevant hormone and, just as important, why proper folding is essential for interactions with coactivator proteins in the nucleus.

## Microarray Technology

Many important molecular mechanisms have been revealed using the candidate gene approach, in which the effects of variants of a suspected gene are measured in an association study. The number of candidate genes is always growing, and the number of candidate genes is limited only by a lack of imagination and/or a lack of reagents. By studying these genes one by one, we get little sense of their relative importance. In contrast, microarray technology has an advantage in that large numbers of transcripts and their responses to steroid hormones and the environment can be screened at one time without reference to the experimenter's theoretical biases (Mong et al., 2002). For example, one microarray compared gene expression in the preoptic area of ovariectomized females treated with estradiol with gene expression in those treated with vehicle. It was discovered that in estradiol-treated females there was downregulation of the mRNA levels for prostaglandin D-synthetase (PgDS), a transcript found mainly in glial cells, with a product that promotes sleep (Mong et al., 2003). This might explain a previous mystery: i.e., how, working through preoptic cells, estrogens can enhance brain arousal and locomotor activity. This is just one example of how microarrays can reveal both predicted and unpredicted downstream effects of steroid hormones.

In addition, microarray technology has been essential for discerning differences in gene expression between neonatal males and females during the most critical period for sexual differentiation of the mouse brain (Gagnidze et al., 2010). Virtually every category of chemical messenger yielded at least one sex difference in its gene expression in hypothalamus or the preoptic area. These results provided evidence beyond doubt that gene expression patterns differ between male and female brains. Furthermore, these results uncovered many new avenues of exploration.

Microarray technologies continue to improve and include whole-genome analyses based on Affymetrix GeneChips and Illumina bead arrays. Software for analyzing the results has also improved, so that analyses can encompass thousands of transcripts and provide pathway analysis of the genetic and neural networks involved in controlling behavior. Furthermore, the RNA sequencing methodologies and the ability to determine the transcriptomes of individual neurons has lifted this domain of neuroendocrinology to an entirely new level.

## Next-Generation Sequencing

Imagine that you need to analyze behaviorally relevant hormone effects on the regulation of gene transcription in a group of neurons located in a particular nucleus "A" that have axonal projections to another brain region "B." It is possible to isolate mRNAs in just those neurons and perform high-throughput whole-genome sequencing. It also is possible to distinguish between the transcriptomes of neurons that project to "B" and the transcriptomes of nearby neurons that do not project to "B" (Ekstrand et al., 2014). Likewise, in populations with large sample sizes, behavioral expression and hormone levels can be correlated with single-nucleotide polymorphisms, which are small changes in DNA sequences. We have advanced from initial analyses of hormone-regulated gene transcription in specific brain regions to large, mechanistic searches of how those changes in transcription come about.

### Interim Summary 20.1

1. Two ERs, ERα and ERβ, have effects on different but overlapping sets of downstream chemical messengers.
2. Different ERs, e.g., ERα and ERβ, interact to control behavior in different ways throughout the life span:
   a. They can act sequentially.
   b. They can complement each other.
   c. They can have opposing effects.

# OTHER NUCLEAR PROTEINS: COACTIVATORS AND COREPRESSORS

Nuclear hormone receptors do not sit in isolation upon the DNA of neurons and glia, facilitating or repressing transcription. Rather, they participate in assemblies of substantial numbers of proteins that mediate their genomic effects. Such nuclear proteins either foster or block the formation of a complex bridge between the hormone-dependent enhancer sequence and the basal transcriptional machinery. Under circumstances in which those nuclear proteins enhance hormone action, they are referred to as *coactivators*; when they block transcription, they are called *corepressors*. Nuclear protein assemblies surrounding hormone-influenced enhancers have complex interactions that determine a cell's sensitivity to hormones. It appears that a particular cell's sensitivity to a particular hormone at a particular time is determined by a combinatorial code among the protein enhancers (Fig. 20.3). Coactivator functions have been described for a number of major hormonal systems, and they are important in explaining hormone actions on behavior.

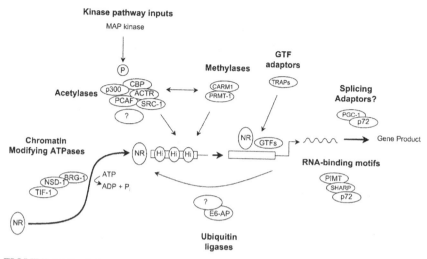

**FIGURE 20.3** A dominant mode of nuclear hormone receptor facilitation of transcription is to bind to DNA response elements; however, a large number of nuclear proteins other than receptors are necessary to complete the reaction. The proteins that surround and shape the DNA (chromatin) can be modified by adenosine triphosphatases (ATPases), acetylases, kinases, and methylases. In addition, especially for hormone receptor response elements on gene promoters far from the transcription start site, adaptor proteins may be necessary to bridge the ligand-bound hormone receptor to general transcription factors (GTFs). Because all of these complexities are presumed to be just as important in nerve cells as in other hormone-responsive cells, they provide mechanisms for developmental and environmental events to alter brain tissue sensitivity to any given hormone. *ACTR*, acid tolerance regulatory protein; *ADP*, adenosine diphosphate; *BRG-1*, brahma homolog 1; *CARM-1*, coactivator-associated arginine methyltransferase 1; *CBP*, cyclical adenosine monophosphate response element binding protein–binding protein; *E6-AP*, E6-associate protein; *MAP*, mitogen-activated protein; *NR*, nuclear receptor; *NSD-1*, nuclear receptor-binding, su (var), enhancer-of-zeste and trithorax domain-containing protein 1; *PCAF*, P300/cyclical adenosine monophosphate response element binding protein–binding protein–associated factor; *PGC-1*, peroxisome proliferator-activated receptor-γ coactivator 1; *PIMT-PRMT-1*, protein arginine N-methyltransferase-1; *SHARP*, Silencing-Mediator for Retinoid/Thyroid/Histone deacetylase 1–associated repressor protein; *SRC-1*, steroid receptor coactivator-1; *TIF-1*, transcriptional intermediary factor 1; *TRAP*, thyroid hormone–associated protein. *From McKenna, N.J., O'Malley, B., 2000. From ligand to response: generating diversity in nuclear receptor coregulator function. J. Steroid Biochem. Mol. Biol. 74 (5), 351–356. With permission.*

## Estrogens

More than 10 nuclear proteins have been shown to influence the efficacy of estrogens bound to ERα in facilitating transcription in a wide variety of cell types. It is exciting to see that different estrogens impose different requirements on the interaction between the ER and the gene's promoter for coactivators to do their job. The precise nucleotide sequence on a particular gene's promoter influences the recruitment of coactivators to the ER, and thus the ability for a particular estrogen to be effective. The protein structural bases for these phenomena are partly understood: Pure estrogens cause a conformational change in ERs, such that a specific part of the receptor protein, helix 12, is available for coactivators to bind. An estrogen antagonist, e.g., of the sort used to retard the growth of breast cancer, causes a different conformational shift, such that the essential coactivator binding surface (helix 12) is blocked.

The nuclear protein story is not limited to positive interactions. A negative regulatory surface within ERα has been identified (Huang et al., 2002). ER is known to interact with at least two types of corepressors: nuclear receptor corepressor (NcoR) and silencing mediator of retinoic acid and TRs. For example, mutation of a single DNA codon, substitution of a leucine for an arginine within the ER, abolished the ability of the receptor to interact with NcoR and dramatically increased transcriptional activity after estradiol administration. Thus, molecular endocrinologists believe that estrogenic signaling in cells includes a regulated release from repression.

The importance of coactivator and corepressor proteins for hormone action is clear. One example occurs in estrogen-sensitive breast cancer cell proliferation. Three coactivator proteins critical for this process were discovered by the use of antisense technology. Antisense sequences are sequences of DNA that are complementary to a target gene. An antisense sequence introduced into the cell can block translation by binding to the mRNA that has been transcribed from that gene. Thus, no protein will be synthesized in the presence of the antisense sequence. The three antisense sequences that were constructed were designed to block the function of each of three coactivator gene products. Estrogen-dependent cell proliferation and DNA synthesis in breast cancer cells were inhibited by treatment with two antisense sequences: those complementary to the sequence of coactivators steroid receptor coactivator-1 (SRC-1) and transcriptional mediator/intermediary factor 2. In contrast, treatment with antisense DNA against the third coactivator, steroid receptor RNA activator, did not affect cell proliferation (Smith and O'Malley, 1999). These results demonstrate that cellular effects of estrogens are specific to particular coactivators.

## Androgens

Androgens affect transcriptional activity of specific genes in a large number of tissues, including the brain. AR effectiveness depends on a large number of nuclear proteins, some of which are coactivators. This set of proteins overlaps with, but is not identical to, the set of proteins important for ER action. Some of these proteins are thought to modify the conformation of the AR whereas others work indirectly by acting on chromatin, the protein that coats the DNA.

Both the amino terminus and the carboxyl terminus of ARs are required for the receptor to interact with coactivators. Likewise, specific amino acid sequences within coactivators are crucial for their interactions with nuclear receptors. For example, deletions of portions of the coactivator, AR-associated protein ARA-70 led to the discovery of a 145–amino acid domain effective for assisting transcriptional activation of ARs (Zhou et al., 2002). One mechanism for these interactions was revealed with three-dimensional imaging methods. The coactivator, cAMP response element–binding protein, helps steroid receptors form specific spatial foci within the cell nucleus. The physical distribution of receptors is closely associated with transcriptional effectiveness (reviewed by Baumann et al., 1999). These are only a few mechanisms by which coactivators facilitate androgen-stimulated transcription.

In addition to coactivators, repressors and corepressors have a part in regulating androgenic hormonal signaling, as they do with estrogens.

The AR contains a target for the repressor, dosage-sensitive sex reversal, adrenal hypoplasia critical region, on chromosome X, gene 1 (DAX-1). The DAX-1 domain involved in its interaction with ARs has been mapped (Holter et al., 2002). DAX-1 might repress binding of the ligand–receptor complex to the androgen response element or it might recruit corepressors in the nucleus, which might either (1) act on ARs to block transcription or (2) negate the effect of coactivators (Fig. 20.4). It is interesting and novel that initiation of this particular repressor–corepressor interaction does not occur exclusively in the cell nucleus (Fig. 20.4).

There are variants of the AR with slightly different nucleotide sequences; the exact nucleotide sequence of the AR makes a difference in androgen efficacy. The most obvious examples comprise AR variants that lack parts of the ligand binding domain, the portion of the AR responsible for accepting testosterone or dihydrotestosterone (Azoitei et al., 2016). Those variants can result from DNA splicing errors or the insertion of "nonsense mutations," which disrupt hormone binding. The ligand binding domain is in the middle of the AR's DNA coding region. In addition, the extreme ends of the protein, the N-terminal and C-terminal domains, have been related to receptor action that does not require hormone binding, leading to altered hormone sensitivity.

## Other Hormonal Systems

Based on a great deal of accumulated information about the roles of nuclear coactivators and repressors in sex hormone signaling, the same principles appear to apply to other hormones with nuclear receptors.

Data on thyroid hormones are impressive. On the one hand, with an apparent molecular weight of about 220 Da, thyroid receptor–associated protein (TRAP) as well as SRC-1 can enhance the transcriptional activating ability of thyroid hormones. On the other hand, NcoR causes the

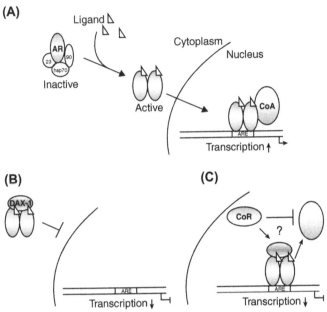

**FIGURE 20.4**    The importance of corepressors: a model applied to androgen receptors (ARs) based on current molecular data. (A) Normally, the inactive AR surrounded by heat-shock proteins (hsp) in the cytoplasm will be activated by the hormonal ligand (dropping off the heat-shock proteins) and will be translocated to the nucleus. There, ARs not only bind to androgen response elements (AREs) on DNA but also achieve the required bonding to nuclear proteins called coactivators (CoA), thus facilitating transcription. (B) How does the protein DAX-1 antagonize androgenic activation of transcription? One possible way is to repress binding to AREs on DNA. (C) A second way in which the protein DAX-1 could work is to call in corepressors in the nucleus, which will act either on ARs themselves or on their coactivators. *From Holter, E., Kotaja, N., Makela, S., Strauss, L., Kietz, S., Janne, O.A., Gustafsson, J.A., Palvimo, J.J., Treuter, E., 2002. Inhibition of androgen receptor (AR) function by the reproductive orphan nuclear receptor DAX-1. Mol. Endocrinol. 16 (3), 515–528. With permission.*

unliganded TRβ to repress basal transcriptional activity of target genes. Thus, relief from this repression upon thyroxine binding to TRβ would elevate the transcriptional rate.

Two points should be made about these TR coregulators: First, vitamin D is another ligand that works through a nuclear receptor. There is considerable overlap among the TRAPs and nuclear proteins associated with vitamin D and those associated with thyroid hormones (e.g., Yuan et al., 1998). That is, some of the vitamin D receptor–interacting proteins are identical to some of the TRAPs. Whereas these similarities may derive from the near-identity of their DNA response elements, many other explanations are possible. Second, with regard to the clinical importance of these coregulators, the syndrome of resistance to thyroid hormone is associated with specific modifications in TR genes. The pathogenicity of these mutations might be rooted in the disturbed relationships between TRs

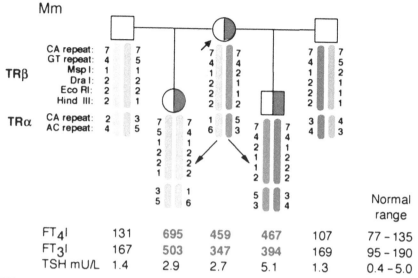

FIGURE 20.5  Haplotypes and thyroid function tests in part of a family in which there is familial resistance to thyroid hormones but no mutation in the coding region of the thyroid hormone-β receptor gene. The high levels of thyroid-stimulating hormone (TSH), free thyroxine (FT$_4$), and free triiodothyronine (FT$_3$) for three affected individuals (the middle three columns) result from a lack of neuroendocrine negative feedback, which normally is mediated through the thyroid receptor (TRβ gene product). Because this gene has no apparent mutation in its coding region, there may be a mutation elsewhere in the gene, some of the problem may be caused by mutations in other nuclear proteins, or both may be true. Because of the effects of thyroid hormones on mood and cognition, these nuclear protein chemistries are likely important to hormone–behavior relations in affected patients. *From Pohlenz, J., Weiss, R.E., et al., 1999. 5 new families with resistance to thyroid hormone not caused by mutations in the thyroid hormone receptor beta gene. J. Clin. Endocrinol. Metab. 84 (11), 3919–3928. With permission.*

and nuclear coactivators and repressors. Over several years, a substantial number of families were identified with inherited resistance to thyroid hormones, although they do not have mutations in TRα or TRβ genes (Fig. 20.5) (Refetoff et al., 2014). An ongoing search for abnormalities in nuclear corepressors or coactivators associated with TRs may uncover the mechanisms of hormone resistance in these patients.

## Studies With Behavioral Measurements

The most exciting studies approach the question of whether nuclear receptor coactivators are actually important for behavior. First among these was a series of experiments from the laboratory of McCarthy et al., who worked with the nuclear protein cyclical adenosine monophosphate response element binding protein–binding protein (CBP) (Auger et al., 2002). Early in postnatal life, the male hypothalamus has more of this protein than does the female hypothalamus. Does this protein participate in defeminization of the brain? That is, might it participate in the causal route by which testicular androgens, affecting the brain neonatally, decrease the ability to perform female-typical behaviors as adults? Antisense DNA oligodeoxynucleotides directed against the mRNA for CBP were infused directly into the hypothalamus of neonatal female rats, which also received testosterone injections. Blocking the gene function of CBP interfered with the defeminizing but not the masculinizing actions of testosterone on adult sex behavior (Fig. 20.6). Data from these experiments provided strong evidence for the role of nuclear receptor coactivators in behavior.

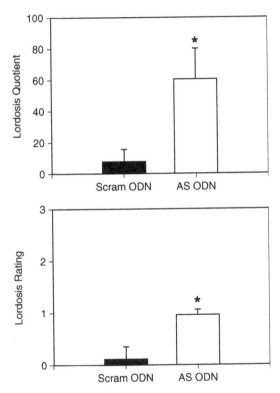

FIGURE 20.6   For this experiment, Auger et al. treated female rats neonatally with testosterone so that they would not show the female-typical sex behavior, lordosis, when they were adults. Rats treated intracerebrally with antisense DNA oligomers interfering with mRNA for the nuclear protein CBP (cyclical adenosine monophosphate response element binding protein–binding protein) had higher lordosis quotients and ratings than did control rats treated intracerebrally with scrambled DNA sequences; therefore, the nuclear protein CBP must be essential for the suppressive effects of testosterone on female sexual behavior. *AS ODN*, antisense oligodeoxynucleotides; *Scram*, scrambled sequence controls. *From Auger, A.P., Perrot-Sinal, T.S., Auger, C.J., Ekas, L.A., Tetel, M.J., McCarthy, M.M., 2002. Expression of the nuclear receptor coactivator, cAMP response element-binding protein, is sexually dimorphic and modulates sexual differentiation of neonatal rat brain. Endocrinology 143 (8), 3009–3016. With permission.*

SRC-1, a nuclear protein that influences the transcriptional effectiveness of many nuclear hormone receptors, is a powerful coactivator. Apostolakis et al. (2002) employed the same type of antisense DNA technique as the McCarthy laboratory used, but the infusion was into the third ventricle of the adult female rat brain. Acute disruption of the function of the SRC-1 gene, but not disruption of other similar gene products, inhibited the ability of estrogens working through ERα to stimulate lordosis behavior (Fig. 20.7).

These pioneering studies established beyond doubt that nuclear coactivator proteins participate within hypothalamic neurons to influence reproductive behaviors.

## SOME OUTSTANDING QUESTIONS

1. Under what circumstances do the nuclear receptors influence behavior, even without hormones bound?
2. Modern sequencing technology reveals that large numbers of genes in the forebrain are hormone-responsive. Do their transcript levels influence nerve cell activity and hormone-dependent behaviors?
3. How do rapid hormone effects (see Chapter 18) operate together with (or against) nuclear, genomic effects to influence behavior?
4. For each hormone that has effects on behavior, what are the crucial coactivator assemblies that allow the cognate nuclear receptor to be effective?
5. In the central nervous system (CNS), in cells contributing to hormonal influences on behavior, how do alterations in receptor protein folding affect neuronal (or glial) sensitivity to hormones?
6. At what point do corepressors have important roles in the behavioral effects of hormones? That is, in the history of behavioral endocrinology, two major influences were studied: hormone levels and "tissue responsiveness" to those hormones. Perhaps corepressor levels help to determine the latter.
7. The discovery of hormone effects on PgDS mentioned in this chapter raised the possibility of hormone-dependent neuronal–glial cooperation. Can this idea be proven?
8. How can changes in genomic structure exerted by hormones be brought into the analysis of CNS mechanisms? Over the long term, perhaps developmentally, nucleotide methylation; over the short term, histone acetylation, methylation, or phosphorylation. Indeed, these molecular states may influence tissue responsiveness to the hormones mentioned earlier.

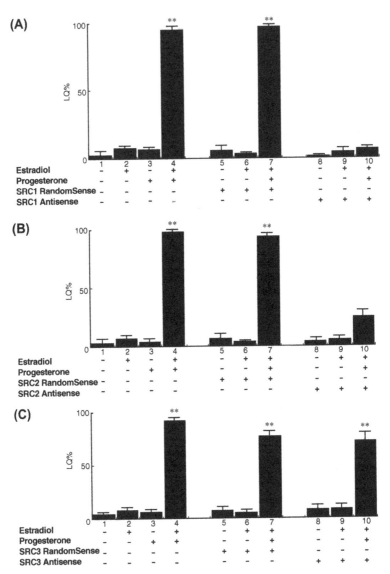

FIGURE 20.7   Coactivators steroid receptor coactivator-1 (SRC-1) (A) and SRC-2 (B) but not SRC-3 (C) participate in the facilitating effects of estradiol and progesterone on lordosis behavior (measured by the lordosis quotient [LQ%]). Their participation is shown by the ability of antisense oligomers against SRC-1 or SRC-2 messenger RNA to block estradiol plus progesterone–facilitated lordosis behavior. *From Apostolakis, E.M., Ramamurphy, M., Zhou, D., Oñate, S., O'Malley, B.W., July 2002. Acute disruption of select steroid receptor coactivators prevents reproductive behavior in rats and unmasks genetic adaptation in knockout mice. Mol. Endocrinol. 16 (7), 1511–1523. With permission.*

# ENVIRONMENT: ENVIRONMENTAL VARIABLES INFLUENCE HORMONE/BEHAVIOR RELATIONS

# 21

# Hormone Effects on Behavior Depend Upon Context

In *Orgasms, Incorporated*, a documentary film about the race to discover a treatment for female sexual dysfunction, filmmaker Liz Canner asks Emory University Professor Kim Wallen to name the factor most important for female sexual motivation. His answer is surprising, given Wallen's long career as a behavioral endocrinologist: The factor most important for sexual motivation is *context*. To illustrate his perspective, he notes that for any particular couple, the best predictor of whether they will be having sex on any given night is not their circulating hormone levels, but rather an array of environmental factors related to the couples' past experience and present exposure to stressors and competing demands for time and energy. For example, in a healthy adult human couple spending their honeymoon in an expensive tropical resort, we do not need to know their sex hormone levels to make a good guess about whether they will be having sex. The abundance of relaxing sunshine, enticing meals, refreshing beverages, privacy, and other honeymooning couples, not to mention the absence of needy young children, meddling parents, geriatric illness, and extreme poverty all combine to ensure that sexual and romantic interactions will be frequent. Whether the couple has never engaged in sexual activity or are familiar sex partners, the influence of the preceding wedding ceremony, the generous gifts, and the warm societal blessings are likely to facilitate marital coupling. Anthropologists tell us that these wedding rituals developed in human societies to promote procreation.

Other significant determinants of sexual motivation and satisfaction in humans are the partners' devotion to each other, romantic imagination, knowledge of erotic zone anatomy, lovemaking skills, and enthusiasm. As illustrated in Chapter 6, hormones are not overriding triggers for behavior. In ways that sometimes are subtle and at other times are obvious, hormones influence the likelihood of particular behaviors in their appropriate context and hormone action differs depending on context. In wild animals, important contexts include the broad geographic location (e.g., tropics, temperate zones, arctic), ambient temperature, microclimate, season,

*Principles of Hormone/Behavior Relations*
http://dx.doi.org/10.1016/B978-0-12-802629-8.00021-8

401

predators, food availability, water availability, the presence of commen-
sal species, and the presence and allure of potential mating partners.
Furthermore, in the wild, hormone action often differs from hormone
action in the laboratory. This chapter goes beyond the initial examples in
Chapters 6 and 7 to illustrate that environmental context and hormones
interact to influence behavior.

# BASIC EXPERIMENTAL EXAMPLES

## Hormones and Context in the Early Vertebrates

Even in some of the early vertebrates, the magnitude of hormone effects
on particular behaviors can be affected by the context in which the hormone is secreted. In the round goby (*Neogobius melanostomus*), an invasive
fish species of the North American Great Lakes and native of the Caspian
and Baltic Seas, male-typical behavior includes increased ventilation, a
rhythmic pumping movement made by the fish's gill structures that pulls
in fresh water. Water that passes over the gills contains vital oxygen and
important chemicals excreted from the bodies and urinary tracts of other
nearby fish. These chemicals include steroid-based pheromones, chemical
messengers secreted from one individual that have effects on the behavior
of other animals. Female pheromones bind to the olfactory receptors of
males and increase their ventilation rates.

Pheromone-induced ventilation behavior is sexually dimorphic; males
have higher ventilation rates as a result of higher circulating androgen
levels. For example, treatment of male gobies with methyltestosterone
increases the ventilation rate. The effect is primarily activational (not organizational) because the ventilation rate can be increased by implanting
methyltestosterone into adult females. In Fig. 21.1, the dark circles represent the higher ventilation rates in methyltestosterone-treated females
compared with untreated ones (open circles) on each of 70 days of testing.
This figure shows that the ventilation rate is further increased by signals
that emanate from fertile females. Normal, mature, intact male round
gobies and methyltestosterone-treated female gobies increase the rate of
ventilation markedly when exposed to nanomolar concentrations of three
putative pheromones: etiocholanolone (ETIO), estrone (E1), and estradiol-
glucuronide (E2-g) (Fig. 21.1).

The underlying mechanism of pheromone action involves neurons in
the part of the brain involved in processing sensory information about
pheromones, the olfactory bulb. Investigators recorded the electrical activity of neurons in the olfactory bulbs of male gobies and found that electrical activity was enhanced by pheromonal cues from females (ETIO, E1,
and E2-g). Increases in olfactory bulb electrical activity occurred only in

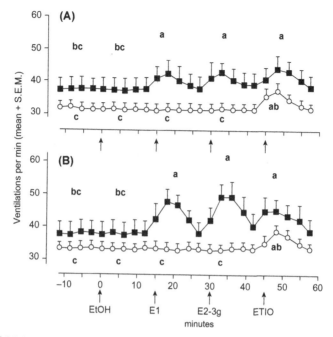

**FIGURE 21.1** Even in fish, environmental context influences hormone-dependent behaviors. These experiments measured the ability of the androgenic hormone methyltestosterone to induce male-like behavior in female round gobies. Experimenters counted the ventilation response by females treated subcutaneously with methyltestosterone (*squares*) or blank capsules (*circles*). (A) Measurements 8–10 days after implantation. (B) Measurements 16–18 days after implantation. Whereas the methyltestosterone-treated animals did not respond to the control (ethanol [EtOH]) added to their water environment, they responded to added estrogens: E1 (estrone), E2 (estradiol-glucuronide), and to a lesser extent, etiocholanolone (ETIO). *S.E.M.*, standard error of the mean. *From Murphy, C.A., Stacey, N.E., 2002. Methyl-testosterone induces male-typical ventilatory behavior in response to putative steroidal pheromones in female round gobies* (Neogobius melanostomus). Horm. Behav. 42 (2), 109–115. With permission.

mature adult males with intact olfactory receptors and not in adults with the olfactory sensors blocked or in immature males with unblocked olfactory receptors (Belanger et al., 2006, 2007). Methyltestosterone-enhanced adult behavior could be further augmented by the presence of olfactory-mediated social cues (pheromones) from opposite-sex conspecifics. These experiments demonstrated that behavioral effects of methyltestosterone depend on two types of context: sexual maturity and the presence of sexually mature females and their pheromones.

The effects of social context on brain and behavior can occur in three or more ways: via steroid receptors, steroid receptor–independent actions, or both. In the brown ghost knife fish (*Apteronotus leptorhynchus*), communication between males includes weak discharges from the electric organ, "chirps" that signal aggression. Male–male social interactions cause a rise in blood levels of the stress hormone cortisol, which correlates with higher rates of electric discharges during bouts of aggression. Long-term male–male interaction and electrical discharge lead to the growth of new neurons (neurogenesis). Treatment with an antagonist to the glucocorticoid receptor (GR) indicated that glucocorticoid binding to its receptor is necessary for male-typical aggressive electric organ discharge, but the presence of a conspecific male can partially increase neurogenesis even if the GR is blocked. Social context thus has two parallel modes of action on neurogenesis: one is GR dependent and the other is GR independent (Dunlap et al., 2011).

Many different species have the natural ability to change their sex, but this occurs only when increases in certain hormones coincide with specific environmental contexts. In some species of fish, such as the bluebanded goby (*Lythrypnus dalli*), social groups include multiple subordinate females and one dominant male. Among the many females in the group, some are dominant over each other and the dominant females increase reproductive success by laying more eggs in the male's nest. The male builds the nest and regulates which females are allowed to lay eggs in the nest for spawning. Dominant males have high levels of the androgen 11-keto-testosterone, but only when there are eggs in the nest. During this time, the male is aggressive with any fish that attempts to eat the eggs, and this aggression requires high levels of 11-keto-testosterone.

In contrast to the egg-guarding behavior of males, females tussle with each other for access to the nest and attempt to cannibalize the eggs laid by other females. This all changes when the male dies or is removed. Within 24 h of the male's removal, the dominant female of the group becomes a male, including a change in color, external genitalia, aggressive behavior, transformation from an egg consumer to an egg protector, and a sharp increase in levels of 11-keto-testosterone in the brain. The most obvious change in the newly transformed male is its high level of male-typical aggressive behavior. In these newly dominant egg protectors, 11-keto-testosterone levels are elevated compared with those of subordinate, egg-eating females. Meanwhile, the subordinate females continue to behave as females. It might be predicted that treatment of a subordinate female with 11-keto-testosterone would transform her into a dominant male, but this is not completely true. Injections of 11-keto-testosterone transform a subordinate female into a male with respect to physical morphology, but in the continued presence of a dominant female, her behavior resembles that of a subordinate. Thus, context prevails over levels of 11-keto-testosterone. The importance of species survival to the newly formed increase in 11-keto-testosterone in a formerly dominant female is that it increases the level of egg guarding versus egg cannibalizing.

The male's reproductive fitness depends on guarding his eggs from multiple females from egg predators, including fish of other species and other gobies. In support of this idea, treatments that decrease the male's synthesis of 11-keto-testosterone prevent him from guarding eggs, with no effect on his other behaviors. When a male is pharmacologically prevented from egg guarding, the dominant female steps up appropriately, i.e., she changes her behavior to protect the nest. When the experimentally manipulated dominant male slacks on his parental behavior, the dominant female changes from a potential egg consumer to an egg protector. This is another example of the importance of context and reciprocal hormone–behavior interactions (reviewed by Pradhan et al., 2015). Additional complex, reciprocal hormone–behavior interactions and their neuroendocrine mechanisms are discussed in Chapter 7.

### Interim Summary 21.1

1. The efficacy of testosterone's influence on male-typical behaviors in round gobies can be influenced by the level of maturity and the presence or absence of pheromonal cues from females.
2. In weakly electric fish, social context affects male–male aggression and neurogenesis via both GR-dependent and GR-independent mechanisms.
3. In sex-changing fish such as the bluebanded goby, behavioral sex change requires an increase in 11-keto-testosterone. The increase in hormone secretion is necessary and sufficient for morphological changes but it is insufficient for behavioral sex change in the presence of a mature male. Paternal behavior (egg guarding) in this fish depends on the levels of 11-keto-testosterone, but this steroid is effective only in the appropriate context (the absence of another individual that is guarding the nest).

## Hormones and Context in Mammals

In laboratory rats, the male's haste and eagerness for mating depend on many external factors, one of which is familiarity with the environment. If an attractive, sexually receptive female is placed in the male's home cage, mating happens quickly. If both animals are put in a strange environment, the male will spend a great deal of time exploring the new sights and smells before attempting to mount the female. Obviously, an animal upset by novel surroundings will be too aroused for optimal mating behavior, whereas an animal that receives no sexual stimulation will be insufficiently aroused. Thus, reproductive motivation and behavior require an optimal or "Goldilocks" level of arousal. The level of arousal cannot be too low or too high; rather, it must be "just right," as illustrated by the Yerkes–Dodson law (Fig. 21.2).

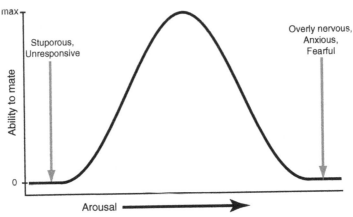

FIGURE 21.2    There is an optimal level of arousal for mating, as there is for a variety of other emotional and cognitive behaviors. Many causes of changes in arousal are environmental or contextual. Animals, including people, might be not aroused enough (e.g., stuporous) or might be too aroused to the point of paralyzing fear and anxiety. This figure represents the classical Yerkes–Dodson law applied to natural behaviors. *max*, maximum.

The Yerkes-Dodson law was first proposed in 1908 to explain the relation between arousal and performance, specifically with regard to performance in intellectual problem solving. The more arousal, the better the performance, up to a point. Too much arousal (stress) is detrimental to problem solving, learning, and memory. A similar inverted-U-shaped curve represents the relation between performance on learning tasks and circulating glucocorticoid and glucose concentrations.

How would generalized arousal affect sexual behavior? Histamine is a neurotransmitter that drives arousal of brain and behavior, and histamine interacts with estradiol. The highest levels of electrical activity in hypothalamic neurons that govern sex behavior in female rodents are achieved by increased histamine in the presence of estrogenic hormones. Conversely, an anesthetic suppresses the estrogen effect, even as it suppresses sex behavior. The combination of sex hormone and arousal-related neurotransmitter appears to turn on both hypothalamic neuron electrical activity and female sex behavior.

The effects of stress hormones come into play when arousal levels become extreme and induce the hypothalamic–pituitary–adrenal system. Corticotropin-releasing hormone (CRH) and glucocorticoids interact with estrogens. In unstressed, ovariectomized female laboratory rats, treatment with estradiol increases locomotor behavior. In contrast, under conditions associated with fear or anxiety, estradiol has the opposite effect. Estradiol and progesterone facilitate female sexual motivation and behavior, but these hormones are less effective in animals that have recently undergone physical or psychological stressors. Rats experience restraint as stressful: After restraint stress, rats paired with opposite-sex conspecifics show increased levels of stress hormones, elevated signs of anticipatory anxiety, inhibited sex behavior, and greater levels of cellular activation in the amygdala, a brain region associated with fear and anxiety.

In ovariectomized hamsters, female consummatory sex behavior is induced by treatment with estradiol followed by progesterone, but food deprivation or treatment with high levels of the hypothalamic stress hormone CRH or the related hormone urocortin inhibits sexual behavior in estradiol- and progesterone-treated females. Conversely, treatments that block CRH or urocortin binding to the brain CRH receptors facilitate female sex behavior, even in energetically stressed females (Jones et al., 2002). These stress–ovarian hormone interactions might have adaptive advantages in the wild. In the absence of danger from predators, levels of estradiol increase arousal to facilitate the locomotion involved in mate searching, courtship, and mating. In the presence of predators, famine, or other dangers, however, arousal will be elevated, and increases in estradiol tend to inhibit sexual motivation, sexual exploration, and locomotion, preventing emergence from the safety of the home or burrow.

Likewise, prior mating experience influences the intensity of sex behavior in many species. In the laboratory, highly experienced stud male rats (males used as breeders to grow a colony) have normal circulating levels of testosterone and mate rapidly and vigorously. Virgin males at any age are more tentative, even if their testosterone levels are the same as those of the stud male. This might be because an initial encounter with a female is perceived as a threat, and only after experience is the encounter perceived as desirable. Effects of experience are not restricted to males. In female rats, sex partner preferences are influenced by previous sexual experience. Furthermore, partner preference manifests differently in different contexts. In the traditional laboratory context, i.e., one male and one female housed together in an arena from which they cannot escape, sexually experienced rats appear to be promiscuous, mating with almost any novel, sexually receptive opposite-sex conspecific introduced into the arena. In nontraditional experiments, mating is observed in two groups of females that receive different types of sexual experience. One group of females, the "paced" group, receives a series of copulatory experiences that are highly rewarding to the female. These copulatory experiences occur in a special chamber that allows the female to escape and return to the male alternately at her own preferred interval. This is called "paced" mating because it allows the female to increase or decrease the time interval between intromissions (insertions of penis into vagina); i.e., pacing allows her to receive penile penetrations at a rate of her own choosing. The other experimental group, the unpaced group, receives the same number of copulatory experiences, except these experiences are not rewarding; i.e., the female cannot escape from the male or pace the timing of intromissions. After this initial phase, in subsequent tests, both groups of females are tested for their preference for either their partner or a strange male.

Females that receive paced matings develop a significant preference for their partner, whereas females that receive unpaced matings fail to develop a preference (they approach and show receptivity to the partner and stranger with equal frequency). Furthermore, females that develop a preference for a particular male actively block their partner's attempts to copulate with competitive females. This is interesting because "mate guarding" behavior is a hallmark of a monogamous species, which suggests that the traditional characterization of laboratory rats as promiscuous (no partner preference) has been premature. Like other monogamous species, rats display mate guarding, but only in the rewarding context created by paced mating. An important lesson comes from this: Only when experimenters test a behavior in an appropriate context can they understand complex social behaviors (Holley et al., 2014).

The hormonal mechanisms involved in prairie vole monogamy are described in more detail in other chapters. In some ways monogamy in primates, including humans, is similar to that in voles, but in other ways it is different. Monogamy in marmoset monkeys, like in voles, involves affiliative behaviors performed in close proximity and a strong preference for mating with only one partner. However, in marmosets, monogamy is more flexible and depends highly on context. After a particular male–female couple have mated, they tend to stay together as they develop an ever-stronger preference for mating with the partner rather than a strange female. The preference is clear when both the partner and stranger are in proximity. In this context, under the watchful gaze of the female partner, the male will show affiliative and courtship behaviors toward his mate and retreat from a stranger. Unlike monogamous prairie voles, monogamous marmosets do not show strong fidelity in the absence of their partner. Minutes after the familiar marmoset partner is out of site, monogamy goes out the window and sexual overtures will be made to the unfamiliar female. Philandering in a marmoset colony can be increased by placing a rock or tree into the cage to allow couples to hide.

A similar flexible pattern is found in many human societies. Many long-term human couples show a willingness to form monogamous pair bonds, and in experiments designed to examine their monogamy, they behave differently when their partner is absent than when their partner is present. When their partner is present, most individuals will avoid social interactions with a novel and highly attractive, opposite-sex adult. They tend to position themselves at a safe distance from the new stranger. When their long-term partner is absent, however, the same individuals show a shorter latency to initiate interactions with a novel, highly attractive opposite-sex adult. Anthropologists note that an inflexible monogamous mating system limits reproductive success because no births occur after the death, exit, or infertility of a long-term partner. Flexible monogamy allows for increased reproductive success if individuals form new pair bonds after the death or exit of a former partner.

Monogamy in marmosets, as in voles, is influenced by oxytocin. Blocking oxytocin binding to its receptor in both male and female marmosets reduces the level of affiliative behavior, including the degree of social proximity and the amount of food sharing with a new mate. Treatment with oxytocin facilitates fidelity with a long-term partner by decreasing time spent in proximity to a stranger. For example, the longer the mating partner is absent, the higher the incidence is of sexual infidelity, and treatment of the long-term partner with oxytocin increases the latency to exhibit sexual solicitation behavior toward a stranger. That is, oxytocin promotes affiliation with a partner and delays affiliation toward a stranger. Furthermore, the effect of exogenous oxytocin on fidelity is greater in long-term partners compared with newly formed partners (Mustoe et al., 2015). Oxytocin-induced effects on affiliative behavior differ, depending on whether the partner is a long-term partner, a new partner, or a stranger.

Hormones associated with reproduction have behavioral effects far beyond sex. For example, estrogens influence the growth and architecture of dendrites in the hippocampus, a brain region associated with memory (Woolley and Cohen, 2002). Of importance, the effects of estrogen on hippocampal morphology depend not only on glutamatergic neurotransmission but also on inputs from basal forebrain cholinergic neurons (Fig. 21.3). These inputs reveal information about environmental context.

A more complicated example of context effects on mammalian behavior is offered by the report of Williamson et al. (2017). In a most interesting manner, the researchers analyzed the dependence of the relations among the sex steroid testosterone, the stress hormone corticosterone, and social status on the social environment of mice. For social hierarchies with despotic alpha males who beat up on subordinate males, testosterone levels in the latter were low. However, in other social environments where subordinate males had lots of "fair fight" competitive interactions, their testosterone levels were higher. Also, in the highly despotic hierarchies, subordinate males had high levels of corticosterone compared with alpha males, but in pair-housed mice the opposite was true. A positive correlation between testosterone and dominance in social hierarchies has been previously shown, but the inclusion of corticosterone and the social context of the dominant–subordinate interaction as additional variables offer a comprehensive view of the determinants of hormone–behavior relations in hierarchical social situations.

Other important contexts for nonreproductive behaviors arise from aspects of the social environment, e.g., the number of males and females in groups of juvenile spotted hyenas. In this species, adult females dominate

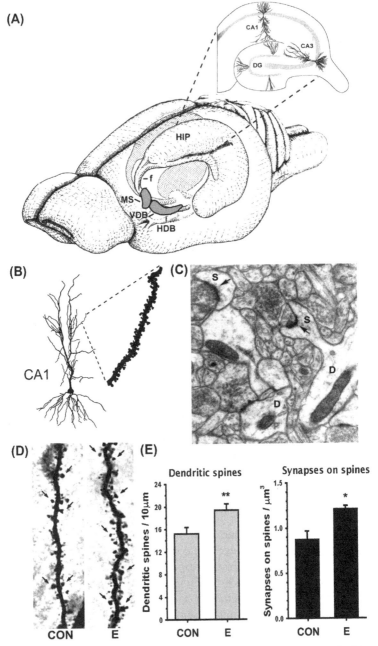

FIGURE 21.3    Hormonal state regulates excitatory synaptic connections in the hippocampus (HIP) of adult female rats. (A) The hippocampus in the rat is a large, curved structure that is oriented septotemporally just beneath the corpus callosum. In cross-section, three types of principal cells are apparent: pyramidal cells of the CA1 and CA3 regions and granule cells

males, and the sex difference in aggressive behavior first appears in juvenile play behavior. For example, when females are observed in all-female groups, aggressive play interactions are of a higher frequency than are those observed in all-male groups. This difference is apparent only when tested in same-sex groups. When females are housed together with the males, female aggressive play interactions decrease whereas male aggressive interactions increase. As a result, the sex difference disappears.

Analyses of genetic influences on behavior revealed clear gene–hormone–behavior interactions that illustrate the effects of context (Figs. 21.3 and 21.4). The increased aggressiveness of female mice with a knockout of the gene that encodes oxytocin depends on the level of stress in the environment; a stressful environment brings out the full oxytocin phenotype (see Chapters 1, 2, and 7). Likewise, the effect of the gene for an estrogen receptor (ERβ) on aggression is a function of the situation in which aggression is being assayed. Indeed, hormone–gene interactions depend exquisitely on details of the experimental protocol. In female mice, testosterone-fueled fighting is reduced by the loss of a functional ERβ gene, but maternal aggression (e.g., defending the nest) is increased (Ogawa et al., 2004) (Fig. 21.4).

Even as there are environmental influences on gene–hormone–behavior relations, strong environmental dependencies are evident in the control of hormone-dependent genes. Consider the conditions that control oxytocin gene expression. As noted previously, physiological conditions include development, puberty, plasmid osmolality, and circadian rhythms; therefore, the developmental context, the amount of water and

---

of the dentate gyrus (DG). Subcortical brain regions, such as the medial septum (MS) and the vertical and horizontal limbs of the diagonal band of Broca (VDB and HDB, respectively) in the basal forebrain, project to the hippocampus via the fornix (f) and the fimbria, which extends as a sheet of fibers along the surface of the hippocampus. (B) The dendrites of each principal cell type in the hippocampus are densely studded with small protrusions termed dendritic spines. A CA1 pyramidal cell is shown with its dendrites covered by spines. Each CA1 pyramidal cell may have as many as 20,000 dendritic spines covering its dendrites. (C) Dendritic spines are the sites of the vast majority of excitatory synaptic input to CA1 pyramidal cells. An electron micrograph is shown in which the *arrows* indicate asymmetric (excitatory) synapses on dendritic spines (S) that arise from the shafts of dendrites (D). In adult female rats, removal of endogenous ovarian hormones by ovariectomy decreases the density of dendritic spines and synapses formed on spines on CA1 pyramidal cells; these decreases can be either prevented or reversed by systemic treatment with estradiol (E) (Woolley and McEwen, 1992, 1993). The density of spines and synapses also fluctuates naturally as estradiol and progesterone levels fluctuate across the estrus cycle; spine/synapse density is greatest during proestrus, when estradiol and progesterone levels are high, and lowest during estrus, when estradiol and progesterone levels are low (Woolley et al., 1990). (D) Light micrographs of CA1 pyramidal cell dendrites from a control (CON) rat that was ovariectomized and treated with oil vehicle and from an ovariectomized animal that was treated with systemic estradiol (E). Note that the density of dendritic spines is greater in the E-treated animal (E).

| | Wildtype Control Females | β-Estrogen Receptor Gene Knocked out |
|---|---|---|
| Postpartum Maternal Aggression | Normal | **Increased** |
| Testosterone-induced Aggression | Normal | **Reduced** |

FIGURE 21.4   The effect of a gene disruption on a class of natural behaviors can depend on the context in which those behaviors are assayed. Here, knocking out the estrogen receptor-β gene increases one form of aggression but reduces another form.

salt in the environment, and the time of day all can influence gene expression for this neuropeptide, which is behaviorally important. Likewise, for expression of the opioid peptide gene encoding enkephalin, gender and hormone state interact with stress level (depending on the environment, of course) to determine the amount of enkephalin transcript in hypothalamic neurons.

## Seasons of the Year

Endocrine effects on brain, body, and behavior are influenced by the time of year. Important cues are generated by features in the environment that change with the seasons, including day length, ambient temperature, rainfall, and food availability. Understanding the importance of these cues requires an understanding of how animals are affected by seasonal changes. For example, in some seasons there is less snow cover, more unfrozen water, more edible vegetation, and a higher availability of prey animals. Reproductive success is higher when offspring births coincide with greater availability of food and water, warm ambient temperatures, and low levels of snow cover. Reproduction is energetically expensive, and reproductive success is higher when energy availability is high and other energetic demands are low.

To ensure that births coincide with favorable energetic conditions, many species use cues from the environment to initiate territorial defense, courtship, and mating in anticipation of energetically favorable conditions. In addition to reproductive processes, many other physiological, morphological, and behavioral processes are synchronized with seasonal changes in the environment. For example, in a wide variety of species, molting, foraging, body weight, body fat accumulation, hibernation, and migration are all synchronized with the seasons to maximize chances of survival. Many of these behaviors also are indirectly related to reproductive success.

As discussed in Chapter 15, there are many seasonal rhythms that differ from one another, and they can be characterized by component properties such as their amplitude and period. The period is the time it takes for the rhythm to complete one cycle. The frequency of the rhythm is the inverse of period, i.e., the number of complete cycles per unit of time. Chapter 15 is concerned with circadian rhythms with a period of about 24 h. Many seasonally breeding animals have rhythms that occur with a frequency of once per year (i.e., these annual rhythms have a period of about 12 months). There are three general types of annual rhythms: (1) those generated by external cues, (2) those generated by an endogenous annual oscillator, and (3) those generated by an interval timer set each year by exogenous seasonal cues such as day length. The simplest of these rhythms are generated by cues that occur annually in the environment and are sensed by the individual. These externally generated rhythms do not persist if and when environmental cues are removed. For example, many of us experience allergic symptoms each year in the spring. The symptoms (sneezing, watery eyes, and runny nose) are triggered by pollen that is produced by spring flowers, and we can escape the annual sneeze-fest by removing the pollen-producing plants or moving away from the geographic area where the pollen is produced. The rhythm disappears when the external stimulus disappears.

Another type of rhythm is the endogenous circannual rhythm, which is produced by an intrinsic cellular oscillator with a period of just under 12 months. We know that internal circannual oscillators exist because they persist under constant conditions. For example, when rhythms are observed for several consecutive years under constant conditions, such as unchanging day length and ambient temperature, the periods of these rhythms persist; i.e., they *free run* with roughly the same period, usually just under 12 months. An example is the circannual rhythm of reproduction in starlings (*Sturnus vulgaris*) kept on constant photoperiods (e.g., 12 h of light followed by 12 h of dark each day). These birds exhibit repeated circannual cycles in gonadal maturation and regression. In addition, starlings show circannual rhythms in other phenotypic traits, including body mass and feather molt. Furthermore, migratory birds show circannual rhythms in a quantifiable behavior known as *zugunruhe*, i.e., migratory restlessness. Circannual rhythms also occur in mammals. A well-studied mammalian circannual clock controls the yearly rhythm in body weight, hibernation, and reproduction in golden-mantled ground squirrels, *Spermophilus lateralis*. The annual rhythm is self-generated by an internal oscillator analogous to a clock. The internal clock is typically shorter than 12 months (e.g., 11.5 months) and is entrained by day length.

Whether mechanical or biological, clocks keep time automatically; they have their own endogenous periods and require nothing from the outside to start the cycle each period. In contrast, timers must be initiated each year by an external cue (think of an egg timer or stopwatch). Some annual biological rhythms are controlled by intrinsic timers instead of endogenous clocks. In animals with circannual interval timers, the timer is initiated each year by a change in day length. For example, in reproductively active Syrian hamsters (*Mesocricetus auratus*) exposed to long summer day lengths, suppression of the hypothalamic–pituitary–gonadal (HPG) system is initiated by a decrease in day length, and release of the HPG system from inhibition occurs at a predetermined interval after the onset of reproductive suppression. Decreasing day lengths initiate gonadal regression in the late summer and initiate the interval timer. When a species-specific duration of time has passed (e.g., 20 weeks), the animal spontaneously becomes refractory to short days and the HPG function returns. *Refractory* means that the individual is no longer sensitive to the reproductive suppression of short days. The return of HPG function is termed spontaneous *gonadal recrudescence*. Exposure to long day lengths in the spring and summer breaks refractoriness and resensitizes the neuroendocrine axis to the short-day signal that initiates gonadal regression and the interval timer each year (reviewed by Prendergast et al., 2002).

An important hormonal mechanism controls annual reproductive cycles in animals such as Syrian and Siberian (*Phodopus sungorus*) hamsters. Experiments with these species show that annual cycles in reproduction and body weight are linked to circadian oscillators that control pineal melatonin secretion. A brief description of the system is as follows: In a hamster housed in long days (e.g., 14 h of light followed by 10 h of dark), the external cue that initiates winter responses is decreasing day length detected by the eye. When a day length of less than 11.5 h of light is detected, a neural signal is sent to the circadian oscillator in the suprachiasmatic nucleus. Next, the signal is transmitted through a circuitous route to the pineal gland. The pineal gland secretes melatonin, a monoamine derived from serotonin, during the dark phase of the photoperiod (the nighttime). In all mammals studied thus far, whether nocturnal or diurnal, melatonin secretion is suppressed by light and increases during the dark phase of the 24-h photoperiod. The longer the dark period (night), the longer the duration is of the nighttime melatonin pulse. In photoperiodic rodents such as Siberian hamsters housed in constant long days or constant light, the reproductive consequences of short day lengths can be mimicked by infusions of melatonin that created the short-day melatonin pulse. Thus, long-duration melatonin pulses create the context for suppressed reproductive hormone secretion and other winter adaptations. Long-duration melatonin pulses increase the hypothalamic sensitivity to negative feedback from the gonadal steroids, thus inhibiting the HPG system.

Long-duration melatonin pulses also alter orexigenic ("feeding") neuropeptide secretion and sensitivity, leading to changes in food intake and body weight. As well, alterations in body weight and adiposity in response to winter-like short days are orchestrated by multiple peripheral changes such as changes in sympathetic tone and the innervation of adipose tissue. During the long days of summer, Siberian hamsters are obese (i.e.,

50% of their body weight is made of fat); in the short days of winter only 20% of their body weight is made of fat. These marked changes in lipid mass are mimicked in the laboratory by decreasing day length while holding all other environmental factors constant. Short-day body weight loss is achieved by increases in lipolysis (fat burning). Melatonin and many other hormones (thyroid hormones, insulin, leptin, and epinephrine) have been ruled out as mediators of these winter increases in lipolysis. Rather, the effects of short-day (long-duration) melatonin pulses are mediated by increased secretion of norepinephrine by the sympathetic nervous system (reviewed by Bartness et al., 2014). These experiments show how seasonal context alters hormone secretion (e.g., melatonin secretion), which affects both reproductive hormones (HPG hormones) and nonreproductive hormones (norepinephrine and so-called "feeding" neuropeptides).

Seasonality, via day length, has significant effects on behavior. For example, in female meadow voles, preference for the odors of other conspecifics is different in the spring breeding season compared with the winter. During the spring and summer, female voles prefer odors from males over those of females. During the winter, they switch their preference. When brought into the laboratory and placed on long-day photoperiods typical of summer, female voles prefer male vole odors. In contrast, during short-day photoperiods (only 10h of light per day), they prefer the odors of females. In the winter, when the day length is short, survival is facilitated by huddling behavior in same-sex female pairs. Unlike intact long day–housed voles, intact female voles housed in short-day photoperiods prefer the odors of other females and live in close proximity to other females with whom they huddle for warmth. Estradiol treatment of short day–housed voles significantly reduces time spent huddling with female cage mates. In contrast to short day–housed females, in long day–housed females, neither ovariectomy nor estradiol treatment significantly alters same-sex affiliation, which suggests that low circulating concentrations of estradiol represent a necessary but not sufficient condition for same-sex affiliation: Low levels of estradiol must be accompanied by long days. Thus, ovarian hormones interact with day length to affect social behavior.

## Interim Summary 21.2

1. The relation between arousal and performance is described by an inverted U-shaped function, i.e., too much or too little arousal inhibits performance. This relation holds for stress hormones and reproduction.

2. Increases in central oxytocin secretion and oxytocin binding to its receptors tend to increase affiliative behaviors, but the effects differ with the phase of pair-bonding. In a long-term pair-bonded individual, oxytocin can increase affiliation with the long-term partner and also can delay affiliation with a stranger.

3. Seasonal cues provide the context for hormone action. Cues such as day length alter the duration of melatonin secretion, which has multiple physiological effects including altered secretion of other hormones and sensitivity of hormone receptors.

## CLINICAL EXAMPLES

The relation between hormones and mood in human subjects highly depend on context. For example, premenstrual dysphoric disorder (PMDD) is an affective disorder characterized by the appearance of negative mood symptoms (irritability, sadness, depression, and mood swings) in a menstrual cycle phase-specific fashion. Symptoms appear during the luteal phase and disappear at or soon after the onset of menstruation. Consequently, abnormalities of ovarian steroid (estradiol and progesterone) secretion have been sought as an explanation for PMDD (Fig. 21.5). Under some conditions, estrogens can increase responses to stress in postmenopausal women. This implies a context dependency of the hormone effect: Estrogens can heighten arousal whereas progesterone can dampen it. In high concentrations, progesterone can be analgesic or even anesthetic. Under neutral or positive conditions, high levels of estradiol in the late follicular phase may heighten positive affect; under stressful conditions, the same estrogen treatment may worsen mood and lead to agitation or even mania. During the menstrual cycle, when these hormones fluctuate, agitation in the follicular phase may be followed by depression in the luteal phase, when levels of progesterone are high. These hormones show even more marked fluctuations over the course of pregnancy and lactation.

An analysis of human gene transcription in cultured cell lines revealed a large gene complex overexpressed in women with menstrually related mood disorder, the Extra Sex Combs/Enhancer of Zeste. This complex regulates gene transcription in response to sex hormones and environmental stressors. Expression of some genes is increased by progesterone whereas the expression of some is decreased by estradiol (Dubey et al., 2017). This work has promise for understanding the inherited context of PMDD and how it interacts with ovarian steroid secretion during different phases of reproductive life. Thus it could be important for the development of treatments for PMDD and related conditions. It is another demonstration of the altered effects of hormones in different genetic and environmental contexts.

In a more general social context, sex hormones, stress hormones, and serotonin interact with social factors such as socioeconomic status, school size, and social environment to affect agonistic interactions, including aggressive behavior in teenage boys (see also Chapter 12). Large discrepancies in socioeconomic status (which increase the chance of personal

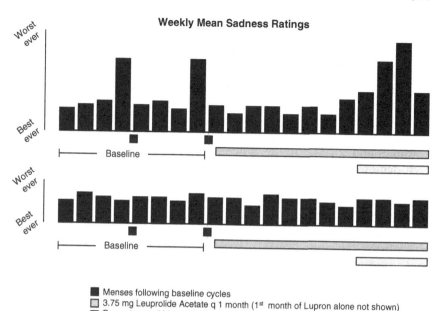

**Weekly Mean Sadness Ratings**

■ Menses following baseline cycles
☐ 3.75 mg Leuprolide Acetate q 1 month (1st month of Lupron alone not shown)
☐ Progesterone 200mg bid by suppository

**FIGURE 21.5** Ten women with premenstrual dysphoric disorder (PMDD) (top row) (see baseline cycles) and 15 controls (bottom row) had minimal mood and behavioral symptoms during administration of leuprolide, a long-acting, gonadotropin-releasing hormone agonist. Women with PMDD, but not the control women, had a significant increase in sadness during either estradiol or progesterone (P) administration. Histograms represent means of seven daily scores on the Daily Rating Form Sadness scale for each of the 8 baseline weeks, 8 weeks preceding hormone replacement (leuprolide alone), and 4 weeks of leuprolide plus P replacement. The role of ovarian steroids in PMDD was evaluated by first suppressing ovarian function (predicted to prevent symptoms) and then adding back physiologic levels of P or estradiol (predicted to precipitate symptoms). In the top row of the figure, symptoms can be seen to appear during the premenstrual week in the two baseline cycles, to disappear (both symptom severity and cyclicity) during the second and third months of ovarian suppression (achieved with the gonadotropin-releasing hormone agonist leuprolide acetate; first month of leuprolide alone not shown), and to be precipitated during replacement with either P or estradiol. However, identical hormone manipulations have no effect on mood in a group of comparison women with no history of PMDD (bottom row); that is, women with PMDD have a differential response to gonadal steroids such that levels or changes in hormones that have little or no impact on mood in normal women are capable of destabilizing mood in women with PMDD. Therefore, the effects of the stimulus are entirely contextual (presence or absence of a history of PMDD) with the physiologic underpinnings of the differential response being the focus of current investigations. *bid*, twice a day; *q*, once a day. *From Schmidt, P.J., Nieman, L.K., Danaceau, M.A., Adams, L.F., Rubinow, D.R., 1998. Differential behavioral effects of gonadal steroids in women with and in those without premenstrual syndrome. N. Engl. J. Med. 338, 209–216. With permission.*

humiliation of an adolescent boy), schools with large numbers of students (which increase the sense of anonymity), and an absence of formal initiation rites (which fail to provide a positive view of a teenage boy's role in adult society) all increase the chance of violent behavior. These studies suggest that future research should focus on interactions among olfactory, visual, auditory, and tactile cues that communicate information about the surrounding population, safety in the environment, and social bonds, and how these cues are interpreted in individuals with different brain levels of neuropeptides such as vasopressin and oxytocin, serotonin, and the gonadal steroids, especially androgens.

We are all aware that stress affects our behavior. Although it is difficult to prove quantitatively, some stress is useful in that it drives us to meet deadlines and succeed in life, whereas high stress can be unhealthy and even debilitating. The concept of stress encompasses a variety of forms but does not always differentiate the effects. The different effects of stress are important, and we cannot assume that the mechanisms that produce physiological responses are the same as those that produce our experience of stress. Hemorrhage, for example, is a stress that elicits release of adrenocorticotropic hormone (ACTH), corticosteroids, and other hormones. We tend to assume that these hormonal changes produce specific emotions. Immobilization stress activates the release of ACTH and corticosteroids, but so does forced exercise (i.e., exercise that we do even when we do not feel like it). The dogma suggests that our autonomic system responds to stress by preparing us for what has been traditionally termed "fight or flight." The accumulated data, however, suggest that a variety of stressors first cause us to halt and assess the situation. The emotions we experience at that moment depend on many contextual factors. Most likely, there is an area of unconscious processing happening at that moment. Eventually a decision might be made to either fight or flee, but most often the response is to do nothing, followed by worry, anxiety, and fear. Under a barrage of catecholamine release from autonomic nerves of the sympathetic nervous system, heart rate increases, blood pressure rises, bronchi of the lungs open for maximum ventilation, and blood flows more to the brain and muscles and away from the gut and skin, but these same responses occur when you fall precipitously in love! In addition, for some humans, love, bonding, and sexual attraction are heightened in dangerous and/or painful situations. In other humans, the opposite is true.

These and other considerations suggest that the mechanisms that underlie fear and anxiety are not the same as those that cause the innate stress response. The stress responses, the neuroendocrine actions of epinephrine and the HPA system, react to any extra requirement for oxygen and metabolic fuel. When we need to survive a truly dangerous threat, the association between fear and the stress response is matched. In many other contexts, however, the stress response is activated but fear and anxiety are not necessarily present or appropriate. In humans, emotions such as fear and anxiety depend on brain areas involved with learning, memory, language, past and present experience, and the elusive concept of consciousness. Many of these are absent in other vertebrate organisms, but the neuroendocrine stress response still occurs in their absence. When we watch a mouse in an open field, its behavior may or may not be accompanied by fear and anxiety regardless of our assumptions. It is critical to understand that the neural substrates for fear and anxiety are different from those of the neuroendocrine stress response, although there is some overlap (LeDoux, 2015). Conversely, the physiological impact of a particular neuroendocrine stress response and the individual's experience of the neuroendocrine stress response can vary considerably, depending on the species and many contextual factors, including the level of consciousness, intellectual ability, past experience, development, metabolic state, surrounding conspecifics and predators, ambient temperature, and season.

# SOME OUTSTANDING NEW CLINICAL OR BASIC QUESTIONS

1. What neural and molecular steps are involved when an environmental manipulation alters hormone effects on gene expression in a hypothalamic neuron?
2. Why do hormone effects in humans apparently depend more on social context compared with lower mammals?
3. Hypertension is a word that conjures up the notion of stress, but what does that really mean? Can lasting hypertension truly be induced by stress?

# Behavioral–Environmental Context Alters Hormone Release

Almost every hormone–behavior relation is a two-way street. In Chapter 7, we illustrated how reciprocal hormone–behavior relations can underlie the interactions between two organisms and create long, multi-step sequences in behavior. In the current chapter, we turn our attention to specific effects of the environment on hormone secretion.

Some clear examples are behaviors initiated by environmental stressors. A *stressor* is an external stimulus, including chemical and biological agents, environmental conditions (e.g., cold temperatures), thoughts, memories, or psychological triggers that increase the physiological stress response. Examples of stressors employed in the laboratory to study stress responses include high-volume sounds, restraint, overcrowding, and food restriction. Real-life stressors with important health consequences in humans include physical and mental abuse, physical injury, blood loss, malnutrition, and prolonged high levels of physical activity, such as those that occur in athletic or military training. As explained in the introduction to this book, the physiological stress response includes the release of epinephrine (adrenaline) by the autonomic (sympathetic) nervous system and activation of the hypothalamic–pituitary–adrenal (HPA) system. The HPA system includes corticotropin-releasing hormone (CRH) secreted from the cells in the hypothalamus into the pituitary portal blood circulation, which reaches the anterior pituitary to stimulate the secretion of adrenocorticotropic hormone (ACTH). ACTH is secreted into the general circulation, reaches the adrenal cortex, and stimulates the secretion of glucocorticoids, including cortisol. In the short term, the hormone action of the stress response increases the chances of overcoming threats to survival, but in the long term, a chronically elevated stress response can lead to mental and physical illness.

The stress response includes almost all other neuroendocrine systems, such as the hypothalamic–pituitary–gonadal (HPG) system, the hypothalamic–pituitary–thyroid (HPT) system, and the posterior pituitary hormones vasopressin and oxytocin. Environmental stressors therefore alter the secretion and action of a wide array of chemical messengers that affect behavioral perceptions and reactions. We begin by discussing how environmental stressors alter the HPG system, and then dig deeper into physiological and psychological aspects of the stress response.

## BASIC EXPERIMENTAL EXAMPLES

Reproductive processes are often inhibited under environmental conditions that are not conducive to the successful rearing of offspring. Offspring survival requires safety from stressors, including predation, extreme temperatures, and famine. Under chronically stressful conditions, reproductive efforts are often inhibited before mating, preventing conception and pregnancy. These stressful environmental stimuli inhibit the HPG system, which controls ovulation and sex behavior. Some of these inhibitory effects on reproduction occur when aversive stimuli are sensed and an inhibitory signal is relayed to the HPG system. A clear example is the need for an adequate food supply to allow for normal reproduction in adult males and females. A low food supply can delay pubertal development in young animals and inhibit reproductive behavior and fertility in adults.

Energy availability is the most important environmental factor that controls mammalian reproduction; examples can be found in every mammalian order (Bronson, 1989). Energy for cellular life and organismal activity comes from the oxidation of metabolic fuels such as glucose and free fatty acids. A sufficient pool of metabolic fuels is required for adult maintenance of the reproductive system, as well as for pubertal development of the reproductive system. Most organisms get their energy by eating food, and if food is not available, they derive fuel from molecules stored as lipids in adipose tissue. This is true in all mammalian species studied, with very few exceptions. A sufficient level of metabolic energy is particularly important for female mammals because after the birth of offspring, reproductive success depends on the energetically costly process of lactation. Thus, in many mammalian species, reproduction is sensitive to the availability of metabolic fuels.

Ovulatory cycles are inhibited by many environmental stressors that decrease the availability of fuels, including food restriction or deprivation, living in cold ambient temperatures, and excessive exercise (reviewed by Bronson, 1989; Wade and Schneider, 1992). Ovulatory cycles are known as estrous cycles in nonprimates and menstrual cycles in primates, because primates show menstrual bleeding whereas most nonprimates do not. In all species with spontaneous ovulatory cycles, a period of follicular development and growth is followed by ovulation, a requirement for fertilization of the egg by the sperm. Follicular development requires pulsatile secretion of gonadotropin-releasing hormone (GnRH), which stimulates the pulsatile secretion of follicle-stimulating hormone (FSH) and luteinizing hormone (LH). In turn, LH and FSH stimulate ovarian steroid secretion. Levels of circulating estradiol gradually increase as the follicle matures. When the follicle is mature and circulating levels of estradiol are high, a surge in GnRH and LH is required for ovulation. After ovulation, the ruptured follicle transforms into the corpus luteum, which secretes high levels of progesterone during the luteal phase of the ovulatory cycle. Thus, ovarian steroids fluctuate over the ovulatory cycle, with low levels of estradiol in the early follicular phase, increasing levels as ovulation approaches, and high levels of progesterone in the luteal phase.

The onset of these ovulatory cycles around the time of puberty and the maintenance of ovulatory cycles are affected by the overall level of nutrition and availability of calories (Foster and Olster, 1985; Foster et al., 1985). Overall calorie availability is a function of the amount of food being eaten, the potential metabolic fuel supply stored as body fat, and the amount of energy being expended. In most organisms, including humans, energy is expended on resting metabolism, growth, immune function, exercise, nonexercise thermogenesis (heat production), and cognition, and in response to stressors in the environment. In girls, a sufficient supply of fuels (derived from food eaten and stored as fat) is critical for the pubertal onset of menstrual cycles (menarche); in women, a sufficient fuel supply is critical for the maintenance of normal menstrual cycles and for sexual motivation and performance.

In girls and women, stressors that influence menarche and the menstrual cycle include trauma, abuse, and peer pressure, especially peer pressure regarding body image. The HPG system of both males and females is inhibited by low food availability, famine, and the eating disorder anorexia *nervosa*, which occurs considerably more frequently in females (Roze et al., 2007; van Noord and Kaaks, 1991). Environmental factors can interact to influence ovulatory cycles: Fuel availability can interact with cold ambient temperatures and excessive exercise to hinder HPG function during pubertal development and adulthood. As well, excessive exercise and the psychological stress of competition can exaggerate the effects of low food intake to delay the onset of puberty (Warren and Perlroth, 2001) (Fig. 22.1).

Energy availability affects all levels of the HPG system in both males and females. The easiest way to monitor metabolic effects on the HPG system is to take frequent blood samples and measure LH.

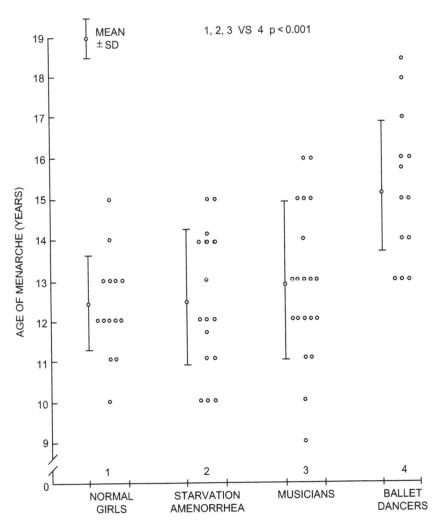

**FIGURE 22.1**   Effect of context on the ability to raise pituitary follicle-stimulating hormone and luteinizing hormone release levels so as to initiate menarche (the onset of menstrual cycles) in young women. "Normal girls" were neither adhering to food-restricting diets nor engaging in high-intensity activity and had regular menstrual cycles. Girls with starvation amenorrhea had disrupted menstrual cycles linked to the eating disorder anorexia nervosa, but were not engaged in high-intensity activity. Musicians were not adhering to a food-restricting diet but were engaged in competitive, high-intensity nonathletic activity. Ballet dancers were engaged in a food-restricting diet plus high-intensity, highly competitive athletic training. The professional and peer pressure to maintain a very low body weight is particularly intense among ballet dancers. In this data set, the delay in menarche occurred only in those with the combined stressors (Column 4) compared with the other groups. Neither the musical environment alone (Column 3) nor food restriction alone delayed menarche (Column 2). *SD*, standard deviation. *From Warren, M.P., Perlroth, N.E., 2001. The effects of intense exercise on the female reproductive system. J. Endocrinol. 170 (1), 3–11. With permission.*

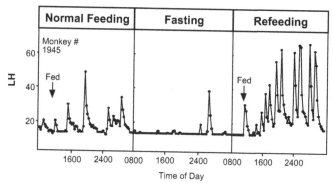

FIGURE 22.2   Changes in the environment can lead to changes in pituitary hormone secretion. Here, a lack of food in the experimental environment (fasting) causes a remarkable decline in pulses of luteinizing hormone (LH) secreted from a male monkey's anterior pituitary gland (Helmreich et al., 1993).

In mammalian species allowed to eat as much food as they want, LH levels fluctuate in regular pulses that occur every 1–2 h, depending on the species (Fig. 22.2). Normally in the well-fed mammal, each pulse of LH is stimulated by a pulse of GnRH. It is difficult to measure the tiny levels of GnRH secreted into the pituitary portal plexus, but it is easier to measure LH in the peripheral blood; thus, most early studies focused on LH pulses. In food-restricted males and females, these pulses decrease in frequency. If restriction continues, LH pulses cease altogether (Fig. 22.2).

If we observe a cessation of LH pulses in the food-deprived animal, how do we know whether the initial inhibition occurred in pituitary or brain? The primary locus of effect of food restriction on the HPG system is thought to be inhibition of the GnRH pulse generator. Early evidence for this came from studies that administered GnRH to rats in which puberty had been delayed by chronic food restriction (Bronson, 1986). In these rats, pulsatile LH secretion, estrous cycles, and ovulation were fully reinstated by administration of GnRH into the pituitary at the species-specific frequency. This demonstrated that in food-restricted animals, the rest of the HPG system is functional and can respond to GnRH. Later, after technology was developed to measure GnRH in the brain and pituitary portal blood, it was confirmed that food restriction inhibits the pulsatile secretion of GnRH.

However, if food restriction is severe and long-lasting, the pituitary becomes less responsive to GnRH and the gonads become less responsive to LH and FSH. This has been shown for a variety of mammalian species. Total food deprivation in lean animals can also block behavioral responsiveness to gonadal steroids. Thus, if ovariectomized, food-deprived female Syrian hamsters are treated with estrus-inducing levels of ovarian steroids, they fail to show mating behavior in response to an adult male of the same species (reviewed by Wade and Jones, 2004). This is thought to be related to a food deprivation–induced decrease in levels of estrogen receptor-α in the ventromedial hypothalamus, one brain area where estradiol facilitates sex behavior. Even levels of food restriction that are insufficient to decrease levels of ovarian steroids can decrease sexual motivation (reviewed by Schneider et al., 2013).

Decreases in the availability of metabolic fuels such as glucose and fatty acids initiate changes in the HPG axis. Starvation is accompanied by low levels of leptin secreted from adipocytes and high levels of ghrelin secreted from the gut. Food restriction–induced inhibition of the HPG system can be reversed by central or peripheral treatment with leptin or antagonism of the receptors for ghrelin (Budak et al., 2006). These peripheral hormones also increase fuel oxidation and energy expenditure, which might explain all or part of their effect on the HPG system (Schneider and Zhou, 1999). Understanding metabolic control of reproduction requires us to understand how deficits in metabolic fuel availability are sensed (Schneider et al., 2012).

Metabolic fuels and peripheral hormones might act directly on GnRH-producing cells, but they might act in other brain areas as well as in the periphery. The signal for energy deficit is then sent by neural projections to the GnRH pulse generator cells. Two important controllers of the GnRH pulse generator are the RFamide peptides, kisspeptin and gonadotropin-inhibiting hormone (GnIH). GnIH was first discovered in quail; the mammalian version of this peptide is also known as RFamide-related peptide-3 (RFRP-3) (reviewed by Clarke and Arbabi, 2016). Kisspeptin acts as a stimulator and GnIH/RFRP-3 acts as an inhibitor of GnRH secretion. Depending on the species, deficits in energy availability decrease the activity of kisspeptin cells and increase the activity of GnIH/RFRP-3 cells (e.g., Schneider et al., 2017).

Seasonal breeding is another example of environmental effects on neuroendocrine systems that control reproduction as well as energy balance. As discussed in other chapters, the inhibition of the reproductive system is controlled by both internal and external cues. In some species, seasonality is orchestrated by an endogenous annual clock influenced by day length, such that the decreasing day length in autumn inhibits the reproductive system. In addition, the endogenous clock is influenced by food availability in some species. Day length, clocks, and food availability all interact to affect kisspeptin and GnIH/RFRP-3, the on and off switches for GnRH (e.g., Paul et al., 2009).

Another important principle of hormone–behavior interactions is that stressors in the environment have multiple effects on hormone secretion (Fig. 22.3). CRH is exquisitely responsive to a variety of environmental stressors. CRH release in the median eminence acts as a major driver

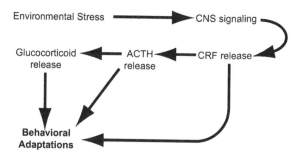

FIGURE 22.3 Classical examples of the effects of environmental context on hormone secretion, which have to do with stress. Illustrated here are three routes that comprise the minimum number by which stress-related hormonal signaling could influence behavior. *ACTH*, adrenocorticotropic hormone; *CNS*, central nervous system; *CRF*, corticotropin-releasing factor.

for ACTH. In turn, ACTH can have behavioral effects on its own (for example, on certain forms of learning and memory) and elevate corticosterone/cortisol levels, thus altering behavior indirectly. The primary role of ACTH is to stimulate the glucocorticoids, the hormones that break down stored metabolic fuels into oxidizable metabolic fuels. In this way, the HPA system interacts with the systems that use and affect metabolic fuels such as the HPG, HPT, and posterior pituitary hormones, oxytocin and vasopressin. The HPG system, for example, is inhibited by chronically elevated glucocorticoids as well as by high levels of CRH. The glucocorticoids, in particular, have a key role in delaying the energetically costly process of reproduction to conserve energy during various energetic emergencies. In many wild species, the reproductive system is inhibited by high levels of glucocorticoids in response to environmental stressors such as intense rain, heavy winds, flooding, tornados, hurricanes, and forest fires. In these species, termination of the breeding season is a natural part of life and is often accompanied by a period of dispersal or migration. These changes in the seasonal activities of animals often have alterations in the levels of circulating glucocorticoids (Angelier and Wingfield, 2013).

In Chapters 1 and 2, we pointed out that hormones act in concert with one another to modulate metabolism and behavior and that they are produced and secreted in an orderly fashion over time in response to stressors. This was demonstrated in the laboratory by restraining animals in enclosed spaces. For example, the HPA response is elicited in monkeys when they are restrained in a chair and trained to avoid a mild electric shock applied to the foot. Plasma and urinary hormones and their metabolites are measured before, during, and after the application of mild shock. Shock avoidance is immediately followed by the secretion of catabolic hormones, including adrenal medullary catechol-amines (epinephrine and norepinephrine) and the adrenal glucocorti-coids (17-hydroxycorticosteroid [17-OHCS]) that promote energy use by mobilizing energy stores in support of "fight or flight" behavior (Mason, 1968). Catabolic hormones stimulate the breakdown of larger molecules such as triglycerides into smaller molecules such as free fatty acids that can be oxidized in a process that yields adenosine triphosphate (the energy currency of the cell). Release of catecholamines is followed by secretion of anabolic hormones (hormones that promote chemical reac-tions involved in constructing larger molecules from smaller molecules), including female and male gonadal steroids (estrone and testosterone) and insulin, which rebuilds tissues previously catabolized for their energy-rich substrates. Fig. 22.4 also shows that during the immediate catabolic hormone response, anabolic hormones are suppressed below baseline, and conversely, during the later anabolic hormone response, the earlier-secreted catabolic hormones are suppressed below baseline. These data highlight the finely orchestrated physiological reciprocity between these two groups of hormones. This hormonal reciprocity parallels the reciprocity between the sympathetic and parasympathetic nervous sys-tems: the former is activated in response to an immediate threat (leading to increasing heart rate and muscle activity, for example) and the latter is more restorative in function.

Furthermore, repeated environmental stressors of the same type result in adaptation of hormonal responses. In many instances, hormonal responses can become habituated to repeated exposure to the same stressor. The chair-restrained monkey paradigm illustrates this. Weekly sessions of 72-h foot-shock avoidance initially results in clear HPA axis activation, as reflected by elevated 24-h urinary 17-OHCS excretion. However, this response rapidly habituates, such that the second 72-h session produces a much smaller response, and subsequent sessions produced little or no HPA axis activation.

## CLINICAL EXAMPLES

Family environment is important. One sensitive measure of social effects on the release of reproductive hormones is the age at which girls enter menarche: that is, the age at which LH is released in a quantity suf-ficient to initiate the menstrual cycle. Familial factors, including the pres-ence of a male family member, basic approval of the girl by the family, and the absence of conflict all have significant effects on the timing of men-arche. Less-positive family relations were associated with earlier matura-tion of pubertal LH secretion patterns (Graber et al., 1995).

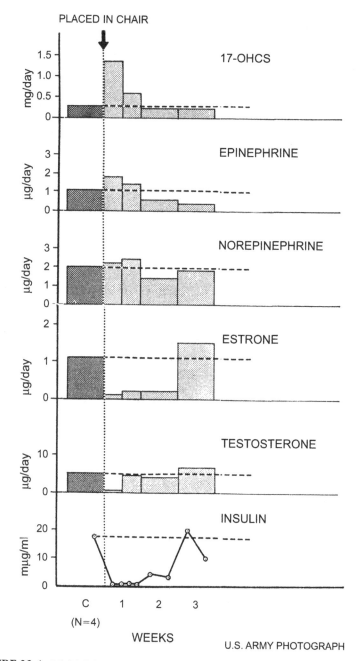

**FIGURE 22.4** Multiple hormone responses during 3-week adaptation to restraining chair in a monkey. 17-Hydroxycorticosteroid (17-OHCS) is a measure of adrenal glucocorticoids. *Modified from Mason, J.W., 1968. Organization of the multiple endocrine responses to avoidance in the monkey. Psychosom. Med. 30 (Suppl. 5), 774–790.*

VI. ENVIRONMENTAL VARIABLES INFLUENCE HORMONE/BEHAVIOR RELATIONS

A second kind of influence on age of menarche in girls could be considered to be the effect of family, stress, nutrition, or all three. Warren and Perlroth (2001) found that trained ballet dancers enter menarche at a significantly later age than do other girls. Of course, beginning and persisting with ballet could result in part from family pressures, and dancing itself could represent a source of environmental stress. However, a dominating factor is that the aesthetics of ballet encourage dancers to maintain a very low percentage of body fat, which can interact with low food intake, high levels of energy expended in training, and the psychological stress of competition to delay reproductive hormonal maturation.

Oxytocin secretion is sensitive to strong environmental signals that are relevant for a woman's emotional life. Positive emotions can be induced by recalling pleasant experiences from the past, which lowers oxytocin levels, increases prolactin levels, but has no effect on ACTH.

The opposite occurs when distress interrupts the neuroendocrine event known as the milk let-down reflex. Under normal circumstances, olfactory, tactile, auditory, and visual stimuli from the infant trigger the release of oxytocin, which stimulates milk ejection from the breast. Nursing mothers are familiar with the tendency for lactation to be disrupted by distressing situations or even by distressing thoughts and memories. This effect of stressors is thought to be mediated by opioid synaptic inputs inhibiting electrical activity in oxytocinergic neurons.

A surprising example of contextual influences over hormone secretion came from an international team of researchers (Hirschenhauser et al., 2002; Hirschenhauser et al., 2008). They studied 27 volunteer men during young adulthood. First, intense sexual activity tended to be associated with testosterone peaks. Second, higher hormone levels tended to occur around weekends. Finally, some men who reported a current wish for children had a 28-day (mensual) pattern of testosterone levels. Although these findings are complex and require further examination, they remind us that even in a simple hormone–behavior causal relation, subtleties may occur when humans are under study.

## Stress

Some environmental circumstances alter the degree to which stressors govern CRH and ACTH release. In some situations, children and adults under stress might be dismissively encouraged to "suck it up" or "forget about it," and be expected to recover quickly from the stress. Plasma glucocorticoids might not get the message (the HPA axis remains activated), and levels of these stress hormones might remain chronically high. Without time for healing and recovery, the effects of chronic stress are damaging. As McEwen (2002) pointed out, stress management is medically important because chronic stress without adequate recovery can lead not only to physical disturbances but also to mental changes, including depression.

Many studies in human subjects have illustrated the importance of central nervous system control of the HPA system. The frontal lobes of the cerebral cortex are the critical brain areas for higher executive function, and the demand for decision making, mastery, and control over novel situations (i.e., learning and coping) is a powerful influence on the HPA axis. These areas come into play when an individual responds to a stressful experience, and afterward.

The military setting is in many respects ideal for human stress studies. Personnel are mostly young and healthy; they are available for repeat studies and follow-up; they are frequently willing to be volunteer subjects; and they often experience extreme stressors in "naturalistic" settings (i.e., as part of their training and in combat). In the US Navy, underwater demolition team (UDT; Navy Sea Air and Land [SEAL]) training represents such a situation. Trainees who enter this program may have little knowledge of swimming and may be in poor physical condition. The training program requires the development of expert swimming skills, outstanding physical endurance, and expertise in handling potentially dangerous situations (e.g., underwater explosives), certainly highly stressful circumstances.

Fig. 22.5 portrays mean early morning serum cortisol concentrations in 20 young, healthy Navy men during the first 2 months of UDT training.

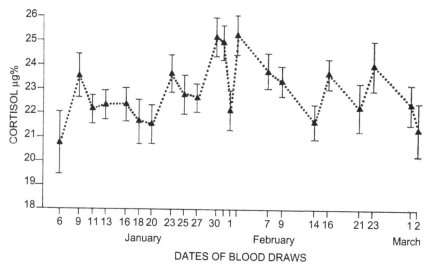

FIGURE 22.5   Mean (standard error of the mean) serum cortisol concentrations for 20 US Navy underwater demolition team (Sea Air and Land) candidates during their first 2 months of training. *From Rubin, R.T., Rahe, R.H., Arthur, R.J., Clark, B.R., 1969. Adrenal cortical activity changes during underwater demolition team training. Psychosom. Med. 31, 557. With permission.*

The first week's training was primarily long-distance running, and the trainees felt confident of their abilities. Serum cortisol was in the normal morning range. The first peak in cortisol occurred coincident with the start of swimming practice (Jan. 9); at this time, most of the men were better runners than swimmers.

Over the next 2 weeks, swimming increased in distance and shifted from a heated pool to the ocean (water temperature, 56°F); nevertheless, cortisol values declined as the men adapted to the swimming training. A significant cortisol increase occurred coincident with the introduction of the face mask back in the heated pool (Jan. 23). Most trainees had little underwater diving experience, and this phase was met with overt anxiety. Over the next week, long ocean swims were conducted with the face mask and the mean cortisol level again trended lower.

In the fifth week of training ("Hell Week"), the men were kept constantly on the move for 5 days and nights, with little or no sleep. The objective was to teach them that no matter how tired they felt, they could always manage further effort. Mean serum cortisol peaked significantly on the first day of Hell Week (Jan. 30) and fluctuated considerably over the next 5 days. During the week, when the men realized they could not keep up such a pace, their esprit de corps quickly diminished, and by the end of the week their tolerance for the grueling schedule had been depleted. During this week, cortisol values were the highest compared with values at any other time during the study.

For the 2 weeks after Hell Week, the schedule was relatively easy. Classroom reconnaissance work began and mean cortisol values decreased significantly. Cortisol then peaked again, coincident with the introduction of diving fins and weapons (February 16). During the next week, cortisol levels declined somewhat as ocean swimming with fins was practiced. An increase in serum cortisol again occurred on the first day of ocean drops and pickups by helicopter (February 23). Values trended lower during the next week of helicopter maneuvers and with the beginning of night ocean swims.

At the end of these 2 months, the group was divided for the demolition phase of training. Cortisol concentrations for the two new groups followed similar patterns of increase in relation to new activities, such as the introduction of the self-contained underwater breathing apparatus, mine searching, and long night compass swims.

An important principle of environmental activation of the HPA axis is illustrated by this UDT/SEAL training study. Mean serum cortisol levels increased coincident with novel experiences that evoked anticipatory anxiety. As each new technique was practiced and became familiar, cortisol levels trended lower, even though instructors made increasing demands for the use of the new technique. The determinant of the transient increases in HPA axis activity was thus the *anticipation* of an unknown situation more than the inherent difficulty of the situation itself. There was a fairly rapid adaptation of the HPA axis once the novelty had passed and mastery of the technique had begun, even though the need for the newly learned technique was increasing. These findings highlight not only the stress reactivity but also the adaptability of the HPA axis to real-life situations. Many additional studies in both military and civilian subjects have corroborated these principles.

Another example illustrates activation of the HPA axis in response to decision making. If a pair of chair-restrained monkeys is subjected to a noxious stimulus in an avoidance situation, the monkey permitted to press a lever to avoid the noxious stimulus to both animals develops gastrointestinal stress lesions, whereas the other monkey, which has no control over the stimulus, does not. This led to the concept of the "executive" monkey: An animal in a situation requiring decision making undergoes more physiological stress than does an animal experiencing the same adversity but having little or no control over its fate.

Several studies have tested this concept in humans, including another investigation of Navy personnel. Naval aviators in training during the Vietnam conflict had blood drawn for serum cortisol determinations shortly after they exited their aircraft after a series of landing practices of increasing difficulty. These included night mirror landing practice on shore and both daytime and nighttime aircraft carrier landing practice. In addition, serum cortisol was measured on a control, nonflying day.

The aircraft in use during these landing practices was the F-4B Phantom, which carried two aviators: the pilot in the front seat, who flew the aircraft, and the radar intercept officer (RIO) in the back seat, who had no control over the aircraft. During landings, the RIO called out the airspeed to the pilot but in all other respects was a passive partner. Thus, the pilot was the "executive" naval aviator, whose split-second decisions during the most hazardous aspect of naval aviation, carrier landings, affected the fate of both himself and his RIO partner. In an emergency, both individuals could eject from the aircraft, although not necessarily safely during a final landing approach.

Fig. 22.6 shows mean serum cortisol concentrations for 9 pilots and 10 RIOs on the 4 test days. Mean cortisol on the control, nonflying day was similar between the two groups. In contrast, during all three landing practices, the pilots had highly significantly increased cortisol concentrations, whereas the RIOs had no significant increases in cortisol compared with their control days. Similar to the observations in monkeys, these findings indicate that a person in a situation requiring decision making undergoes more physiological stress than does a person experiencing the same potential adversity but having little or no control over his fate; this extends the concept of "executive" stress to humans. Other studies

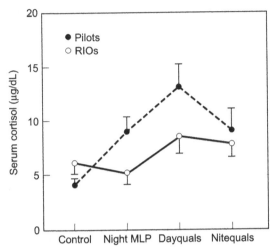

**FIGURE 22.6**   Mean (standard error of the mean) serum cortisol concentrations for 9 US Navy pilots and 10 radar intercept officers (RIOs) on a nonflying control day compared with land-based (night mirror landing practice [MLP]) and aircraft carrier–based (daytime aircraft carrier landing practice [DAYQUALS], nighttime carrier landing practice [NITEQUALS]) landing practice. *Modified from Miller, R.G., Rubin, R.T., Clark, B.R., Crawford, W.R., Arthur, R.J., 1970. The stress of aircraft carrier landings. I. Corticosteroid responses in naval aviators. Psychosom. Med. 32, 585.*

in military personnel (e.g., B-52 aircraft commanders versus their crews, and Special Forces officers versus men under their command) support this concept.

Modern warfare is increasingly being conducted at a distance through computers and remotely guided weapons. Military pilots are being trained more frequently on flight simulators and in the art of computer-guided drones, including pilotless reconnaissance aircraft that can be armed. One concern that has been expressed by "old-school," combat-hardened senior military personnel is that if they are needed in an actual combat situation, this new generation of simulator-trained aviators might have "inexperienced" physiological anxiety reactions, including stress-hormone responses that could be deleterious to optimal functioning when faced with a real threat of bodily harm. On the other hand, we are learning that even pilots who command drones in combat from a distance may develop the same psychological problems, e.g., posttraumatic stress disorder (PTSD), as those who have experienced the traumas of actual combat. Do drone pilots have HPA axis activation during training when new tasks are introduced? The lesson is that continual study of real-life stress situations, as new ones arise, is important for fully understanding the effects of environmental stress on physical and psychological well-being.

Another important clinical example of HPA axis adaptation is that of patients subjected to the stress of hospitalization. As much as 30% to 50% of patients with major depression have increased HPA axis activity, as indicated by increased circulating ACTH and cortisol concentrations and resistance of the HPA axis to suppression by the synthetic glucocorticoid, dexamethasone. The more severe the depression, the more likely there will be increased activity of this endocrine axis. Successful treatment results in return of HPA activity to normal, as well as relief of depressive symptoms.

Even patients with severe major depression remain responsive to environmental stressors that can contribute to increased HPA axis activity. One indicator of this increase is the dexamethasone suppression test (DST), which involves administration of low-dose dexamethasone at about midnight and measurement of circulating cortisol at intervals thereafter. Dexamethasone, which acts primarily at the pituitary, normally strongly suppresses ACTH and cortisol secretion for the next 24 h. If CRH driving of the HPA axis is increased, ACTH and cortisol escape from this suppression and their circulating levels increase.

Fig. 22.7 shows plasma cortisol before and during repeated DSTs in a hospitalized depressed woman before and during treatment. All of the midnight blood samples were taken immediately before dexamethasone administration (2 mg orally), and all but the last sample (May 15, after

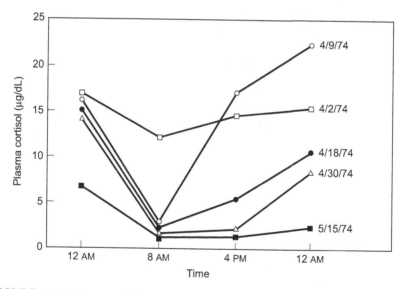

FIGURE 22.7 Plasma cortisol concentrations prior to (12 a.m.) and following administration of dexamethasone (2 mg) at 12 a.m. on the first test day to a hospitalized patient with major depression. April 2 and 9 are prior to antidepressant treatment; April 18, 30, and May 15 are during treatment. *From Carroll, B.J., Curtis, G.C., Mendels, J., 1976. Neuroendocrine regulation in depression. I. Limbic system-adrenocortical dysfunction. Arch. Gen. Psychiatry 33, 1041.*

5 weeks of antidepressant treatment) show elevated cortisol levels. On Apr. 2, just after hospital admission, the DST was strongly abnormal, with almost no suppression of postdexamethasone cortisol. One week later, on Apr. 9, there was normal suppression of cortisol at 8 a.m. but clear escape afterward, which indicated some lessening of HPA axis activation.

Of relevance to this lesson about HPA axis activation and adaptation, during the week between April 2 and April 9, the patient was treated only with placebo medication, so that the weakening of the abnormal DST on April 9 compared with April 2 almost certainly reflected the patient's increasing familiarity with and adaptation to the hospital environment. After active treatment was instituted, the DST levels gradually improved, becoming fully normal (cortisol suppression for a full 24 h) on May 15. Thus, another important principle is that both internal factors (assumed neurotransmitter abnormalities related to the disease process) and external factors (environmental stressors) can be simultaneous activators of a stress-responsive hormone axis.

When and why might the stress response fail to habituate? Early work focused on glucocorticoid negative feedback in hypothalamic areas where cortisol normally inhibits CRH secretion. Modern work focuses on forebrain areas involved in learning and memory. Habituation is essentially a type of learning. The normal, healthy individual learns that when a formerly stress-inducing signal (e.g., a loud bang) is repeated, it no longer indicates danger. However, if the stress-responsive HPA axis fails to habituate in an individual with hypersensitive neuroendocrine systems and/or a poor coping response, chronic exposure to catabolic hormones can damage tissues of the heart, kidney, pancreas, liver, and other organs, and decrease life span. One line of research explores the possibility that certain stress disorders such as PTSD arise from a learning disability in which the individual cannot separate specific kinds of stressors, which bring specific outcomes, from other types of stressors with different outcomes. The lack of ability to associate specific stressors with their outcomes hypothetically prevents habituation (Kheirbek et al., 2012). This line of investigation is promising because it might link a widespread psychiatric syndrome with molecular mechanisms of neurogenesis involved in learning and memory.

# SOME OUTSTANDING QUESTIONS

1. How plastic are hormonal responses to stressors, i.e., which ones are reversible when an animal is removed from a stressful environment, and how quickly do they normalize? Are stressful influences on hormone secretion ameliorated when a subject is placed in a hygienic or therapeutic environment? How does this vary with age (e.g., prepubertal versus postpubertal comparisons and studies of the aged)?

2. What severity and duration of a given stressor are required to produce irreversible changes in endocrine function and associated metabolic effects?
a. What are the characteristics of stressors in early life that lead to increased HPA axis activity in adulthood?
b. Which factors can mitigate the effects of stressors in early life such that irreversible changes in endocrine function are not produced?
3. How do competing forces from the environment become integrated to form a net force of defined direction, amplitude, and time course on the release of a given hormone?
4. To what extent have certain environmental–endocrine relations remained intact during evolution from lower mammals to nonhuman primates to humans? To what extent have they changed in an understandable and lawful manner?
5. Which ascending neural pathways to the basomedial hypothalamus from the brain stem influence environmental–endocrine causal sequences? Likewise, which pathways descending from the cerebral cortex and limbic system do the same?

# EVOLUTION

# Neuroendocrine Mechanisms Have Been Conserved to Provide Biologically Adaptive Body–Brain–Behavior Coordination

Evolution is the change in heritable phenotypes over time. As simple as it seems, this is the process by which all life forms on earth came about. There are many kinds of evolutionary forces, one of which is natural selection, the differential survival and reproductive success of individuals with propitious, heritable phenotypes. Many hormone-controlled behaviors, especially social, courtship, sex, and parental, are means by which individuals increase their fitness (i.e., their chances of reproductive success) and pass on their heritable phenotypes to the next generation. Evolutionary theorists from Charles Darwin to Ernst Mayr have seen such behaviors as the "leading edge of evolutionary change" (Mayr, 1974). Understanding hormonal control of behavior helps us understand the process of evolution. At first glance, some phenotypes might appear to be adaptive. For example, some traits increase life span, which should allow more time to reproduce and raise more offspring to reproductive age. However, if the activation of one particular hormone receptor increases longevity but decreases fitness, the alleles that increase longevity will not be selected, and longevity will not increase over successive generations. Similarly, mutations in some alleles might increase longevity, but those alleles are unlikely to be selected if they decrease reproductive success. Selection for longevity is thus constrained by the deleterious effects of these genes on reproduction. For example, genes that encode gonadal steroid receptors or enzymes involved in steroid synthesis are necessary for normal reproductive function, but these steroid hormones have other effects that

*Principles of Hormone/Behavior Relations*
http://dx.doi.org/10.1016/B978-0-12-802629-8.00023-1

441

can decrease survival and longevity, e.g., the effects of sustained hypo-thalamic–pituitary–adrenal axis activation by stress. Longevity cannot be increased without deleterious effects on the reproductive system; gonadal steroids, as well as adrenal steroids (glucocorticoids), carry a cost in lon-gevity. Survival and longevity are constrained (cannot increase) unless new adaptations arise that decrease the cost of reproductive hormones (Cox et al., 2010; Jenkins et al., 2004; Mukhopadhyay and Tissenbaum, 2007; Wingfield et al., 2001).

There are many ways in which variations in hormonal systems can affect the evolutionary paths followed for a particular species. Part of this chapter will discuss the role of hormones and receptors in the evolution of life history strategies. The evolution of hormonal mechanisms enriches our understanding of hormone function. If each organism evolved from preexisting organisms, all of these organisms would be expected to share many genes, biochemical processes, cellular structures, and organ systems. We share about 98% of our DNA with chimpanzees, 85% with zebra fish, and over a third of our DNA with fruit flies. Many of these genes encode hormones, neuropeptides, neurotransmitters, and their receptors. It has been argued (Pfaff, 1999) (Chapter 8) that many neuroendocrine mecha-nisms were conserved from earlier mammalian brains to the human brain. Thus, to some degree, we can learn about hormone–behavior relations in our own species by studying those systems in laboratory rodents with shorter life spans, smaller size, and the ability to complete many environ-mentally controlled experiments within weeks or months.

In addition to mammals, birds have been particularly useful in an understanding of basic mechanisms that define hormone–behavior rela-tions. In fact, many firsts in the fields of animal behavior, behavioral endocrinology, and neuroscience were achieved using birds as the model system, perhaps because avian species are often diurnal (awake in the day when we tend to be awake), more conspicuous in the wild, and eas-ily captured; and many species can be raised in captivity with success (Wingfield, 2005). As mentioned in the first chapters of this book, the first experimental evidence for the effects of glandular secretions on behavior were performed by Berthold in the 1800s using roosters as experimental subjects. Other firsts include experiments performed in 1924–25 demon-strating that increasing day lengths stimulate the reproductive systems of the dark-eyed junco and American crow (Rowan, 1930, 1932) and the landmark experiments of Hinde and Lehrman in the 1960s using canar-ies and ring doves to show that hormones affect behavior, but changes in behavior also lead to more changes in hormone secretion (Chapter 7). Experimenters such as those of Fernando Nottebohm and Masami Konishi made pivotal contributions to an understanding of neural plasticity and neurogenesis, processes that underlie the ability of birds to learn species-specific songs. They and their academic offspring were pioneers in the field of the development of sex differences, especially as they pertain to social behaviors. Also using birds, Barney Schlinger and Eliot Brenowitz began an influential line of research on the role of the enzyme aromatase in rapid hormonal effects on behavior. All of the work in birds was relevant to understanding these processes in mammals, including humans, which reiterates the fact that many functions of the neuroendocrine system have been conserved during evolution.

In other chapters, we describe the use of other vertebrates, such as fish, amphibians, and reptiles such as snakes, in the field of behavioral neuroendocrinology. Even the earliest invertebrates have much to tell us. Mice and rats have short life spans, but those of fruit flies (*Drosophila melanogaster*) and nematode worms (such as *Caenorhabditis elegans*) are even shorter, which allows many experiments to be performed rapidly. Members of these two species have surprisingly complex social behaviors despite having tractable nervous systems. The nervous system of *C. elegans* consists of about 300 neurons and has been used to study simple behaviors such as chemotaxis, avoidance, and oxygen sensing, as well as more complex behaviors including foraging, mating behavior, alcohol intoxication, learning, habituation, and adaptation. Another advantage of this roundworm species is that all of its genes and neural synapses are known. These invertebrates have simpler organ systems; yet because all animals share basic components, they have relevance for an understanding of mammals, including *Homo sapiens*. Thus, studies of hormone–brain–behavior mechanisms in many simpler species are likely to contribute to a medical understanding of disorders involving the same neurons and same biochemical reactions in humans. More generally, the principle that "hormones evolve for specific functions" can be illustrated by several hormones and their receptors.

Throughout this book, we have seen many examples of how hormones affect specific behaviors. Pari passu, the material has illustrated (1) how hormones have evolved to provide essential physiological functions for survival, and (2) that behavioral mechanisms comprise one aspect of that evolution. The most extensive examples quoted in this short text include steroid and peptide hormones for feeding, drinking, stress, and sex, which are clearly basic for any organism. It is unsurprising, therefore, that several important families of hormones have been conserved for millions of years to serve a specific function.

The two main animal lineages, the Protostomia and Deuterostomia, split from each other about 700 million years ago. Most invertebrates belong to the Protostomia, including flatworms, earthworms and leeches, molluscs, roundworms, and arthropods (insects, crustaceans, spiders, and ticks). All vertebrates belong to the Deuterostomia, including fishes, amphibians, nonavian reptiles, birds, and mammals. Before this early split in evolutionary lineages, the sponges and cnidarians existed, including the hydra, sea anemones, corals, and jellyfishes. The sponges have no nervous system, so any molecules that resemble mammalian neuropeptides would have a different function in sponges. Similarly, sponges have no circulatory system, so if you define a hormone as a chemical messenger that travels in the circulation, sponges have no true hormones. However, sponges have insulin-like receptors, and these influence growth and metabolism just as they do in other organisms. In fact, the appearance of the insulin-like receptor in sponges and its effect on metabolism, growth, and differentiation made possible the appearance of subsequent metazoan species. The cnidarians are important because the common ancestor of these organisms had the first components of a rudimentary nervous system. Do cnidarians have hormones, neuropeptides, and neurotransmitters? Yes, yes, and no. Cnidarians have the insulin–insulin receptor signaling system, and their nervous system is strongly peptidergic. Despite having a nervous system, most fast neurotransmitters (glutamate, glycine, γ-aminobutyric acid, and acetylcholine) have not been found in cnidarians. The exception is the transmitter serotonin, which has been located in neurons of the sea pansy, *Renilla koellikeri*. In most cnidarians, neuropeptides act via G protein–coupled receptors. In addition, in some cnidarians, neuropeptides activate ligand-gated ion channels, thus allowing fast responses (as if they had neurotransmitters). The gonadotropin-releasing hormone (GnRH) receptor originated in cnidarians, i.e., before the split in the Protostomia and Deuterostomia over 700 million years ago. The neuropeptides oxytocin and vasopressin have been conserved almost perfectly in many different species from both Protostomia and Deuterostomia. Similarly, neuropeptide Y (NPY) is highly conserved in both lineages. This prolonged conservation might be nature's way of saying, "If it ain't broke, don't fix it." Hormones are synthesized from DNA-encoding specific sequences of a gene. Over time, mutations occur in individual nucleotide bases within these sequences, and different species may develop different variants of primordial hormones. Nevertheless, as a principle, the biologically active part of the hormone is the most highly conserved sequence. Thus, we now explore examples in which the general physiological function often remains the same over time.

## INGESTION

As we have seen, ingestive behavior is a complex behavioral event affected by many hormones (recounted in Chapters 1 and 2). The effects of these hormones are almost never limited to food intake; they also influence energy intake, storage, and expenditure, including the energy expended on reproduction. The main function of mechanisms that increase and decrease food intake is to maintain homeostasis in energy availability.

Animals need energy to remain alive, move around, forage and hunt for food, grow, fight, mate, and, in some species, care for offspring. Energy also generates heat, and in poikilothermic animals, feeding and eating are necessary for thermogenesis. Most animals evolved in environments where food is not continuously available. Food sources fluctuate with seasons and with unexpected events (e.g., famine, flood, fire, tornados, hurricanes, and typhoons). Thus, most organisms engage in anticipatory overeating so that they can store energy on the body, which can be broken down later for oxidizable metabolic fuels (glucose and free fatty acids), which are metabolized to maintain energy homeostasis when there is no food to eat or when animals are busy with other activities incompatible with eating. Leptin, insulin, and NPY, for example, are not only important hormones in the control of ingestive behavior, they also regulate energy storage and expenditure. These multiple functions are evident throughout the animal kingdom and are managed by many receptors for individual hormones and neuropeptides.

The potent orexigenic peptide NPY and its receptors are present in most vertebrates. There are at least six receptors for NPY in mammals. In species in which NPY increases food intake (e.g., rats, mice, hamsters, and humans), its action is primarily via Y5. In mice, specific pharmacological agonists to the Y5 receptor increase food intake, and antagonists to the Y5 receptor decrease food intake. Molecules with structural similarity to NPY and the Y1 receptor are found in some invertebrates, including the gastropod mollusk, *Lymnaea stagnalis* (a pond snail). Even in the roundworm, *C. elegans*, a seven-transmembrane protein has been identified as the critical receptor in ingestive behavior and is remarkably similar to the NPY receptor family. In *D. melanogaster*, ingestion is stimulated by neuropeptide F (NPF) and its receptor NPF-R, a ligand–receptor pair with striking similarities to vertebrate NPY and the Y2 receptor. During development, the fly larva burrow into and eat food, grow continuously until they reach maturity, stop eating food, and begin to wander in search of a place to pupate. The shift in ingestive behavior is accompanied by a significant downregulation of *npf* expression in the brain.

Larvae that are fed high-quality food *ad libitum* will refuse to eat food of a lower quality, but acceptance of low-quality food increases with the duration of prior food deprivation. Food deprivation–induced ingestion of low-quality food can be mimicked in *ad libitum*–fed larvae by overexpression of the gene for the NPF receptor, *npf-r* (reviewed by Nassel and Wegener, 2011). From peptide sequencing, molecular cloning, and comparative genomics, it has been shown that the NPY family has a common ancestral gene (NYY), and by gene duplications has produced NPY, PYY, pancreatic polypeptide, NPF, and others. By the time the first vertebrates such as the lamprey evolved, they had NPY and PYY. The pleiotropic effects of NPY and Y receptor genes on ingestive and reproductive behavior are ancient. In the pond snail, *L. stagnalis*, NPY fails to increase food intake, but instead conserves energy by inhibiting reproduction. In this species, NPY's effects are all mediated via the Y1 receptor. Thus, in the pond snail, NPY functions mainly to inhibit the allocation of energy to reproduction and is not involved in controlling food intake *per se* (reviewed by de Jong-Brink et al., 2001). The reproduction–energy metabolism link is thus ancient.

Insulin and insulin-like molecules appear in the earliest organisms and function in growth and metabolism. The unicellular ciliates, such as Tetrahymena, as well as other protozoa, are known to synthesize, store, secrete, and take up insulin. The plasma membrane and nuclear envelope of Tetrahymena have functional insulin receptors structurally similar to those in mammals. These single-celled organisms possess the entire intracellular insulin-signaling pathway (Csaba, 2013). In vertebrates, insulin typically controls tissue uptake of glucose, whereas insulin-like peptides control growth. Insulin is critical for the uptake of glucose into peripheral cells, and insulin acts in brain to signal energy availability. Also, insulin-mediated glucose uptake is critical for energetically costly processes related to reproduction. Reproduction is inhibited in diabetes and in mutant mice with deficits in insulin or the insulin receptor. In invertebrate species, only one insulin-like family of molecules controls both growth and metabolism. In *D. melanogaster*, insulin-like peptide has a role in carbohydrate homeostasis and growth. In insects such as *D. melanogaster*, juvenile hormone, which promotes larval growth and delays metamorphosis, is a member of the insulin-like family. Relaxin is also found in the fly in the form of an insulin/relaxin-like peptide. The preservation of function is shown by the fact that flies with mutations that reduce insulin signaling grow up to be small and infertile, as do mice with insulin deficiency (reviewed by Nassel and Winther, 2010).

# REPRODUCTION

## Gonadotropin-Releasing Hormone

GnRH, a decapeptide, is the master neuropeptide for reproduction in every vertebrate that has been studied, and GnRH is found in invertebrates including the cnidarians. All 16 chemical forms of GnRH are involved in reproduction, including mating behavior. In mammals, the two forms in the brain are GnRH I and GnRH II. The latter is the most conserved during evolution and has been shown to coordinate reproductive behavior, but only when the nutritional state is appropriate. Most exquisitely, in the female, GnRH signaling coordinates mating behavior with ovulation, so that when the female takes risks to encounter the male, mating is likely to be productive.

One prominent aspect of the conservation of GnRH among vertebrates has to do with the life history of GnRH neurons. Unlike other neurons in the central nervous system, GnRH neurons are not born in the brain, but in the olfactory epithelium. During the embryonic period, they migrate into the brain, arriving at their final functional positions in the preoptic area and anterior hypothalamus. This unique developmental migration from the nose into the brain holds true for all vertebrates studied, from fish through humans.

In terms of brain and behavior, we know that systemic injections of GnRH can enhance the performance of sex behavior in the female and male. In the case of female behavior, lordosis, GnRH is not acting through the pituitary gland (as it would to cause ovulation), because it works in females from which the pituitary has been surgically removed. GnRH facilitates mating behavior in male rats, as well. One mechanism by which GnRH works is to act as a neuromodulator enhancing norepinephrine signaling in the hypothalamus. Another mechanism is to enhance the effect of nitric oxide. GnRH is secreted in a pulsatile manner; in fact, in men whose criminal behavior requires them by law to be "chemically castrated," long-term, constant GnRH exposure shuts down their reproductive system.

The GnRH pulse generator is inhibited by environmental factors such as low food availability and short winter days. In populations of wild mammals, there is variation in response to short day lengths. Some animals are highly responsive, completely ceasing function of the NPG system during day lengths typical of the autumn. The variation in response to short days is linked to variations in the gene for GnRH. Natural populations of rodents contain significant genetic variation in the number and location of GnRH neurons, which might contribute to the functional variation in fertility that occurs in short photoperiods (Kaugars et al., 2016).

## STEROID HORMONE RECEPTORS

Virtually all vertebrates have versions of estrogen, androgen, progestin, and glucocorticoid receptors, although it is not clear whether they exist in invertebrates. In lampreys, jawless fish that diverged from the rest of the fish line about 450 million years ago, steroid receptors include the estrogen, progestin, and glucocorticoid receptors, but not androgen receptor. In lampreys, pubertal development is mediated by estradiol in both males and females. Various statistical methods have been used to "resurrect" the ancestral steroid hormone receptor, which was clearly an estrogen receptor. As you will remember from the first chapter, estradiol is the last in a long chain of prohormones synthesized from cholesterol. Thus, it has been inferred that estradiol was the first steroid hormone ligand with functional binding to the estrogen receptor (Thornton, 2001). In this hypothesized scenario, other steroid receptors came later via duplication and mutation in the gene for the estrogen receptor. Before the evolution of the progestin, glucocorticoid, and androgen receptors, the only purpose of progesterone, cortisol, and androgens was to serve as substrates from which to form estradiol. It was only later that progesterone became a functional hormone acting via progestin receptors (Thornton, 2001). Viewing steroid hormone receptors as examples of gene transcription factors, Thornton argued that small permissive substitutions lead the way for a steroid hormone receptor "to tolerate large-effect mutations that cause a shift in (hormone receiving) specificity." There is an abundance of variety in these receptors that, say, work with estrogen effects in brain does not lead us to appreciate. Even among fish, for example, amino acid sequences are not identical from species to species. Future work using protein alignment programs currently available from several companies may explain why some sequences have been conserved and others have undergone frequent mutations.

# STRESS HORMONES

## Corticotropin-Releasing Factor

Corticotropic releasing factor (CRF) and a series of related peptides, most notably, the urocortins, can be found throughout the vertebrate phyla. CRF is best known for its stimulation of adrenocorticotropic hormone and its involvement in behavioral stress responses in both vertebrates and invertebrates, but osmoregulation and thermoregulation also may have depended on CRH or related peptides.

## Vasopressin

Consider the conserved role of the nine–amino acid (nona)peptide produced in certain magnocellular neuronal groups in the anterior hypothalamus. When animals moved out of the sea and onto land, they had to use hormones that already existed to maintain their water balance. The amount of water ingested must be balanced by the quantity and concentration of urine produced to regulate fluid volume and osmolarity within a normal range. As we have seen in Chapter 5, the principal hormone for controlling urine output is vasopressin. Vasopressin and oxytocin are nonapeptides evolved from the same gene. Conserved sequences of vasopressin have been found almost unaltered in cnidarians such as Hydra, an animal that has existed for over 700 million years. In fish, isotocin is the equivalent of oxytocin, whereas vasotocin is the equivalent of vasopressin. Although fish and humans have had separate lineages for 400 million years, the hormones are similarly involved in fluid and electrolyte balance. Thus, from an ancestral, ancient prohormone, one evolutionary line, through fish and reptiles to mammals, eventually yielded mammalian vasopressin and another evolutionary line eventually yielded mammalian oxytocin (see subsequent discussion).

## Oxytocin

The nonapeptide oxytocin is widely considered to facilitate prosocial behaviors in laboratory animals, mammals, and humans. What is less appreciated is the role of arginine vasotocin (AVT) in teleost fish. AVT neurons in the preoptic area project to both the pituitary gland and extrahypothalamic areas. Although the behavioral effects of AVT are not identical in all fish species, a common theme seems to emphasize behaviors related to reproduction. Similarly, AVT stimulates the courtship behavior, "advertising calling," by frogs. In birds, the data support a role for vasopressin in control of so-called "courtship song," although in some species under some conditions singing mainly serves aggressive or territorial functions as a prelude to reproduction. Biologists performing this comparative work are interested not only in the enduring roles for a given ancestor to oxytocin, but also in the different adaptations of behavioral function for this family of neuropeptides.

# HORMONAL EFFECTS ON EVOLUTIONARY PATHS FOLLOWED

In earlier chapters we noted that the same hormone can have opposite effects on different behaviors. The opposing pleiotropic effects of hormones impose constraints on the evolution of individual phenotypes that are affected by these hormones (Finch and Rose, 1995; Hau and Wingfield, 2011; Ketterson and Nolan, 1992). For example, in many bird species, high levels of androgens promote territorial aggression, and winners of aggressive encounters receive benefits in greater access to females, but there is a cost to high levels of androgens. Chronically elevated plasma androgens inhibit immune function, decrease survival rate, and suppress behaviors related to care and feeding of offspring (reviewed by Hau and Wingfield, 2011; Wingfield et al., 1990). In the brown anole lizard, *Anolis sagrei*, high levels of fecundity are at odds with maximum longevity. This was demonstrated when longevity was significantly increased by the removal of ovarian hormones by ovariectomy. Conversely, high levels of estradiol and progesterone decrease (and ovariectomy increases) growth, body fat content, hematocrit, immune function, and parasite tolerance (Cox et al., 2010). These examples show how hormones force animals to make physiological trade-offs. Individuals must trade their own longevity for greater reproductive success.

In other species, animals trade dispersal for reproductive success. Wing-polymorphic *Gryllus* crickets contain morphs that trade fecundity for dispersal capabilities related to large wing size and highly developed flight muscles. A long-winged morph has the ability to fly. Flight is supported by storage of triglycerides in flight muscles, from which free fatty acids can be broken down and used as flight fuel. In females of the long-winged morph, ovarian growth and egg production are delayed, whereas in short-wing females, ovarian development is rapid, resulting in greater fecundity associated with higher levels of insulin-like peptides and ecdysteroids (reviewed by Zera, 2016a,b). At first glance, this suggests that the short-wing flightless morph might be more successful. However, the long-wing morph enjoys some advantages, including higher levels of acoustic sensitivity (they are better equipped to learn and make courtship songs) coupled with circadian rhythms in levels of juvenile hormone that underlie daily dispersals. These landmark studies demonstrate the important role of hormones in microevolutionary trade-offs.

Although this text is devoted to identifying and explaining principles of hormone relations in mammals, the comparative approach that includes nonmammals and invertebrates gives us a perspective: For any given neuropeptide or hormone receptor, we may realize that they have carrying out similar physiological and behavioral functions for millions of years of evolution.

# Epilogue: A Look to the Future

OK, if we understand how this happens, we can maybe manipulate it. *Sierra* (2015).

Two of the greatest challenges facing mankind are determining the nature of the smallest elements of matter and determining the size and extent of our universe. These are perhaps the grandest questions we can ask about our existence. Similarly, with regard to hormone-behavior relations, two of the greatest challenges we face are determining their smallest components, their genetic foundations and molecular processes, and their largest components, the brain networks related to endocrine function and the environmental factors that affect their plasticity in the human brain. Advances in both small-scale and large-scale technologies, such as whole-genome sequencing and functional magnetic resonance imaging, are enabling scientists to expand their understanding of the smallest and largest components of hormone-behavior relations.

It is said about acquiring real estate that the three most important considerations are location, location, location. We, similarly, can say about understanding hormone-behavior relations that the three most important considerations are mechanisms, mechanisms, mechanisms. For example, as discussed in Chapter 22, what are the mechanisms involved in the hypothalamo–pituitary–adrenal cortical (HPA) response to the organism's experience of novel environmental stressors, as well as later reduction of the HPA response when the environmental stressors are reexperienced and become familiar? What happens in the brain when the organism realizes the experience is new and perhaps anxiety-provoking, versus the experience being a ho-hum repetition of something already experienced many times (e.g., in high-level training for dangerous activities)? What happens in the hypothalamus? Is there a smaller net secretion of CRH and/or AVP to stimulate ACTH secretion from the pituitary? If the adrenal cortex has been sensitized and perhaps hypertrophied by earlier stimulation from high levels of secreted ACTH, does the adrenal cortex take longer to reduce its output of corticosteroids than it takes the brain, hypothalamus, and pituitary to reduce their stimulation? Overall, novel stressors lead to strongly increased HPA axis activity, and familiarity with stressors leads to adaptation and normalization of the HPA axis hormonal cascade. What are the mechanisms?

Other hormonal systems work in the opposite way: consider the growth hormone (GH) response to severe psychosocial deprivation in infants and young children, as discussed in Chapter 5. GH secretion and IGF-1 production become profoundly suppressed for long periods of time, resulting in failure to grow and thrive, the psychosocial dwarfism syndrome. What happens in the brain when the infant, craving interpersonal contact, finally despairs and gives up psychologically? Is there chronic hypersecretion of somatostatin from the hypothalamus that impairs GH secretion from the pituitary? Why is there no adaptation to the chronicity of the experience? How much does physical malnutrition add to the hormonal disruption? And, of great clinical importance, there is a rapid reversal of the hormonal shutdown with the introduction of interpersonal caring and nurturing. Indeed, if the interpersonal deprivation is interrupted early enough, there can be a complete reversal of the failure to thrive physically. What are the brain and hormonal mechanisms that control this process? And, of considerable importance, do the higher brain functions of the deprived infant, i.e., its psychology, develop and mature normally? If not, in what aspects of higher brain function (thinking, feeling, acting) do abnormalities continue to be manifest? (It also is of interest that, at the other end of life, these same GH and IGF-1 deficits may play key roles in the delay of aging, as shown in long-lived mice with mutations in GH signaling.)

These are just two examples of hormone-behavior relations, but they are stark and contrasting in their direction: hormone activation by stressors in the first example, and hormone suppression by stressors in the second. For both, however, understanding their mechanisms is the critical undertaking for the future, especially for refining our treatments of pathologies related to these processes. How are we to do this? We need to push every advance in experimental methodology and bring it to bear on our work.

For example, computers lie at the heart of many of our current technological advances: they perform complex calculations with superb accuracy and at lightning speed. For the handling of large data sets such as are found in gene sequencing and neuroimaging, high-speed computational analysis is essential; the human brain is so sluggish in its calculating ability that it cannot compare with even the simplest hand-held computer. In contrast, the human brain is supreme in pattern recognition. We take for granted that we can understand complex, rapidly changing emotional expressions in both friends and strangers, and we consider those who have difficulty in ascertaining emotions in others to be handicapped in some way. In pattern recognition and its interpretation, even the most powerful computer still lags behind the human brain, with its lifetime of stored images and memories. And, the human brain is startlingly more efficient (20-watt power requirement) than the fastest supercomputer yet invented (10-million-watt power requirement) (Fischetti, 2011). Many computer scientists are trying to develop neural network algorithms as a basis for pattern recognition and for emulating other higher-order brain functions; their success has been impressive but incremental, given the enormity of the task (Service, 2014). Future application of these algorithms to hormone-behavior interactions should lead to important insights into the mechanisms of neural network control of endocrine functions and, conversely, the effect of hormones on neural networks, with important therapeutic implications.

Put this all together, and smart students may conclude that hormone actions lead us to at least three frontiers: (1) molecular mechanisms of hormone-sensitive gene transcription; (2) behavior-related environmental problems, e.g., toxins that are endocrine disruptors; and (3) modern individualized medicine for treating hormone abnormalities that affect mental function in patients. As a result, the authors firmly believe that, when a third edition of *Principles of Hormone Behavior Relations* is contemplated some years hence, unimagined scientific advances will have been reported, and the authors will be able to chronicle these discoveries in the future volume. Our strong hope is that, in the intervening years, readers of this current edition will be sufficiently excited by these possibilities to make their own lasting contributions to the deeper understanding of hormone-behavior relations and their potential therapeutic benefits.

# Further Reading

This is not a comprehensive bibliography; instead, we have tried to provide enough literature citations to support the figures and tables in this text and to lead the reader to some of the primary data in our field. Note that a comprehensive reference source, Hormones, Brain and Behavior, has been published recently in five volumes by Academic Press, third edition (2017). As well, two other excellent texts treat neuroendocrinology from different points of view: Behavioral Endocrinology by J. Becker, M. Breedlove, D. Crews, and M. McCarthy (MIT Press, 2002), and An Introduction to Behavioral Endocrinology by R. Nelson (Sinauer Associates, 2000).

## Introduction

Ahima, R.S., 2004. Body fat, leptin, and hypothalamic amenorrhea. N. Engl. J. Med. 351, 959–962.

Albers, H.E., 2015. Species, sex and individual differences in the vasotocin/vasopressin system: relationship to neurochemical signaling in the social behavior neural network. Front. Neuroendocrinol. 35, 49–71.

Andrews, R.C., Walker, B.R., 1999. Glucocorticoids and insulin resistance: old hormones, new targets. Clin. Sci. 96, 513–523.

Andrews, R.C., Herlihy, O., Livingstone, D.E., Andrew, R., Walker, B.R., 2002. Abnormal cortisol metabolism and tissue sensitivity to cortisol in patients with glucose intolerance. J. Clin. Endocrinol. Metab. 87, 5587–5593.

Balthazart, J., Ball, G., 2006. Is brain estradiol a hormone or a neurotransmitter? Trends Neurosci. 29 (5), 241–249.

Calisi, R., 2014. An integrative overview of the role of gonadotropin-inhibitory hormone in behavior: applying Tinbergen's four questions. Gen. Comp. Endocrinol. 203, 95–105.

Campese, V.D., Kim, J., Lazaro-Munoz, G., Pena, L., LeDoux, J.E., Cain, C.K., 2014. Lesions of lateral or central amygdala abolish aversive Pavlovian-to-instrumental transfer in rats. Front. Behav. Neurosci. 8, 161.

Carter, S., 2014. Oxytocin pathways and the evolution of human behavior. Annu. Rev. Psychol. 65, 17–39.

Chehab, F.F., 2014. 20 years of leptin: leptin and reproduction: past milestones, present undertakings, and future endeavors. J. Endocrinol. 223, T37–T48.

Clark, J.H., Schrader, W.T., O'Malley, B.W., 1992. Mechanism of action of steroid hormones. In: Wilson, J.D., Foster, D.W. (Eds.), Williams Textbook of Endocrinology, eighth ed. Saunders, Philadelphia, pp. 35–90.

Clarke, I.J., Cummins, J.T., 1982. The temporal relationship between gonadotropin releasing hormone (GnRH) and luteinizing hormone (LH) secretion in ovariectomized ewes. Endocrinology 111, 1737–1739.

Clarke, I.J., Smith, J.T., Henry, B.A., Oldfield, B.J., Stefanidis, A., Millar, R.P., Sari, I.P., Chng, K., Fabre-Nys, C., Caraty, A., Ang, B.T., Chan, L., Fraley, G.S., 2012. Gonadotropin-inhibitory hormone is a hypothalamic peptide that provides a molecular switch between reproduction and feeding. Neuroendocrinology 95, 305–315.

Conlon, M., Lightfoot, N., Kreiger, N., 2007. Rotating shift work and risk of prostate cancer. Epidemiology 18, 182–183.

Cottier, F., Mühlschlegel, F.A., 2012. Communication in fungi. Int. J. Microbiol. http://dx.doi.org/10.1155/2012/351832.

Dallman, M.F., 2010. Stress-induced obesity and the emotional nervous system. Trends Endocrinol. Metab. TEM 21, 159–165.

Deutch, A.Y., Roth, R.H., 2003. Neurotransmitters. In: Squire, L.R., Bloom, F.E., McConnell, S.K., Roberts, J.L., Spitzer, N.C., Zigmond, M.J. (Eds.), Fundamental Neuroscience, second ed. Academic Press, Amsterdam, pp. 163–196.

Epel, E.S., McEwen, B., Seeman, T., Matthews, K., Castellazzo, G., Brownell, K.D., Bell, J., Ickovics, J.R., 2000. Stress and body shape: stress-induced cortisol secretion is consistently greater among women with central fat. Psychosom. Med. 62, 623–632.

Gan, Y., Yang, C., Tong, X., Sun, H., Cong, Y., Yin, X., Li, L., Cao, S., Dong, X., Gong, Y., Shi, O., Deng, J., Bi, H., Lu, Z., 2015. Shift work and diabetes mellitus: a meta-analysis of observational studies. Occup. Environ. Med. 72, 72–78.

Gore, A.C., 2008. Developmental programming and endocrine disruptor effects on reproductive neuroendocrine systems. Front. Neuroendocrinol. 29, 358–374.

Grill, H.J., Hayes, M.R., 2009. The nucleus tractus solitarius: a portal for visceral afferent signal processing, energy status assessment and integration of their combined effects on food intake. Int. J. Obes. (Lond.) 33 (Suppl. 1), S11–S15.

Grill, H.J., 2006. Distributed neural control of energy balance: contributions from hindbrain and hypothalamus. Obes. (Silver Spring) 14 (Suppl. 5), 216S–221S.

Ha, M., Park, J., 2005. Shiftwork and metabolic risk factors of cardiovascular disease. J. Occup. Health 47, 89–95.

Hansen, J., 2006. Risk of breast cancer after night- and shift work: current evidence and ongoing studies in Denmark. Cancer Causes Control 17, 531–537.

Herman, J.P., 2013. Neural control of chronic stress adaptation. Front. Behav. Neurosci. 7, 61.

Kahn, C.R., Smith, R.J., Chin, W.W., 1992. Mechanism of action of hormones that act at the cell surface. In: Wilson, J.D., Foster, D.W. (Eds.), Williams Textbook of Endocrinology, eighth ed. Saunders, Philadelphia, pp. 91–134.

Kandel, E.R., Schwartz, J.H., Jessel, T.M., 2000. Principles of Neural Science, fourth ed. McGraw-Hill, New York.

Kelly, A.M., Goodson, J.L., 2014. Social functions of individual vasopressin–oxytocin cell groups in vertebrates: what do we really know? Front. Neuroendocrinol. 35, 512–529.

Kivimaki, M., Virtanen, M., Elovainio, M., Vaananen, A., Keltikangas-Jarvinen, L., Vahtera, J., 2006. Prevalent cardiovascular disease, risk factors and selection out of shift work. Scand. J. Work Environ. Health 32, 204–208.

Klingerman, C.M., Williams 3rd, W.P., Simberlund, J., Brahme, N., Prasad, A., Schneider, J.E., Kriegsfeld, L.J., 2011. Food restriction-induced changes in gonadotropin-inhibiting hormone cells are associated with changes in sexual motivation and food hoarding, but not sexual performance and food intake. Front. Endocrinol. 2 (Article 101), 1–15.

Krieger, D.T., 1980. The hypothalamus and neuroendocrinology. In: Krieger, D.T., Hughes, J.C. (Eds.), Neuroendocrinology. Sinauer Associates, Sunderland, MA, pp. 3–12.

Kriegsfeld, L.J., Gibson, E.M., Williams 3rd, W.P., Zhao, S., Mason, A.O., Bentley, G.E., Tsutsui, K., 2010. The roles of RFamide-related peptide-3 in mammalian reproductive function and behaviour. J. Neuroendocrinol. 22, 692–700.

Kuo, L.E., Czarnecka, M., Kitlinska, J.B., Tilan, J.U., Kvetnansky, R., Zukowska, Z., 2008. Chronic stress, combined with a high-fat/high-sugar diet, shifts sympathetic signaling toward neuropeptide Y and leads to obesity and the metabolic syndrome. Ann. N.Y. Acad. Sci. 1148, 232–237.

Kust, 1932. Die feststellung der trachtigkeit der stute durch hormonnachweis im blute. Berl. Tierarztl. Wschr. 48, 817–818.

Marquie, J.C., Tucker, P., Folkard, S., Gentil, C., Ansiau, D., 2014. Chronic effects of shift work on cognition: findings from the VISAT longitudinal study. Occup. Environ. Med. 72, 258–264.

Micheel, C.M., Nass, S.J., Omenn, G.S., 2012. Evolution of Translational Omics. Institute of Medicine of the National Academies, National Academies Press, Washington, DC, pp. 1–274.

Paredes, S., Ribeiro, L., 2014. Cortisol: the villain in metabolic syndrome? Rev. Assoc. Med. Bras. 60, 84–92.

Poole, C.J., Evans, G.R., Spurgeon, A., Bridges, K.W., 1992a. Effects of a change in shift work on health. Occup. Med. (Lond.) 42, 193–199.

Poole, C.J., Wright, A.D., Nattrass, M., 1992b. Control of diabetes mellitus in shift workers. Br. J. Ind. Med. 49, 513–515.

Schneider, J.E., Wise, J.D., Benton, N.A., Brozek, J.M., Keen-Rhinehart, E., 2013. When do we eat? Ingestive behavior, survival, and reproductive success. Horm. Behav. 64, 702–728.

Schneider, J.E., Brozek, J.M., Keen-Rhinehart, E., 2014. Our stolen figures: the interface of sexual differentiation, endocrine disruptors, maternal programming, and energy balance. Horm. Behav. 66, 104–119.

Simon, N., 2002. Hormonal processes in the development and expression of aggressive behavior. In: Pfaff, D.W., Arnold, A.P., Fahrbach, S., Etgen, A.M., Rubin, R.T. (Eds.), Hormones, Brain, and Behavior. Elsevier.

Simpson, E.R., Davis, S.R., 2001. Minireview: aromatase and the regulation of estrogen biosynthesis–some new perspectives. Endocrinology 142, 4589–4594.

Simpson, E.R., Jones, M.E., 2006. Of mice and men: the many guises of estrogens. Ernst Scher. Found. Symp. Proc. 1, 45–67.

Simpson, E.R., Clyne, C., Speed, C., Rubin, G., Bulun, S., 2001. Tissue-specific estrogen biosynthesis and metabolism. Ann. N.Y. Acad. Sci. 949, 58–67.

Simpson, E.R., Misso, M., Hewitt, K.N., Hill, R.A., Boon, W.C., Jones, M.E., Kovacic, A., Zhou, J., Clyne, C.D., 2005. Estrogen–the good, the bad, and the unexpected. Endocr. Rev. 26, 322–330.

Simpson, E.R., 2003. Sources of estrogen and their importance. J. Steroid Biochem. Mol. Biol. 86, 225–230.

Taylor, S.E., Klein, L.C., et al., 2000. Female responses to stress: tend and befriend; not fight or flight. Psychol. Rev. 107, 411–429.

Temple, J.L., Millar, R.P., Rissman, E.F., 2003. An evolutionarily conserved form of gonadotropin-releasing hormone coordinates energy and reproductive behavior. Endocrinology 144, 13–19.

Vicennati, V., Garelli, S., Rinaldi, E., Di Dalmazi, G., Pagotto, U., Pasquali, R., 2014. Cross-talk between adipose tissue and the HPA axis in obesity and overt hypercortisolemic states. Horm. Mol. Biol. Clin. Investig. 17, 63–77.

Vienberg, S.G., Bjornholm, M., 2014. Chronic glucocorticoid treatment increases de novo lipogenesis in visceral adipose tissue. Acta Physiol. (Oxf.) 211, 257–259.

Weitzman, E.D., 1980. Biologic rhythms and hormone secretion patterns. In: Krieger, D.T., Hughes, J.C. (Eds.), Neuroendocrinology. Sinauer Associates, Sunderland, MA, pp. 85–92.

Zondek, 1934. Oestrogenic hormone in the urine of the stallion. Nature 133, 494–495.

## Chapter 1: *Hormones Can Facilitate or Suppress Behaviors*

Abizaid, A., Horvath, T.L., 2012. Ghrelin and the central regulation of feeding and energy balance. Indian J. Endocrinol. Metab. 16 (Suppl. 3), S617–S626.

Alexander, J.L., Kotz, K., Dennerstein, L., Kutner, S.J., Wallen, K., Notelovitz, M., 2004. The effects of postmenopausal hormone therapies on female sexual functioning: a review of double-blind, randomized controlled trials. Menopause 11, 749–765.

Ammar, A.A., Sederholm, F., Saito, T.R., Scheurink, A.J., Johnson, A.E., Sodersten, P., 2000. NPY-leptin: opposing effects on appetitive and consummatory ingestive behavior and sexual behavior. Am. J. Physiol. Regul. Integr. Comp. Physiol. 278 (6), R1627–R1633.

Andrews, Z.B., 2011. Central mechanisms involved in the orexigenic actions of ghrelin. Peptides 32 (11), 2248–2255.

Barbaccia, M.L., Serra, M., Purdy, R.H., Biggio, G., 2001. Stress and neuroactive steroids. Int. Rev. Neurobiol. 46, 243–272.

Barbarich, N., 2002. Is there a common mechanism of serotonin dysregulation in anorexia nervosa and obsessive compulsive disorder? Eat. Weight Disord. 7 (3), 221–231.

Barsh, G.S., Schwartz, M.W., 2002. Genetic approaches to studying energy balance: perception and integration. Nat. Rev. Genet. 3 (8), 589–600.

Bartness, T.J., Keen-Rhinehart, E., Dailey, M.J., Teubner, B.J., 2011. Neural and hormonal control of food hoarding. Am. J. Physiol. Regul. Integr. Comp. Physiol. 301 (3), R641–R655.

Baskin, D.G., Blevins, J.E., Schwartz, M.W., 2001. How the brain regulates food intake and body weight: the role of leptin. J. Pediatr. Endocrinol. Metab. 14 (Suppl. 6), 1417–1429.

Berthold, O., 1849. Transplantation der Hoden. In: Archiv für Anatomie, physiologie und wissenschaftliche Medicin, pp. 42–46 Trans. by Quiring, D.P., 1944. Transplantation of Testis. In: Bulletin of the History of Medicine vol. 16, 399–401.

Blaustein, J.D., 2008. Neuroendocrine regulation of feminine sexual behavior: lessons from rodent models and thoughts about humans. Annu. Rev. Psychol. 59, 93–118.

Blevins, J.E., Schwartz, M.W., Baskin, D.G., 2002. Peptide signals regulating food intake and energy homeostasis. Can. J. Physiol. Pharmacol. 80 (5), 396–406.

Branson, R., Potoczna, N., Kral, J.G., Lentes, K.U., Hoehe, M.R., Horber, F.F., 2003. Binge eating as a major phenotype of melanocortin 4 receptor gene mutations. N. Engl. J. Med. 348 (12), 1096–1103.

Carter, C.S., Keverne, E.B., 2002. The neurobiology of social affiliation and pair bonding. In: Pfaff, D.W., Arnold, A.P., Etgen, A.M., Fahrbach, S.E., Rubin, R.T. (Eds.), Hormones, Brain and Behavior, vol. 1. Academic Press, San Diego, CA, pp. 299–337.

Clark, J.T., Kalra, P.S., Kalra, S.P., 1984. Neuropeptide Y stimulates feeding but inhibits sexual behavior in rats. Endocrinology 117 (6), 2435–2442.

Clark, J.T., Kalra, P.S., Kalra, S.P., 1985. Neuropeptide Y stimulates feeding but inhibits sexual behavior in rats. Endocrinology 117 (6), 2435–2442.

Cummings, J.A., Becker, J.B., 2012. Quantitative assessment of emale sexual motivation in the rat: hormonal control of motivation. J. Neurosci. Methods 204 (2), 227–233.

Cummings, D.E., Schwartz, M.W., 2003. Genetics and pathophysiology of human obesity. Annu. Rev. Med. 54, 453–471.

Cummings, D.E., Clement, K., Purnell, J.Q., Vaisse, C., Foster, K.E., Frayo, R.S., Schwartz, M.W., Basdevant, A., Weigle, D.S., 2002a. Elevated plasma ghrelin levels in Prader-Willi syndrome. Nat. Med. 7, 643–644.

Cummings, D.E., Weigle, D.S., Frayo, R.S., Breen, P.A., Ma, M.K., Dellinger, E.P., Purnell, J.Q., 2002b. Plasma ghrelin levels after diet-induced weight loss or gastric bypass surgery. N. Engl. J. Med. 346 (21), 1623–1630.

Dailey, M.J., Moran, T.H., 2013. Glucagon-like peptide 1 and appetite. Trends Endocrinol. Metab. 24, 85–91.

Damassa, D.A., Smith, E.R., Tennent, B., Davidson, J.M., 1977. The relationship between circulating testosterone levels and male sexual behavior in rats. Horm. Behav. 8, 275–286.

DelParigi, A., Tschop, M., Heiman, M.L., Salbe, A.D., Vozarova, B., Sell, S.M., Bunt, J.C., Tataranni, P.A., 2002. High circulating ghrelin: a potential cause of hyperphagia and obesity in Prader-Willi syndrome. J. Clin. Endocrinol. Metab. 87 (12), 5461–5464.

Denis, R.G., Joly-Amado, A., Webber, E., Langlet, F., Schaeffer, M., Padilla, S.L., Cansell, C., Dehouck, B., Castel, J., Delbes, A.S., Martinez, S., Lacombe, A., Rouch, C., Kassis, N., Fehrentz, J.A., Martinez, J., Verdie, P., Hnasko, T.S., Palmiter, R.D., Krashes, M.J., Guler, A.D., Magnan, C., Luquet, S., 2015. Palatability can drive feeding independent of AgRP neurons. Cell Metab. 22 (4), 646–657.

DiPatrizio, N.V., Astarita, G., Schwartz, G., Li, X., Piomelli, D., 2011. Endocannabinoid signal in the gut controls dietary fat intake. Proc. Natl. Acad. Sci. U.S.A. 108, 12904–12908.

Ellacott, K.L., Cone, R.D., 2006. The role of the central melanocortin system in the regulation of food intake and energy homeostasis: lessons from mouse models. Philos. Trans. R. Soc. Lond. B. Biol. Sci. 361, 1265–1274.

Engel, S.R., Grant, K.A., 2001. Neurosteroids and behavior. Int. Rev. Neurobiol. 46, 321–348.

Flier, J.S., May 1, 1998. What's in a name? In search of leptin's physiologic role. J. Clin. Endocrinol. Metab. 83 (5), 1407–1413.

Freeman, M.E., 1994. The neuroendocrine control of the ovarian cycle of the rat. In: Knobil, E., Neill, J.D. (Eds.), The Physiology of Reproduction, vol. 2. second ed. Raven Press, New York.

Grill, H.J., Schwartz, M.W., Kaplan, J.M., Foxhall, J.S., Breininger, J., Baskin, D.G., 2002. Evidence that the caudal brainstem is a target for the inhibitory effect of leptin on food intake. Endocrinology 143 (1), 239–246.

Grunt, J.A., Young, W.C., 1952. Differential reactivity of individuals and the response of the male Guinea pig to testosterone propionate. Endocrinology 51 (3), 237–248.

Gunst, A., Jern, P., Westberg, L., Johansson, A., Salo, B., Burri, A., Spector, T., Eriksson, E., Sandnabba, N.K., Santtila, P., 2015. A study of possible associations between single nucleotide polymorphisms in the estrogen receptor 2 gene and female sexual desire. J. Sex. Med. 12 (3), 676–684.

Halaas, J.L., Gajiwala, K.S., Maffei, M., Cohen, S.L., Chait, B.T., Rabinowitz, D., Lallone, R.L., Burley, S.K., Friedman, J.M., 1995. Weight-reducing effects of the plasma protein encoded by the obese gene. Science 269 (5223), 543–546.

Hewson, A.K., Tung, L.Y., Connell, D.W., Tookman, L., Dickson, S.L., 2002. The rat arcuate nucleus integrates peripheral signals provided by leptin, insulin, and a ghrelin mimetic. Diabetes 51 (12), 3412–3419.

Hull, E., 2017. Male sexual behavior. In: Pfaff, D., Joels, M. (Eds.), Hormones, Brain and Behavior, vol. 1, third ed., pp. 1–45.

Inaba, A., Komori, Y., Muroi, Y., Kinoshita, K., Ishii, T., 2016. Neuropeptide Y signaling in the dorsal raphe nucleus inhibits male sexual behavior in mice. Neuroscience 320, 140–148.

Insel, T.R., Young, L.E., 2000. Neuropeptides and the evolution of social behavior. Curr. Opin. Neurobiol. 10, 784–789.

Jacoangeli, F., Zoli, A., Taranto, A., Staar-Mezzasalma, F., Ficoneri, C., Pierangeli, S., Menzinger, G., Bollea, M.R., 2002. Osteoporosis and anorexia nervosa: relative role of endocrine alterations and malnutrition. Eat. Weight Disord. 7 (3), 190–195.

Kaufman, J., Plotsky, P.M., Nemeroff, C.B., Charney, D.S., 2000. Effects of early adverse experiences on brain structure and function: clinical implications. Biol. Psychiatry 48, 778–790.

Keen-Rhinehart, E., Ondek, K., Schneider, J.E., 2013. Neuroendocrine regulation of ingestive behavior. Front. Neurosci. 7, 213. http://dx.doi.org/10.3389/fnins.2013.00213.

Kowalska, I., Karczewska-Kupczewska, M., Straczkowski, M., Rubin, R.T., 2016. Anorexia nervosa, bulimia nervosa, and other eating disorders. In: Jameson, J.L., DeGroot, L.J. (Eds.), Endocrinology Adult and Pediatric, seventh ed. Elsevier/Saunders, Philadelphia, pp. 498–514.

Lal, S., Kirkup, A.J., Brunsden, A.M., Thompson, D.G., Grundy, D., 2001. Vagal afferent responses to fatty acids of different chain length in the rat. Am. J. Physiol. Gastrointest. Liver Physiol. 281 (4), G907–G915.

MacDougald, O.A., Hwang, C.S., Fan, H., Lane, M.D., 1995. Regulated expression of the obese gene product (leptin) in white adipose tissue and 3T3-L1 adipocytes. Proc. Natl. Acad. Sci. U.S.A. 92 (20), 9034–9037.

Madhuranath, B.N., Yajuryedi, H.N., 2011. Progesterone prevents corticosterone mediated inhibition of estrous behavior in rats. Indian J. Exp. Biol. 49 (5), 313–318.

Maffei, M., Fei, H., Lee, G.H., Dani, C., Leroy, P., Zhang, Y., Proenca, R., Negrel, R., Ailhaud, G., Friedman, J.M., 1995. Increased expression in adipocytes of *ob* RNA in mice with lesions of the hypothalamus and with mutations at the *db* locus. Proc. Natl. Acad. Sci. U.S.A. 92 (15), 6957–6960.

Margetic, S., Gazzola, C., Pregg, G.G., Hill, R.A., 2002. Leptin: a review of its peripheral actions and interactions. Int. J. Obes. Relat. Metab. Disord. 26 (11), 1407–1433.

Masbad, S., 2014. The role of glucagon-like peptide-1 impairment in obesity and potential therapeutic implications. Diabetes, Obes. Metab. 16, 9–21.

Mason, J.W., 1968. Organization of the multiple endocrine responses to avoidance in the monkey. Psychosom. Med. 30, 774–790.

McMinn, J.E., Baskin, D.G., Schwartz, M.W., 2002. Neuroendocrine mechanisms regulation food intake and body weight. Obes. Rev. 26 (11), 1407–1433.

Meisel, R.L., Sterner, M.R., 1990. Progesterone inhibition of sexual behavior is accompanied by an activation of aggression in female Syrian hamsters. Physiol. Behav. 47 (3), 415–417.

Mercer, R.E., Chee, M.J.S., Colmers, W.F., 2011. The role of NPY in hypothalamic mediated food intake. Front. Neuroendocrinol. 32 (4), 398–415.

Morton, G.J., Schwartz, M.W., 2001. The NPY/AgrP neuron and energy homeostasis. Int. J. Obes. Relat. Metab. Disord. 5 (Suppl.), S56–S62.

Morton, G.J., Niswender, K.D., Rhodes, C.J., Myers Jr., M.G., Blevins, J.E., Baskin, D.G., Schwartz, M.W., 2003. Arcuate nucleus-specific leptin receptor gene therapy attenuates the obesity phenotype of koletsky (fa(k)/fa(k)) rats. Endocrinology 144 (5), 2016–2024.

Nijenhusiu, W.A., Garner, K.M., VanRozen, R.J., Adan, R.A., 2003. Poor cell surface expression of human melanocortin-4 receptor mutations associated with obesity. J. Biol. Chem. 278.

Pacak, K., Palkovits, M., 2001. Stressor specificity of central neuroendocrine responses: implications for stress-related disorders. Endocr. Rev. 22, 502–548.

Parry, B.L., Berga, S.L., 2002. Premenstrual dysphoric disorder. In: Pfaff, D.W., Arnold, A.P., Etgen, A.M., Fahrbach, S.E., Rubin, R.T. (Eds.), Hormones, Brain and Behavior, vol. 5. Academic Press, San Diego, CA, pp. 531–552.

Pastor, Z., Holla, K., Chmel, R., 2013. The influence of combined oral contraceptives on female sexual desire: a systematic review. Eur. J. Contracept. Reprod. Health Care 18 (1), 27–43.

Phillips, M.I., Olds, J., 1969. Unit activity: motivation-dependent responses from midbrain neurons. Science 165 (899), 1269–1271.

Phillips, M.I., Heininger, F., Toffolo, S., 1996. The role of brain angiotensin in thirst and AVP release induced by hemorrhage. Regul. Pept. 66 (1–2), 3–11.

Poggioli, R., Vergoni, A.V., Marrama, D., Giuliani, D., Bertolini, A., 1990. NPY-induced inhibition of male copulatory activity is a direct behavioural effect. Neuropeptides 16.

Reame, N.E., 2001. Premenstrual syndrome. In: DeGroot, L.J., Jameson, J.L. (Eds.), Endocrinology, fourth ed. Saunders, Philadelphia, PA, pp. 2147–2152.

Roney, J.R., Simmons, Z.L., 2013. Hormonal predictors of sexual motivation in natural menstrual cycles. Horm. Behav. 63, 636–645.

Roujeau, C., Jockers, R., Dam, J., 2014. New pharmacological perspectives for the leptin receptor. Front. Endocrinol 5. http://dx.doi.org/10.3389/fendo.2014.00167.

Rubinow, D.R., Schmidt, P.J., Roca, C.A., Daly, R.C., 2002. Gonadal hormones and behavior in women: concentrations versus context. In: Pfaff, D.W., Arnold, A.P., Etgen, A.M., Fahrbach, S.E., Rubin, R.T. (Eds.), Hormones, Brain and Behavior, vol. 5. Academic Press, San Diego, CA, pp. 37–73.

Schneider, J.E., Brozek, J.M., Keen-Rhinehart, E., 2014. Our stolen figures: the interface of sexual differentiation, endocrine distruptors, maternal programming, and energy balance. Horm. Behav. 66 (1), 104–119.

Schneider, J.E., Wise, J.D., Benton, N.A., Brozek, J.M., Keen-Rhinehart, E., 2013. When do we eat? Ingestive behavior, survival, and reproductive success. Horm. Behav. 64, 702–728.

Schwartz, M.W., Morton, G.J., 2002. Obesity: keeping hunger at bay. Nature 418, 595–597.

Schwartz, M.W., Woods, S.C., Seeley, R.J., Barsh, G.S., Baskin, D.G., Leibel, R.L., 2003. Is the energy homeostasis system inherently biased toward weight gain? Diabetes 52 (2), 227–231.

Schwartz, M.W., 2001. Brain pathways controlling food intake and body weight. Exp. Biol. Med. 226 (11), 978–981.

Shibuya, I., Utsunomiya, K., Toyohira, Y., Ueno, S., Tsutsui, M., Cheah, T.B., Ueta, Y., Izuma, F., Yanagihara, N., 2002. Regulation of catecholamine synthesis by leptin. Ann. N.Y. Acad. Sci. 971, 522–527.

Smith, M.S., Freeman, M.E., Neill, J.D., 1975. The control of progesterone secretion during the estrous cycle and early pseudopregnancy in the rat: prolactin, gonadotropin and steroid levels associated with rescue of the corpus luteum of pseudopregnancy. Endocrinology 96 (1), 219–226.

Smith, S.S., 2002. Novel effects of neuroactive steroids in the central nervous system. In: Pfaff, D.W., Arnold, A.P., Etgen, A.M., Fahrbach, S.E., Rubin, R.T. (Eds.), Hormones, Brain and Behavior, vol. 3. Academic Press, San Diego, CA, pp. 747–778.

Snyder, P.J., Bhasin, S., Cunningham, G.R., Matsumoto, A.M., Stephens-Shields, A.J., Cauley, J.A., Gill, T.M., Barrett-Connor, E., Swerdloff, R.S., Wang, C., Ensrud, K.E., Lewis, C.E., Farrar, J.T., Cella, D., Rosen, R.C., Pahor, M., Crandall, J.P., Molitch, M.E., Cifelli, D., Dougar, D., Fluharty, L., Resnick, S.M., Storer, T.W., Anton, S., Basaria, S., Diem, S.J., Hou, X., Mohler III, E.R., Parsons, J.K., Wenger, N.K., Zeldow, B., Landis, J.R., Ellenberg, S.S., for the Testosterone Trials Investigators, 2016. Effects of testosterone treatment in older men. N. Engl. J. Med. 374 (7), 611–624.

Swart, I., Jahng, J.W., Overton, J.M., Houpt, T.A., 2002. Hypothalamic NPY, AgrP, and POMC mRNA responses to leptin and refeeding mice. Am. J. Physiol. Regul. Integ. Comp. Physiol. 283 (5), R1020–R1026.

Wallen, K., Winston, L.A., Gaventa, S., Davis-DaSilva, M., Collins, D.C., 1984. Periovulatory changes in female sexual behavior and patterns of ovarian steroid secretion in group-living rhesus monkeys. Horm. Behav. 18, 431–450.

Wallen, K., 2013. Women are not as unique as thought by some: comment on "hormonal predictors of sexual motivation in natural menstrual cycles," by Roney and Simmons. Horm. Behav. 63 (4), 634–635.

Wang, C., Swerdloff, R.S., Iranmanesh, A., Dobs, A., Snyder, P.J., Cunningham, G., Matsumoto, A.M., Weber, T., Berman, N., The Testosterone Gel Study Group, 2000. Transdermal testosterone gel improves sexual function, mood, muscle strength, and body composition parameters in hypogonadal men. J. Clin. Endocrinol. Metab. 85, 2839–2853.

Wilson, J.D., Roehrborn, C., 1999. Long-term consequences of castration in men: lessons from the Skoptzy and the eunuchs of the Chinese and Ottoman courts. J. Clin. Endocrinol. Metab. 84, 4324–4331.

Wisse, B.E., Schwartz, M.W., 2001. Role of melanocortins in control of obesity. Lancet 358, 857–859.

Wisse, B.E., Frayo, R.S., Schwartz, M.W., Cummings, D.E., 2001. Reversal of cancer anorexia by blockade of central melanocortin receptors in rats. Endocrinology 142 (8), 3292–3301.

Zhou, Q.Y., Palmiter, R.D., 1995. Dopamine-deficient mice are severely hypoactive, adipsic, and aphagic. Cell 83 (7), 1197–1209.

Zucker, I., 1968. Biphasic effects of progesterone on sexual receptivity in the female Guinea pig. J. Comp. Physiol. Psychol. 65 (3), 472–478.

## Chapter 2: Hormone Combinations Can Be Important for Behavior

Abiko, T., Kimura, Y., 2001. Syntheses and effect of bombesin-fragment 6–14 and its four analogues on food intake in rats. Curr. Pharm. Biotechnol. 2 (2), 201–207.

Blackburn, R.E., Samson, W.K., Fulton, R.J., Stricker, E.M., Verbalis, J.G., 1995. Central oxytocin and ANP receptors mediate osmotic inhibition of salt appetite in rats. Am. J. Physiol. 269, R245–R251.

Blair-West, J.R., Carey, K.D., Denton, D.A., Weisinger, R.S., Shade, R.E., 1998. Evidence that brain angiotensin II is involved in both thirst and sodium appetite in baboons. Am. J. Physiol. 275 (5, Pt. 2), R1639–R1646.

Blaustein, J.D., Erskine, M.S., 2002. Feminine sexual behavior: cellular integration of hormonal and afferent information in the rodent forebrain. In: Pfaff, D.W., Arnold, A.P., Etgen, A.M., Fahrbach, S.E., Rubin, R.T. (Eds.), Hormones, Brain and Behavior, vol. 1. Academic Press, San Diego, CA, pp. 139–214.

Briggs, D.I., Enriori, P.J., Lemus, M.B., Cowley, M.A., Andrews, Z.B., 2010. Diet-induced obesity causes ghrelin resistance in arcuate npy/agrp neurons. Endocrinology 151 (10), 4745–4755.

Brown, M., Allen, R., Villarreal, J., Rivier, J., Vale, W., 1978. Bombesin-like activity: radioimmunologic assessment in biological tissues. Life Sci. 23 (27–28), 2721–2728.

Buggy, J., Jonklaas, J., 1984. Sodium appetite decreased by central angiotensin blockade. Physiol. Behav. 32 (5), 737–742.

Burrell, L.M., Lambert, H.J., Baylis, P.H., 1991. Effect of atrial natriuretic peptide on thirst and arginine vasopressin release in humans. Am. J. Physiol. 260, R475–R479.

Choleris, E., Gustafsson, J.A., et al., 2003. An estrogen-dependent four-gene micronet regulating social recognition: a study with oxytocin and estrogen receptor-alpha and -beta knockout mice. Proc. Natl. Acad. Sci. U.S.A. 100 (10), 6192–6197.

Cummings, D.E., Overduin, J., 2007. Gastrointestinal regulation of food intake. J. Clin. Invest. 117 (1), 13–23.

Daniels, D., Flanagan-Cato, L.M., 2000. Functionally-defined compartments of the lordosis neural circuit in the ventromedial hypothalamus in female rats. J. Neurobiol. 45, 1–3.

Daniels, D., Miselis, R.R., Flanagan-Cato, L.M., 1999. Central neuronal circuit innervating the lordosis muscles defined by transneuronal transport of pseudorabies virus. J. Neurosci. 19, 2823–2833.

de Bold, A.J., November 15, 1985. Atrial natriuretic factor: a hormone produced by the heart. Science 230 (4727), 767–770.

Denton, D., 1984. The Hunger for Salt: An Anthropological, Physiological and Medical Analysis. Springer, Berlin.

De Wied, D., Jolles, J., 1982. Neuropeptides derived from pro-opiocortin: behavioral, physiological, and neurochemical effects. Physiol. Rev. 62, 976–1059.

Dellovade, T.L., Zhu, Y.-S., Krey, L., Pfaff, D.W., 1996. Thyroid hormone and estrogen interact to regulate behavior. Proc. Natl. Acad. Sci. U.S.A. 93, 12581–12586.

Dhingra, H., Roongsritong, C., Kurtzman, N.A., 2002. Brain natriuretic peptide: role in cardiovascular and volume homeostasis. Semin. Nephrol. 22 (5), 423–437.

Dohanich, G., 2002. Gonadal steroids, learning, and memory. In: Pfaff, D.W., Arnold, A.P., Etgen, A.M., Fahrbach, S.E., Rubin, R.T. (Eds.), Hormones, Brain and Behavior, vol. 2. Academic Press, San Diego, CA, pp. 265–327.

Ehrlich, K.J., Fitts, D.A., 1990. Atrial natriuretic peptide in the subfornical organ reduces drinking induced by angiotensin or in response to water deprivation. Behav. Neurosci. 104 (2), 365–372.

Epstein, A.N., Fitzsimons, J.T., Rolls, B.J., 1970. Drinking induced by injection of angiotensin into the brain of the rat. J. Physiol. 210 (2), 457–474.

Figlewicz, D.P., Nadzan, A.M., Sipols, A.J., Green, P.K., Liddle, R.A., Porte Jr., D., Woods, S.C., 1992. Intraventricular CCK-8 reduces single meal size in the baboon by interaction with type-A CCK receptors. Am. J. Physiol. 263 (4, Pt. 2), R863–R867.

Fitts, D.A., Thunhorst, R.L., Simpson, J.B., 1985. Diuresis and reduction of salt appetite by lateral ventricular infusions of atriopeptin II. Brain Res. 348, 118–124.

Fitzsimons, J.T., 1998. Angiotensin, thrist, and sodium appetite. Physiol. Rev. 78, 583–686.

Fitzsimons, J.T., Stricker, E.M., 1971. Sodium appetite and the renin–angiotensin system. Nat. New Biol. 231 (19), 58–60.

Fletcher, S.W., Colditz, G.A., 2002. Failure of estrogen plus progestin therapy for prevention [editorial]. JAMA 288, 366–367.

Fluharty, S.J., Epstein, A.N., 1983. Sodium appetite elicited by intracerebroventricular infusion of angiotensin in the rat. II. Synergistic interaction with systemic mineralocorticoids. Behav. Neurosci. 97 (5), 746–758.

Ganesan, R., Sumners, C., 1989. Glucocorticoids potentiate the dipsogenic action of angiotensin II. Brain Res. 499 (1), 121–130.

Geerling, J.C., Lowey, 2008. Central regulation of sodium appetite. Exp. Physiol. 93 (2), 177–209.

Gibbs, J., Fauser, D.J., Row, E.A., Rolls, B.J., Rolls, E.T., Maddison, S.P., 1979. Bombesin suppresses feeding in rats. Nature 282 (5735), 208–210.

Gibbs, J., Young, R.C., Smith, G.P., 1997. Cholecystokinin decreases food intake in rats. Obes. Res. 5 (3), 284–290.

Gonzalez-Mariscal, G., Poindron, P., 2002. Parental care in mammals: immediate internal and sensory factors of control. In: Pfaff, D.W., Arnold, A.P., Etgen, A.M., Fahrbach, S.E., Rubin, R.T. (Eds.), Hormones, Brain and Behavior, vol. 1. Academic Press, San Diego, CA, pp. 215–298.

Grill, H.J., Hayes, M.R., 2012. Hindbrain neurons as an essential hub in the neuroanatomically distributed control of energy balance. Cell Metab. 16, 296–309.

Hamilton, R.B., Norgren, R., 1984. Central projections of gustatory nerves in the rat. J. Comp. Neurol. 222 (4), 560–577.

Holsboer, F., 1999. The rationale for corticotropin-releasing hormone receptor (CRH-R) antagonists to treat depression and anxiety. J. Psychiatr. Res. 33, 181–214.

Horvath, T.L., 2006. Synaptic plasticity in energy balance regulation. Obesity 5, 228S–233S.

Huda, M.S., Dovey, T., Wong, S.P., English, P.J., Halford, J., McCulloch, P., et al., 2009. Ghrelin restores 'lean-type' hunger and energy expenditure profiles in morbidly obese subjects but has no effect on postgastrectomy subjects. Int. J. Obes. 33 (3), 317–325.

Insel, T.R., Gingrich, B.S., Young, L.J., 2001. Oxytocin: who needs it? Prog. Brain Res. 133, 59–66.

Johnson, A.K., Thunhorst, R.L., 1997. The neuroendocrinology of thirst and salt appetite: visceral sensory signals and mechanisms of central integration. Front. Neuroendocrinol. 18 (3), 292–353.

Johnson, Z.V., Young, L.J., 2015. Neurobiological mechanisms of social attachment and pair bonding. Curr. Opin. Behav. Sci. 3, 38–44.

Jovanovic, Z., Yeo, G.S., 2010. Central leptin signaling: beyond the arcuate nucleus. Auton. Neurosci. 156, 8–14.

Kaufman, J., Plotsky, P.M., Nemeroff, C.B., Charney, D.S., 2000. Effects of early adverse experiences on brain structure and function: clinical implications. Biol. Psychiatry 48, 778–790.

Kirchner, H., Heppner, K.M., Tschop, M.H., 2012. The role of ghrelin in the control of energy balance. Handb. Exp. Pharmacol. 209, 161–184.

Kissileff, H.R., 2000. Ingestive behavior microstructure, basic mechanisms and clinical applications. Neurosci. Biobehav. Rev. 24, 171–172.

Kojima, M., Hosoda, H., Date, Y., Nakazato, M., Matsuo, H., Kangawa, K., 1999. Ghrelin is a growth-hormone-releasing acylated peptide from stomach. Nature 402 (6762), 656–660.

Lewko, B., Stepinski, J., May 15, 2002. Cyclic GMP signaling in podocytes. Microsc. Res. Tech. 574 (4), 232–235.

Li, Z., Ferguson, A.V., 1993. Subfornical organ efferents to paraventricular nucleus utilize angiotensin as a neurotransmitter. Am. J. Physiol. 265 (2, Pt. 2), R302–R309.

Lim, M.M., Young, L.J., 2004. Vasopressin-dependent neural circuits underlying pair bond formation in the monogamous prairie vole. Neuroscience 125, 35–45.

Ma, L.Y., McEwen, B.S., Sakai, R.R., Schulkin, J., 1993. Glucocorticoids facilitate mineralocorticoid-induced sodium intake in the rat. Horm. Behav. 27 (2), 240–250.

McCann, S.M., Gutkowska, J., Antunes-Rodrigues, J., 2003. Neuroendocrine control of body fluid homeostasis. Braz. J. Med. Biol. Res. 36, 165–181.

McCarthy, M.M., Kleopoulos, S.P., Mobbs, C.V., Pfaff, D.W., 1994. Infusion of antisense oligodeoxynucleotides to the oxytocin receptor in the ventromedial hypothalamus reduces estrogen-induced sexual receptivity and oxytocin receptor binding in the female rat. Neuroendocrinology 59, 432–440.

Meisel, R.L., Dohanich, G.P., McEwen, B.S., Pfaff, D.W., 1987. Antagonism of sexual behavior in female rats by ventromedial hypothalamic implants of antiestrogen. Neuroendocrinology 45 (3), 201–207.

Meisel, R.L., Pfaff, D.W., 1984. RNA and protein synthesis inhibitors: effects on sexual behavior in female rats. Brain Res. Bull. 12, 1870193.

Meisel, R.L., Pfaff, D.W., 1985. Specificity and neural sites of action of anisomycin in the reduction or facilitation of female sexual behavior in rats. Horm. Behav. 19, 237–251.

Melis, M.R., Arigolas, A., 1995. Nitric oxide donors in penile erection and yawing when infected in the central nervous system of male rats. Eur. J. Pharm. 294 (1), 9.

Mohon, M.A., 2001. An airman with Schmidt's syndrome. Fed. Air Surgeon's Med. Bull. 1–3.

Moran, T.H., Sawyer, T.K., Seeb, D.H., Arriglio, P.J., Lombard, M.A., McHugh, P.R., 1992. Potent and sustained satiety actions of a cholecystokinin octapeptid analogue. Am. J. Clin. Nutr. 2865–2905.

Morgan, M., Dellovade, T.L., Pfaff, D.W., 2000. Effect of thyroid hormone administration on estrogen-induced sex behavior in female mice. Horm. Behav. 37, 15–22.

Niswender, K.D., Schwartz, M.W., 2003. Insulin and leptin revisited: adiposity with overlapping physiological and intracellular signaling capabilities. Front. Neuroendcrinol. 24 (1), 1–10.

Odell, W.D., Burger, H.G., 2001. Menopause and hormone replacement. In: DeGroot, L.J., Jameson, J.L. (Eds.), Endocrinology, fourth ed. Saunders, Philadelphia, PA, pp. 2153–2162.

Parry, B.L., Berga, S.L., 2002. Premenstrual dysphoric disorder. In: Pfaff, D.W., Arnold, A.P., Etgen, A.M., Fahrbach, S.E., Rubin, R.T. (Eds.), Hormones, Brain and Behavior, vol. 5. Academic Press, San Diego, CA, pp. 531–552.

Pfaff, D.W., 1999. Drive: Neurobiological and Molecular Mechanisms of Sexual Motivation. MIT Press, Cambridge, MA.

Phillips, M.I., Gyurko, R., 1995. In vivo applications of antisense oligonucleotides for peptide research. Regul. Pept. 59 (2), 131–141.

Phillips, M.I., Heininger, F., Toffolo, S., 1996. The role of brain angiotensin in thirst and AVP release induced by hemorrhage. Regul. Pept. 66 (1–2), 3–11.

Phillips, M.I., 1987. Functions of angiotensin in the central nervous system. Annu. Rev. Physiol. 49, 413–435.

Porte Jr., D., Baskin, D.G., Schwartz, M.W., 2002. Leptin and insulin action in the central nervous system. Nutr. Rev. 60 (10, Pt. 2), S20–S29.

Reame, N.E., 2001. Premenstrual syndrome. In: DeGroot, L.J., Jameson, J.L. (Eds.), Endocrinology, fourth ed. Saunders, Philadelphia, PA, pp. 2147–2152.

Rubin, R.T., Reinisch, J.M., Haskett, R.F., 1981. Postnatal gonadal steroid effects on human behavior. Science 211, 1318–1324.

Rudolph, L.M., Bentley, G.E., Calandra, R.S., Paredes, A.H., Tesone, M., Wu, T.J., Micevych, P.E., 2016. Peripheral and central mechanisms involved in the hormonal control of male and female reproduction. J. Neuroendocrinol. 28.

Sakai, R.R., Ma, L.Y., Zhang, D.M., McEwen, B.S., Fluharty, S.J., 1996. Intracerebral administration of mineralocorticoid receptor antisense oligonucleotides attenuate adrenal steroid-induced salt appetite in rats. Neuroendocrinology 64 (6), 425–429.

Sakai, R.R., McEwen, B.S., Fluharty, S.J., Ma, L.Y., 2000. The amygdala: site of genomic and nongenomic arousal of aldosterone-induced sodium intake. Kidney Intern. 57 (4), 1337–1345.

Schulkin, J., 2002. Hormonal modulation of central motivational states. In: Pfaff, D.W., Arnold, A.P., Etgen, A.M., Fahrbach, S.E., Rubin, R.T. (Eds.), Hormones, Brain and Behavior, vol. 1. Academic Press, San Diego, CA, pp. 633–657.

Schulkin, J., 2003. Rethinking Homeostasis: Allostatic Regulation in Physiology and Pathophysiology. MIT Press, Cambridge, MA.

Schumacher, M., Robert, F., 2002. Progesterone: synthesis, metabolism, mechanisms of action, and effects in the nervous system. In: Pfaff, D.W., Arnold, A.P., Etgen, A.M., Fahrbach, S.E., Rubin, R.T. (Eds.), Hormones, Brain and Behavior, vol. 3. Academic Press, San Diego, CA, pp. 683–745.

Seibel, M.M., 2001. Ovulation induction and assisted reproduction. In: DeGroot, L.J., Jameson, J.L. (Eds.), Endocrinology, fourth ed. Saunders, Philadelphia, PA, pp. 2138–2145.

Simon, N.G., 2002. Hormonal processes in the development and expression of aggressive behavior. In: Pfaff, D.W., Arnold, A.P., Etgen, A.M., Fahrbach, S.E., Rubin, R.T. (Eds.), Hormones, Brain and Behavior, vol. 1. Academic Press, San Diego, CA, pp. 339–392.

Sinchak, K., Wagner, E., 2012. Front. Neuroendocrinol. 3, 342–363.

Stewart, W.K., Fleming, L.W., 1973. Features of a successful therapeutic fast of 382 days' duration. Postgrad. Med. J. 49 (569), 203–209.

Sumners, C., Fregly, M.J., 1989. Modulation of angiotensin II binding sites in neuronal cultures by mineralocorticoids. Am. J. Physiol. 256 (1, Pt. 1), C121–C129.

Sumners, C., Myers, L.M., 1991. Angiotensin II decreases cGMP levels in neuronal cultures from rat brain. Am. J. Physiol. 260 (1, Pt. 1), C79–C87.

Sutton, A.K., Myers Jr., M.G., Olson, D.P., 2016. The role of PVH circuits in leptin action and energy balance. Annu. Rev. Physiol. 78, 207–221.

Takeda, R., Takayama, Y., Tagawa, S., Komel, L., 1999. Schmidt's syndrome: autoimmune polyglandular disease of the adrenal and thyroid glands. Isr. Med. Assoc. J. 1, 285–286.

Thomas, C.J., Head, G.A., Woods, R.L., 1998. ANP and bradycardic reflexes in hypertensive rats: influence of cardiac hypertrophy. Hypertension 32, 548–555.

Vasudevan, N., Davidkova, G., Zhu, Y.S., Koibuchi, N., Chin, W.W., Pfaff, D.W., 2001a. Differential interactions of estrogen receptor and thyroid hormone receptor isoforms on the rat oxytocin receptor promoter leads to differences in transcriptional regulation. Neuroendocrinology 74 (5), 309–324.

Vasudevan, N., Koibuchi, N., Chin, W.W., Pfaff, D.W., 2001b. Differential crosstalk between estrogen receptor (ER)α and ERβ and the thyroid hormone receptor isoforms results in flexible regulation of the consensus ERE. Brain Res. Mol. Brain Res. 95 (1–2), 9–17.

Vasudevan, N., Zhu, Y.S., Daniels, S., Koibuchi, N., Chin, W.W., Pfaff, D.W., 2001c. Crosstalk between oestrogen receptors and thyroid hormone receptor isoforms results in differential regulation of the preproenkephalin gene. J. Neuroendocrinol. 13.

Vink, S., Jin, A.H., Poth, K.J., Head, G.A., Alewood, P.F., 2012. Natriuretic peptide drug leads from snake venom. Toxicon 59, 434–445.

Wilson, W.L., Starbuck, E.M., Fitts, D.A., 2002. Salt appetite of adrenalectomized rats after a lesion of the SFO. Behav. Brain Res. 136 (2), 449–453.

Wingfield, J.C., Romero, L.M., 2001. Adrenocortical responses to stress and their modulation in free-living vertebrates. In: McEwew, B.S. (Ed.), Coping with the Environment. Neural and Endocrine Mechanisms. Oxford University Press, Oxford.

Woods, R.L., 2004. Cardioprotective functions of atrial natriuretic peptide and B-type natriuretic peptide: a brief review. Clin. Exp. Pharmacol. Physiol. 31, 791–794.

Young, L.J., Lim, M.M., Gingrich, B., Insel, T.R., 2001. Cellular mechanisms of social attachment. Horm. Behav. 40, 133–138.

Zhang, D.M., Stellar, E., Epstein, A.N., 1984. Together intracranial angiotensin and systemic mineralocorticoid produce avidity for salt in the rat. Physiol. Behav. 32, 677–681.

## Chapter 3: One Hormone Can Have Many Effects

Abiko, T., Kimura, Y., 2001. Syntheses and effect of bombesin-fragment 6-14 and its four analogues on food intake in rats. Curr. Pharm. Biotechnol. 2 (2), 201–207.

Ahima, R.S., Prabakaran, D., Mantzoros, C., Qu, D., Lowell, B., Maratos-Flier, E., Flier, J.S., 1996. Role of leptin in the neuroendocrine response to fasting. Nature 382, 250–252.

Ahima, R.S., Dushay, J., Flier, S.N., Prabakaran, D., Flier, J.S., 1997. Leptin accelerates the onset of puberty in normal female mice. J. Clin. Invest. 99, 391–395.

Alde, S., Celis, M.E., 1980. Influence of alpha-melanotropin on LH release in the rat. Neuroendocrinology 31, 116–120.

Ammar, A.A., Sederholm, F., Saito, T.R., Scheurink, A.J., Johnson, A.E., Sodersten, P., 2000. NPY-leptin: opposing effects on appetitive and consummatory ingestive behavior and sexual behavior. Am. J. Physiol. Regul. Integr. Comp. Physiol. 278, R1627–R1633.

Babaei-Balderlou, F., Khazali, H., 2016. Effects of ghrelin on sexual behavior and luteinizing hormone beta-subunit gene expression in male rats. J. Reprod. Infertil. 17, 88–96.

Backholer, K., Smith, J., Clarke, I.J., 2009. Melanocortins may stimulate reproduction by activating orexin neurons in the dorsomedial hypothalamus and kisspeptin neurons in the preoptic area of the ewe. Endocrinology 150, 5488–5497.

Balon, R., Segraves, R.T., 2008. Survey of treatment practices for sexual dysfuntion(s) associated with anti-depressants. J. Sex. Marital Ther. 34, 353–365.

Beak, S.A., Heath, M.M., Small, C.J., Morgan, D.G., Ghatei, M.A., Taylor, A.D., Buckingham, J.C., Bloom, S.R., Smith, D.M., 1998. Glucagon-like peptide-1 stimulates luteinizing hormone-releasing hormone secretion in a rodent hypothalamic neuronal cell line. J. Clin. Invest. 101, 1334–1341.

Bellefontaine, N., Elias, C.F., 2014. Minireview: metabolic control of the reproductive physiology: insights from genetic mouse models. Horm. Behav. 66, 7–14.

Blair-West, J.R., Carey, K.D., Denton, D.A., Weisinger, R.S., Shade, R.E., 1998. Evidence that brain angiotensin II is involved in both thirst and sodium appetite in baboons. Am. J. Physiol. 275 (5, Pt. 2), R1639–R1646.

Brindley, G.S., 1983. Cavernosal alpha-bockade: a new technique for the investigating and treating erectile impotence. Br. J. Psychiatry 143, 332–337.

Brown, M., Allen, R., Villarreal, J., Rivier, J., Vale, W., 1978. Bombesin-like activity: radioimmunologic assessment in biological tissues. Life Sci. 23 (27–28), 2721–2728.

Calisi, R.M., 2014. An integrative overview of the role of gonadotropin-inhibitory hormone in behavior: applying Tinbergen's four questions. Gen. Comp. Endocrinol. 203, 95–105.

Choleris, E., Gustafsson, J.A., et al., 2003. An estrogen-dependent four-gene micronet regulating social recognition: a study with oxytocin and estrogen receptor-alpha and -beta knockout mice. Proc. Natl. Acad. Sci. U.S.A. 100 (10), 6192–6197.

Clark, J.T., Kalra, P.S., Kalra, S.P., 1985. Neuropeptide Y stimulates feeding but inhibits sexual behavior in rats. Endocrinology 117, 2435–2442.

Clarke, I.J., Smith, J.T., Henry, B.A., Oldfield, B.J., Stefanidis, A., Millar, R.P., Sari, I.P., Chng, K., Fabre-Nys, C., Caraty, A., Ang, B.T., Chan, L., Fraley, G.S., 2012. Gonadotropin-inhibitory hormone is a hypothalamic peptide that provides a molecular switch between reproduction and feeding. Neuroendocrinology 95, 305–315.

Clarke, I.J., 2014. Interface between metabolic balance and reproduction in ruminants: focus on the hypothalamus and pituitary. Horm. Behav. 66, 15–40.

Corp, E.S., Greco, B., Powers, J.B., Marin Bivens, C.L., Wade, G.N., 2001. Neuropeptide Y inhibits estrous behavior and stimulates feeding via separate receptors in Syrian hamsters. Am. J. Physiol. Regul. Integr. Comp. Physiol. 280, R1061–R1068.

Csoka, A.B., Bahrick, A., Mehtonen, O.P., 2008. Persistent sexual dysfunction after discontinuation of selective serotonin reuptake inhibitors. J. Sexual Med. 5, 227–233.

Dellovade, T.L., Zhu, Y.-S., Krey, L., Pfaff, D.W., 1996. Thyroid hormone and estrogen interact to regulate behavior. Proc. Natl. Acad. Sci. U.S.A. 93, 12581–12586.

Dohanich, G., 2002. Gonadal steroids, learning, and memory. In: Pfaff, D.W., Arnold, A.P., Etgen, A.M., Fahrbach, S.E., Rubin, R.T. (Eds.), Hormones, Brain and Behavior, vol. 2. Academic Press, San Diego, CA, pp. 265–327.

Dorr, R.T., Lines, R., Levine, N., Brooks, C., Xiang, L., Hruby, V.J., Hadley, M.E., 1996. Evaluation of melanotan-II, a superpotent cyclic melanotropic peptide in a pilot phase-I clinical study. Life Sci. 58, 1777–1784.

Dungan, H.M., Clifton, D.K., Steiner, R.A., 2006. Minireview: kisspeptin neurons as central processors in the regulation of gonadotropin-releasing hormone secretion. Endocrinology 147, 1154–1158.

Elias, C.F., 2012. Leptin action in pubertal development: recent advances and unanswered questions. Trends Endocrinol. Metab. 23, 9–15.

Elias, C.F., 2014. A critical view of the use of genetic tools to unveil neural circuits: the case of leptin action in reproduction. Am. J. Physiol. Regul. Integr. Comp. Physiol. 306, R1–R9.

Ferguson, A.V., Donevan, S.D., Papas, S., Smith, P.M., 1990. Circumventricular structures: CNS sensors of circulating peptides and autonomic control centres. Endocrinol. Exp. 24, 19–27.

Figlewicz, D.P., Nadzan, A.M., Sipols, A.J., Green, P.K., Liddle, R.A., Porte Jr., D., Woods, S.C., 1992. Intraventricular CCK-8 reduces single meal size in the baboon by interaction with type-A CCK receptors. Am. J. Physiol. 263 (4, Pt. 2), R863–R867.

Fitts, D.A., Thurnhorst, R.L., Simpson, J.B., 1985. Diuresis and reduction of salt appetite by lateral ventricular infusions of atriopeptin II. Brain Res. 348, 118–124.

Fitzsimons, J.T., 1998. Angiotensin, thirst, and sodium appetite. Physiol. Rev. 78 (3), 583–686.

Fitzsimons, J.T., Stricker, E.M., 1971. Sodium appetite and the renin-angiotensin system. Nat. New Biol. 231 (19), 58–60.

Fitzsimons, J.T., 1969. The role of renal thirst factor in drinking induced by extracellular stimuli. J. Physiol. 201 (2), 349–368.

Fletcher, S.W., Colditz, G.A., 2002. Failure of estrogen plus progestin therapy for prevention [editorial]. JAMA 288, 366–367.

Gibbs, J., Fauser, D.J., Row, E.A., Rolls, B.J., Rolls, E.T., Maddison, S.P., 1979. Bombesin suppresses feeding in rats. Nature 282 (5735), 208–210.

Gibbs, J., Young, R.C., Smith, G.P., 1997. Cholecystokinin decreases food intake in rats. Obes. Res. 5 (3), 284–290.

Hamilton, R.B., Norgren, R., 1984. Central projections of gustatory nerves in the rat. J. Comp. Neurol. 222 (4), 560–577.

Hull, E., 2017. Male sexual behavior. In: Pfaff, D., Joels, M. (Eds.), Hormones, Brain and Behavior, vol. 1, third ed., pp. 1–45.

Johnson, A.K., Thunhorst, R.L., 1997. The neuroendocrinology of thirst and salt appetite: visceral sensory signals and mechanisms of central integration. Front. Neuroendocrinol. 18 (3), 292–353.

Jones, J.E., Pick, R.R., Dettloff, S.L., Wade, G.N., 2004. Metabolic fuels, neuropeptide Y, and estrous behavior in Syrian hamsters. Brain Res. 1007, 78–85.

Kalra, S.P., Clark, J.T., Sahu, A., Dube, M.G., Kalra, P.S., 1988. Control of feeding and sexual behaviors by neuropeptide Y: physiological implications. Synapse 2, 254–257.

Keene, A.C., Jones, J.E., Wade, G.N., Corp, E.S., 2003. Forebrain sites of NPY action on estrous behavior in Syrian hamsters. Physiol. Behav. 78, 711–716.

Kiyokawa, M., Matsuzaki, T., Iwasa, T., Ogata, R., Murakami, M., Kinouchi, R., Yoshida, S., Kuwahara, A., Yasui, T., Irahara, M., 2011. Neuropeptide Y mediates orexin A-mediated suppression of pulsatile gonadotropin-releasing hormone secretion in ovariectomized rats. J. Med. Invest. 58, 11–18.

Kleitz-Nelson, H.K., Domingues, J.M., Klingerman, C.M., Krishnamoorthy, K., Patel, K., Spiro, A.B., Struby, C., Patel, A., Schneider, J.E., 2010. Energetic challenges unmask the role of ovarian hormones in orchestrating ingestive and sex behaviors. Horm. Behav. 58, 563–574.

Klingerman, C.M., Krishnamoorthy, K., Patel, K., Spiro, A.B., Struby, C., Patel, A., Schneider, J.E., 2010. Energetic challenges unmask the role of ovarian hormones in orchestrating ingestive and sex behaviors. Horm. Behav. 58, 563–574.

Klingerman, C.M., Williams 3rd, W.P., Simberlund, J., Brahme, N., Prasad, A., Schneider, J.E., Kriegsfeld, L.J., 2011. Food restriction-induced changes in gonadotropin-inhibiting hormone cells are associated with changes in sexual motivation and food hoarding, but not sexual performance and food intake. Front. Endocrinol. (Lausanne) 2, 101.

Kojima, M., Hosoda, H., Date, Y., Nakazato, M., Matsuo, H., Kangawa, K., 1999. Ghrelin is a growth-hormone-releasing acylated peptide from stomach. Nature 402 (6762), 656–660.

Kriegsfeld, L.J., 2006. Driving reproduction: RFamide peptides behind the wheel. Horm. Behav. 50, 655–666.

Leon, S., Garcia-Galiano, D., Ruiz-Pino, F., Barroso, A., Manfredi-Lozano, M., Romero-Ruiz, A., Roa, J., Vazquez, M.J., Gaytan, F., Blomenrohr, M., van Duin, M., Pinilla, L., Tena-Sempere, M., 2014. Physiological roles of gonadotropin-inhibitory hormone signaling in the control of mammalian reproductive axis: studies in the NPFF1 receptor null mouse. Endocrinology 155, 2953–2965.

Limone, P., Calvelli, P., Altare, F., Ajmone-Catt, P., Lima, T., Molinatti, G.M., 1997. Evidence for an interaction between alpha-MSH and opioids in the regulation of gonadotropin secretion in man. J. Endocrinol. Invest. 20, 207–210.

Marin-Bivens, C.L., Kalra, S.P., Olster, D.H., 1998. Intraventricular injection of neuropeptide Y antisera curbs weight gain and feeding, and increases the display of sexual behaviors in obese Zucker female rats. Regul. Pept. 75–76, 327–334.

Meisel, R.L., O'Hanlon, J.K., Sachs, B.D., March 1984. Differential maintenance of penile responses and copulatory behavior by gonadal hormones in castrated male rats. Horm. Behav. 18 (1), 56–64.

Mohon, M.A., 2001. An airman with Schmidt's syndrome. Fed. Air Surgeon's Med. Bull. 1–3.

Molinoff, P.B., Shadiack, A.M., Earle, D., Diamond, L.E., Quon, C.Y., 2003. PT-141: a melanocortin agonist for the treatment of sexual dysfunction. Ann. N.Y. Acad. Sci. 994, 96–102.

Moran, T.H., Sawyer, T.K., Seeb, D.H., Arriglio, P.J., Lombard, M.A., McHugh, P.R., 1992. Potent and sustained satiety actions of a cholecystokinin octapeptid analogue. Am. J. Clin. Nutr. 2865–2905.

Morgan, M., Dellovade, T.L., Pfaff, D.W., 2000. Effect of thyroid hormone administration on estrogen-induced sex behavior in female mice. Horm. Behav. 37, 15–22.

Niswender, K.D., Schwartz, M.W., 2003. Insulin and leptin revisited: adiposity with overlapping physiological and intracellular signaling capabilities. Front. Neuroendcrinol. 24 (1), 1–10.

Odell, W.D., Burger, H.G., 2001. Menopause and hormone replacement. In: DeGroot, L.J., Jameson, J.L. (Eds.), Endocrinology, fourth ed. Saunders, Philadelphia, PA, pp. 2153–2162.

Parry, B.L., Berga, S.L., 2002. Premenstrual dysphoric disorder. In: Pfaff, D.W., Arnold, A.P., Etgen, A.M., Fahrbach, S.E., Rubin, R.T. (Eds.), Hormones, Brain and Behavior, vol. 5. Academic Press, San Diego, CA, pp. 531–552.

Pfaff, D.W., 1999. Drive: Neurobiological and Molecular Mechanisms of Sexual Motivation. MIT Press, Cambridge, MA.

Pfaus, J.G., 2011. Physiology of libido. In: Mulhall, J., Incrocci, L., Goldstein, I., Rosen, R. (Eds.), Cancer and Sexual Health. Current Clinical Urology. Humana Press.

Pfaus, J.G., Shadiack, A., Van Soest, T., Tse, M., Molinoff, P., 2004. Selective facilitation of sexual solicitation in the female rat by a melanocortin receptor agonist. Proc. Natl. Acad. Sci. U.S.A. 101, 10201–10204.

Phillips, M.I., 1987. Functions of angiotensin in the central nervous system. Annu. Rev. Physiol. 49, 413–435.

Piekarski, D.J., Zhao, S., Jennings, K.J., Iwasa, T., Kriegsfeld, L.J., 2012. Gonadotropin-Inhibitory Hormone Reduces Sexual Motivation and Paracopulatory Behaviors but Not Copulatory Behavior in Female Syrian Hamsters. Society for Behavioral Neuroendocrinology. Hormones and Behavior. Madison, Wisconsin.

Porte Jr., D., Baskin, D.G., Schwartz, M.W., 2002. Leptin and insulin action in the central nervous system. Nutr. Rev. 60 (10, Pt. 2), S20–S29.

Rasmussen, N., 2008. America's first amphetamine epidemic 1929-1971: a quantitative and qualitative retrospective with implications for the present. Am. J. Public Health 98, 974–985.

Reame, N.E., 2001. Premenstrual syndrome. In: DeGroot, L.J., Jameson, J.L. (Eds.), Endocrinology, fourth ed. Saunders, Philadelphia, PA, pp. 2147–2152.

Romero, L.M., Wingfield, J.C., 2016. Tempests, Poxes, Predators, and People: Stress in Wild Animals and How They Cope. Oxford University Press, NY, p. 614.

Sakai, R.R., Ma, L.Y., Zhang, D.M., McEwen, B.S., Fluharty, S.J., 1996. Intracerebral administration of mineralocorticoid receptor antisense oligonucleotides attenuate adrenal steroid-induced salt appetite in rats. Neuroendocrinology 64 (6), 425–429.

Schneider, J.E., Goldman, M.D., Tang, S., Bean, B., Ji, H., Friedman, M.I., 1998. Leptin indirectly affects estrous cycles by increasing metabolic fuel oxidation. Horm. Behav. 33, 217–228.

Schneider, J.E., Casper, J.F., Barisich, A., Schoengold, C., Cherry, S., Surico, J., DeBarba, A., Fabris, F., Rabold, E., 2007. Food deprivation and leptin prioritize ingestive and sex behavior without affecting estrous cycles in Syrian hamsters. Horm. Behav. 51, 413–427.

Schneider, J.E., Wise, J.D., Benton, N.A., Brozek, J.M., Keen-Rhinehart, E., 2013. When do we eat: ingestive behavior, survival, and reproductive success. Horm. Behav. 64 (4), 702–728.

Schneider, J.E., 2004. Energy balance and reproduction. Physiol. Behav. 81, 289–317.

Schreihofer, D.A., Golden, G.A., Cameron, J.L., 1993. Cholecystokinin (CCK)-induced stimulation of luteinizing hormone (LH) secretion in adult male rhesus monkeys: examination of the role of CCK in nutritional regulation of LH secretion. Endocrinology 132, 1553–1560.

Schumacher, M., Robert, F., 2002. Progesterone: synthesis, metabolism, mechanisms of action, and effects in the nervous system. In: Pfaff, D.W., Arnold, A.P., Etgen, A.M., Fahrbach, S.E., Rubin, R.T. (Eds.), Hormones, Brain and Behavior, vol. 3. Academic Press, San Diego, CA, pp. 683–745.

Seibel, M.M., 2001. Ovulation induction and assisted reproduction. In: DeGroot, L.J., Jameson, J.L. (Eds.), Endocrinology, fourth ed. Saunders, Philadelphia, PA, pp. 2138–2145.

Stahl, S.M., Sommer, B., Allers, K.A., 2011. Multifunctional pharmacology of flibanserin: possible mechanism of therapeutic action in hypoactive sexual desire disorder. J. Sexual Med. 8, 15–27.

Takeda, R., Takayama, Y., Tagawa, S., Komel, L., 1999. Schmidt's syndrome: autoimmune polyglandular disease of the adrenal and thyroid glands. Isr. Med. Assoc. J. 1, 285–286.

Tena-Sempere, M., 2010. Kisspeptins and the metabolic control of reproduction: physiologic roles and physiopathological implications. Ann. D'Endocrinol. 71, 201–202.

Valentino, R.J., Bangasser, D., Van Bockstaele, E.J., 2013. Sex-biased stress signaling: the corticotropin-releasing factor receptor as a model. Mol. Pharmacol. 83 (4), 737–745.

Vasudevan, N., Davidkova, G., Zhu, Y.S., Koibuchi, N., Chin, W.W., Pfaff, D.W., 2001a. Differential interactions of estrogen receptor and thyroid hormone receptor isoforms on the rat oxytocin receptor promoter leads to differences in transcriptional regulation. Neuroendocrinology 74 (5), 309–324.

Vasudevan, N., Koibuchi, N., Chin, W.W., Pfaff, D.W., 2001. Differential crosstalk between estrogen receptor (ER)α and ERß and the thyroid hormone receptor isoforms results in flexible regulation of the consensus ERE. Brain Res. Mol. Brain Res. 95 (1–2), 9–17.

Vasudevan, N., Zhu, Y.S., Daniels, S., Koibuchi, N., Chin, W.W., Pfaff, D.W., 2001. Crosstalk between oestrogen receptors and thyroid hormone receptor isoforms results in differential regulation of the preproenkephalin gene. J. Neuroendocrinol. 13.

Volkow, N.D., Wang, G.J., Fowler, J.S., Telang, F., Jayne, M., Wong, C., 2007. Stimulant-induced enhanced sexual desire as a potential contributing factor in HIV transmission. Am. J. Psychiatry 164, 157–160.

Vulliemoz, N.R., Xiao, E., Xia-Zhang, L., Germond, M., Rivier, J., Ferin, M., 2004. Decrease in luteinizing hormone pulse frequency during a five-hour peripheral ghrelin infusion in the ovariectomized rhesus monkey. J. Clin. Endocrinol. Metab. 89, 5718–5723.

Vulliemoz, N.R., Xiao, E., Xia-Zhang, L., Wardlaw, S.L., Ferin, M., 2005. Central infusion of agouti-related peptide suppresses pulsatile luteinizing hormone release in the ovariectomized rhesus monkey. Endocrinology 146, 784–789.

Wessells, H., Hruby, V.J., Hackett, J., Han, G., Balse-Srinivasan, P., Vanderah, T.W., 2003. MT-II induces penile erection via brain and spinal mechanisms. Ann. N.Y. Acad. Sci. 994, 90–95.

Wilson, W.L., Starbuck, E.M., Fitts, D.A., 2002. Salt appetite of adrenalectomized rats after a lesion of the SFO. Behav. Brain Res. 136 (2), 449–453.

Wingfield, J.C., Romero, L.M., 2001. Adrenocortical responses to stress and their modulation in free-living vertebrates. In: McEwen, B.S., Goodman, H.M. (Eds.), Handbook of Physiology; Section 7: The Endocrine System. Coping with the Environment: Neural and Endocrine Mechanisms, vol. IV. Oxford University Press, New York, pp. 211–234.

## Chapter 4: Hormone Metabolites Can Be the Behaviorally Active Compounds

Ahmad, S., Varagic, J., Groban, L., Dell'Italia, L.J., Nagata, S., Kon, N.D., Ferrario, C.M., 2014. Angiotensin-(1-12): a chymase-mediated cellular angiotensin II substrate. Curr. Hypertens. Rep. 16, 429.

Baulieu, E.-E., Robel, P., 1996. Dehydroepiandrosterone (DHEA) and dehydroepiandrosterone sulfate (DHEAS) as neuroactive steroids. J. Endocrinol. 150, 5221–5239.

Baulieu, E.-E., Robel, P., 1998. Dehydroepiandrosterone (DHEA) and dehydroepiandrosterone sulfate (DHEAS) as neuroactive steroids [commentary]. Proc. Natl. Acad. Sci. U.S.A. 95, 4089–4091.

Beach, F.A., 1942. Copulatory behavior in prepuberally castrated male rats and its modification by estrogen administration. Endocrinology 31, 679–683.

Brown, N.J., Kumar, S., Painter, C.A., Vaughan, D.E., 2002. ACE inhibition versus angiotensin type 1 receptor antagonism: differential effects on PAI-1 over time. Hypertension 40 (6), 859–865.

Carey, R.M., 2013. Newly discovered components and actions of the renin-angiotensin system. Hypertension 62, 818–822.

Culman, J., Blume, A., Gohlke, P., Unger, T., 2002. The renin-angiotensin system in the brain: possible therapeutic implications for AT(1)-receptor blockers. J. Hum. Hypertens. 16, S64–S70.

Davidson, J.M., Bloch, G.J., 1969. Neuroendocrine aspects of male reproduction. Biol. Reprod. 1, 67–92.

De Wied, D., Jolles, J., 1982. Neuropeptides derived from pro-opiocortin: behavioral, physiological, and neurochemical effects. Physiol. Rev. 62, 976–1059.

Ebling, F.J.P., 2014. On the value of seasonal mammals for identifying mechanisms underlying the control of food intake and body weight. Horm. Behav. 66 (1), 56–65.

Ferrario, C.M., 2003. Contribution of angiotensin-(1-7) to cardiovascular physiology and pathology. Curr. Hypertens. Rep. 5 (2), 129–134.

Ferreira, P.M., Souza Dos Santos, R.A., Campagnole-Santos, M.J., 2007. Angiotensin-(3-7) pressor effect at the rostral ventrolateral medulla. Regul. Pept. 141, 168–174.

Forest, M.G., 2001. Diagnosis and treatment of disorders of sexual development. In: DeGroot, L.J., Jameson, J.L. (Eds.), Endocrinology, fourth ed. Saunders, Philadelphia, pp. 1974–2010.

Frye, 2001a. The role of neurosteroids and nongenomic effects of progestins and androgens in mediating sexual receptivity of rodents. Brain Res. Rev. 37 (1–3), 201–222.

Frye, C.A., 2001b. The role of neurosteroids and nongenomic effects of progestins in the ventral tegmental area in mediating sexual receptivity of rodents. Horm. Behav. 40 (2), 226–233.

Gooren, L.J.G., 2001. Gender identity and sexual behavior. In: DeGroot, L.J., Jameson, J.L. (Eds.), Endocrinology, fourth ed. Saunders, Philadelphia, PA, pp. 2033–2042.

Gorski, R.A., 2000. Sexual differentiation of the nervous system. In: Kandel, E.R., Schwartz, J.H., Jessel, T.M. (Eds.), Principles of Neural Science, fourth ed. McGraw-Hill, New York, pp. 1131–1148.

Green, R., 2002. Sexual identity and sexual orientation. In: Pfaff, D.W., Arnold, A.P., Etgen, A.M., Fahrbach, S.E., Rubin, R.T. (Eds.), Hormones, Brain and Behavior, vol. 4. Academic Press, San Diego, CA, pp. 463–485.

Harno, E., White, A., 2016. In: Jameson, J.L., DeGroot, L.J. (Eds.), Endocrinology. Elsevier Saunders, Philadelphia, PA.

Kasckow, J., Geracioti, J.T.D., 2002. Neuroregulatory peptides of central nervous system origin: from bench to bedside. In: Pfaff, D.W., Arnold, A.P., Etgen, A.M., Fahrbach, S.E., Rubin, R.T. (Eds.), Hormones, Brain and Behavior, vol. 5. Academic Press, San Diego, CA, pp. 153–208.

McCarthy, M.M., 1994. Molecular aspects of sexual differentiation in the rodent brain. Psychoneuroendocrinology 19, 5–7.

McCarthy, M.M., 2016. Multifaceted origins of sex differences in the brain. Philos. Trans. R. Soc. B 371, 20150106.

McKinley, M.J., Johnson, A.K., 2004. The physiological regulation of thirst and fluid intake. News Physiol. Sci. 19 (1), 1–6.

Ohinata, K., Fujiwara, Y., Fukumoto, S., Iwai, M., Horiuchi, M., Yoshikawa, M., 2008. Angiotensin II and III suppress food intake via angiotensin AT(2) receptor and prostaglandin EP(4) receptor in mice. FEBS Lett. 582, 773–777.

Okeigwe, I., Kuohung, W., 2014. 5-Alpha reductase deficiency: a 40-year retrospective review. Curr. Opin. Endocrinol. Diabetes, Obes. 21 (6), 483–487.

Pavlides, C., McEwen, B.S., 1999. Effects of mineralocorticoid and glucocorticoid receptors on long-term potentiation in the CA3 hippocampal field. Brain Res. 851 (1–2), 204–214.

Pavlides, C., Kimura, A., et al., 1995a. Hippocampal homosynaptic long-term depression/depotentiation induced by adrenal steroids. Neuroscience 68 (2), 379–385.

Pavlides, C., Watanabe, Y., Magarinos, A.M., McEwen, B.S., 1995b. Opposing roles of type I and type II adrenal steroid receptors in hippocampal long-term potentiation. Neuroscience 68, 387–394.

Petrulis, A., 2013. Chemosignals and hormones in the neural control of mammalian sexual behavior. Front. Neuroendocrinol. 34, 255–267.

Phoenix, C.H., Goy, R.W., Gerral, A.A., Young, W.C., 1959. Organizing action of prenatally administered testosterone propionate on the tissues mediating mating behavior in the female Guinea pig. Endocrinology. 1959 (65), 369–382.

Reaux, A., Fournier-Zaluski, M.C., David, C., Zini, S., Roques, B.P., Corvol, P., Llorens-Cortes, C., 1999. Aminopeptidase A inhibitors as potential central antihypertensive agents. Proc. Natl. Acad. Sci. U.S.A. 96, 13415–13420.

Reddy, D.S., Estes, W.A., 2016. Clinical potential of neurosteroids for CNS disorders. Trends Pharmacol. Sci. S0165–S6147 (16)30004-9.

Rubin, R.T., Reinisch, J.M., Haskett, R.F., 1981. Postnatal gonadal steroid effects on human behavior. Science 211, 1318–1324.

Soma, K.K., Rendon, N.M., Boonstra, R., Albers, H.E., Demas, G.E., January 2015. DHEA effects on brain and behavior: insights from comparative studies of aggression. J. Steroid Biochem. Mol. Biol. 145, 261–272. http://dx.doi.org/10.1016/j.jsbmb.2014.05.011. Epub June 11, 2014.

Styne, D.M., Grumbach, M.M., 2002. Puberty in boys and girls. In: Pfaff, D.W., Arnold, A.P., Etgen, A.M., Fahrbach, S.E., Rubin, R.T. (Eds.), Hormones, Brain and Behavior, vol. 4. Academic Press, San Diego, CA, pp. 661–716.

Sumners, C., Horiuchi, M., Widdop, R.E., McCarthy, C., Unger, T., Steckelings, U.M., 2013. Protective arms of the renin-angiotensin-system in neurological disease. Clin. Exp. Pharmacol. Physiol. 40, 580–588.

Toran-Allerand, C.D., Guan, X., MacLusky, N.J., Horvath, T.L., Diano, S., Singh, M., Connolly Jr., E.S., Nethrapalli, I.S., Tinnikov, A.A., 2002. ER-X: a novel, plasma membrane-associated, putative estrogen receptor that is regulated during development and after ischemic brain injury. J. Neurosci. 22, 8391–8401.

White, A., Ray, D.W., 2001. Adrenocorticotropic hormone. In: DeGroot, L.J., Jameson, J.L. (Eds.), Endocrinology, fourth ed. Saunders, Philadelphia, PA, pp. 221–233.

Wilson, W.L., Roques, B.P., Llorens-Cortes, C., Speth, R.C., Harding, J.W., Wright, J.W., 2005. Roles of brain angiotensins II and III in thirst and sodium appetite. Brain Res. 1060, 108–117.

Wright, J.W., Tamura-Myers, E., Wilson, W.L., Roques, B.P., Llorens-Cortes, C., Speth, R.C., Harding, J.W., 2003. Conversion of brain angiotensin II to angiotensin III is critical for pressor response in rats. Am. J. Physiol. Regul. Integr. Comp. Physiol. 284 (3), R725–R733.

Xu, P., Sriramula, S., Lazartigues, E., 2011. ACE2/ANG-(1-7)/Mas pathway in the brain: the axis of good. Am. J. Physiol. Regul. Integr. Comp. Physiol. 300, R804–R817.

## Chapter 5: There Are Optimal Hormone Concentrations: Too Little or Too Much Can Be Damaging

Alatzoglou, K.S., Dattani, M.T., 2015. Growth hormone deficiency in children. In: Jameson, J.L., DeGroot, L.J. (Eds.), Endocrinology: Adult and Pediatric, seventh ed. Saunders, Philadelphia, PA, pp. 418–440.

Andela, C.D., van Haalen, F.M., Ragnarsson, O., Papakokkinou, E., Johansson, G., Santos, A., Webb, S.M., Biermasz, N.R., van der Wee, H.J.A., Pereira, A.M., 2015. Cushing's syndrome causes irreversible effects on the human brain: a systematic review of structural and functional MRI studies. Eur. J. Endocrinol. 173 (1), R1–R14.

Barthel, A., Willenberg, H.S., Gruber, M., Bornstein, S.R., 2015. Adrenal insufficiency. In: Jameson, J.L., DeGroot, L.J. (Eds.), Endocrinology, seventh ed. Saunders, Philadelphia, PA, pp. 1763–1774.

Bauer, M., Whybrow, P.C., 2000. Thyroid hormone, brain, and behavior. In: Fink, G. (Ed.), Encyclopedia of Stress, vol. 2. Academic Press, San Diego, CA, pp. 239–264.

Bornstein, S.R., Stratakis, C.A., Chrousos, G.P., 2000. Cushing's syndrome: medical aspects. In: Fink, G. (Ed.), Encyclopedia of Stress, vol. 1. Academic Press, San Diego, CA, pp. 615–621.

Burt, M.G., Ho, K.K.Y., 2015. Hypopituitarism and growth hormone deficiency. In: Jameson, J.L., DeGroot, L.J. (Eds.), Endocrinology: Adult and Pediatric, seventh ed. Saunders, Philadelphia, PA, pp. 188–208.

Chiovato, L., Barbesino, G., Pinchera, A., 2001. Graves' disease. In: DeGroot, L.J., Jameson, J.L. (Eds.), Endocrinology, fourth ed. Saunders, Philadelphia, PA, pp. 1422–1449.

Harno, E., White, A., 2015. Adrenocorticotropic hormone. In: Jameson, J.L., DeGroot, L.J. (Eds.), Endocrinology: Adult and Pediatric, seventh ed. Saunders, Philadelphia, PA, pp. 129–146.

Joffe, R.T., 2000. Hypothyroidism. In: Fink, G. (Ed.), Encyclopedia of Stress, vol. 2. Academic Press, San Diego, CA, pp. 496–499.

Joffe, R.T., 2009. Hypothalamic-pituitary-thyroid axis. In: Rubin, R.T., Pfaff, D.W. (Eds.), Hormone/Behavior Relations of Clinical Importance: Endocrine Systems Interacting with Brain and Behavior. Academic Press, San Diego, CA, pp. 69–83.

Juszczak, A., Morris, D.G., Grossman, A.B., Nieman, L.K., 2015. Cushing's syndrome. In: Jameson, J.L., DeGroot, L.J. (Eds.), Endocrinology: Adult and Pediatric, seventh ed. Saunders, Philadelphia, PA, pp. 227–255.

Laron, Z., 2000. Growth hormone and insulin-like growth factor I: effects on the brain. In: Fink, G. (Ed.), Encyclopedia of Stress, vol. 5. Academic Press, San Diego, CA, pp. 75–96.

Laron, Z., 2009. Growth hormone and insulin-like growth factor I: effects on the brain. In: Rubin, R.T., Pfaff, D.W. (Eds.), Hormone/Behavior Relations of Clinical Importance: Endocrine Systems Interacting with Brain and Behavior. Academic Press, San Diego, CA, pp. 373–394.

Lechan, R.M., Fekete, C., Toni, R., 2009. Thyroid hormone in neural tissue. In: Pfaff, D.W., Arnold, A.P., Etgen, A.M., Fahrbach, S.E., Rubin, R.T. (Eds.), Hormones, Brain and Behavior, second ed. Academic Press, San Diego, CA, pp. 1289–1328.

Lesser, I.M., 2000. Hyperthyroidism. In: Fink, G. (Ed.), Encyclopedia of Stress, vol. 2. Academic Press, San Diego, CA, pp. 439–440.

Lindsay, R.S., Funahashi, T., Hanson, R.L., Matsuzawa, Y., Tanaka, S., Tataranni, P.A., Knowler, W.C., Krakoff, J., 2002. Adiponectin and development of type 2 diabetes in the Pima Indian population. Lancet 360 (9326), 57–58.

Loriaux, D.L., McDonald, W.J., 2001. Adrenal insufficiency. In: DeGroot, L.J., Jameson, J.L. (Eds.), Endocrinology, fourth ed. Saunders, Philadelphia, PA, pp. 1683–1690.

Marino, M., Vitti, P., Chiovato, L., 2015. Graves' disease. In: Jameson, J.L., DeGroot, L.J. (Eds.), Endocrinology: Adult and Pediatric, seventh ed. Saunders, Philadelphia, PA, pp. 1437–1464.

McEwen, B.S., 2002. The End of Stress as We Know It. Joseph Henry Press, Washington, DC, p. 64.

Melmed, S., 2001. Acromegaly. In: DeGroot, L.J., Jameson, J.L. (Eds.), Endocrinology, fourth ed. Saunders, Philadelphia, PA, pp. 300–312.

Melmed, S., 2015. Acromegaly. In: Jameson, J.L., DeGroot, L.J. (Eds.), Endocrinology: Adult and Pediatric, seventh ed. Saunders, Philadelphia, PA, pp. 209–226.

Netter, F.H., 1965a. The suprarenal glands (adrenal glands). In: Forsham, P.H. (Ed.), Endocrine System and Selected Metabolic Diseases: The Ciba Collection of Medical Illustrations, vol. 4. Ciba Pharmaceutical Co., New York, pp. 75–108.

Netter, F.H., 1965b. The thyroid gland; the parathyroid glands. In: Forsham, P.H. (Ed.), Endocrine System and Selected Metabolic Diseases: The Ciba Collection of Medical Illustrations, vol. 4. Ciba Pharmaceutical Co., New York, pp. 39–74.

Niemann, L.K., 2001. Cushing's syndrome. In: DeGroot, L.J., Jameson, J.L. (Eds.), Endocrinology, fourth ed. Saunders, Philadelphia, PA, pp. 1691–1715.

Quax, R.A., Manenschijn, L., Koper, J.W., Hazes, J.M., Lamberts, S.W.J., van Rossum, E.F.C., Feelders, R.A., 2013. Glucocorticoid sensitivity in health and disease. Nat. Rev. Neurosci. 9 (11), 670–686.

Raff, H., Carroll, T., 2015. Cushing's syndrome: from physiological principles to diagnosis and clinical care. J. Physiol. 593 (3), 493–506.

Rhodes, M.E., McKlveen, J.M., Ripepi, D.R., Gentile, N.E., 2009. Hypothalamic-pituitary-adrenal cortical axis. In: Rubin, R.T., Pfaff, D.W. (Eds.), Hormone/Behavior Relations of Clinical Importance: Endocrine Systems Interacting with Brain and Behavior. Academic Press, San Diego, CA, pp. 47–67.

Rosenfeld, R.G., 2001. Growth hormone deficiency in children. In: DeGroot, L.J., Jameson, J.L. (Eds.), Endocrinology, fourth ed. Saunders, Philadelphia, PA, pp. 503–519.

Rosenfield, R.L., Cuttler, L., 2001. Somatic growth and maturation. In: DeGroot, L.J., Jameson, J.L. (Eds.), Endocrinology, fourth ed. Saunders, Philadelphia, PA, pp. 477–502.

Starkman, M., 2000. Cushing's syndrome, neuropsychiatric aspects. In: Fink, G. (Ed.), Encyclopedia of Stress, vol. 1. Academic Press, San Diego, CA, pp. 621–625.

Starkman, M., 2007. Cushing's syndrome, neuropsychiatric aspects. In: Fink, G. (Ed.), Encyclopedia of Stress, vol. 1. Academic Press, San Diego, CA, pp. 688–692.

Stumvoll, M., Haring, H., 2002. Glitazones: clinical effects and molecular mechanisms. Am. Med. 34 (3), 217–224.

Visser, T.J., Fliers, E., 2000. Thyroid hormones. In: Fink, G. (Ed.), Encyclopedia of Stress, vol. 3. Academic Press, San Diego, CA, pp. 605–612.

Walker, C.-D., Welberg, L.A.M., Plotsky, P.M., 2002. Glucocorticoids, stress, and development. In: Pfaff, D.W., Arnold, A.P., Etgen, A.M., Fahrbach, S.E., Rubin, R.T. (Eds.), Hormones, Brain and Behavior, vol. 4. Academic Press, San Diego, CA, pp. 487–534.

White, A., Ray, D.W., 2001. Adrenocorticotropic hormone. In: DeGroot, L.J., Jameson, J.L. (Eds.), Endocrinology, fourth ed. Saunders, Philadelphia, PA, pp. 221–233.

Wiersinga, W.M., 2001. Hypothyroidism and myxedema coma. In: DeGroot, L.J., Jameson, J.L. (Eds.), Endocrinology, fourth ed. Saunders, Philadelphia, PA, pp. 1491–1506.

Wiersinga, W.M., 2015. Hypothyroidism and myxedema coma. In: Jameson, J.L., DeGroot, L.J. (Eds.), Endocrinology: Adult and Pediatric, seventh ed. Saunders, Philadelphia, PA, pp. 1540–1556.

Willenberg, H.S., Bornstein, S.R., Chrousos, G.P., 2000. Adrenal insufficiency. In: Fink, G. (Ed.), Encyclopedia of Stress, vol. 1. Academic Press, San Diego, CA, pp. 58–63.

## Chapter 6: Hormones Do Not "Cause" Behavior; They Alter Probabilities of Responses to Given Stimuli in the Appropriate Context

Abdulhay, A., Benton, N.A., Klingerman, C.M., Krishnamoorthy, K., Brozek, J., Schneider, J.E., 2014. Estrous cycle fluctuations in sex and ingestive behavior are accentuated by exercise and cold ambient temperature. Horm. Behav. 66, 135–147.

Abeck, D.S., McKittrick, C.R., Blanchard, D.C., Blanchard, R.J., Nikulina, J., McEwen, B.S., Sakai, R.S., 1997. Chronic social stress alters expression of corticotrophin-releasing factor and arginine vasopressin mRNA expression in rat brain. J. Neurosci. 12, 4895–4903.

American Psychiatric Association, 1994. Diagnostic and Statistical Manual of Mental Disorders, fourth ed. American Psychiatric Press, Washington, DC.

Bauer, M., Whybrow, P.C., 2000. Thyroid hormone, brain, and behavior. In: Fink, G. (Ed.), Encyclopedia of Stress, vol. 2. Academic Press, San Diego, CA, pp. 239–264.

Blanchard, R.J., Blanchard, D.C., 1989. Anti-predator defensive behaviors in a visible burrow system. J. Comp. Psychol. 103, 70–82.

Blanchard, D.C., McKittrick, C.R., Hardy, M.P., Blanchard, R.J., 2002. Social stress effects on hormones, brain and behavior. In: Pfaff, D.W., Arnold, A.P., Etgen, A.M., Fahrbach, S.E., Rubin, R.T. (Eds.), Hormones, Brain and Behavior, vol. 1. Academic Press, San Diego, CA, pp. 735–772.

Bronson, F.H., 1989. Mammalian Reproductive Biology, first ed. The University of Chicago Press, Chicago and London.

Carroll, B.J., Feinberg, M., Greden, J.F., Tarika, J., Albala, A.A., Haskett, R.F., James, N.M., Kronfol, Z., Lohr, N., Steiner, M., de Vigne, J.P., Young, E., 1981. A specific laboratory test for the diagnosis of melancholia: standardization, validation, and clinical utility. Arch. Gen. Psychiatry 38, 15–22.

Chao, H.M., Blanchard, D.C., Blanchard, R.J., McEwan, B.S., Sakai, R.M., 1993. The effect of social stress on hippocampal gene expression. Mol. Cell. Neurosci. 4, 543–548.

Chiovato, L., Barbesino, G., Pinchera, A., 2001. Graves' disease. In: DeGroot, L.J., Jameson, J.L. (Eds.), Endocrinology, fourth ed. Saunders, Philadelphia, PA, pp. 1422–1449.

Figlewicz, D.P., 2003. Adiposity signals and food reward: expanding the CNS roles of insulin and leptin. Am. J. Physiol. Regul. Integr. Comp. Physiol. 284 (4), R882–R892.

Holsboer, F., 1999. The rationale for corticotropin-releasing hormone receptor (CRH-R) antagonists to treat depression and anxiety. J. Psychiatr. Res. 33, 181–214.

Klingerman, C.M., Krishnamoorthy, K., Patel, K., Spiro, A.B., Struby, C., Patel, A., Schneider, J.E., 2010. Energetic challenges unmask the role of ovarian hormones in orchestrating ingestive and sex behaviors. Horm. Behav. 58, 563–574.

Klingerman, C.M., Williams 3rd, W.P., Simberlund, J., Brahme, N., Prasad, A., Schneider, J.E., Kriegsfeld, L.J., 2011. Food restriction-induced changes in gonadotropin-inhibiting hormone cells are associated with changes in sexual motivation and food hoarding, but not sexual performance and food intake. Front. Endocrinol. (Lausanne) 2, 101.

Kow, L.M., Pfaff, D.W., 1979. Responses of single units in sixth lumbar dorsal root ganglion of female rats to mechanostimulation relevant for lordosis reflex. J. Neurophysiol. 42, 203–213.

Kow, L.M., Montgomery, M.O., Pfaff, D.W., 1979. Triggering of lordosis reflex in female rats with somatosensory stimulation: quantitative determination of stimulus parameters. J. Neurophysiol. 42, 195–202.

Kow, L.M., Zemlan, F.P., Pfaff, D.W., 1980. Responses of lumbosacral spinal units to mechanical stimuli related to analysis of lordosis reflex in female rats. J. Neurophysiol. 43, 27–45.

Lesser, I.M., 2000. Hyperthyroidism. In: Fink, G. (Ed.), Encyclopedia of Stress, vol. 2. Academic Press, San Diego, CA, pp. 439–440.

McClintock, C.J., McKittrick, C.R., Blanchard, D.C., Blanchard, R.J., McEwen, B.S., Sakai, R.R., 1995. Serotonin receptor binding in a colony model of chronic social stress. Biol. Psychiatry 37, 383–393.

McKittrick, C.R., Magarinos, A.M., Blanchard, D.C., Blanchard, R.J., McEwen, B.S., Sakai, R.R., 2000. Chronic social stress reduces dendritic arbors in CA3 of hippocampus and decreases binding to serotonin transporter sites. Synapse 36 (2), 85–94.

Moffatt, C.A., October 2003. Steroid hormone modulation of olfactory processing in the context of socio-sexual behaviors in rodents and humans. Brain Res. Rev. 43 (2), 192–206.

Petrulis, A., 2013. Chemosignals, hormones, and mammalian reproduction. Horm. Behav. 63 (5), 723–741.

Reding, K., Michopoulos, V., Wallen, K., Sanchez, M., Wilson, M.E., Toufexis, D., 2012. Social status modifies estradiol activation of sociosexual behavior of female rhesus monkeys. Horm. Behav. 62 (5), 612–620.

Rubin, R.T., Dinan, T.G., Scott, L.V., 2002. Affective disorders. In: Pfaff, D.W., Arnold, A.P., Etgen, A.M., Fahrbach, S.E., Rubin, R.T. (Eds.), Hormones, Brain and Behavior, vol. 5. Academic Press, San Diego, CA, pp. 467–514.

Schneider, J.E., Wise, J.D., Benton, N.A., Brozek, J., Keen-Rinehart, E., 2013. When do we eat? Ingestive behavior, survival, and reproductive success. Horm. Behav. 64 (4), 702–728.

Spencer, R.L., Miller, A.H., Moday, H., McEwen, B.S., Blanchard, R.J., Blanchard, D.C., Sakai, R.R., 1996. Chronic social stress produces reductions in available splenic type II corticosteroid receptor binding and plasma corticosteroid binding globulin levels. Psychoneuroendocrinology 21, 95–109.

Starkman, M., 2000. Cushing's syndrome, neuropsychiatric aspects. In: Fink, G. (Ed.), Encyclopedia of Stress, vol. 1. Academic Press, San Diego, CA, pp. 621–625.

Vaidya, B., Pearce, S.H., August 21, 2014. Diagnosis and management of thyrotoxicosis. Br. Med. J. 349, g5128.

Wallen, K., 1990. Desire and ability. Neurosci. Biobehav. Rev. 14 (2), 233–241.

Wood, R., Newman, S., 1995. Integration of chemosensory and hormonal cues is essential for mating in Syrian hamsters. J. Neurosci. 15 (11), 7261–7269.

Yoest, K.E., Cummings, J.A., Becker, J.B., 2014. Estradiol, dopamine and motivation. Cent. Nerv. Syst. Agents Med. Chem. 14 (2), 83–89.

Zobel, A.W., Nickel, T., Künzel, H.E., Ackl, N., Sonntag, A., Ising, M., Holsboer, F., 2000. Effects of the high-affinity corticotropin-releasing hormone receptor 1 antagonist R121919 in major depression: the first 20 patients treated. J. Psychiatr. Res. 34, 171–181.

## Chapter 7: Hormone-Behavior Relations Are Reciprocal

Amorin, N., Calisi, R.M., 2015. Measurement of neuronal soma size and estimated peptide concentrations in addition to cell abundance offer a high resolution of seasonal and reproductive influences on GnRH-I and GnIH in European starlings. J. Integr. Comp. Biol. 55, 332–342.

Anisman, H., Zaharia, M.D., Meaney, M.J., Merali, Z., 1998. Do early-life events permanently alter behavioral and hormonal responses to stressors? Int. J. Dev. Neurosci. 16, 149–164.

Aragona, B.J., Wang, Z., 2009. Dopamine regulation of social choice in a monogamous rodent species. Front. Behav. Neurosci. 3, 15.

Archer, J., 2006. Testosterone and human aggression: an evaluation of the challenge hypothesis. Neurosci. Biobehav. Rev. 30 (3), 319–345.

Atkins-Regan, E., 2005. Hormones and Animal Social Behavior. Princeton University Press, Princeton.

Bakker, J., Kelliher, K.R., Baum, M.J., 2001. Mating induces gonadotropin-releasing hormone neuronal activation in anosmic female ferrets. Biol. Reprod. 64 (4), 1100–1105.

Balthazart, J., Ball, G.F., 2006b. Is brain estradiol a hormone or a neurotransmitter? Trends Neurosci. 29 (5), 241–249.

Belle, M.D., Tsutsui, K., Lea, R.W., 2003. Sex steroid communication in the ring dove brain during courtship. Can. J. Physiol. Pharmacol. 81 (4), 359–370.

Belle, M.D., Sharp, P.J., Lea, R.W., 2005. Aromatase inhibition abolishes courtship behaviours in the ring dove (Streptopelia risoria) and reduces androgen and progesterone receptors in the hypothalamus and pituitary gland. Mol. Cell. Biochem. 276 (1–2), 193–204.

Baulieu, E.E., 1991. Neurosteroids: a new function in the brain. Biol. Cell 71, 3–10.

Bennis-Taleb, N., Remacle, C., Hoett, J.J., Reusens, B., 1999. A low-protein isocaloric diet during gestation affects brain development and alters permanently cerebral cortex blood vessels in rat offspring. J. Nutr. 129 (8), 1613–1619.

Bridges, R.S., 2016. Long-term alterations in neural and endocrine processes induced by motherhood in mammals. Horm. Behav. 77, 193–203.

Brockway, B.F., 1965. Stimulation of ovarian development and egg laying by male courtship vocalization in budgerigars (Melopsittacus undulates). Anim. Behav. 13 (4), 575–578.

Calisi, R.M., Diaz-Munoz, S.I.., Wingfield, J.C., Bentley, G.E., 2011. Social and breeding status are associated with the expression of GnIH. Genes, Brain Behav. 10, 557–564.

Carré, Olmstead, 2015. Social neuroendocrinology of human aggression: examining the role of competition-induced testosterone dynamics. Neuroscience 286, 171–186.

Cheng, M.-F., Silver, R., 1975. Estrogen-progesterone regulation of nest-building and incubation behavior in ovariectomized ring doves (Streptopelia risoria). J. Comp. Physiol. Psychol. 88 (1), 256–263.

Cheng, M.-F., Peng, J.P., Johnson, P., 1998. Hypothalamic neurons preferentially respond to female nest coo stimulation: demonstration of direct acoustic stimulation of luteinizing hormone release. J. Neurosci. 18, 5477–5489.

Cheng, M.-F., 1979. Progress and prospects in ring dove research: a personal view. In: Advances in the Study of Behavior, vol. 9. Academic Press, New York, pp. 97–131.

Cheng, M.-F., 1986. Female cooing promotes ovarian development in ring doves. Physiol. Behav. 37, 371–374.

Cho, M.M., DeVries, A.C., Williams, J.R., Carter, C.S., 1999. The effects of oxytocin and vasopressin on partner preferences in male and female prairie voles (Microtus ochrogaster). Behav Neurosci. 113 (5), 1071–1079.

Cornil, C.A., Ball, G.F., Balthazart, J., December 18, 2006. Functional significance of the rapid regulation of brain estrogen action: where do the estrogens come from? Brain Res. 1121 (1), 2–26.

Cornil, C.A., Seredynski, A.L., de Bournonville, C., Dickens, M.J., Charlier, T.D., Ball, G.F., Balthazart, J., 2013. Rapid control of reproductive behaviour by locally synthesised oestrogens: focus on aromatase. J. Neuroendocrinol. 25, 1070–1078.

Dellovade, T.L., Hunter, E., Rissman, E.F., 1995. Interactions with males promote rapid changes in gonadotropin-releasing hormone immunoreactive cells. Neuroendocrinology 62 (4), 385–395.

Fabre-Nys, C., Kendrick, K.M., Scaramuzzi, R.J., 2015. The "ram effect": new insights into neural modulation of the gonadotropic axis by male odors and socio-sexual interactions. Front. Neurosci. 9, 111.

Feder, H.H., Storey, A., Goodwin, D., Reboulleau, C., Silver, R., 1977. Testosterone and "5alpha-dihydrotestosterone" levels in peripheral plasma of male and female ring doves (Streptopelia risoria) during a reproductive cycle. Biol. Reprod. 16, 666–677.

Fernald, R.D., 2015. Social behaviour: can it change the brain? Anim. Behav. 103, 259–265.

Fernald, R.D., Maruska, K.P., 2012. Social information changes the brain. Proc. Natl. Acad. Sci. U.S.A. 109 (Suppl 2), 17194–17199.

Forest, M.G., 2001. Diagnosis and treatment of disorders of sexual development. In: DeGroot, L.J., Jameson, J.L. (Eds.), Endocrinology, fourth ed. Saunders, Philadelphia, pp. 1974–2010.

Forlano, P.M., Schlinger, B.A., Bass, A.H., 2006. Brain aromatase: new lessons from non-mammalian model systems. Front. Neuroendocrinol. 27 (3), 247–274.

Francis, D.D., Young, L.J., Meaney, M.J., Insel, T.R., 2002. Naturally occurring differences in maternal care are associated with the expression of oxytocin and vasopressin (V1a). J. Neuroendocrinol. 14 (5), 349–353.

Gleason, E.D., Fuxjager, M.J., Oyegbile, T.O., Marler, C.A., 2009. Testosterone release and social context: when it occurs and why. Front. Neuroendocrinol. 30 (4), 460–469.

Glover, V., 1999. Maternal stress or anxiety during pregnancy and development of the baby. Pract. Midwife 2 (5), 20–22.

Glushakov, A.V., Dennis, D.M., Morey, T.E., Sumners, C., Cucchiara, R.F., Seubert, C.N., Martynuk, A.E., 2002. Specific inhibition of N-methyl-d-aspartate receptor function in rat hippocampal neurons by l-phenylalanine at concentrations observed during phenylketonuria. Mol. Psychiatry 7 (4), 359–367.

Goldsmith, A.R., Edwards, C., Koprucu, M., Silver, R., 1981. Concentrations of prolactin and luteinizing hormone in plasma of doves in relation to incubation and development of the crop gland. J. Endocrinol. 90, 437–443.

Gooren, L.J.G., 2001. Gender identity and sexual behavior. In: DeGroot, L.J., Jameson, J.L. (Eds.), Endocrinology, fourth ed. Saunders, Philadelphia, PA, pp. 2033–2042.

Gore, A.C., 2013. GnRH: The Master Molecule of Reproduction. Kluwer Academic Publishers, Boston..

Green, R., 2002. Sexual identity and sexual orientation. In: Pfaff, D.W., Arnold, A.P., Etgen, A.M., Fahrbach, S.E., Rubin, R.T. (Eds.), Hormones, Brain and Behavior, vol. 4. Academic Press, San Diego, CA, pp. 463–485.

Greer, J.M., Capecchi, M.R., 2002. Hoxb8 is required for normal grooming behavior in mice. Neuron 33 (1), 23–43.

Hamilton, L.D., Carre, J.M., Mehta, P.H., Olmstead, N., Whitaker, J.D., 2015. Social neuroendocrinology of status: a review and future directions. Adapt. Hum. Behav. Physiol. 1 (2), 202–230.

Hayden-Hixson, D.M., Ferris, C.F., 1991. Steroid-specific regulation of agonistic responding in the anterior hypothalamus of male hamsters. Physiol. Behav. 50 (4), 793–799.

Heimovics, S.A., Trainor, B.C., Soma, K.K., 2015. Rapid effects of estradiol on aggression in birds and mice: the fast and the furious. Integr. Comp. Biol. 55 (2), 281–293.

Herbison, A.E., 2006. Physiology of the gonadotropin-releasing hormone neuronal network. In: Neill, J.C. (Ed.), Knobil and Neill's Physiology of Reproduction. Academic Press, New York, NY.

Hinde, R.A., Steele, E., 1978. The influence of daylength and male vocalizations on the estrogen-dependent behavior of female canaries and budgerigars, with discussion of data from other species. In: Rosenblatt, J. (Ed.), Advances in the Study of Behavior, 8. Academic Press, pp. 39–73.

Hinde, R.A., 1965. Interaction of internal and external factors in integration of canary reproduction. In: Beach, F.A. (Ed.), Sex and Behavior. John Wiley Sons, New York, pp. 381–415.

Hines, M., 2002. Sexual differentiation of human brain and behavior. In: Pfaff, D.W., Arnold, A.P., Etgen, A.M., Fahrbach, S.E., Rubin, R.T. (Eds.), Hormones, Brain and Behavior, vol. 4. Academic Press, San Diego, CA, pp. 425–462.

Hirschenhauser, K., Oliveira, R.F., 2006. Social modulation of androgens in male vertebrates: meta-analyses of the challenge hypothesis. Anim. Behav. 71, 265–277.

Insel, T.R., 2010. The challenge of translation in social neuroscience: a review of oxytocin, vasopressin, and affiliative behavior. Neuron 65 (6), 768–779.

Johnson, Z.V., Walum, H., Jamal, Y.A., Xiao, Y., Keebaugh, A.C., Inoue, K., Young, L.J., 2016. Central oxytocin receptors mediate mating-induced partner preferences and enhance correlated activation across forebrain nuclei in male prairie voles. Horm. Behav. 79, 8–17.

Kaufman, J., Plotsky, P.M., Nemeroff, C.B., Charney, D.S., 2000. Effects of early adverse experiences on brain structure and function: clinical implications. Biol. Psychiatry 48, 778–790.

Korzan, W.J., Summers, C.H., 2007. Behavioral diversity and neurochemical plasticity: selection of stress coping strategies that define social status. Brain Behav. Evol. 70 (4), 257–266.

Kroodsma, D.E., 1976. Reproductive development in a female songbird: differential stimulation by quality of male song. Science 192 (4239), 574–575.

Lehrman, D.S., 1964. The reproductive behavior of ring doves. In: Eisner, T., Wilson, E.O. (Eds.), Animal Behavior: Readings from Scientific American. W. H. Freeman and Co., San Francisco.

Lehrman, D.S., 1965. Interaction between internal and external environments in the regulation of the reproductive cycle of the ring dove. In: Beach, F.A. (Ed.), Sex and Behavior. John Wiley Sons, New York, pp. 355–380.

Leo, C.P., Hsu, S.Y., Hsueh, A.J.W., 2002. Hormonal genomics. Endocr. Rev. 23, 369–381.

Leon, D.A., Johanson, M., Rasmussen, F., 2000. Gestational age growth rate of fetal mass are inversely associated with systolic blood pressure in young adults: an epidemiologic study of 165,136 Swedish men aged 18 years. Am. J. Epidemiol. 152 (7), 597–604.

Lim, M.M., Young, L.J., 2006. Neuropeptidergic regulation of affiliative behavior and social bonding in animals. Horm. Behav. 50 (4), 506–517.

Lipina, S.J., Posner, M.I., 2012. The impact of poverty on the development of brain networks. Front. Hum. Neurosci. 6, 238.

Maney, D.L., Goode, C.T., Lake, J.I., Lange, H.S., O'Brien, S., 2007. Rapid neuroendocrine responses to auditory courtship signals. Endocrinology 148 (12), 5614–5623.

Mantei, K.E., Ramakrishnan, S., Sharp, P.J., Buntin, J.D., November 2008. Courtship interactions stimulate rapid changes in GnRH synthesis in male ring doves. Horm. Behav. 54(5), 669–675. http://dx.doi.org/10.1016/j.yhbeh.2008.07.005. Epub July 29, 2008.

Matthews, L.H., 1939. Visual stimulation and ovulation in pigeons. Proc. R. Soc. Lond. 126, 557–560.

McCarton, C.M., Brooks-Gunn, J., Wallace, I.F., Bauer, C.R., Bennett, F.C., Bernbaum, J.C., Broyles, R.S., Casey, P.H., McCormick, M.C., Scott, D.T., Tyson, J., Tonascia, J., Meinert, C.L., 1997. Results at age 8 years of early intervention for low-birth-weight premature infants: the Infant Health and Development Program. JAMA 277 (2), 126–132.

McEwen, B.S., Tucker, P., 2011. Critical biological pathways for chronic psychosocial stress and research opportunities to advance the consideration of stress in chemical risk assessment. Am. J. Public Health 101 (Suppl. 1), S131–S139.

McEwen, B.S., 1999. Behavioral control of hormonal secretion. In: Siegel, G.J., Agranoff, B.W., Albers, R.W., et al. (Eds.), Basic Neurochemistry: Molecular, Cellular and Medical Aspects, sixth ed. Lippencot-Raven, Philadelphia.

McEwen, B.S., 2010. Stress, sex, and neural adaptation to a changing environment: mechanisms of neuronal remodeling. Ann. N.Y. Acad. Sci. 1204 (Suppl.), E38–E59.

McEwen, B.S., 2016. The Brain on Stress: How Behavior and the Social Environment "Get under the Skin". Agency for Healthcare Research and Quality, Rockville, MD. http://www.ahrq. gov/professionals/education/curriculum-tools/population-health/mcewen.html.

McKinney, W.T., Suomi, S.J., Harlow, H.F., 1971. Depression in primates. Am. J. Psychiatry 127, 1313–1320.

Müller, M.B., Keck, M.E., Steckler, T., Holsboer, F., 2002. Genetics of endocrine-behavior interactions. In: Pfaff, D.W., Arnold, A.P., Etgen, A.M., Fahrbach, S.E., Rubin, R.T. (Eds.), Hormones, Brain and Behavior, vol. 5. Academic Press, San Diego, CA, pp. 263–301.

Oliveira, R.F., Lopes, M., Carneiro, L.A., Canário, A.V., 2001. Watching fights raises fish hormone levels. Nature 409 (6819), 475.

Patel, M.D., 1936. The physiology of the formation of pigeon's milk. Physiol. Zool. 9, 129–152.

Pedersen, C.A., Boccia, M.L., 2002. Oxytocin links mothering received, mothering bestowed and adult stress responses. Stress 5 (4), 259–267.

Petersen, C.L., Hurley, L.M., 2017. Putting it in context: linking auditory processing with social behavior circuits in the vertebrate brain. Integr. Comp. Biol. 57, 865–877.

Pohlenz, J., Weiss, R.E., et al., 1999. 5 new families with resistance to thyroid hormone not caused by mutations in the thyroid hormone receptor beta gene. J. Clin. Endocrinol. Metab. 84 (11), 3919–3928.

Ramsey, S.M., Goldsmith, A.R., Silver, R., 1985. Stimulus requirements for prolactin and luteinizing hormone secretion in incubating ring doves. Gen. Comp. Endocrinol. 59 (2), 246.

Refetoff, S., 2000. Resistance to thyroid hormone. In: Braverman, L.E., Utiger, R.E. (Eds.), Werner & Ingbar's the Thyroid: A Fundamental and Clinical Text, eighth ed. Lippincott Williams & Wilkins, Philadelphia, PA, pp. 1028–1043.

Remage-Healey, L., Maidment, N.T., Schlinger, B.A., 2008. Forebrain steroid levels fluctuate rapidly during social interactions. Nat. Neurosci. 11, 1327–1334.

Reutrakul, S., Sadow, P.M., et al., 2000. Search for abnormalities of nuclear corepressors, coactivators, and a coregulator in families with resistance to thyroid hormone without mutations in thyroid hormone receptor beta or alpha genes. J. Clin. Endocrinol. Metab. 85 (10), 3609–3617.

Richardson, H.N., Gore, A.C., Venier, J., Romeo, R.D., Sisk, C.L., 2004. Increased expression of forebrain GnRH mRNA and changes in testosterone negative feedback following pubertal maturation. Mol. Cell Endocrinol. 214, 63–70.

Rissman, E.F., 1997. Behavioral regulation of the GnRH system. In: Pahar, I.S., Sakuma, Y. (Eds.), GnRH Neurons: Gene to Behavior. Brain Shuppan, Tokyo, pp. 325–342.

Sadow, P., Reutrakul, Weiss, Refetoff, 2000. Resistance to thyroid hormone in the absence of mutations in the thyroid hormone receptor genes. Curr. Opin. Endocrinol. Diabetes 7, 253–259.

Saldanha, C.J., Remage-Healey, L., Schlinger, B.A., 2011. Synaptocrine signaling: steroid synthesis and action at the synapse. Endocr. Rev. 32, 532–549.

Sapolsky, R.M., 2003. Stress and plasticity in the limbic system. Neurochem. Res. 28 (11), 1735–1742.

Seminara, S.B., Crowley, W.F., 2002. Genetic approaches to unraveling reproductive disorders: examples of bedside to bench research in the genomic era. Endocr. Rev. 23, 382–392.

Silver, R., Goldsmith, A.R., Follett, B.K., 1980. Plasma luteinizing hormone in male ring doves during the breeding cycle. Gen. Comp. Endocrinol. 42, 19–24.

Snoeck, A., Remacle, C., Reusens, B., Hoett, J.J., 1990. Effect of a low protein diet during pregnancy on the fetal rat endocrine pancreas. Biol. Neonate 57 (2), 107–118.

Soma, K.K., 2006. Testosterone and aggression: Berthold, birds and beyond. J. Neuroendocrinol. 18, 543–551.

Stevenson, T.J., Ball, G.F., 2009. Anatomical localization of the effects of reproductive state, castration, and social milieu on cells immunoreactive for gonadotropin-releasing hormone-I in male European starlings (Sturnus vulgaris). J. Comp. Neurol. 517 (2), 146–155.

Styne, D.M., Grumbach, M.M., 2002. Puberty in boys and girls. In: Pfaff, D.W., Arnold, A.P., Etgen, A.M., Fahrbach, S.E., Rubin, R.T. (Eds.), Hormones, Brain and Behavior, vol. 4. Academic Press, San Diego, CA, pp. 661–716.

Swann, J.M., Fabre-Nys, C., Barton, R., 2009. Hormonal and pheromonal modulation of the extended amygdala: implications for social behavior. In: Pfaff, D.W., Arnold, A.P., Fahrbach, S.E., Etgen, A.M., Rubin, R.T. (Eds.), Hormones, Brain and Behavior. Elsevier, New York (Chapter 12).

Thompson, E.B., 2002. The impact of genomics and proteomics on endocrinology [editorial]. Endocr. Rev. 23, 366–368.

Tobari, Y., Son, Y.L., Ubuka, T., Hasegawa, Y., Tsutsui, K., 2014. A new pathway mediating social effects on the endocrine system: female presence acting via norepinephrine release stimulates gonadotropin-inhibitory hormone in the paraventricular nucleus and suppresses luteinizing hormone in quail. J. Neurosci. 34, 9803–9811.

Trainor, B.C., Marler, C.A., 2002. Testosterone promotes paternal behaviour in a monogamous mammal via conversion to oestrogen. Proc. Biol. Sci. 269 (1493), 823–829.

Ubuka, T., Bentley, G.E., Ukena, K., Wingfield, J.C., Tsutsui, K., 2005. Melatonin induces the expression of gonadotropin-inhibitory hormone in the avian brain. Proc. Natl. Acad. Sci. 102, 3052–3057.

Van Anders, S.M., Watson, N.V., 2006. Social neuroendocrinology: effects of social contexts and behaviors on sex steroids in humans. Hum. Nat. 17 (2), 2112–2237.

Van Anders, S.M., Steiger, J., Goldey, K.L., 2015. Effects of gendered behavior on testosterone in women and men. Proc. Natl. Acad. Sci. 112, 13805–13810.

Vasudevan, N., Pfaff, D.W., 2008. Non-genomic actions of estrogens and their interaction with genomic actions in the brain. Front. Neuroendocrinol. 29 (2), 238–257.

Walker, C.-D., Welberg, L.A.M., Plotsky, P.M., 2002. Glucocorticoids, stress, and development. In: Pfaff, D.W., Arnold, A.P., Etgen, A.M., Fahrbach, S.E., Rubin, R.T. (Eds.), Hormones, Brain and Behavior, vol. 4. Academic Press, San Diego, CA, pp. 487–534.

Weaver, S.A., Aherne, F.X., Meaney, M.J., Schaefer, A.L., Dixon, W.T., 2000. Neonatal handling permanently alters hypothalamic-pituitary-adrenal axis function, behavior, and body weight in boars. J. Endocrinol. 164, 349–359.

Welberg, L.A., Seckl, J.R., Holmes, M.C., 2001. Prenatal glucocorticoid programming of brain corticosteroid receptors and corticotrophin-releasing hormone: possible implications for behaviour. Neuroscience 104 (1), 71–79.

White, S.A., Nguyen, T., Fernald, R.D., 2002. Social regulation of gonadotropin-releasing hormone. J. Exp. Biol. 205, 2567–2581.

Wilczynski, W., Lynch, K.S., 2006. Social regulation of plasma estradiol concentration in a female anuaran. Horm. Behav. 50 (1), 101–106.

Wilczynski, W., Lynch, K.S., 2011. Female sexual arousal in amphibians. Horm. Behav. 59, 630–636.

Wingfield, J.C., Hegner, R.E., Dufty, A.M., Ball, G.F., 1990. The challenge hypothesis: theoretical implications for patterns of testosterone secretion, mating systems, and breeding strategies. Am. Nat. 136, 829–846.

Wood, R.I., Stanton, S.J., 2012. Testosterone and sport: current perspectives. Horm. Behav. 61 (1), 147–155.

Yoon, H., Enquist, L.W., Dulac, C., 2005. Olfactory inputs to hypothalamic neurons controlling reproduction and fertility. Cell 123, 669–682.

Zilioli, S., Watson, N.V., 2014. Testosterone across successive competitions: evidence for a 'winner effect' in humans? Psychoneuroendocrinol 47, 1–9.

## Chapter 8: Familial/Genetic Dispositions to Hormone Responsiveness Can Influence Behavior

Arnold, A.P., Chen, X., 2009. What the "four core genotypes" mouse model tell us about sex difference in the brain and other tissue? Front. Neuroendocrinol. 1, 1–9.

Caspi, A., Sugden, K., Moffitt, T.E., Taylor, A., Craig, I.W., Harrington, H.L., McClay, J., Mill, J., Martin, J., Braithwaite, A., Poulton, R., 2003. Influence of life stress on depression: moderation by a polymorphisminthe 5-HTT gene. Science 301, 386.

Cavalcante, J.C., Bittencourt, J.C., Elias, C.F., 2014. Distribution of the neuronal inputs to the ventral premammalary nucleus of male and female rats. Brain Res. 1582, 77–90.

Chodorow, N., 1999. The Reproduction of Motherhood: Psychoanalysis and the Sociology of Gender. University of California Press, Berkeley.

DeVries, G.J., Simerly, R.B., 2002. Anatomy, development, and function of sexually diorphic neural circuits in the mammalian brain. In: Pfaff, D.W., Arnold, A.P., Etgen, A.M., Fahrbach, S.E., Rubin, R.T. (Eds.), Hormones, Brain and Behavior, vol. 4. Academic Press, San Diego, CA, pp. 137–191.

Francis, D.D., Meaney, M.J., 1999. Maternal care and development of stress responses. Curr. Opin. Neurobiol. 9, 128–134.

Francis, S.M., Kistner-Griffin, E., Yan, Z., Guter, S., Cook, E.H., Jacob, S., May 12, 2016. Variants in adjacent oxytocin/vasopressin gene region and associations with ASD diagnosis and other autism related endophenotypes. Front. Neurosci. 10, 195.

Glushakov, A.V., Dennis, D.M., Morey, T.E., Sumners, C., Cucchiara, R.F., Seubert, C.N., Martynuk, A.E., 2002. Specific inhibition of N-methyl-d-aspartate receptor function in rat hippocampal neurons by l-phenylalanine at concentrations observed during phenylketonuria. Mol. Psychiatry 7 (4), 359–367.

Gonzalez-Mariscal, 2002. In: Pfaff, D.W., Arnold, A.P., Etgen, A.M., Fahrbach, S.E., Rubin, R.T. (Eds.), Hormones, Brain and Behavior. Academic Press, San Diego, CA.

Goy, R.W., McEwen, B.S., et al., 1980. Sexual Differentiation of the Brain: Based on a Work Session of the Neurosciences Research Program. MIT Press, Cambridge, MA.

Greer, J.M., Capecchi, M.R., 2002. Hoxb8 is required for normal grooming behavior in mice. Neuron 33 (1), 23–43.

Hamann, S., Stevens, J., Vick, J.H., Bryk, K., Quigley, C.A., Berenbaum, S.A., Wallen, K., 2014. Brain responses to sexual images in 46,XY women with complete androgen insensitivity syndrome are female-typical. Horm. Behav. 66 (5), 724–730.

Handa, R.J., McGivern, R.F., 2000. Gender and stress. In: Fink, G. (Ed.), Encyclopedia of Stress, vol. 2. Academic Press, San Diego, CA, pp. 196–204.

Imperato-McGinley, 2002. In: Pfaff, D.W., Arnold, A.P., Etgen, A.M., Fahrbach, S.E., Rubin, R.T. (Eds.), Hormones, Brain and Behavior. Academic Press, San Diego, CA.

Landgraf, R., Wigger, A., 2002. High vs low anxiety-related behavior rats: an animal model of extremes in trait anxiety. Behav. Genet. 32, 301–314.

Liu, D., Diorio, J., Tannenbaum, B., Caldji, C., Francis, D., Freedman, A., Sharma, S., Pearson, D., Plotsky, P.M., Meaney, M.J., 1997. Maternal care, hippocampal glucocorticoid receptors, and hypothalamic-pituitary-adrenal responses to stress. Science 277, 1659.

McCormick, C.M., Linkroum, W., et al., 2002. Peripheral and central sex steroids have differential effects on the HPA axis of male and female rats. Stress 5 (4), 235–247.

McGowan, P.O., Sasaki, A., D'Alessio, A.C., Dymov, S., Labonté, B., Szyf, M., Turecki, G., Meaney, M.J., 2009. Epigenetic regulation of the glucocorticoid receptor in human brain associates with childhood abuse. Nat. Neurosci. 12 (3), 342–348.

Mogil, J.S., Wilson, S.G., et al., 2003. The melanocortin-1 receptor gene mediates female- specific mechanisms of analgesia in mice and humans. Proc. Natl. Acad. Sci. U.S.A. 100 (8), 4867–4872.

Ogawa, S., Taylor, J., et al., 1996. Reversal of sex roles in genetic female mice by disruption of estrogen receptor gene. Neuroendocrinology 64, 467–470.

Ogawa, S., Eng, V., et al., 1998a. Roles of estrogen receptor-alpha gene expression in reproduction-related behaviors in female mice. Endocrinology 139, 5070–5081.

Ogawa, S., Washburn, T., et al., 1998b. Modifications of testosterone-dependent behaviors by estrogen receptor-alpha gene disruption in male mice. Endocrinology 139, 5058–5069.

Qayyum, A., Zai, C.C., Hirata, Y., Tiwari, A.K., Cheema, S., Nowrouzi, B., Beitchman, J.H., Kennedy, J.L., 2015. The role of the catechol-o-methyltransferase (COMT) gene Val158Met in aggressive behaviour, a review of genetic studies. Curr. Neuropharmacol. 13, 802–814.

Refetoff, S., 2000. Resistance to thyroid hormone. In: Braverman, L.E., Utiger, R.E. (Eds.), Werner & Ingbar's the Thyroid: A Fundamental and Clinical Text, eighth ed. Lippincott Williams & Wilkins, Philadelphia, PA, pp. 1028–1043.

Rubin, R.T., O'Toole, S.M., Rhodes, M.E., Sekula, L.K., Czambel, R.K., 1999. Hypothalamo-pituitary-adrenal cortical responses to low-dose physostigmine and arginine vasopressin administration: sex differences between major depressives and matched control subjects. Psychiatry Res. 89, 1–20.

Simerly, R.B., 1990. Hormonal control of neuropeptide gene expression in sexually dimorphic olfactory pathways. Trends Neurosci. 13, 104–110.

Simerly, R.B., 2002. Wired for reproduction: organization and development of sexually dimorphic circuits in the mammalian forebrain. Annu. Rev. Neurosci. 25, 507–5036.

Simon, N., 2002. In: Pfaff, D.W., Arnold, A.P., Etgen, A.M., Fahrbach, S.E., Rubin, R.T. (Eds.), Hormones, Brain and Behavior. Academic Press, San Diego, CA.

Suzuki, S., Lund, T.D., Price, R.H., Handa, R.J., 2001. Sex differences in the hypothalamo-pituitary-adrenal axis: novel roles for androgen and estrogen receptors. Recent Res. Dev. Endocrinol. 69–86.

Wilhelm, K., Parker, G., Dewhurst, J., 1998. Examining sex differences in the impact of anticipated and actual life events. J. Affect. Disord. 48, 37–45.

## Chapter 9: Sex Differences Can Influence Behavioral Responses

Aloisi, A.M., Steenbergen, J.L., van De Poll, N.E., Farabollini, F., 1994. Sex-dependent effects of restraint on nociception and pituitary–adrenal hormones in the rat. Physiol. Behav. 55, 789–793.

Arnold, A.P., Burgoyne, P.S., 2004. Are XX and XY brain cells intrinsically different? Trends Endocrinol. Metab. 15, 6–11.

Babb, J.A., Masini, C.V., Day, H.E., Campeau, S., 2013. Sex differences in activated corticotropin-releasing factor neurons within stress-related neurocircuitry and hypothalamic-pituitary-adrenocortical axis hormones following restraint in rats. Neuroscience 234, 40–52.

Balthazart, J., 2011a. Hormones and human sexual orientation. Endocrinology 152 (8), 2937–2947.

Balthazart, J., 2011b. The Biology of Homosexuality. Oxford University Press, London, England.

Barbazanges, A., Piazza, P.V., et al., 1996. Maternal glucocorticoid secretion mediates long-term effects of prenatal stress. J. Neurosci. 16 (12), 3943–3949.

Beery, A.K., Zucker, I., 2011. Sex bias in neuroscience and biomedical research. Neurosci. Biobehav. Rev. 35, 565–572.

Belz, E.E., Kennell, J.S., Czambel, R.K., Rubin, R.T., Rhodes, M.E., 2003. Environmental enrichment lowers stress-responsive hormones in singly-housed male and female rats. Pharm. Biochem. Behav. 76, 481–486.

Berenbaum, S.A., Meyer-Bahlburg, H.F., 2015. Gender development and sexuality in disorders of sex development. Horm. Metab. Res. 47 (5), 361–366.

Berenbaum, S.A., Duck, S.C., Bryk, K., 2000. Behavioral effects of prenatal versus postnatal androgen excess in children with 21-hydroxylase-deficient congenital adrenal hyperplasia. J. Clin. Endocrinol. Metab. 85 (2), 727–733.

Blaustein, J.D., Erskine, M.S., 2002. Feminine sexual behavior: cellular integration of hormonal and afferent information in the rodent forebrain. In: Pfaff, D.W., Arnold, A.P., Etgen, A.M., Fahrbach, S.E., Rubin, R.T. (Eds.), Hormones, Brain and Behavior, vol. 1. Academic Press, San Diego, CA, pp. 139–214.

Bosch, O.J., Nair, H.P., Ahern, T.H., Neumann, I.D., Young, L.J., 2009. The CRF system mediates increased passive stress-coping behavior following the loss of a bonded partner in a monogamous rodent. Neuropsychopharmacology 34, 1406–1415.

Breedlove, S.M., 2010. Minireview: organizational hypothesis: instances of the fingerpost. Endocrinology 151, 4116–4122.

Bridges, R.S., 2016. Long-term alterations in neural and endocrine processes induced by motherhood in mammals. Horm. Behav. 77, 193–203.

Chodorow, 1999. In: The Reproduction of Mothering.

De Vries, G.J., Forger, N.G., 2015. Sex differences in the brain: a whole body perspective. Biol. Sex. Differ. 6, 15–20.

DeVries, G.J., Simerly, R.B., 2002. Anatomy, development, and function of sexually diorphic neural circuits in the mammalian brain. In: Pfaff, D.W., Arnold, A.P., Etgen, A.M., Fahrbach, S.E., Rubin, R.T. (Eds.), Hormones, Brain and Behavior, vol. 4. Academic Press, San Diego, CA, pp. 137–191.

Dowling, A.L.S., Zoeller, R.T., 2000. Thyroid hormone of maternal origin regulates the expression of RC3/Neurogranin mRNA in the fetal rat brain. Brain Res. 82, 126–132.

Dowling, A.L.S., Martz, G.U., Leonard, J.L., Zoeller, R.T., 2000. Acute changes in maternal thyroid hormone induce rapid and transient changes in specific gene expression in fetal rat brain. J. Neurosci. 20, 2255–2265.

Dowling, A.L.S., Iannacone, E.A., Zoeller, R.T., 2001. Maternal hypothyroidism selectively affects the expression of neuroendocrine-specific protein-A messenger ribonucleic acid in the proliferative zone of the fetal rat brain cortex. Endocrinology 142, 390–399.

Forest, M.G., 2001. Diagnosis and treatment of disorders of sexual development. In: DeGroot, L.J., Jameson, J.L. (Eds.), Endocrinology, fourth ed. Saunders, Philadelphia, pp. 1974–2010.

Gonzalez-Mariscal, 2002. Hormones, Brain and Behavior.

Gooren, L.J., Byne, W., 2009. Sexual orientation in men and women. In: Rubin, R.T., Pfaff, D.W. (Eds.), Hormone/Behavior Relations of Clinical Importance. Elsevier, Amsterdam, pp. 291–310.

Gooren, L.J.G., 2001. Gender identity and sexual behavior. In: DeGroot, L.J., Jameson, J.L. (Eds.), Endocrinology, fourth ed. Saunders, Philadelphia, PA, pp. 2033–2042.

Gorski, R.A., 2000. Sexual differentiation of the nervous system. In: Kandel, E.R., Schwartz, J.H., Jessel, T.M. (Eds.), Principles of Neural Science, fourth ed. McGraw-Hill, New York, pp. 1131–1148.

Green, R., 2002. Sexual identity and sexual orientation. In: Pfaff, D.W., Arnold, A.P., Etgen, A.M., Fahrbach, S.E., Rubin, R.T. (Eds.), Hormones, Brain and Behavior, vol. 4. Academic Press, San Diego, CA, pp. 463–485.

Hines, M., Constantinescu, M., Spencer, D., 2015. Early androgen exposure and human gender development. Biol. Sex Differ. 6, 3.

Hines, M., 2002. Sexual differentiation of human brain and behavior. In: Pfaff, D.W., Arnold, A.P., Etgen, A.M., Fahrbach, S.E., Rubin, R.T. (Eds.), Hormones, Brain and Behavior, vol. 4. Academic Press, San Diego, CA, pp. 425–462.

Juraska, J.M., Sisk, C.L., DonCarlos, L.L., 2013. Sexual differentiation of the adolescent rodent brain: hormonal influences and developmental mechanisms. Horm. Behav. 64 (2), 203–210.

Kamphuis, P.J., Bakker, J.M., et al., 2002. Enhanced glucocorticoid feedback inhibition of hypothalamo-pituitary-adrenal responses to stress in adult rats neonatally treated with dexamethasone. Neuroendocrinology 76 (3), 158–169.

Lenz, K.M., Wright, C.L., Martin, R.C., McCarthy, M.M., April 7, 2011. Prostaglandin $E_2$ regulates AMPA receptor phosphorylation and promotes membrane insertion in preoptic area neurons and glia during sexual differentiation. PLoS One 6 (4), e18500.

Loyd, D.R., Murphy, A.Z., 2014. The neuroanatomy of sexual dimorphism in opioid analgesia. Exp. Neurol. 0, 57–63.

McCarthy, M.M., Pickett, L.A., VanRyzin, J.W., Kight, K.E., 2015. Surprising origins of sex differences in the brain. Horm. Behav. 76, 3–10.

McCormick, C.M., et al., 2002. Stress 5 (4), 235–324.

Meaney, M.J., Diorio, J., et al., 1994. Environmental regulation of the development of glucocorticoid receptor systems in the rat forebrain: the role of serotonin. Ann. N.Y. Acad. Sci. 746, 260–273 Discussion: 274, 289–293.

Mogil, J.S., Wilson, S.G., et al., 2003. The melanocortin-1 receptor gene mediates female- specific mechanisms of analgesia in mice and humans. Proc. Natl. Acad. Sci. U.S.A. 100 (8), 4867–4872.

Mohammed, A.H., Henriksson, B.G., Södeström, S., Ebendal, T., Olsson, T., Seckl, J.R., 1993. Environmental influences on the central nervous system and their implications for the aging rat. Behav. Brain Res. 57, 183–192.

Nathanielsz, P.W., 1999. Life in the Womb: The Origin of Health and Disease. Promethean Press, Ithaca, NY.

Nohara, K., Zhang, Y., Waraich, R.S., Laque, A., Tiano, J.P., Tong, J., Munzberg, H., Mauvais-Jarvis, F., 2011. Early-life exposure to testosterone programs the hypothalamic melanocortin system. Endocrinology 152, 1661–1669.

Oakley, A., 2015. Sex, Gender, and Society. Dorset Press, Dorchester, England.

Ogawa, S., Taylor, J., et al., 1996. Reversal of sex roles in genetic female mice by disruption of estrogen receptor gene. Neuroendocrinology 64, 467–470.

Ogawa, S., Eng, V., et al., 1998a. Roles of estrogen receptor-alpha gene expression in repro-duction-related behaviors in female mice. Endocrinology 139, 5070–5081.

Ogawa, S., Washburn, T., et al., 1998b. Modifications of testosterone-dependent behaviors by estrogen receptor-alpha gene disruption in male mice. Endocrinology 139, 5058–5069.

Rubin, R.T., Carroll, B.J., 2009. Mood disorders. In: Rubin, R.T., Pfaff, D.W. (Eds.), Hormone/ Behavior Relations of Clinical Importance. Elsevier, Amsterdam, pp. 593–620.

Rubin, R.T., 1982. Testosterone and aggression in men. In: Beumont, P.J.V., Burrows, G.D. (Eds.), Handbook of Psychiatry and Endocrinology. Elsevier, Amsterdam, pp. 355–366.

Rubin, R.T., O'Toole, S.M., Rhodes, M.E., Sekula, L.K., Czambel, R.K., 1999. Hypothalamo-pituitary-adrenal cortical responses to low-dose physostigmine and arginine vasopressin administration: sex differences between major depressives and matched control subjects. Psychiatry Res. 89, 1–20.

Schneider, J.E., Brozek, J.M., Keen-Rhinehart, E., 2014. Our stolen figures: the interface of sexual differentiation, endocrine disruptors, maternal programming, and energy balance. Horm. Behav. 66 (1), 104–119.

Simerly, R.B., 1990. Hormonal control of neuropeptide gene expression in sexually dimor-phic olfactory pathways. Trends Neurosci. 13, 104–110.

Simerly, R.B., 2002. Wired for reproduction: organization and development of sexually dimorphic circuits in the mammalian forebrain. Annu. Rev. Neurosci. 25, 507–5036.

Simon, N.G., 2002. Hormonal processes in the development and expression of aggressive behavior. In: Pfaff, D.W., Arnold, A.P., Etgen, A.M., Fahrbach, S.E., Rubin, R.T. (Eds.), Hormones, Brain and Behavior, vol. 1. Academic Press, San Diego, CA, pp. 339–392.

Styne, D.M., Grumbach, M.M., 2002. Puberty in boys and girls. In: Pfaff, D.W., Arnold, A.P., Etgen, A.M., Fahrbach, S.E., Rubin, R.T. (Eds.), Hormones, Brain and Behavior, vol. 4. Academic Press, San Diego, CA, pp. 661–716.

Wallen, K., Baum, M.J., 2002. Masculinization and defeminization in altricial and precocial mammals: comparative aspects of steroid hormone action. In: Pfaff, D.W., Arnold, A.P., Etgen, A.M., Fahrbach, S.E., Rubin, R.T. (Eds.), Hormones, Brain and Behavior, vol. 4. Academic Press, San Diego, CA, pp. 385–423.

Welberg, L.A., Seckl, J.R., 2001. Prenatal stress, glucocorticoids and the programming of the brain. J. Neuroendocrinol. 13 (2), 113–128.

Zoeller, T.R., Dowling, A.L., et al., 2002. Thyroid hormone, brain development, and the envi-ronment. Environ. Health Perspect. 110 (Suppl. 3), 355–361.

Zubieta, J.-K., Smith, Y.R., Bueller, J.A., Xu, Y., Kilbourn, M.R., Jewett, D.M., Meyer, C.R., Koeppe, R.A., Stohler, C.S., 2001. Science 293, 311–315.

## Chapter 10: Hormone Actions Early in Development Can Influence Hormone Responsiveness in the CNS During Adulthood

Asarian, L., Geary, N., 2013. Sex differences in the physiology of eating. Am. J. Physiol. Regul. Integr. Comp. Physiol. 305, R1215–R1267.

Barbazanges, A., Piazza, P.V., et al., 1996. Maternal glucocorticoid secretion mediates long-term effects of prenatal stress. J. Neurosci. 16 (12), 3943–3949.

Barker, D., Osmond, C., 1986. Infant mortality, childhood nutrition, and ischaemic heart dis-ease in England and Wales. Lancet 1077–1081.

Berenbaum, S.A., Duck, S.C., Bryk, K., 2000. Behavioral effects of prenatal versus postnatal androgen excess in children with 21-hydroxylase-deficient congenital adrenal hyperpla-sia. J. Clin. Endocrinol. Metab. 85 (2), 727–733.

Blaustein, J.D., Erskine, M.S., 2002. Feminine sexual behavior: cellular integration of hor-monal and afferent information in the rodent forebrain. In: Pfaff, D.W., Arnold, A.P., Etgen, A.M., Fahrbach, S.E., Rubin, R.T. (Eds.), Hormones, Brain and Behavior, vol. 1. Academic Press, San Diego, CA, pp. 139–214.

Bouret, S.G., Draper, S.J., Simerly, R.B., 2004. Trophic action of leptin on 18 hypothalamic neurons that regulate feeding. Science 304, 108–110.

Clark, P.A., Iranmanesh, A., et al., 1997. Comparison of pulsatile luteinizing hormone secretion between prepubertal children and young adults: evidence for a mass/amplitude-dependent difference without gender or day/night contrasts. J. Clin. Endocrinol. Metab. 82 (9), 2950–2955.

Dowling, A.L.S., Zoeller, R.T., 2000. Thyroid hormone of maternal origin regulates the expression of RC3/neurogranin mRNA in the fetal rat brain. Brain Res. 82, 126–132.

Dowling, A.L.S., Martz, G.U., Leonard, J.L., Zoeller, R.T., 2000. Acute changes in maternal thyroid hormone induce rapid and transient changes in specific gene expression in fetal rat brain. J. Neurosci. 20, 2255–2265.

Dowling, A.L.S., Iannacone, E.A., Zoeller, R.T., 2001. Maternal hypothyroidism selectively affects the expression of neuroendocrine-specific protein-A messenger ribonucleic acid in the proliferative zone of the fetal rat brain cortex. Endocrinology 142, 390–399.

Ducy, P., Amling, M., Takeda, S., Priemel, M., Schilling, A.F., Beil, F.T., Shen, J., Vinson, C., Rueger, J.M., Karsenty, G., 2000. Leptin inhibits bone formation through a hypothalamic relay: a central control of bone mass. Cell 100 (2), 197–207.

Erickson-Schroth, L., 2016. Psychological and biological influences on gender roles. In: Pfaff, D., Volkow, N. (Eds.), Neuroscience in the 21st Century. Springer, Heidelberg.

Forest, M.G., 2001. Diagnosis and treatment of disorders of sexual development. In: DeGroot, L.J., Jameson, J.L. (Eds.), Endocrinology, fourth ed. Saunders, Philadelphia, pp. 1974–2010.

Gooren, L.J.G., 2001. Gender identity and sexual behavior. In: DeGroot, L.J., Jameson, J.L. (Eds.), Endocrinology, fourth ed. Saunders, Philadelphia, PA, pp. 2033–2042.

Gorski, R.A., 2000. Sexual differentiation of the nervous system. In: Kandel, E.R., Schwartz, J.H., Jessel, T.M. (Eds.), Principles of Neural Science, fourth ed. McGraw-Hill, New York, pp. 1131–1148.

Green, R., 2002. Sexual identity and sexual orientation. In: Pfaff, D.W., Arnold, A.P., Etgen, A.M., Fahrbach, S.E., Rubin, R.T. (Eds.), Hormones, Brain and Behavior, vol. 4. Academic Press, San Diego, CA, pp. 463–485.

Hines, M., 2002. Sexual differentiation of human brain and behavior. In: Pfaff, D.W., Arnold, A.P., Etgen, A.M., Fahrbach, S.E., Rubin, R.T. (Eds.), Hormones, Brain and Behavior, vol. 4. Academic Press, San Diego, CA, pp. 425–462.

Jones, A.P., Friedman, M.I., 1982. Obesity and adipocyte abnormalities in offspring of rats undernourished during pregnancy. Sciences 215, 1518–1519.

Jones, A.P., Simon, E.L., Friedman, M.I., 1986. Gestational undernutrition and the development of obesity in rats. J. Nutr. 1484–1492.

Kamphuis, P.J., Bakker, J.M., et al., 2002. Enhanced glucocorticoid feedback inhibition of hypothalamo–pituitary–adrenal responses to stress in adult rats neonatally treated with dexamethasone. Neuroendocrinology 76 (3), 158–169.

Lumey, L.H., Van Poppel, F.W., August 1994. The Dutch famine of 1944-45: mortality and morbidity in past and present generations. Soc. Hist. Med. 7 (2), 229–246.

Maccari, S., Darnaudery Jr., M., Morely-Fletcher, S., Van Reeth, O., 2003. Prenatal stress and long-term consequences: implications of glucocorticoid hormones. Neurosci. Biobehav. Rev. 27, 119–127.

Marshall, W.A., Tanner, J.M., 1970. Variations in the pattern of pubertal changes in boys. Arch. Dis. Child. 45 (239), 13–23.

Marshall, W.A., Tanner, J.M., 1999. Variations in pattern of pubertal changes in girls. Arch. Dis. Child. 44 (235), 291–303.

Mauras, N., Blizzard, R.M., et al., 1987. Augmentation of growth hormone secretion during puberty: evidence for a pulse amplitude-modulated phenomenon. J. Clin. Endocrinol. Metab. 64 (3), 596–601.

Meaney, M.J., Diorio, J., et al., 1994. Environmental regulation of the development of glucocorticoid receptor systems in the rat forebrain: the role of serotonin. Ann. N.Y. Acad. Sci. 746, 260–273 Discussion: 274, 289–293.

Mohammed, A.H., Henriksson, B.G., Södeström, S., Ebendal, T., Olsson, T., Seckl, J.R., 1993. Environmental influences on the central nervous system and their implications for the aging rat. Behav. Brain Res. 57, 183–192.

Nathanielsz, P.W., 1999. Life in the Womb: The Origin of Health and Disease. Promethean Press, Ithaca, NY.

Nugent, B.M., Wright, C.L., Shetty, A.C., Hodes, G.E., Lenz, K.M., Mahurkar, A., Russo, S.J., Devine, S.E., McCarthy, M.M., May 2015. Brain feminization requires active repression of masculinization via DNA methylation. Nat. Neurosci. 18 (5), 690–697.

Olney, J.W., 1969. Brain lesions, obesity, and other disturbances in mice treated with monosodium glutamate. Science 164 (880), 719–721.

Phoenix, C.H., Goy, R.W., Gerral, A.A., Young, W.C., 1959. Organizing action of prenatally administered testosterone propionate on the tissues mediating mating behavior in the female Guinea pig. Endocrinology. 1959 (65), 369–382.

Ravelli, G.P., Stein, Z., Susser, M., 1976. Obesity in young men after famine exposure in utero and early infancy. N. Engl. J. Med. 295, 349–353.

Rogol, A.D., Roemmich, J.N., Clark, P.A., 2002. Growth at puberty. J. Adolesc. Health 31 (Suppl. 6), 192–200.

Rubin, R.T., 1982. Testosterone and aggression in men. In: Beumont, P.J.V., Burrows, G.D. (Eds.), Handbook of Psychiatry and Endocrinology. Elsevier, Amsterdam, pp. 355–366.

Schneider, M.L., Roughton, E.C., Koehler, A.J., Lubach, G.R., 1999. Growth and development following prenatal stress exposure in primates: an examination of ontogenetic vulnerability. Child. Dev. 70, 263–274.

Schneider, J.E., Brozek, J.M., Keen-Rhinehart, E., 2014. Our stolen figures: the interface of sexual differentiation, endocrine disruptors, maternal programming, and energy balance. Horm. Behav. 66 (1), 104–119.

Shi, H., Clegg, D.J., 2009. Sex differences in the regulation of body weight. Physiol. Behav. 97, 199–204.

Shi, H., Seeley, R.J., Clegg, D.J., 2009. Sexual differences in the control of energy homeostasis. Front. Neuroendocrinol. 30, 396–404.

Simon, N.G., 2002. Hormonal processes in the development and expression of aggressive behavior. In: Pfaff, D.W., Arnold, A.P., Etgen, A.M., Fahrbach, S.E., Rubin, R.T. (Eds.), Hormones, Brain and Behavior, vol. 1. Academic Press, San Diego, CA, pp. 339–392.

Styne, D.M., Grumbach, M.M., 2002. Puberty in boys and girls. In: Pfaff, D.W., Arnold, A.P., Etgen, A.M., Fahrbach, S.E., Rubin, R.T. (Eds.), Hormones, Brain and Behavior, vol. 4. Academic Press, San Diego, CA, pp. 661–716.

Takeda, S., Elefteriou, F., Levasseur, R., Liu, X., Zhao, L., Parker, K.L., Armstrong, D., Ducy, P., Karenty, G., 2002. Leptin regulates bone formation via the sympathetic nervous system. Cell 111, 305–317.

Wallen, K., Baum, M.J., 2002. Masculinization and defeminization in altricial and precocial mammals: comparative aspects of steroid hormone action. In: Pfaff, D.W., Arnold, A.P., Etgen, A.M., Fahrbach, S.E., Rubin, R.T. (Eds.), Hormones, Brain and Behavior, vol. 4. Academic Press, San Diego, CA, pp. 385–423.

Welberg, L.A., Seckl, J.R., 2001. Prenatal stress, glucocorticoids and the programming of the brain. J. Neuroendocrinol. 13 (2), 113–128.

Zoeller, T.R., Dowling, A.L., et al., 2002. Thyroid hormone, brain development, and the environment. Environ. Health Perspect 110 (Suppl. 3), 355–361.

## Chapter 11: Epigenetic Changes Mediate Effects of Hormones on Behavior

Abiko, T., Kimura, Y., 2001. Syntheses and effect of bombesin-fragment 6–14 and its four analogues on food intake in rats. Curr. Pharm. Biotechnol. 2 (2), 201–207.

Anawalt, B.D., Merriam, G.R., 2001. Neuroendocrine aging in men. Endocrinol. Metab. Clin. North Am. 30, 647–669.

Berkley, K.J., Hoffman, G.E., Murphy, A.Z., Holdcroft, A., 2002. Pain: sex/gender differences. In: Pfaff, D.W., Arnold, A.P., Etgen, A.M., Fahrbach, S.E., Rubin, R.T. (Eds.), Hormones, Brain and Behavior, vol. 5. Academic Press, San Diego, CA, pp. 409–442.

Blair-West, J.R., Carey, K.D., Denton, D.A., Weisinger, R.S., Shade, R.E., 1998. Evidence that brain angiotensin II is involved in both thirst and sodium appetite in baboons. Am. J. Physiol. 275 (5, Pt. 2), R1639–R1646.

Briggs, D.I., Enriori, P.J., Lemus, M.B., Cowley, M.A., Andrews, Z.B., 2010. Diet-induced obesity causes ghrelin resistance in arcuate npy/agrp neurons. Endocrinology 151 (10), 4745–4755.

Brown, M., Allen, R., Villarreal, J., Rivier, J., Vale, W., 1978. Bombesin-like activity: radioimmunologic assessment in biological tissues. Life Sci. 23 (27–28), 2721–2728.

Choleris, E., Gustafsson, J.A., et al., 2003. An estrogen-dependent four-gene micronet regulating social recognition: a study with oxytocin and estrogen receptor-alpha and -beta knockout mice. Proc. Natl. Acad. Sci. U.S.A. 100 (10), 6192–6197.

Crews, D., 2008. Epigenetics and its implications for behavioral neuroendocrinology. Front. Neuroendocrinol. 29 (3), 344–357.

Crowe, M.J., Frosling, M.L., Rolls, B.J., Phillips, P.A., Ledingham, J.G., Smith, R., 1987. Altered water excretion in healthy elderly men. Age Ageing 16 (5), 285–293.

Cummings, D.E., Overduin, J., 2007. Gastrointestinal regulation of food intake. J. Clin. Invest. 117 (1), 13–23.

Dellovade, T.L., Zhu, Y.-S., Krey, L., Pfaff, D.W., 1996. Thyroid hormone and estrogen interact to regulate behavior. Proc. Natl. Acad. Sci. U.S.A. 93, 12581–12586.

Dohanich, G., 2002. Gonadal steroids, learning, and memory. In: Pfaff, D.W., Arnold, A.P., Etgen, A.M., Fahrbach, S.E., Rubin, R.T. (Eds.), Hormones, Brain and Behavior, vol. 2. Academic Press, San Diego, CA, pp. 265–327.

Dubal, D.B., Pettigrew, L.C., Kashon, M., Ren, J.M., Finklestein, S.P., Rau, S.W., Wise, P.M., 1998. Estradiol protects against ischemia-induced brain injury. J. Cereb. Blood Flow Metab. 18, 1253–1258.

Dubal, D.B., Shughrue, P.J., Wilson, M.E., Merchenthaler, I., Wise, P.M., 1999. Estradiol modulates bcl-2 in cerebral ischemia: a potential role for estrogen receptors. J. Neurosci. 19, 6385–6393.

Dubal, D.B., Zhu, H., Yu, J., Rau, S.W., Shughrue, P.J., Merchenthaler, I., Kindy, M.S., Wise, P.M., 2001. Estrogen receptor α, not ß, is a critical link in estradiol-mediated protection against brain injury. Proc. Natl. Acad. Sci. U.S.A. 98, 1952–1957.

Endocrine Society, 2001. In: 2nd Annual Andropause Consensus Meeting, Beverly Hills, CA.

Figlewicz, D.P., Nadzan, A.M., Sipols, A.J., Green, P.K., Liddle, R.A., Porte Jr., D., Woods, S.C., 1992. Intraventricular CCK-8 reduces single meal size in the baboon by interaction with type-A CCK receptors. Am. J. Physiol. 263 (4, Pt. 2), R863–R867.

Fitts, D.A., Thurnhorst, R.L., Simpson, J.B., 1985. Diuresis and reduction of salt appetite by lateral ventricular infusions of atriopeptin II. Brain Res. 348, 118–124.

Fitzsimons, J.T., Stricker, E.M., 1971. Sodium appetite and the renin–angiotensin system. Nat. New Biol. 231 (19), 58–60.

Fletcher, S.W., Colditz, G.A., 2002. Failure of estrogen plus progestin therapy for prevention [editorial]. JAMA 288, 366–367.

Foster, T.C., 1999. Involvement of hippocampal synaptic plasticity in age-related memory decline. Brain Res. Rev. 30, 236–249.

Gagnidze, K., Weil, Z.M., Faustino, L.C., Schaafsma, S.M., Pfaff, D.W., 2013. Early histone modifications in the ventromedial hypothalamus and preoptic area following oestradiol administration. J. Neuroendocrinol. 25, 939–955.

Geerling, J.C., Lowey, 2008. Central regulation of sodium appetite. Exp. Physiol. 93 (2), 177–209.

Gibbs, J., Fauser, D.J., Row, E.A., Rolls, B.J., Rolls, E.T., Maddison, S.P., 1979. Bombesin suppresses feeding in rats. Nature 282 (5735), 208–210.

Gibbs, J., Young, R.C., Smith, G.P., 1997. Cholecystokinin decreases food intake in rats. Obes. Res. 5 (3), 284–290.

Golden, G.A., Mason, R.P., Tulenko, T.N., Zubenko, G.S., Rubin, R.T., 1999. Rapid and opposite effects of cortisol and estradiol on human erythrocyte Na⁺, K⁺ ATPase activity: relationship to steroid intercalation into the cell membrane. Life Sci. 65, 1247–1255.

Gore, A.C., Patisaul, H.B., 2010. Neuroendocrine disruption: historical roots, current progress, questions for the future. Front. Neuroendocrinol. 31, 395–399.

Guttman, M., Rinn, J.L., 2012. Modular regulatory principles of large non-coding RNAs. Nature 482, 339–346.

Hamilton, R.B., Norgren, R., 1984. Central projections of gustatory nerves in the rat. J. Comp. Neurol. 222 (4), 560–577.

Henderson, V.W., Reynolds, D.W., 2002. Protective effects of estrogen on aging and damaged neural systems. In: Pfaff, D.W., Arnold, A.P., Etgen, A.M., Fahrbach, S.E., Rubin, R.T. (Eds.), Hormones, Brain and Behavior, vol. 4. Academic Press, San Diego, CA, pp. 821–837.

Huda, M.S., Dovey, T., Wong, S.P., English, P.J., Halford, J., McCulloch, P., et al., 2009. Ghrelin restores 'lean-type' hunger and energy expenditure profiles in morbidly obese subjects but has no effect on postgastrectomy subjects. Int. J. Obes. 33 (3), 317–325.

Hunter, R., Gagnidze, K., McEwen, B., Pfaff, D., 2015. Stress and the dynamic genome: steroids, epigenetics and the transposome. Proc. Natl. Acad. Sci. 112, 6828–6833.

Jensen Peña, C., Monk, C., Champagne, F.A., 2012. Epigenetic effects of prenatal stress on 11β-hydroxysteroid dehydrogenase-2 in the placenta and fetal brain. PLoS One 7, e39791.

Johnson, A.K., Thunhorst, R.L., 1997. The neuroendocrinology of thirst and salt appetite: visceral sensory signals and mechanisms of central integration. Front. Neuroendocrinol. 18 (3), 292–353.

Keverne, E.B., Tabansky, I., Pfaff, D., 2015. Epigenetic changes in the developing brain: effects on behavior. Proc. Natl. Acad. Sci. 112, 6789–6795.

Kirchner, H., Heppner, K.M., Tschop, M.H., 2012. The role of ghrelin in the control of energy balance. Handb. Exp. Pharmacol. 209, 161–184.

Kojima, M., Hosoda, H., Date, Y., Nakazato, M., Matsuo, H., Kangawa, K., 1999. Ghrelin is a growth-hormone-releasing acylated peptide from stomach. Nature 402 (6762), 656–660.

Kumar, A., Foster, T.C., 2002. 17β-estradiol benzoate decreases the AHP amplitude in CA1 pyramidal neurons. J. Neurophysiol. 88, 621–626.

Kundakovic, M., Gudsnuk, K., Franks, B., Madrid, J., Miller, R.L., Perera, F.P., Champagne, F.A., 2013. Sex-specific epigenetic disruption and behavioral changes following low-dose in utero bisphenol A exposure. Proc. Natl. Acad. Sci. U.S.A. 110, 9956–9961.

Meaney, M.J., 2001. Maternal care gene expression and the transmission of individual differences in stress reactivity across generations. Annu. Rev. Neurosci. 24, 161–192.

Meisel, R.L., Dohanich, G.P., McEwen, B.S., Pfaff, D.W., 1987. Antagonism of sexual behavior in female rats by ventromedial hypothalamic implants of antiestrogen. Neuroendocrinology 45 (3), 201–207.

Mohon, M.A., 2001. An airman with Schmidt's syndrome. Fed. Air Surgeon's Med. Bull. 1–3.

Moran, T.H., Sawyer, T.K., Seeb, D.H., Arriglio, P.J., Lombard, M.A., McHugh, P.R., 1992. Potent and sustained satiety actions of a cholecystokinin octapeptid analogue. Am. J. Clin. Nutr. 2865–2905.

Morgan, M., Dellovade, T.L., Pfaff, D.W., 2000. Effect of thyroid hormone administration on estrogen-induced sex behavior in female mice. Horm. Behav. 37, 15–22.

Niswender, K.D., Schwartz, M.W., 2003. Insulin and leptin revisited: adiposity with overlapping physiological and intracellular signaling capabilities. Front. Neuroendcrinol. 24 (1), 1–10.

Odell, W.D., Burger, H.G., 2001. Menopause and hormone replacement. In: DeGroot, L.J., Jameson, J.L. (Eds.), Endocrinology, fourth ed. Saunders, Philadelphia, PA, pp. 2153–2162.

Ojeda, S.R., Lomniczi, A., 2014. Puberty in 2013: unravelling the mystery of puberty. Nat. Rev. Endocrinol. 10, 67–69.

Parry, B.L., Berga, S.L., 2002. Premenstrual dysphoric disorder. In: Pfaff, D.W., Arnold, A.P., Etgen, A.M., Fahrbach, S.E., Rubin, R.T. (Eds.), Hormones, Brain and Behavior, vol. 5. Academic Press, San Diego, CA, pp. 531–552.

Pfaff, D.W., 1999. Drive: Neurobiological and Molecular Mechanisms of Sexual Motivation. MIT Press, Cambridge, MA.

Phillips, P.A., Bretherton, M., Johnston, C.I., Gray, L., 1991a. Reduced osmotic thirst in healthy elderly men. Am. J. Physiol. 261 (1, Pt. 2), R166–R171.

Phillips, P.A., Hodsman, G.P., Johnston, C.I., 1991b. Neuroendocrine mechanisms and cardiovascular homeostasis in the elderly. Cardiovasc. Drugs Ther. 6 (Suppl.), 1209–1213.

Phillips, P.A., Bretherton, M., Risvanis, J., Casley, D., Johnston, C., Gray, L., 1993a. Effects of drinking on thirst and vasopressin in dehydrated elderly men. Am. J. Physiol. 264 (5, Pt. 2), R877–R881.

Phillips, P.A., Johnston, C.I., Gray, L., 1993b. Disturbed fluid and electrolyte homeostasis following dehydration in elderly people. Age Ageing 22 (1), S26–S33.

Porte Jr., D., Baskin, D.G., Schwartz, M.W., 2002. Leptin and insulin action in the central nervous system. Nutr. Rev. 60 (10, Pt. 2), S20–S29.

Raskind, M.A., Wilkinson, C.W., Peskind, E.R., 2002. Aging and Alzheimer's disease. In: Pfaff, D.W., Arnold, A.P., Etgen, A.M., Fahrbach, S.E., Rubin, R.T. (Eds.), Hormones, Brain and Behavior, vol. 5. Academic Press, San Diego, CA, pp. 637–664.

Reame, N.E., 2001. Premenstrual syndrome. In: DeGroot, L.J., Jameson, J.L. (Eds.), Endocrinology, fourth ed. Saunders, Philadelphia, PA, pp. 2147–2152.

Rinn, J.L., et al., 2014. Spatiotemporal expression and transcriptional perturbations by long noncoding RNAs in the mouse brain. Proc. Natl. Acad. Sci. 112, 6855–6863.

Roth, J., Koch, C.A., Rother, K.I., 2001. Aging, endocrinology, and the elderly patient. In: DeGroot, L.J., Jameson, J.L. (Eds.), Endocrinology, fourth ed. Saunders, Philadelphia, PA, pp. 529–555.

Rubinow, D.R., Schmidt, P.J., Roca, C.A., Daly, R.C., 2002. Gonadal hormones and behavior in women: concentrations versus context. In: Pfaff, D.W., Arnold, A.P., Etgen, A.M., Fahrbach, S.E., Rubin, R.T. (Eds.), Hormones, Brain and Behavior, vol. 5. Academic Press, San Diego, CA, pp. 37–73.

Sadow, T.F., Rubin, R.T., 1992. Effects of hypothalamic peptides on the aging brain. Psychoneuroendocrinology 17, 293–314.

Sakai, R.R., Ma, L.Y., Zhang, D.M., McEwen, B.S., Fluharty, S.J., 1996. Intracerebral administration of mineralocorticoid receptor antisense oligonucleotides attenuate adrenal steroid-induced salt appetite in rats. Neuroendocrinology 64 (6), 425–429.

Sapolsky, R.M., 2000. Glucocorticoids and hippocampal atrophy in neuropsychiatric disorders. Arch. Gen. Psychiatry 57 (10), 925–935.

Schaaf, M.J., Cidlowski, J.A., 2002. Molecular mechanisms of glucocorticoid action and resistance. J. Steroid Biochem. Mol. Biol. 83 (1–5), 37–48.

Schneider, J.E., Brozek, J.M., Keen-Rhinehart, E., 2014. Our stolen figures: the interface of sexual differentiation, endocrine disruptors, maternal programming, and energy balance. Horm. Behav. 66, 104–119.

Schumacher, M., Robert, F., 2002. Progesterone: synthesis, metabolism, mechanisms of action, and effects in the nervous system. In: Pfaff, D.W., Arnold, A.P., Etgen, A.M., Fahrbach, S.E., Rubin, R.T. (Eds.), Hormones, Brain and Behavior, vol. 3. Academic Press, San Diego, CA, pp. 683–745.

Seibel, M.M., 2001. Ovulation induction and assisted reproduction. In: DeGroot, L.J., Jameson, J.L. (Eds.), Endocrinology, fourth ed. Saunders, Philadelphia, PA, pp. 2138–2145.

Sharrow, K.M., Kumar, A., Foster, T.C., 2002. Calcineurin as a potential contributor in estradiol regulation of hippocampal synaptic function. Neuroscience 113, 89–97.

Sherwin, B.B., 2003. Estrogen and cognitive functioning in women. Endocrine Rev. 24, 133–151.

Simon, N.G., 2002. Hormonal processes in the development and expression of aggressive behavior. In: Pfaff, D.W., Arnold, A.P., Etgen, A.M., Fahrbach, S.E., Rubin, R.T. (Eds.), Hormones, Brain and Behavior, vol. 1. Academic Press, San Diego, CA, pp. 339–392.

Sinchak, K., Wagner, E., 2012. Front. Neuroendocrinol. 3, 342–363.

Snyder, P.J., 2001. Effect of age on testicular function and consequences of testosterone treatment. J. Clin. Endocrinol. Metab. 86, 2369–2372.

Van Coevorden, A., Mockel, J., Laurent, E., Kerkhofs, M., L'Hermite-Baleriaux, M., Decoster, C., Neve, P., Van Cauter, E., 1991. Neuroendocrine rhythms and sleep in aging men. Am. J. Physiol. 260, E651–E661.

Vasudevan, N., Davidkova, G., Zhu, Y.S., Koibuchi, N., Chin, W.W., Pfaff, D.W., 2001. Differential interactions of estrogen receptor and thyroid hormone receptor isoforms on the rat oxytocin receptor promoter leads to differences in transcriptional regulation. Neuroendocrinology 74 (5), 309–324.

Vasudevan, N., Koibuchi, N., Chin, W.W., Pfaff, D.W., 2001. Differential crosstalk between estrogen receptor (ER) and ER and the thyroid hormone receptor isoforms results in flexible regulation of the consensus ERE. Brain Res. Mol. Brain Res. 95 (1–2), 9–17.

Vasudevan, N., Zhu, Y.S., Daniels, S., Koibuchi, N., Chin, W.W., Pfaff, D.W., 2001. Crosstalk between oestrogen receptors and thyroid hormone receptor isoforms results in differential regulation of the preproenkephalin gene. J. Neuroendocrinol. 13.

Vermeulen, A., 2001. Androgen replacement therapy in the aging male: a critical evaluation. J. Clin. Endocrinol. Metab. 86, 2380–2390.

Waddington, C.H., 1942. Canalization of development and the inheritance of acquired characters. Nature 150, 563–565.

Wilson, M.E., Dubal, D.B., Wise, P.M., 2000. Estradiol protects injury-induced cell death in cortical explant cultures: a role for estrogen receptors. Brain Res. 873, 235–242.

Wilson, W.L., Starbuck, E.M., Fitts, D.A., 2002. Salt appetite of adrenalectomized rats after a lesion of the SFO. Behav. Brain Res. 136 (2), 449–453.

Wilson, M.E., Liu, Y., Wise, P.M., 2002. Estradiol enhances Akt activation in cortical explant cultures following neuronal injury. Mol. Brain Res. 102, 88–94.

Wise, P.M., Dubal, D., Wilson, M.E., Rau, S.W., Liu, Y., 2001. Estradiol: a trophic and protective factor in the adult brain. Front. Neuroendocrinol. 22, 33–66.

Wise, P.M., 2000. Neuroendocrine correlates of aging. In: Conn, P.M., Freeman, M.E. (Eds.), Neuroendocrinology in Physiology and Medicine. Humana Press, Totowa, NJ.

## Chapter 12: Puberty Alters Hormone Secretion and Hormone Responsivity and Heralds Sex Differences

Bellefontaine, N., Elias, C.F., 2014. Minireview: metabolic control of the reproductive physiology: insights from genetic mouse models. Horm. Behav. 66, 7–14.

Bronson, F.H., 1989. Mammalian Reproductive Biology, first ed. The University of Chicago Press, Chicago and London.

Clark, P.A., Iranmanesh, A., et al., 1997. Comparison of pulsatile luteinizing hormone secretion between prepubertal children and young adults: evidence for a mass/amplitude-dependent difference without gender or day/night contrasts. J. Clin. Endocrinol. Metab. 82 (9), 2950–2955.

De Bond, J.A., Smith, J.T., 2014. Kisspeptin and energy balance in reproduction. Reproduction 147, R53–R63.

Juraska, J.M., Sisk, C.L., DonCarlos, L.L., 2013. Sexual differentiation of the adolescent rodent brain: hormonal influences and developmental mechanisms. Horm. Behav. 64, 203–210.

Marshall, W.A., Tanner, J.M., 1969. Variations in pattern of pubertal changes in girls. Arch. Dis. Child. 44 (235), 291–303.

Marshall, W.A., Tanner, J.M., 1970. Variations in pattern of pubertal changes in boys. Arch. Dis. Child. 45 (239), 13–23.

Mitsushima, D., Hei, D.L., Terasawa, E., 1994. gamma-Aminobutyric acid is an inhibitory neurotransmitter restricting the release of luteinizing hormone-releasing hormone before the onset of puberty. Proc. Natl. Acad. Sci. U.S.A. 91 (1), 395–399.

Ojeda, S.R., Lomniczi, A., 2014. Puberty in 2013: unravelling the mystery of puberty. Nat. Rev. Endocrinol. 10, 67–69.

Ojeda, S.R., Ma, Y.J., Lee, B.J., Prevot, V., 2000. Glia-to-neuron signaling and the neuroendo-crine control of female puberty. Recent Prog. Horm. Res. 55, 197–223.

Parsons, B., MacLusky, N.J., Krieger, M.S., McEwen, B.S., Pfaff, D.W., 1979. The effects of long-term estrogen exposure on the induction of sexual behavior and measurements of brain estrogen and progestin receptors in the female rat. Horm. Behav. 13, 301–313.

Parsons, B., Rainbow, T.C., Pfaff, D.W., McEwen, B.S., 1981. Oestradiol, sexual receptivity and cytosol progestin receptors in rat hypothalamus. Nature 292, 58–59.

Parsons, B., McEwen, B.S., Pfaff, D.W., 1982a. A discontinuous schedule of estradiol treat-ment is sufficient to activate progesterone-facilitated feminine sexual behavior and to increase cytosol receptors for progestins in the hypothalamus of the rat. Endocrinology 110, 613–619.

Parsons, B., Rainbow, T.C., Pfaff, D.W., McEwen, B.S., 1982b. Hypothalamic protein synthe-sis essential for the activation of the lordosis reflex in the female rat. Endocrinology 110, 620–624.

Plant, et al., 1989. Puberty in monkeys is triggered by chemical stimulation of the hypothala-mus. Proc. Natl. Acad. Sci. U.S.A. 86 (7), 2506–2510.

Reaves, P.Y., Gelband, C.H., Wang, H., Yang, H., Lu, D., Berecek, K.H., Katovich, M.J., Raizada, M.K., 1999. Permanent cardiovascular protection from hypertension by the AT(1) recep-tor antisense gene therapy in hypertensive rat offspring. Circ. Res. 85 (10), 44–50.

Sanchez-Garrido, M.A., Tena-Sempere, M., 2013. Metabolic control of puberty: roles of leptin and kisspeptins. Horm. Behav. 64, 187–194.

Santoro, N., Filicori, M., et al., 1986. Hypogonadotropic disorders in men and women: diagnosis and therapy with pulsatile gonadotropin-releasing hormone. Endocrine Rev. 7 (1), 11–23.

Seminara, S.B., Hayes, F.J., et al., 1998. Gonadotropin-releasing hormone deficiency in the human (idiopathic hypogonadotropic hypogonadism and Kallmann's syndrome): patho-physiological and genetic considerations. Endocrine Rev. 19 (5), 521–539.

Sisk, C.L., Zehr, J.L., 2005. Pubertal hormones organize the adolescent brain and behavior. Front. Neuroendocrinol. 26, 163–174.

Tannenbaum, G.S., Epelbaum, J., Bowers, C.Y., 2002. Ghrelin and the growth hormone neuro-endocrine axis. In: Kordon, et al. (Ed.), Brain Somatic Cross-Talk and the Central Control of Metabolism. Springer-Verlag, Berlin.

Valk, T.W., Corley, K.P., et al., 1980. Hypogonadotropic hypogonadism: hormonal responses to low dose pulsatile administration of gonadotropin-releasing hormone. J. Clin. Endocrinol. Metab. 51 (4), 730–738.

Wilson, K.M., Margargal, W., Berecek, K.H., 1988. Long-term captopril treatment: angioten-sin II receptors and responses. Hypertension 11 (2, Pt. 2), 148–152.

Wilson, M., Daly, M., Pound, N., 2002. An evolutionary psychological perspective on the modulation of competitive confrontation and risk-taking. In: Pfaff, D.W. (Ed.), Hormones, Brain, and Behavior, vol. V. Elsevier, San Diego, p. 389.

## Chapter 13: Changes in Hormone Levels and Responsiveness During Aging Affect Behavior

Aminoff, M.J., 2000. Brown-Sequard: selected contributions of a nineteenth-century neuro-scientist. Neuroscientist 6 (1), 60–65.

Aminoff, M.J., 2011. Brown-Sequard: An Improbable Genius Who Transformed Medicine. Oxford Univeristy Press, Oxford, England.

Asakawa, A., Inui, A., Kaga, T., Yuzuriha, H., Nagata, T., Ueno, N., Makino, S., Fujimiya, M., Niijima, A., Fujino, M.A., Kasuga, M., 2001. Ghrelin is an appetite-stimulatory signal from stomach with structural resemblance to motilin. Gastroenterology 120 (2), 337–345.

Bagnasco, M., Dube, M.G., Kalra, P.S., Kalra, S.P., 2002a. Evidence for the existence of distinct central appetite, energy expenditure, and ghrelin stimulation pathways as revealed by hypothalamic site-specific leptin gene therapy. Endocrinology 143 (11), 4409–4421.

Bagnasco, M., Kalra, P.S., Kalra, S.P., 2002b. Plasma leptin levels are pulsatile in adult rats: effects of gonadectomy. Neuroendocrinology 75 (4), 257–263.

Balsalobre, A., Brown, S.A., Marcacci, L., Tronche, F., Kellendonk, C., Reichardt, H.M., Eschutz, G., Schibler, U., 2000. Resetting of circadian time in peripheral tissues by glucorticoid signaling. Science 289, 2344–2347.

Bartke, A., 2011. Single-gene mutations and healthy ageing in mammals. Philos. Trans. Soc. Lond. B. Biol. Sci. 366, 28–34.

Cohen, P., Zhao, C., Cai, X., Montez, J.M., Rohani, S.C., Feinstein, P., Mombaerts, P., Friedman, J.M., 2001. J. Clin. Invest. 108 (8), 1113–1121.

Czeisler, C.A., Klerman, E.B., 1999. Circadian and sleep-dependent regulation of hormone release in humans. Recent Prog. Horm. Res. 54, 97–132.

Damiola, F., LeMinh, N., Preitner, N., Kornmann, B., Fleury-Olela, F., Schibler, U., 2000. Restricted feeding uncouples circadian oscillators in peripheral tissues from the central pacemaker in the suprachiasmatic nucleus. Genes Dev. 14 (23), 2950–2961.

Darlington, T.K., Wager-Smith, K., et al., 1998. Closing the circadian loop: CLOCK- induced transcription of its own inhibitors per and tim. Science 280 (5369), 1599–1603.

Down, J.L., Wise, P.M., 2009. The role of the brain in female reproductive aging. Mol. Cell. Endocrinol. 299, 32–38.

Duncan, K., 2015. Estrogen Effects on Traumatic Brain Injury: Mechanisms of Neuroprotection and Repair. Elsevier, London.

Ferkin, M.H., Zucker, I., 1991. Seasonal control of odour preferences of meadow voles (Microtus pennsylvanicus) by photoperiod and ovarian hormones. J. Reprod. Fertil. 92 (2), 433–441.

Foster, T.C., 1999. Involvement of hippocampal synaptic plasticity in age-related memory decline. Brain Res. Rev. 30, 236–249.

Friedman, J.M., 2002. The function of leptin in nutrition, weight, physiology. Nutr. Rev. 60 (10, Pt. 2), S1–S14 Discussion: S68–S87.

Golden, G.A., Mason, R.P., Tulenko, T.N., Zubenko, G.S., Rubin, R.T., 1999. Rapid and opposite effects of cortisol and estradiol on human erythrocyte $Na^+$, $K^+$ ATPase activity: relationship to steroid intercalation into the cell membrane. Life Sci. 65, 1247–1255.

Henderson, V.W., Reynolds, D.W., 2002. Protective effects of estrogen on aging and damaged neural systems. In: Pfaff, D.W., Arnold, A.P., Etgen, A.M., Fahrbach, S.E., Rubin, R.T. (Eds.), Hormones, Brain and Behavior, vol. 4. Academic Press, San Diego, CA, pp. 821–837.

Herzog, E.D., Takahashi, J.S., et al., 1998. Clock controls circadian period in isolated suprachiasmatic nucleus neurons. Nat. Neurosci. 1 (8), 708–713.

Kalra, S.P., Dube, M.G., Sahu, A., Phelps, C.P., Kalra, P.S., 1991. Neuropeptide Y secretion increases in the paraventricular nucleus in association with increased appetite for food. Proc. Natl. Acad. Sci. 88 (23), 10931–10935.

Kalra, S.P., Bagnasco, M., Otukonyong, E.E., Dube, M.G., Kalra, P.S., 2003. Rhythmic, reciprocal ghrelin and leptin signaling: new insight in the development of obesity. Regul. Pept. 111 (1–3), 1–11.

Kumar, A., Foster, T.C., 2002. 17β-estradiol benzoate decreases the AHP amplitude in CA1 pyramidal neurons. J. Neurophysiol. 88, 621–626.

LeMinh, N., Damiola, F., Tronche, F., Schutz, G., Schibler, U., 2001. Glucorticoid hormones inhibit food-induced phase-shifting of peripheral circadian oscillators. EMBRO J. 20 (24), 7128–7136.

Low-Zeddies, S.S., Takahashi, J.S., 2001. Chimera analysis of the Clock mutation in mice shows that complex cellular integration determines circadian behavior. Cell 105 (1), 25–42.

Luine, V.N., 2014. Estradiol and cognitive function: past, present and future. Horm. Behav. 66, 602–618.

Sapolsky, R.M., 2000. Glucocorticoids and hippocampal atrophy in neuropsychiatric disorders. Arch. Gen. Psychiatry 57 (10), 925–935.

Schibler, U., Juergen, A., Ripperger, J.A., Brown, S.A., 2001. Circadian rhythms: chronobiology-reducing time. Science 293, 437–438.

Sharrow, K.M., Kumar, A., Foster, T.C., 2002. Calcineurin as a potential contributor in estradiol regulation of hippocampal synaptic function. Neuroscience 113, 89–97.

Song, S.H., McIntyre, S.S., Shah, H., Veldhuis, J.D., Hayes, P.C., Butler, P.C., 2000. Direct measurement of pulsalile insulin secretion from the portal vein in human subjects. J. Clin. Endocrinol. Metab. 85 (12), 4491–4499.

Tissenbaum, H.A., 2012. Genetics, life span, health span, and the aging process in *Caenorhabditis elegans*. J. Gerontol. A. Biol. Sci. Med. Sci. 67A (5), 503–510 (special issue on genetics and aging).

Toh, K.L., Jones, C.R., et al., 2001. An hPer2 phosphorylation site mutation in familial advanced sleep phase syndrome. Science 291 (5506), 1040–1043.

Villeda, S.A., et al., 2011. The ageing systemic milieu negatively regulates neurogenesis and cognitive function. Nature 477 (7362), 90–94.

Vitaterna, M.H., King, D.P., et al., 1994. Mutagenesis and mapping of a mouse gene, *Clock*, essential for circadian behavior. Science 264 (5159), 719–725.

Woolley, C.S., Cohen, R.S., 2002. Sex steroids and neuronal growth in adulthood. In: Pfaff, D.W., Arnold, A., Fahrbach, S.E., Etgen, A.M., Rubin, R.T. (Eds.), Hormones, Brain and Behavior. Academic Press.

Xu, B., Kalra, P.S., Farmerie, W.G., Kalra, S.P., 1999. Daily changes in hypothalamic gene expression of neuropeptide Y, galanin, proopiomelanocortin, and adipocyte leptin gene expression and secretion: effects of food restriction. Endocrinology 140 (6), 2668–2675.

Young, M.W., Kay, S.A., 2001. Time zones: a comparative genetics of circadian clocks. Nat. Rev. Genet. 2 (9), 702–715.

Young, M.W., 2002. Big ben rings in a lesson on biological clocks. Neuron 36 (6), 1001–1005.

Zhai, G., Teumer, A., Stolk, L., Perry, J.R.B., Vandenput, L., Coviello, A.D., Koster, A., Bell, J.T., Bhasin, S., Eriksson, J., Eriksson, A., Ernst, F., Ferrucci, L., Frayling, T.M., Glass, D., Grundberg, E., Haring, R., Hedman, Å.K., Hofman, A., Kiel, D.P., Kroemer, H.K., Liu, Y., Lunetta, K.L., Maggio, M., Lorentzon, M., Mangino, M., Melzer, D., Miljkovic, I., Nica, A., Penninx, B.W.J.H., Vasan, R.S., Rivadeneira, F., Small, K.S., Soranzo, N., Uitterlinden, A.G., Völzke, H., Wilson, S.G., Xi, L., Vivian Zhuang, W., Harris, T.B., Murabito, J.M., Ohlsson, C., Murray, A., de Jong, F.H., Spector, T.D., Wallaschofski, H., 2011. Eight common genetic variants associated with serum DHEAS levels suggest a key role in ageing mechanisms. PLoS Genet. 7 (4), e1002025. http://dx.doi.org/10.1371/journal.pgen.1002025.

Zucker, I., 2002. In: Pfaff, D.W., Arnold, A.P., Etgen, A.M., Fahrbach, S.E., Rubin, R.T. (Eds.), Hormones, Brain and Behavior. Academic Press, San Diego, CA.

# Chapter 14: Duration of Hormone Exposure Can Make a Big Difference: In Some Cases, Longer Is Better; in Other Cases, Brief Pulses Are Optimal for Behavioral Effects

Fan, P., Maximov, P.Y., Curpan, R.F., Abderrahman, B., Jordan, V.C., 2015. The molecular, cellular and clinical consequences of targeting the estrogen receptor following estrogen deprivation therapy. Mol. Cell. Endocrinol. 418, 245–263.

Kow, L.M., Pfaff, D.W., 2016. Rapid estrogen actions on ion channels: a survey in search for mechanisms. Steroids 111, 46–53.

Kow, L.M., Pataky, S., Dupré, C., Phan, A., Martin-Alguacil, N., Pfaff, D.W., 2016. Analyses of rapid estrogen actions on rat ventromedial hypothalamic neurons. Steroids 111, 100–112.

McEwen, B.S., 1998. Protective and damaging effects of stress mediators. Semin. Med. (Beth Isr.) 338, 171–198.

McEwen, B.S., 2002. The End of Stress as We Know It. Joseph Henry Press, Washington, DC.

Moss, R., McCann, S.M., 1973. Induction of mating behavior by LRF. Science 181, 177–179.

Palovcik, R.A., Phillips, M.I., Kappy, M.S., Raizada, M.K., 1984. Insulin inhibits pyramidal neurons in hippocampal slices. Brain Res. 309 (1), 187–191.

Parsons, B., McEwen, B.S., Pfaff, D.W., 1982a. A discontinuous schedule of estradiol treatment is sufficient to activate progesterone-facilitated feminine sexual behavior and to increase cytosol receptors for progestins in the hypothalamus of the rat. Endocrinology 110, 613–619.

Pauls, S.D., Honma, K.-I., Honma, S., Silver, R., 2016. Deconstructing circadian rhythmicity with models and manipulations. TINS 39, 405–419.

Pfaff, D.W., 1973. Luteinizing hormone releasing factor (LRF) potentiates lordosis behavior in hypophysectomized ovariectomized female rats. Science 182, 1148–1149.

Sassarini, J., Lumsden, M.A., 2015. Oestrogen replacement in postmenopausal women. Age Ageing 44, 551–558.

Schulkin, J., 2003. Rethinking Homeostasis: Allostatic Regulation in Physiology and Pathophysiology. MIT Press, Cambridge, MA.

Schwanzel Fukuda, M., Pfaff, D.W., 1989. Origin of luteinizing hormone-releasing hormone neurons. Nature 338, 161–164.

Schwanzel Fukuda, M., Bick, D., Pfaff, D.W., 1989. Luteinizing hormone-releasing hormone (LHRH)-expressing cells do not migrate normally in an inherited hypogonadal (Kallmann) syndrome. Mol. Brain Res. 6, 311–326.

Schwanzel-Fukuda, M., Crossin, K.L., Pfaff, D.W., Bouloux, P.M., Hardelin, J.-P., Petit, C., 1996. Migration of LHRH neurons in early human embryos: association with neural cell adhesion molecules. J. Comp. Neurol. 366, 547–557.

## Chapter 15: Hormonal Secretions and Responses Are Affected by Biological Clocks

Allen, A.M., MacGregor, D.P., McKinley, M.J., Mendelsohn, F.A., 1999. Angiotensin II receptors in the human brain. Regul. Pept. 79 (1), 1–7.

Anke, J., Van Eekelen, M., Phillips, M.I., 1988. Plasma angiotensin II levels at moment of drinking during angiotensin II intravenous infusion. Am. J. Physiol. 255 (3, Pt. 2), R500–R506.

Bailey, C.J., 2001. New pharmacologic agents for diabetes. Curr. Diab. Rep. 1 (2), 119–126.

Bartness, T.J., Liu, Y., Shrestha, Y.B., Ryu, V., 2014. Neural innervation of white adipose tissue and the control of lipolysis. Front. Neuroendocrinol. 35, 473–493.

Bodnar, R., Commons, K., Pfaff, D.W., 2002. Central Neural States Relating Sex and Pain. The Johns Hopkins University Press, Baltimore, MD.

Buggy, J., Hoffman, W.E., Phillips, M.I., Fisher, A.E., Johnson, A.K., 1979. Osmosensitivity of rat third ventricle and interactions with angiotensin. Am. J. Physiol. 236 (1), R75–R82.

Buijs, F.N., Leon-Mercado, L., Guzman-Ruiz, M., Guerrero-Vargas, N.N., Romo-Nava, F., Buijs, R.M., 2016. The circadian system: a regulatory feedback network of periphery and brain. Physiology 31, 170–181.

Challet, E., 2015. Keeping circadian time with hormones. Diabetes Obes. Metab 17 (Suppl. 1), 76–83.

Czeisler, C.A., Klerman, E.B., 1999. Circadian and sleep-dependent regulation of hormone release in humans. Recent Prog. Horm. Res. 54, 97–132.

Erren, T.C., Reiter, R.J., 2015. Melatonin: a universal time messenger. Neuroendocrinol. Lett. 36, 187–192.

Hastings, M.H., Brancaccio, M., Maywood, E.S., 2014. Circadian pacemaking in cells and circuits of the suprachiasmatic nucleus. J. Neuroendocrinol. 26, 2–10.

Hayaishi, O., 2000. Molecular mechanisms of sleep-wake regulation: a role of prostaglandin D2. Philos. Trans. R. Soc. Lond. B Biol. Sci. 355 (1394), 275–280.

He, B., Nohara, K., Park, N., Park, Y.S., Guillory, B., Zhao, Z., Garcia, J.M., Koike, N., Lee, C.C., Takahashi, J.S., Yoo, S.H., Chen, Z., 2016. The small molecule nobiletin targets the molecular oscillator to enhance circadian rhythms and protect against metabolic syndrome. Cell Metab. 23, 610–621.

Hoffman, W.E., Phillips, M.I., Wilson, E., Schmid, P.G., 1977. A pressor response with drinking in rats. Proc. Soc. Biol. Med. 154 (1), 121–124.

Hogarty, D.C., Speakman, E.A., Puig, V., Phillips, M.I., 1992. The role of angiotensin, AT1 and AT2 receptors in the pressor, drinking and vasopressin responses to central angiotensin. Brain Res. 586 (2), 289–294.

Hogarty, D.C., Tran, D.N., Phillips, M.I., 1994. Involvement of angiotensin receptor subtypes in osmotically induced release of vasopressin. Brain Res. 6371 (1–2), 126–132.

Jordan, J., Shannon, J.R., Black, B.K., Ali, Y., Farley, M., Costa, F., Diedrich, A., Robertson, R.M., Biaggioni, I., Robertson, D., 2000. The pressor response to water drinking in humans: a sympathetic reflex? Circulation 101, 504–509.

Kleitman, N., 1939. Sleep and Wakefulness as Alternating Phases in the Cycle of Existence. University of Chicago Press, Chicago, USA.

Kuriyama, R., 1996. Angiotensin converting enzyme inhibitor induced anemia in a kidney transplant recipient. Transplant Proc. 28, 1635.

Maghnie, M., 2003. Diabetes insipidus. Horm. Res. 59 (Suppl. 1), 42–54.

McKinley, M.J., Gerstberger, R., Mathai, M.L., Oldfield, B.J., Schmid, H., 1999. The lamina terminalis and its role in fluid and electrolyte homeostasis. J. Clin. Neurosci. 6 (4), 289–301.

Paul, M.J., Pyter, L.M., Freeman, D.A., Galang, J., Prendergast, B.J., 2009. Photic and nonphotic seasonal cues differentially engage hypothalamic kisspeptin and rfamide-related peptide mRNA expression in Siberian hamsters. J. Neuroendocrinol. 21, 1007–1014.

Pauls, S.D., Honma, K., Honma, S., Silver, R., 2016. Deconstructing circadian rhythmicity with models and manipulations. Trends Neurosci. 39, 405–419.

Phillips, M.I., Quilan, J.T., Weyhenmeyer, J., 1980. An angiotensin-like peptide in the brain. Life Sci. 27 (25–26), 2589–2594.

Phillips, M.I., Hoffman, W.E., Bealer, S.L., 1982. Dehydration and fluid balance: central effects of angiotensin. Fed. Proc. 41 (9), 2520–2527.

Prendergast, B.J., Nelson, R.J., Zucker, I., 2002. Mammalian seasonal rhythms: behavior and neuroendocrine substrates. In: Pfaff, D.W. (Ed.), Hormones, Brain and Behavior, vol. 2. Elsevier Science, pp. 93–156.

Quinlan, J.T., Phillips, M.I., 1981. Immunoreactivity for an angiotensin II-like peptide in human brain. Brain Res. 205 (1), 212–218.

Schibler, U., Juergen, A., Ripperger, J.A., Brown, S.A., 2001. Circadian rhythms: chronobiology-reducing time. Science 293, 437–438.

Schroeder, C., Bush, V.E., Norcliffe, L.J., Luft, F.C., Tank, J., Jordan, J., Hainsworth, R., 2002. Water drinking acutely improves orthostatic tolerance in health subjects. Circulation 106, 2806–2811.

Shannon, J.R., Diedrich, A., Biaggioni, I., Tank, J., Robertson, R.M., Robertson, D., Jordan, J., 2002. Water drinking as a treatment for orthostatic syndromes. Am. J. Med. 112, 355–360.

Silver, R., Kriegsfeld, L.J., 2014. Circadian rhythms have broad implications for understanding brain and behavior. Eur. J. Neurosci. 39 (11), 1866–1880.

Smale, L., Lee, T., Nunez, A.A., 2003. Mammalian diurnality: some facts and gaps. J. Biol. Rhythm. 18, 356–366.

Sunn, N., Egli, M., Burazin, T., Colvill, C., Davern, P., Denton, D.A., Oldfield, B.J., Weisinger, R.S., Rauch, M., Schmid, H.A., McKinley, M.J., 2002. Circulating relaxin acts on subfornical organ neurons to stimulate water drinking in the rat. Proc. Natl. Acad. Sci. U.S.A. 99 (3), 1701–1706.

Tamura, R., Norgren, R., 2003. Intracranial renin alters gustatory neural responses in the nucleus of the solitary tract of rats. Am. J. Physiol. Regul. Integr. Comp. Physiol. 284, R1108–R1118.

Wingfield, J.C., Farner, D.S., 1980. Control of seasonal reproduction in temperate zone birds. In: Reiter, R.J., Follett, B.K. (Eds.), Progress in Reproductive Biology: Seasonal Reproduction in Higher Vertebrates. S. Karger, New York, pp. 62–101.

Wirz-Justice, A., 2007. How to measure circadian rhythms in humans. Medicographia 29, 84–90.

Wong, N.L., Tsui, J.K., 2003. Angiotensin II upregulates the expression of vasopressin V2 mRNA in the medullary collecting duct of the rat. Metabolism 52 (3), 290–295.

Yoshii, T., Hermann-Luibl, C., Helfrich-Förster, C., 2016. Circadian light-input pathways in Drosophila. Commun. Integr. Biol. 9 pp. e1102805–e1102805-8.

Young, M.W., Kay, S.A., 2001. Time zones: a comparative genetics of circadian clocks. Nat. Rev. Genet. 2 (9), 702–715.

Zordan, M.A., Sandrelli, F., 2015. Circadian clock dysfunction and psychiatric disease: could fruit flies have a say? Front Neurol 6, 1–16.

## Chapter 16: Effects of a Given Hormone Can Be Widespread Across the Body; Central Effects Consonant With Peripheral Effects Form Coordinated, Unified Mechanisms

Aikey, J.L., Nyby, J.G., Anmuth, D.N., James, P.J., 2002. Testosterone rapidly reduces anxiety in male house mice (Mus musculus). Horm. Behav. 42 (4), 448–460 .

Backhed, F., Ding, H., Wang, T., Hooper, L.V., Koh, G.Y., Nagy, A., Semenkovich, C.F., Gordon, J.I., 2004. The gut microbiota as an environmental factor that regulates fat storage. Proc. Natl. Acad. Sci. U.S.A. 101, 15718–15723.

Bartness, T.J., Liu, Y., Shrestha, Y.B., Ryu, V., 2014. Neural innervation of white adipose tissue and the control of lipolysis. Front. Neuroendocrinol. 35, 473–493.

Bryson, J.M., Phuyal, J.L., Swan, V., Caterson, I.D., 1999. Leptin has acute effects on glucose and lipid metabolism in both lean and gold thioglucose-obese mice. Am. J. Physiol. 277, E417–E422.

Ceddia, R.B., William Jr., W.N., Curi, R., 1999. Comparing effects of leptin and insulin on glucose metabolism in skeletal muscle: evidence for an effect of leptin on glucose uptake and decarboxylation. Int. J. Obes. Relat. Metab. Disord. 23, 75–82.

Chambliss, K.L., Yuhanna, I.S., et al., 2002a. ERß has nongenomic action in caveolae. Mol. Endocrinol. 16 (5), 938–946.

Clarke, I.J., Arbabi, L., 2016a. New concepts of the central control of reproduction, integrating influence of stress, metabolic state, and season. Domest. Anim. Endocrinol. 56 (Suppl), S165–S179.

Cox-York, K.A., Sheflin, A.M., Foster, M.T., Gentile, C.L., Kahl, A., Koch, L.G., Britton, S.L., Weir, T.L., 2015. Ovariectomy results in differential shifts in gut microbiota in low versus high aerobic capacity rats. Physiol. Rep. 3.

Cunnane, S.C., Crawford, M.A., 2003. Survival of the fattest: fat babies were the key to evolution of the large human brain. Comp. Biochem. Physiol. A Mol. Integr. Physiol. 136, 17–26.

Cunnane, S.C., 2005. Survival of the Fattest: The Key to Human Brain Evolution. World Scientific, London.

De Bond, J.A., Smith, J.T., 2014. Kisspeptin and energy balance in reproduction. Reproduction 147, R53–R63.

Fonken, L.K., Nelson, R.J., 2014. The effects of light at night on circadian clocks and metabolism. Endocr. Rev. 35, 648–670.

Friedman, M.I., Stricker, E.M., 1976. The physiological psychology of hunger: a physiological perspective. Psychol.Rev. 83, 409–431.

Friedman, M.I., 1995. Control of energy intake by energy metabolism. Am. J. Clin. Nutr. 62 (Suppl), 1096S–1100S.

Friedman, M.I., 2008. Food intake: control, regulation and the illusion of dysregulation. In: Harris, R.B., Mattes, R. (Eds.), Appetite and Food Intake: Behavioral and Physiological Considerations. CRC Press, Boca Raton, Florida, USA, pp. 1–19.

Frye, C.A., Vongher, J.M., 1999a. Progestins' rapid facilitation of lordosis when applied to the ventral tegmentum corresponds to efficacy at enhancing GABA(A)receptor activity. J. Neuroendocrinol. 11 (11), 829–837.

Frye, C.A., Bayon, L.E., Vongher, J.M., 2000. Intravenous progesterone elicits a more rapid induction of lordosis in rats than does SKF38393. Psychobiology 28 (1), 99–109.

Garver, W.S., Newman, S.B., Gonzales-Pacheco, D.M., Castillo, J.J., Jelinek, D., Heidenreich, R.A., Orlando, R.A., 2013. The genetics of childhood obesity and interaction with dietary macronutrients. Genes Nutr. 8, 271–287.

Grattan, D.R., 2008. Fetal programming from maternal obesity: eating too much for two? Endocrinology 149, 5345–5347.

Hinney, A., Hebebrand, J., 2008. Polygenic obesity in humans. Obes. Facts 1, 35–42.

James, P.J., Nyby, J.G., 2002. Testosterone rapidly affects the expression of copulatory behavior in house mice (*Mus musculus*). Physiol. Behav. 75, 287–294.

Ludwig, D.S., Friedman, M.I., 2014. Increasing adiposity: consequence or cause of overeating? JAMA 311, 2167–2168.

Martin, B., Ji, S., Maudsley, S., Mattson, M.P., 2010. "Control" laboratory rodents are metabolically morbid: why it matters. Proc. Natl. Acad. Sci. U.S.A. 107, 6127–6133.

Minokoshi, Y., Kim, Y.B., Peroni, O.D., Fryer, L.G., Muller, C., Carling, D., Kahn, B.B., 2002. Leptin stimulates fatty-acid oxidation by activating AMP-activated protein kinase. Nature 415, 339–343.

Minokoshi, Y., Alquier, T., Furukawa, N., Kim, Y.B., Lee, A., Xue, B., Mu, J., Foufelle, F., Ferre, P., Birnbaum, M.J., Stuck, B.J., Kahn, B.B., 2004. AMP-kinase regulates food intake by responding to hormonal and nutrient signals in the hypothalamus. Nature 428, 569–574.

Minokoshi, Y., Shiuchi, T., Lee, S., Suzuki, A., Okamoto, S., 2008. Role of hypothalamic AMP-kinase in food intake regulation. Nutrition 24, 786–790.

Nunez, A.A., Gray, J.M., Wade, G.N., 1980. Food intake and adipose tissue lipoprotein lipase activity after hypothalamic estradiol benzoate implants in rats. Physiol. Behav. 25, 595–598.

Ogden, C.L., Flegal, K.M., Carroll, M.D., Johnson, C.L., 2002. Prevalence and trends in overweight among US children and adolescents, 1999-2000. JAMA 288, 1728–1732.

Ogden, C.L., Carroll, M.D., Kit, B.K., Flegal, K.M., 2013. Prevalence of obesity among adults: United States, 2011-2012. NCHS Data Brief 1–8.

Rettberg, J.R., Yao, J., Brinton, R.D., 2014. Estrogen: a master regulator of bioenergetic systems in the brain and body. Front. Neuroendocrinol. 35, 8–30.

Ridaura, V.K., Faith, J.J., Rey, F.E., Cheng, J., Duncan, A.E., Kau, A.L., Griffin, N.W., Lombard, V., Henrissat, B., Bain, J.R., Muehlbauer, M.J., Ilkayeva, O., Semenkovich, C.F., Funai, K., Hayashi, D.K., Lyle, B.J., Martini, M.C., Ursell, L.K., Clemente, J.C., Van Treuren, W., Walters, W.A., Knight, R., Newgard, C.B., Heath, A.C., Gordon, J.I., 2013. Gut microbiota from twins discordant for obesity modulate metabolism in mice. Science 341, 1241214.

Sajapitak, S., Iwata, K., Shahab, M., Uenoyama, Y., Yamada, S., Kinoshita, M., Bari, F.Y., I'Anson, H., Tsukamura, H., Maeda, K., 2008a. Central lipoprivation-induced suppression of luteinizing hormone pulses is mediated by paraventricular catecholaminergic inputs in female rats. Endocrinology 149, 3016–3024.

Sajapitak, S., Uenoyama, Y., Yamada, S., Kinoshita, M., Iwata, K., Bari, F.Y., I'Anson, H., Tsukamula, H., Maeda, K., 2008b. Paraventricular alpha1- and alpha2-adrenergic receptors mediate hindbrain lipoprivation-induced suppression of luteinizing hormone pulses in female rats. J. Reprod. Dev. 54, 198–202.

Schneider, J.E., Zhou, D., 1999. Interactive effects of central leptin and peripheral fuel oxidation on estrous cyclicity. Am. J. Physiol. 277, R1020–R1024.

Schneider, J.E., Klingerman, C.M., Abdulhay, A., 2012. Sense and nonsense in metabolic control of reproduction. Front. Endocrinol. 3 (3), 26. http://dx.doi.org/10.3389/fendo.2012.00026.

Schneider, J.E., Wise, J.D., Benton, N.A., Brozek, J.M., Keen-Rhinehart, E., 2013. When do we eat? Ingestive behavior, survival, and reproductive success. Horm. Behav. 64, 702–728.

Schneider, J.E., Brozek, J.M., Keen-Rhinehart, E., 2014. Our stolen figures: the interface of sexual differentiation, endocrine disruptors, maternal programming, and energy balance. Horm. Behav. 66.

Schneider, J.E., 2004. Energy balance and reproduction. Physiol. Behav. 81, 289–317.

Schwartz, G.J., Salorio, C.F., Skoglund, C., Moran, T.H., 1999. Gut vagal afferent lesions increase meal size but do not block gastric preload-induced feeding suppression. Am. J. Physiol. 276, R1623–R1629.

Schwartz, G.J., 2006. Integrative capacity of the caudal brainstem in the control of food intake. Philos. Trans. R Soc. Lond. B Biol. Sci. 361, 1275–1280.

Shechter, A., Schwartz, G.J., December 20, 2016. Gut-brain nutrient sensing in food reward. Appetite. Pii: S0195-6663(16)30900-X. doi: 10.1016/j.appet.2016.12.009. [Epub ahead of print].

Shimabukuro, M., Koyama, K., Chen, G., Wang, M.Y., Trieu, F., Lee, Y., Newgard, C.B., Unger, R.H., 1997. Direct antidiabetic effect of leptin through triglyceride depletion of tissues. Proc. Natl. Acad. Sci. U.S.A. 94, 4637–4641.

Singh, D., 1993. Adaptive significance of female physical attractiveness: role of waist-to-hip ratio. J. Pers. Soc. Psychol. 65, 293–307.

Tsutsui, K., Bentley, G.E., Bedecarrats, G., Osugi, T., Ubuka, T., Kriegsfeld, L.J., 2010. Gonadotropin-inhibitory hormone (GnIH) and its control of central and peripheral reproductive function. Front. Neuroendocrinol. 31, 284–295.

Van Pelt, R.E., Jankowski, C.M., Gozansky, W.S., Schwartz, R.S., Kohrt, W.M., 2005. Lower-body adiposity and metabolic protection in postmenopausal women. J. Clin. Endocrinol. Metab. 90, 4573–4578.

Vasudevan, N., Kow, L.-M., Pfaff, D.W., 2001. Early membrane estrogenic effects required for full expression of slower genomic actions in a nerve cell line. PNAS 98 (21), 12267–12271.

Wade, G.N., Jones, J.E., 2004. Neuroendocrinology of nutritional infertility. Am. J. Physiol. Regul. Integr. Comp. Physiol. 287, R1277–R1296.

Wade, G.N., Gray, J.M., Bartness, T.J., 1985. Gonadal influences on adiposity. Int. J. Obes. 9 (Suppl. 1), 83–92.

Wade, G.N., Schneider, J.E., Friedman, M.I., 1991. Insulin-induced anestrus in Syrian hamsters. Am. J. Physiol. 260, R148–R152.

Wade, G.N., Schneider, J.E., Li, H.Y., 1996. Control of fertility by metabolic cues. Am. J. Physiol. 270, E1–E19.

Wang, T., Hung, C.C., Randall, D.J., 2006. The comparative physiology of food deprivation: from feast to famine. Annu. Rev.Physiol. 68, 223–251.

## Chapter 17: Hormones Can Act at All Levels of the Neuraxis to Exert Behavioral Effects; The Nature of the Behavioral Effect Depends on the Site of Action

Bain, D.L., Franden, M.A., et al., 2001. The N-terminal region of human progesterone B- receptors: biophysical and biochemical comparison to A-receptors. J. Biol. Chem. 276 (26), 23825–23831.

Badoer, E., McKinlay, D., 1997. Effect of intravenous angiotensin II on Fos distribution and drinking behavior in rabbits. Am. J. Physiol. Regul. Integr. Comp. Physiol. 272 (5 Pt 2), R1515–R1524.

Burson, J.M., Aguilera, G., Gross, K.W., Sigmund, C.D., 1994. Differential expression of angiotensin receptor 1A and 1B in mouse. Am. J. Physiol. 267 (2, Pt. 1), E260–E267.

Carrasco, G.A., Van de Kar, L.D., 2003. Neuroendocrine pharmacology of stress. Eur. J. Pharmacol. 463 (1–3), 235–272.

Carter, C.S., 2014. Oxytocin pathways and the evolution of human behavior. Annu. Rev. Psychol. 65, 17–39.

Dellovade, T.L., Chan, J., Vennstrom, B., Forrest, D., Pfaff, D.W., 2000. The two thyroid hormone receptor genes have opposite effects on estrogen stimulated sex behaviors. Nat. Neurosci. 3 (5), 472–475.

Evered, M.D., Robinson, M.M., 1984. Increased or decreased thirst caused by inhibition of angiotensin-converting enzyme in the rat. J. Physiol. 348, 573–588.

Fitzsimons, J.T., 1998. Angiotensin, thirst, and sodium appetite. Physiol. Rev. 78 (3), 583–686.

Grill, H.J., 2006. Distributed neural control of energy balance: contributions from hindbrain and hypothalamus. Obesity 5, 216S–221S.

Head, G.A., Quail, A.W., Woods, R.L., 1987. Lesions of a1 noradrenergic cells affect arginine vasopressin release and heart rate during hemorrhage. Am. J. Physiol. Heart Circ. Physiol. 253 (5 Pt. 2), H1012–H1017.

Kakar, S.S., Riel, K.K., Neill, J.D., 1992. Differential expression of angiotensin II receptor subtype mRNAs (AT-1A and AT-1B) in the brain. Biochem. Biophys. Res. Commun. 185 (2), 688–692.

Kuriyama, R., 1996. Angiotensin converting enzyme inhibitor induced anemia in a kidney transplant recipient. Transplant Proc. 28, 1635.

Mavani, G.P., DeVita, M.V., Michelis, M.F., 2015. A review of the nonpressor and nonantidiuretic actions of the hormone vasopressin. Front. Med. (Lausanne) 2, 19.

McKinley, M.J., Johnson, A.K., 2004. The physiological regulation of thirst and fluid intake. News Physiol. Sci. 19 (1), 1–6.

Scott, R.E., Wu-Peng, X.S., et al., 2002. Regulation and expression of progesterone receptor mRNA isoforms A and B in the male and female rat hypothalamus and pituitary following oestrogen treatment. J. Neuroendocrinol. 14 (3), 175–183.

Tung, L., Mohamed, M.K., et al., 1993. Antagonist-occupied human progesterone B- receptors activate transcription without binding to progesterone response elements and are dominantly inhibited by A-receptors. Mol. Endocrinol. 7 (10), 1256–1265.

Vegeto, E., Shahbaz, M.M., et al., 1993. Human progesterone receptor A form is a cell- and promoter-specific repressor of human progesterone receptor B function. Mol. Endocrinol. 7 (10), 1244–1255.

Wen, D.X., Xu, Y.F., et al., 1994. The A and B isoforms of the human progesterone receptor operate through distinct signaling pathways within target cells. Mol. Cell. Biol. 14 (12), 8356–8364.

Yudt, M.R., Cidlowski, J.A., 2001. Molecular identification and characterization of A and B forms of the glucocorticoid receptor. Mol. Endocrinol. 15 (7), 1093–1103.

Yudt, M.R., Cidlowski, J.A., 2002. The glucocorticoid receptor: coding a diversity of proteins and responses through a single gene. Mol. Endocrinol. 16 (8), 1719–1726.

Zell, V., Juif, P.É., Hanesch, U., Poisbeau, P., Anton, F., Darbon, P., 2015. Corticosterone analgesia is mediated by the spinal production of neuroactive metabolites that enhance GABAergic inhibitory transmission on dorsal horn rat neurons. Eur. J. Neurosci. 41, 390–397.

Zubeldia-Brenner, L., Roselli, C.E., Recabarren, S.E., Gonzalez Deniselle, M.C., Lara, H.E., June 4, 2016. Developmental and functional effects of steroid hormones on the neuroendocrine axis and spinal cord. J. Neuroendocrinol. http://dx.doi.org/10.1111/jne.1240.

## Chapter 18: Hormone Receptors Act by Multiple Interacting Mechanisms

Aikey, J.L., Nyby, J.G., Anmuth, D.N., James, P.J., 2002. Testosterone rapidly reduces anxiety in male house mice (Mus musculus). Horm. Behav. 42 (4), 448–460.

Anderson, J.N., Peck, E.J., Clark, J.H., 1974. Nuclear receptor estradiol complex: a requirement for uterotrophic responses. Endocrinology 95, 174.

Apostolakis, E.M., Garai, J., et al., 2000. Epidermal growth factor activates reproductive behavior independent of ovarian steroids in female rodents. Mol. Endocrinol. 14 (7), 1086–1098.

Apostolakis, E.M., Ramamurphy, M., et al., 2002. Acute disruption of select steroid receptor coactivators prevents reproductive behavior in rats and unmasks genetic adaptation in knockout mice. Mol. Endocrinol. 16 (7), 1511–1523.

Auger, A.P., Perrot-Sinal, T.S., et al., 2002. Expression of the nuclear receptor coactivator, cAMP response element-binding protein, is sexually dimorphic and modulates sexual differentiation of neonatal rat brain. Endocrinology 143 (8), 3009–3016.

Camille Melón, L., Maguire, J., 2016. GABAergic regulation of the HPA and HPG axes and the impact of stress on reproductive function. J. Steroid Biochem. Mol. Biol. 160, 196–203.

Chambliss, K.L., Yuhanna, I.S., Anderson, R.G.W., Mendelsohn, M.E., Shaul, P.W., 2002b. ERbeta has nongenomic action in caveolae. Mol. Endocrinol. 16 (5), 938–946.

Frye, C.A., Vongher, J.M., 1999. Progestins' rapid facilitation of lordosis when applied to the ventral tegmentum corresponds to efficacy at enhancing GABA(A) receptor activity. J. Neuroendocrinol. 11, 829–837.

Frye, C.A., Bayon, L.E., Vongher, J.M., 2000. Intravenous progesterone elicits a more rapid induction of lordosis in rats than does SKF38393. Psychobiology 28, 99–109.

Gaub, M.P., Bellard, M., Scheuer, I., Chambon, P., Sassone-Corsi, P., 1990. Activation of the ovalbumin gene by the estrogen receptor involves the fos-jun complex. Cell. 63, 1267–1276.

Golden, G.A., Mason, P.E., Rubin, R.T., Mason, R.P., 1998. Biophysical membrane interactions of steroid hormones: a potential complementary mechanism of steroid action. Clin. Neuropharmacol. 21, 181–189.

Gu, Q., Moss, R.L., 1996. 17 beta-Estradiol potentiates kainate-induced currents via activation of the cAMP cascade. J. Neurosci. 16, 3620–3629.

Hewitt, S.C., Li, Y., Li, L., Korach, K.S., et al., 2010. Estrogen-mediated regulation of Igf1 transcription and uterine growth involves direct binding of estrogen receptor α to estrogen-responsive elements. J. Biol. Chem. 285, 2676–2685.

Holter, E., Kotaja, N., et al., 2002. Inhibition of androgen receptor (AR) function by the reproductive orphan nuclear receptor DAX-1. Mol. Endocrinol. 16 (3), 515–528.

James, P.J., Nyby, J.G., 2002. Testosterone rapidly affects the expression of copulatory behavior in house mice (Mus musculus). Physiol. Behav. 75, 287–294.

Janson, H.T., Popiela, C.L., Jackson, G.L., Iwamoto, G.A., 1993. A re-evaluation of the effects of gonadal steroids on neuronal activity in the male rat. Brain Res. Bull. 31, 217–223.

Jensen, E.V., Jacobson, H.I., 1960. Fate of steroid estrogens in target tissues. In: Pincus, G., Vollmer, E.P. (Eds.), Biological Activities of Steroids in Relation to Cancer. Academic Press, New York, pp. 161–174.

Kelly, M.J., Levin, E.R., 2001. Rapid actions of plasma membrane estrogen receptors. Trends in Endocrinol. Metab. 12, 152–156.

Madak-Erdogan, Z., Kim, S.H., Gong, P., Zhao, Y.C., Zhang, H., Chambliss, K.L., Carlson, K.E., Mayne, C.G., Shaul, P.W., Korach, K.S., Katzenellenbogen, J.A., Katzenellenbogen, B.S., 2016. Design of pathway preferential estrogens that provide beneficial metabolic and vascular effects without stimulating reproductive tissues. Sci. Signal. 9 (429), ra53.

McKenna, N.J., O'Malley, B., 2000. From ligand to response: generating diversity in nuclear receptor coregulator function. J. Steroid Biochem. Mol. Biol. 74 (5), 351–356.

Meitzen, J., Mermelstein, P.G., 2011. Estrogen receptors stimulate brain region specific metabotropic glutamate receptors to rapidly initiate signal transduction pathways. J. Chem. Neuroanat. 42, 236–241.

Mong, J.A., Krebs, C., et al., 2002. Perspective: microarrays and differential display PCR-tools for studying transcript levels of genes in neuroendocrine systems. Endocrinology 143 (6), 2002–2006.

Mong, J.A., Devidze, N., et al., 2003. Estradiol regulation of lipocalin-type prostaglandin D synthase transcript levels in the rodent brain: evidence from high density oligonucleotide arrays and in situ hybridization. PNAS 100 (1), 318–323.

Olesen, K.M., Jessen, H.M., Auger, C.J., Auger, A.P., 2005. Dopaminergic activation of estrogen receptors in neonatal brain alters progestin receptor expression and juvenile social play behavior. Endocrinology 146, 3705–3712.

Park, C.J., Zhao, Z., Glidewell-Kenney, C., Lazic, M., Chambon, P., Krust, A., Weiss, J., Clegg, D.J., Dunaif, A., Jameson, J.L., Levine, J.E., 2011. Genetic rescue of nonclassical ERalpha signaling normalizes energy balance in obese ER-alpha-null mutant mice. J. Clin. Invest. 121, 604–612.

Parsons, B., Rainbow, T.C., Pfaff, D.W., McEwen, B.S., 1981. Oestradiol, sexual receptivity and cytosol progestin receptors in rat hypothalamus. Nature 292, 58–59.

Pfaff, D.W., 1999. Drive; Neurobiological and Molecular Mechanisms of Sexual Motivation. MIT Press, Cambridge, MA.

Pohlenz, J., Weiss, R.E., Macchia, P.E., Pannain, S., Lau, I.T., Ho, H., Refetoff, S., 1999. 5 new families with resistance to thyroid hormone not caused by mutations in the thyroid hormone receptor ß gene. J. Clin. Endocrinol. Metab. 84, 3919–3928.

Rowan, B.G., O'Malley, B.W., 2000. Progesterone receptor coactivators. Steroids 65 (10–11), 545–549.

Santollo, J., Daniels, D., 2015. Multiple estrogen receptor subtypes influence ingestive behavior in female rodents. Physiol. Behav. 152, 431–437.

Segars, J.H., Driggers, P.H., 2002. Estrogen action and cytoplasmic signaling cascades. Part I: membrane-associated signaling complexes. Trends Endocrinol. Metab. 13, 349–354.

Smith, C.L., O'Malley, B.W., 1999. Evolving concepts of selective estrogen receptor action: from basic science to clinical applications. Trends Endocrinol. Metab. 10 (8), 299–300.

Toft, D., Gorski, J., 1966. A receptor molecule for estrogens: isolation from the rat uterus and preliminary characterization. Proc. Natl. Acad. Sci. U.S.A. 55, 1574–1581.

Tsai, M.J., O'Malley, B.W., 1994. Molecular mechanisms of action of steroid/thyroid receptor superfamily members. Annu. Rev. Biochem. 63, 451–486.

Vasudevan, N., Kow, L.-M., Pfaff, D.W., 2001. Early membrane estrogenic effects required for full expression of slower genomic actions in a nerve cell line. Proc. Natl. Acad. Sci. 98, 12267–12271.

Weigel, N.L., Zhang, Y., 1998. Ligand-independent activation of steroid hormone receptors. J. Mol. Med. 76, 469–479.

Weiss, R.E., Hayashi, Y., et al., 1996. Dominant inheritance of resistance to thyroid hormone not linked to defects in the thyroid hormone receptor alpha or beta genes may be due to a defective cofactor. J. Clin. Endocrinol. Metab. 81 (12), 4196–4203.

## Chapter 19: Gene Duplication and Splicing Products for Hormone Receptors in the Central Nervous System Often Have Different Behavioral Effects

Breedlove, M., 2010. Minireview: organizational hypothesis: instances of the fingerpost. Endocrinology 151 (9), 4116–4122.

Carter, C.S., 2014. Oxytocin pathways and the evolution of human behavior. Annu. Rev. Psychol. 65, 1–771.

Craig, I.W., Halton, K.E., 2009. Genetics of human aggressive behaviour. Hum. Genet. 126, 101–113.

Daniel, J.M., Dohanich, G.P., 2001. Acetylcholine mediates the estrogen-induced increase in NMDA receptor binding in CA1 of the hippocampus and the associated improvement in working memory. J. Neurosci. 21, 6949–6956.

Dehm, S.M., Tindall, D.J., 2011. Alternatively spliced androgen receptor variants. Endocr. Relat. Cancer 18, R183–R196.

Horth, L., 2007. Sensory genes and mate choice: evidence that duplications, mutations, and adaptive evolution alter variation in mating cue genes and their receptors. Genomics 90, 159–175.

Horton, B.M., Hudson, W.H., Ortlund, E.A., Shirk, S., Thomas, J.W., Young, E.R., Zinsow-Kramer, W., Maney, D.L., 2014. Estrogen receptor α polymorphism in a species with alternative behavioral phenotypes. Proc. Natl. Acad. Sci. 111 (4), 1443–1448.

Krukoff, T.L., Khalili, P., 1997. Stress-induced activation of nitric oxide producing neurons in the rat brain. J. Comp. Neurol. 377 (4), 509–519.

Lam, T., Leranth, C., 2002. Locally Administered Estradiol into the MSDB of OVX Rats Affects the Density of CA1 Area Pyramidal Cell Spine Synapses and Glial Processes. Program No. 444.8, 2002 Abstract Viewer/Itinerary Planner. Society for Neuroscience, Washington, DC.

Leranth, C., Shanabrough, M., Horvath, T.L., 2000. Hormonal regulation of hippocampal spine synapse density involves subcortical mediation. Neuroscience 101 (2), 349–356.

Morgan, M.A., Pfaff, D.W., 2001. Effects of estrogen on activity and fear-related behaviors in mice. Horm. Behav. 40 (4), 472–482.

Morgan, M.A., Pfaff, D.W., 2002. Estrogen's effects on activity, anxiety, and fear in two mouse strains. Behav. Brain Res. 132 (1), 85–93.

Murphy, C.A., Stacey, N.E., 2002. Methyl-testosterone induces male-typical ventilatory behavior in response to putative steroidal pheromones in female round gobies (Neogobius melanostomus). Horm. Behav. 42 (2), 109–115.

Murphy, D.D., Cole, N.B., Greenberger, V., Segal, M., 1998. Estradiol increases dendritic spine density by reducing GABA neurotransmission in hippocampal neurons. J. Neurosci. 18, 2550–2559.

Rudick, C.N., Woolley, C.S., 2001. Estrogen regulates functional inhibition of hippocampal CA1 pyramidal cells in the adult female rat. J. Neurosci. 21 (17), 6532–6543.

Rudick, C.N., Gibbs, R.B., Woolley, C.S., 2002. Estrogen-Induced Disinhibition of Hippocampal CA1 Pyramidal Cells Depends on Basal Forebrain Cholinergic Neurons. Program No. 740.6, 2002 Abstract Viewer/Itinerary Planner. Society for Neuroscience, Washington, DC.

Sandstrom, N.J., Williams, C.L., 2001. Memory retention is modulated by acute estradiol and progesterone replacement. Behav. Neurosci. 115, 384–393.

Schmidt, P.J., Nieman, L.K., Danaceau, M.A., Adams, L.F., Rubinow, D.R., 1998. Differential behavioral effects of gonadal steroids in women with and in those without premenstrual syndrome. N.Engl. J. Med. 338, 209–216.

Shors, T.J., Chua, C., Falduto, J., 2001. Sex differences and opposite effects of stress on dendritic spine density in the male versus female hippocampus. J. Neurosci. 21 (16), 6292–6297.

Thornton, J.W., Need, E., Crews, D., 2003. Resurrecting the ancestral steroid receptor: ancient origin of estrogen signaling. Science 301, 1714–1717.

Uppari, N., Joseph, V., Bairam, A., 2016. Inhibitory respiratory responses to progesterone and allopregnanolone in newborn rats chronically treated with caffeine. J. Physiol. 594, 373–389.

Vasudevan, N., Ogawa, S., Pfaff, D., 2002. Physiol. Rev. 82 (4), 923–944.

Vasudevan, N., Morgan, M., Pfaff, D., Ogawa, S., 2013. Distinct behavioral phenotypes in male mice lacking the thyroid hormone receptor α1 or β isoforms. Horm. Behav. 63, 742–751.

Woolley, C.S., McEwen, B.S., 1992. Estradiol mediates fluctuation in hippocampal synapse density during the estrous cycle in the adult rat. J. Neurosci. 12, 2549–2554.

Woolley, C.S., McEwen, B.S., 1993. Roles of estradiol and progesterone in regulation of hippocampal dendritic spine density during the estrous cycle in the rat. J. Comp. Neurol. 336, 293–306.

Woolley, C.S., McEwen, B.S., 1994. Estradiol regulates hippocampal dendritic spine density via an NMDA receptor-dependent mechanism. J. Neurosci. 14, 7680–7687.

Woolley, C.S., Gould, E., Frankfurt, M., McEwen, B.S., 1990. Naturally occurring fluctuation in dendritic spine density on adult hippocampal pyramidal neurons. J. Neurosci. 10, 4035–4039.

Woolley, C.S., Weiland, N.G., McEwen, B.S., Schwartzkroin, P.A., 1997. Estradiol increases the sensitivity of hippocampal CA1 pyramidal cells to NMDA receptor-mediated synaptic input: correlation with dendritic spine density. J. Neurosci. 17, 1848–1859.

## Chapter 20: Hormone Receptors Interact With Other Nuclear Proteins to Influence Hormone Responsiveness

Apostolakis, E.M., Ramamurphy, M., Zhou, D., Oñate, S., O'Malley, B.W., July 2002. Acute disruption of select steroid receptor coactivators prevents reproductive behavior in rats and unmasks genetic adaptation in knockout mice. Mol. Endocrinol. 16 (7), 1511–1523.

Auger, A.P., Perrot-Sinal, T.S., Auger, C.J., Ekas, L.A., Tetel, M.J., McCarthy, M.M., 2002. Expression of the nuclear receptor coactivator, cAMP response element-binding protein, is sexually dimorphic and modulates sexual differentiation of neonatal rat brain. Endocrinology 143, 3009–3016.

Azoitei, A., Merseburger, A.S., Godau, B., Hoda, M.R., Schmid, E., Cronauer, M.V., 2016. C-terminally truncated constitutively active androgen receptor variants and their biologic and clinical significance in castration-resistant prostate cancer. J. Steroid Biochem. Mol. Biol. 166, 38–44.

Baumann, C.T., Lim, C.S., Hager, G.L., 1999. Intracellular localization and trafficking of steroid receptors. Cell Biochem. Biophys. 31, 119–127.

Brady, J.V., 1967. Ulcers in executive monkeys. In: McGaugh, J.L., Weinberger, N.W., Whalen, R.E. (Eds.), The Biological Bases of Behavior. Freeman, San Francisco, CA, pp. 189–192.

Carroll, B.J., Curtis, G.C., Mendels, J., 1976. Neuroendocrine regulation in depression. I. Limbic system-adrenocortical dysfunction. Arch. Gen. Psychiatry 33, 1041–1044.

Carroll, B.J., Feinberg, M., Greden, J.F., Tarika, J., Albala, A.A., Haskett, R.F., James, N.M., Kronfol, Z., Lohr, N., Steiner, M., de Vigne, J.P., Young, E., 1981. A specific laboratory test for the diagnosis of melancholia: standardization, validation, and clinical utility. Arch. Gen. Psychiatry 38, 15–22.

Ekstrand, M.I., Nectow, A.R., Knight, Z.A., Latcha, K.N., Pomeranz, L.E., Friedman, J.M., 2014. Molecular profiling of neurons based on connectivity. Cell 157, 1230–1242.

Gagnidze, K., Pfaff, D.W., Mong, J.A., 2010. Gene expression in neuroendocrine cells during the critical period for sexual differentiation of the brain. Prog. Brain Res. 186, 97–111.

Graber, J.A., Brooks-Gunn, J., et al., 1995. The antecedents of menarcheal age: heredity, family environment, and stressful life events. Child Dev. 66 (2), 346–359.

Helmreich, D.L., Mattern, L.G., Cameron, J.L., 1993. Lack of a role of the hypothalamic- pituitary-adrenal axis in the fasting-induced suppression of luteinizing hormone secretion in adult male rhesus monkeys (Macaca mulatta). Endocrinology 132 (6), 2427–2437.

Holter, E., Kotaja, N., Makela, S., Strauss, L., Kietz, S., Janne, O.A., Gustafsson, J.A., Palvimo, J.J., Treuter, E., 2002. Inhibition of androgen receptor (AR) function by the reproductive orphan nuclear receptor DAX-1. Mol. Endocrinol. 16, 515–528.

Huang, H.J., Norris, J.D., McDonnell, D.P., 2002. Identification of a negative regulatory surface within estrogen receptor alpha provides evidence in support of a role for corepressors in regulating cellular responses to agonists and antagonists. Mol. Endocrinol. 16, 1778–1792.

Mason, J.W., 1968. Organization of the multiple endocrine responses to avoidance in the monkey. Psychosom. Med. 30, 774–790.

McEwen, B.S., 2002. The End of Stress as We Know It. Joseph Henry Press, Washington, DC.

Miller, R.G., Rubin, R.T., Clark, B.R., Crawford, W.R., Arthur, R.J., 1970. The stress of aircraft carrier landings. I. Corticosteroid responses in naval aviators. Psychosom. Med. 32, 581–588.

Mong, J.A., Krebs, C., et al., 2002. Perspective: microarrays and differential display PCR-tools for studying transcript levels of genes in neuroendocrine systems. Endocrinology 143 (6), 2002–2006.

Mong, J.A., Devidze, N., et al., 2003. Estradiol regulation of lipocalin-type prostaglandin D synthase transcript levels in the rodent brain: evidence from high density oligonucleotide arrays and in situ hybridization. PNAS 100 (1), 318–323.

Pacak, K., Palkovits, M., 2001. Stressor specificity of central neuroendocrine responses: implications for stress-related disorders. Endocr. Rev. 22, 502–548.

Refetoff, S., Bassett, J.H., Beck-Peccoz, P., Bernal, J., Brent, G., Chatterjee, K., De Groot, L.J., Dumitrescu, A.M., Jameson, J.L., Kopp, P.A., Murata, Y., Persani, L., Samarut, J., Weiss, R.E., Williams, G.R., Yen, P.M., 2014. Classification and proposed nomenclature for inherited defects of thyroid hormone action, cell transport, and metabolism. J. Clin. Endocrinol. Metab. 99, 768–770.

Rubin, R.T., Rahe, R.H., Arthur, R.J., Clark, B.R., 1969. Adrenal cortical activity changes during underwater demolition team training. Psychosom. Med. 31, 553–564.

Smith, C.L., O'Malley, B.W., 1999. Evolving concepts of selective estrogen receptor action: from basic science to clinical applications. Trends Endocrinol. Metab. 10 (8), 299–300.

Warren, M.P., Fried, J.L., 2001. Hypothalamic amenorrhea: the effects of environmental stresses on the reproductive system—a central effect of the central nervous system. Endocrinol. Metab. Clin. North Am. 30 (3), 611–629.

Warren, M.P., Perlroth, N.E., 2001. The effects of intense exercise on the female reproductive system. J. Endocrinol. 170 (1), 3–11.

Yuan, C.X., Ito, M., Fondell, J.D., Fu, Z.Y., Roeder, R.G., 1998. The TRAP220 component of a thyroid hormone receptor- associated protein (TRAP) coactivator complex interacts directly with nuclear receptors in a ligand-dependent fashion. Proc. Natl. Acad. Sci. U.S.A. 95, 7939–7944.

Zhou, Z.X., He, B., Hall, S.H., Wilson, E.M., French, F.S., 2002. Domain interactions between coregulator ARA(70) and the androgen receptor (AR). Mol. Endocrinol. 16, 287–300.

## Chapter 21: Hormone Effects on Behavior Depend Upon Context

Bachman, E.S., Dhillon, H., Zhang, C.-Y., Cinti, S., Bianco, A.C., Kobilka, B.K., Lowell, B.B., 2002. beta-AR signaling required for diet-induced thermogenesis and obesity resistance. Science 297, 843–845.

Bartness, T.J., Liu, Y., Shrestha, Y.B., Ryu, V., 2015. Neural innervation of white adipose tissue and the control of lipolysis. Front. Neuroendocrinol. 35, 473–493.

Bates, S.H., Stearns, W.H., Dundon, T.A., Schubert, M., Tso, A.W., Wang, Y., Banks, A.S., Lavery, H.J., Haq, A.K., Maratos-Flier, E., Neel, B.G., Schwartz, M.W., Myers Jr., M.G., 2003. STAT3 signalling is required for leptin regulation of energy balance but not reproduction. Nature 421 (6925), 856–859.

Beery, A.K., Loo, T.J., Zucker, I., 2008. Day length and estradiol affect same-sex affiliative behavior in the female meadow vole. Horm. Behav. 54, 153–159.

Belanger, R.M., Corkum, L.D., Zielinski, B.S., 2006. Olfactory sensory input increases gill ventilation in male round gobies (Neogobius melanostomus) during exposure to steroids. Comp. Biochem. Physiol. A. Mol. Integr. Physiol. 144, 196–202.

Belanger, R.M., Corkum, L.D., Zielinski, B.S., 2007. Differential behavioral responses by reproductive and non-reproductive male round gobies (Neogobius melanostomus) to the putative pheromone estrone. Comp. Biochem. Physiol. A. Mol. Integr. Physiol. 147, 77–83.

Cerda-Reverter, J.M., Larhammar, D., 2000. Neuropeptide Y family of peptides: structure, anatomical expression, function, and molecular evolution. Biochem. Cell Biol. 78 (3), 371–392.

Conlon, J.M., 2002. The origin and evolution of peptide YY (PYY) and pancreatic polypeptide (PP). Peptides 23 (2), 269–278.

Davisson, R.L., Oliverio, M.I., Coffman, T.M., Sigmund, C.D., 2000. Divergent functions of angiotensin II receptor isoforms in the brain. J. Clin. Invest. 106 (1), 103–106.

Dubey, N., Hoffman, J.F., Schuebel, K., Yuan, Q., Martinez, P.E., Nieman, L.K., Rubinow, D.R., Schmidt, P.J., Goldman, D., 2017. The ESC/E(Z) complex, an effector of response to ovarian steroids, manifests an intrinsic difference in cells from women with premenstrual dysphoric disorder. Mol. Psychiatry 22.

Ducy, P., Karsenty, G., 2000. The family of bone morphogenetic proteins. Kidney Intern. 57, 2207–2214.

Dunlap, K.D., Jashari, D., Pappas, K.M., 2011. Glucocorticoid receptor blockade inhibits brain cell addition and aggressive signaling in electric fish, Apteronotus leptorhynchus. Horm. Behav. 60 (3), 275–283.

Dupré, C., Lovett-Barron, M., Pfaff, D.W., Kow, L.M., 2010. Histaminergic responses by hypothalamic neurons that regulate lordosis and their modulation by estradiol. Proc. Natl. Acad. Sci. U.S.A. 107 (27), 12311–12316.

Elmquist, J.K., Flier, J.S., 2004. Neuroscience. The fat-brain axis enters a new dimension. Science 304 (5667), 108–110.

Galli, S.M., Phillips, M.I., 1996. Interactions of angiotensin II and atrial natriuretic peptide in the brain: fish to rodent. Proc. Soc. Exp. Biol. 213, 128–137.

Garofalo, R.S., 2002. Genetic analysis of insulin signaling in Drosophilia. Trends Endocrinol. Metab. 13 (4), 156–162.

Holley, A., Shaley, S., Bellevue, S., Pfaus, J.G., 2014. Conditioned mate-guarding behavior in the female rat. Physiol. Behav. 131, 136–141.

Jones, J.E., Pick, R.R., Davenport, M.D., Keen, A.C., Corp, E.S., Wade, G.N., 2002. Disinhibition of female sexual behavior by a CRH receptor antagonist in Syrian hamsters. Am. J. Physiol. Regul. Integr. Comp. Physiol. 283, R591–R597.

Kelly, B., Maguire-Herring, V., Rose, C.M., Gore, H.E., Ferrigno, S., Novak, M.A., Lacreuse, A., 2014. Short-term testosterone manipulations do not affect cognition or motor function but differentially modulate emotions in young and older male rhesus monkeys. Horm. Behav. 66 (5), 731–742. http://dx.doi.org/10.1016/j.yhbeh.2014.08.016.

LeDoux, J., 2015. Anxious: Using the Brain to Understand and Treat Fear and Anxiety. Penguin Books Random House, New York.

Manzon, L.A., 2002. The role of prolactin in fish osmoregulation: a review. Gen. Comp. Endocrinol. 125 (2), 291–310.

Mustoe, A.C., Cavanaugh, J., Harnisch, A.M., Thompson, B.E., French, J.A., 2015. Do marmosets care to share? Oxytocin treatment reduces prosocial behavior toward strangers. Horm. Behav. 71, 83–90.

Nasokin, I.O., Alikasifoglu, A., Barrette, T., Cheng, M.M., Thomas, P.M., Nikitin, A.G., 2002. Cloning, characterization, and embryonic expression analysis of the Drosophilia melanogaster gene encoding insulin/relaxin-like peptide. Biochem. Biophys. Res. Commun. 295 (2), 312–318.

Niswender, K.D., Gallis, B., Blevins, J.E., Corson, M.A., Schwartz, M.W., Baskin, D.G., 2003a. Immunocytochemical detection of phosphatidylinositol 3-kinase activation by insulin and leptin. J. Histochem. Cytochem. 51 (3), 275–283.

Niswender, K.D., Morrison, C.D., Clegg, D.J., Olson, R., Baskin, D.G., Myers Jr., M.G., Seeley, R.J., Schwartz, M.W., 2003b. Insulin activation of phosphatidylinositol 3-kinase in the hypothalamic arcuate nucleus: a key mediator of insulin-induced anorexia. Diabetes 52 (2), 227–231.

Ogawa, S., Choleris, E., Pfaff, D., December 2004. Genetic influences on aggressive behaviors and arousability in animals. Ann. N.Y. Acad. Sci. 1036, 257–266.

Pradhan, D.S., Solomon-Lane, T.K., Grober, M.S., 2015. Contextual modulation of social and endocrine correlates of fitness: insights from the life history of a sex changing fish. Front. Neurosci. 9, 8.

Prendergast, B.J., Nelson, R.J., Zucker, I., 2002. Mammalian seasonal rhythms: behavioral and neuroendocrine substrates. In: Pfaff, D.W., et al. (Ed.), Hormones, Brain and Behavior. Elsevier, New York.

Stephens, S.B., Wallen, K., 2013. Environmental and social influences on neuroendocrine puberty and behavior in macaques and other nonhuman primates. Horm. Behav. 64 (2), 226–239. http://dx.doi.org/10.1016/j.yhbeh.2013.05.003.

Takahashi, A., Yasuda, A., Sullivan, C.V., Kawauchi, H., 2003. Identification of proopiomela-nocortin-related peptides in the rostral pars distalis of the pituitary in coelacanth: evolutional implications. Gen. Comp. Endocrinol. 130 (3), 340–349.

Williamson, C.M., Lee, W., Romeo, R.D., Curley, J.P., March 15, 2017. Social context-dependent relationships between mouse dominance rank and plasma hormone levels. Physiol. Behav. 171, 110–119.

Woolley, C.S., Gould, E., Frankfurt, M., McEwen, B.S., 1990. Naturally occurring fluctuation in dendritic spine density on adult hippocampal pyramidal neurons. J. Neurosci. 10, 4035–4039.

Woolley, C.S., Cohen, R.S., 2002. Sex steroids and neuronal growth in adulthood. In: Pfaff, D.W., Arnold, A., Fahrbach, S.E., Etgen, A.M., Rubin, R.T. (Eds.), Hormones, Brain and Behavior. Academic Press.

Woolley, C.S., McEwen, B.S., 1992. Estradiol mediates fluctuation in hippocampal synapse density during the estrous cycle in the adult rat. J. Neurosci. 12, 2549–2554.

Woolley, C.S., McEwen, B.S., 1993. Roles of estradiol and progesterone in regulation of hippocampal dendritic spine density during the estrous cycle in the rat. J. Comp. Neurol. 336, 293–306.

## Chapter 22: Behavioral–Environmental Context Alters Hormone Release

Angelier, F., Wingfield, J.C., 2013. Importance of the glucocorticoid stress response in a changing world: theory, hypotheses and perspectives. Gen. Comp. Endocrinol. 190, 118–128. http://dx.doi.org/10.1016/j.ygcen.2013.05.022.

Bronson, F.H., 1986. Food-restricted, prepubertal, female rats: rapid recovery of luteinizing hormone pulsing with excess food, and full recovery of pubertal development with gonadotropin-releasing hormone. Endocrinology 118 (6), 2483–2487.

Bronson, F.H., 1989. Mammalian Reproductive Biology, first ed. The University of Chicago Press, Chicago and London.

Budak, E., Fernandez Sanchez, M., Bellver, J., Cervero, A., Simon, C., Pellicer, A., 2006. Interactions of the hormones leptin, ghrelin, adiponectin, resistin, and PYY3-36 with the reproductive system. Fertil. Steril. 85 (6), 1563–1581.

Clarke, I.J., Arbabi, L., 2016. New concepts of the central control of reproduction, integrating influence of stress, metabolic state, and season. Domest. Anim. Endocrinol. 56 (Suppl.), S165–S179. http://dx.doi.org/10.1016/j.domaniend.2016.03.001.

Foster, D.L., Olster, D.H., 1985. Effect of restricted nutrition on puberty in the lamb: patterns of tonic luteinizing hormone (LH) secretion and competency of the LH surge system. Endocrinology 116 (1), 375–381.

Foster, D.L., Yellon, S.M., Olster, D.H., 1985. Internal and external determinants of the timing of puberty in the female. J. Reprod. Fertil. 75 (1), 327–344.

Graber, J.A., Brooks-Gunn, J., Warren, M.P., 1995. The antecedents of menarcheal age: heredity, family environment, and stressful life events. Child Dev. 66 (2), 346–359.

Hirschenhauser, K., Frigerio, D., Grammer, K., Magnusson, M.S., 2002. Monthly patterns of testosterone and behavior in prospective fathers. Horm. Behav. 42 (2), 172–181.

Hirschenhauser, K., Wittek, M., Johnston, P., Mostl, E., 2008. Social context rather than behavioral output or winning modulates post-conflict testosterone responses in Japanese quail (Coturnix japonica). Physiol. Behav. 95 (3), 457–463. http://dx.doi.org/10.1016/j.physbeh.2008.07.013.

Kheirbek, M.A., Klemenhagen, K.C., Sahay, A., Hen, R., 2012. Neurogenesis and generalization: a new approach to stratify and treat anxiety disorders. Nat. Neurosci. 15 (12), 1613–1620. http://dx.doi.org/10.1038/nn.3262.

Mason, J.W., 1968. Organization of the multiple endocrine responses to avoidance in the monkey. Psychosom. Med. 30 (Suppl. 5), 774–790.

McEwen, B.S., 2002. The End of Stress as We Know It. Joseph Henry Press, Washington, DC, p. 64.

Paul, M.J., Pyter, L.M., Freeman, D.A., Galang, J., Prendergast, B.J., 2009. Photic and non-photic seasonal cues differentially engage hypothalamic kisspeptin and RFamide-related peptide mRNA expression in Siberian hamsters. J. Neuroendocrinol. 21 (12), 1007–1014. http://dx.doi.org/10.1111/j.1365-2826.2009.01924.x.

Roze, C., Doyen, C., Le Heuzey, M.F., Armoogum, P., Mouren, M.C., Leger, J., 2007. Predictors of late menarche and adult height in children with anorexia nervosa. Clin. Endocrinol. (Oxf.) 67 (3), 462–467. http://dx.doi.org/10.1111/j.1365-2265.2007.02912.x.

Schneider, J.E., Wade, G.N., 1989. Availability of metabolic fuels controls estrous cyclicity of Syrian hamsters. Science 244 (4910), 1326–1328.

Schneider, J.E., Zhou, D., 1999. Interactive effects of central leptin and peripheral fuel oxidation on estrous cyclicity. Am. J. Physiol. 277 (4 Pt. 2), R1020–R1024.

Schneider, J.E., Klingerman, C.M., Abdulhay, A., 2012. Sense and nonsense in metabolic control of reproduction. Front. Endocrinol. 3 (26). http://dx.doi.org/10.3389/fendo.2012.00026.

Schneider, J.E., Wise, J.D., Benton, N.A., Brozek, J.M., Keen-Rhinehart, E., 2013. When do we eat? Ingestive behavior, survival, and reproductive success. Horm. Behav. 64 (4), 702–728. http://dx.doi.org/10.1016/j.yhbeh.2013.07.005.

Schneider, J.E., Benton, N.A., Russo, K.A., Klingerman, C.M., Williams 3rd, W.P., Simberlund, J., Abdulhay, A., Brozek, J.M., Kriegsfeld, L.J., July 27, 2017. RFamide-related peptide-3 and the trade-off between reproductive and ingestive behavior. Integr. Comp. Biol. https://doi.org/10.1093/icb/icx097. [Epub ahead of print] 28985338.

van Noord, P.A., Kaaks, R., 1991. The effect of wartime conditions and the 1944-45 'Dutch famine' on recalled menarcheal age in participants of the DOM breast cancer screening project. Ann. Hum. Biol. 18 (1), 57–70.

Wade, G.N., Jones, J.E., 2004. Neuroendocrinology of nutritional infertility. Am. J. Physiol. Regul. Integr. Comp. Physiol. 287 (6), R1277–R1296.

Wade, G.N., Schneider, J.E., 1992. Metabolic fuels and reproduction in female mammals. Neurosci. Biobehav. Rev. 16 (2), 235–272.

Warren, M.P., Perlroth, N.E., 2001. The effects of intense exercise on the female reproductive system. J. Endocrinol. 170 (1), 3–11.

## Chapter 23: Neuroendocrine Mechanisms Have Been Conserved to Provide Biologically Adaptive Body–Brain–Behavior Coordination

Anderson, D.W., McKeown, A.N., Thornton, J.W., June 15, 2015. Intermolecular epistasis shaped the function and evolution of an ancient transcription factor and its DNA binding sites. Elife 4, e07864.

Brozek, J.M., Schneider, J.E., Rhinehart, E., September 14, 2017. Maternal programming of body weight in syrian hamsters. Integr. Comp. Biol. https://doi.org/10.1093/icb/icx108. [Epub ahead of print] 28992103.

Choleris, E., Pfaff, D., Kavaliers, M. (Eds.), 2013. Oxytocin, Vasopressin and Related Peptides in the Regulation of Behavior. Cambridge University Press, Cambridge, England.

Cox, R.M., Parker, E.U., Cheney, D.M., Lieble, A.L., Martin, L.B., Calsbeek, R., 2010. Experimental evidence for physiological costs underlying the trade-off between reproduction and survival. Func. Ecol. 24, 1262–1269. http://dx.doi.org/10.1111/j.1365-2435.2010.01756.x.

Crespi, E.J., Travis, J.A., 2017. The search for mechanisms underlying evolutionary trade-offs in response to different selection pressures in the least killifish. Integr. Comp. Biol.

Csaba, G., 2013. Insulin at a unicellular eukaryote level. Cell Biol. Intern. 37 (4), 267–275. http://dx.doi.org/10.1002/cbin.10054.

de Jong-Brink, M., ter Maat, A., Tensen, C.P., 2001. NPY in invertebrates: molecular answers to altered functions during evolution. Peptides 22 (3), 309–315.

Finch, C.E., Rose, M.R., 1995. Hormones and the physiological architecture of life history evolution. Q. Rev. Biol. 70 (1), 1–52.

Hau, M., Wingfield, J.C., 2011. Hormonally-regulated trade-offs: evolutionary variability and phenotypic plasticity in testosterone signalling pathways. In: Flatt, T., Heyland, A. (Eds.), Mechanisms of Life History Evolution. Oxford University Press, Oxford, UK, pp. 349–361.

Jenkins, N.L., McColl, G., Lithgow, G.J., 2004. Fitness cost of extended lifespan in *Caenorhabditis elegans*. Proc. R. Soc. Lond. B 271 (1556), 2523–2526. http://dx.doi.org/10.1098/rspb.2004.2897.

Kaugars, K.E., Rivers, C.I., Saha, M.S., Heideman, P.D., 2016. Genetic variation in total number and locations of GnRH neurons identified using in situ hybridization in a wild-source population. J. Exp. Zool. A Ecol. Genet. Physiol. 325 (2), 106–115. http://dx.doi.org/10.1002/jez.2000.

Ketterson, E.D., Nolan, V., 1992. Hormones and life histories: an integrative approach. Am. Nat. 140 (Suppl. 1), S33–S62. http://dx.doi.org/10.1086/285396.

Mayr, E., 1974. Behavior programs and evolutionary strategies. Am. Sci. 62 (6), 650–659.

Mukhopadhyay, A., Tissenbaum, H.A., 2007. Reproduction and longevity: secrets revealed by *C. elegans*. Trends Cell Biol. 17 (2), 65–71. http://dx.doi.org/10.1016/j.tcb.2006.12.004.

Nassel, D.R., Wegener, C., 2011. A comparative review of short and long neuropeptide F signaling in invertebrates: any similarities to vertebrate neuropeptide Y signaling? Peptides 32 (6), 1335–1355. http://dx.doi.org/10.1016/j.peptides.2011.03.013.

Nassel, D.R., Winther, A.M., 2010. Drosophila neuropeptides in regulation of physiology and behavior. Prog. Neurobiol. 92 (1), 42–104. http://dx.doi.org/10.1016/j.pneurobio.2010.04.010.

Pfaff, D.W., 1999c. Drive: Neurobiological and Molecular Mechanisms of Sexual Motivation. MIT Press, Cambridge, MA.

Rowan, W., 1930. Experiments in bird migration. II. Reversed migration. Proc. Natl. Acad. Sci. U.S.A. 16 (7), 520–525.

Rowan, W., 1932. Experiments in bird migration: III. The effects of artificial light, castration and certain extracts on the Autumn Movements of the American Crow(*Corvus brachyrhynchis*). Proc. Natl. Acad. Sci. U.S.A. 18 (11), 639–654.

Stearns, S.C., 1989. Trade-offs in life-history evolution. Func. Ecol. 3, 259–268.

Thornton, J.W., 2001. Evolution of vertebrate steroid receptors from an ancestral estrogen receptor by ligand exploitation and serial genome expansions. Proc. Natl. Acad. Sci. U.S.A. 98 (10), 5671–5676.

Wingfield, J., Hegner, R.E., Dufty Jr., A.M., Ball, G., 1990. The "challenge hypothesis": theoretical implications for patterns of testosterone secretion, mating systems, and breeding strategies. Am. Nat. 136, 829–846.

Wingfield, J.C., Lynn, S., Soma, K.K., 2001. Avoiding the 'costs' of testosterone: ecological bases of hormone-behavior interactions. Brain, Behav. Evol. 57 (5), 239–251.

Wingfield, J.C., 2005. Historical contributions of research on birds to behavioral neuroendocrinology. Horm. Behav. 48 (4), 395–402. http://dx.doi.org/10.1016/j.yhbeh.2005.06.003.

Zera, A.J., 2016a. Evolutionary endocrinology of hormonal rhythms: juvenile hormone titer circadian polymorphism in *Gryllus firmus*. Integr. Comp. Biol. 56 (2), 159–170. http://dx.doi.org/10.1093/icb/icw027.

Zera, A.J., 2016b. Juvenile hormone and the endocrine regulation of wing polymorphism in insects: new insights from circadian and functional-genomic studies in Gryllus crickets. Physiol. Entomol. 41, 313–326.

## *Epilogue: A Look to the Future*

Fischetti, M., 2011. Computers versus brains. Sci. Am. 305, 104.

Service, R.F., 2014. The brain chip. Science 345, 614–616.

Sierra, F., 2015. Quoted in Science. 349, 1277.

Tsutsui, K., 2016. How to contribute to the progress of neuroendocrinology: new insights from discovering novel neuropeptides and neurosteroids regulating pituitary and brain functions. Gen. Comp. Endocrinol. 227, 3–15.

# Index

Printed in the United States
By Bookmasters